T0318581

Assessing and Measuring Environmental Impact and Sustainability

Assessing and Measuring Environmental Impact and Sustainability

Edited by

Jiří Jaromír Klemeš

AMSTERDAM • BOSTON • HEIDELBERG • LONDON
NEW YORK • OXFORD • PARIS • SAN DIEGO
SAN FRANCISCO • SINGAPORE • SYDNEY • TOKYO
Butterworth-Heinemann is an imprint of Elsevier

Butterworth-Heinemann is an imprint of Elsevier
The Boulevard, Langford Lane, Kidlington, Oxford OX5 1GB, UK
225 Wyman Street, Waltham, MA 02451, USA

ISBN: 978-0-12-799968-5

British Library Cataloguing-in-Publication Data
A catalogue record for this book is available from the British Library

Library of Congress Cataloging-in-Publication Data
A catalog record for this book is available from the Library of Congress

For information on all Butterworth-Heinemann publications visit our website at http://store.elsevier.com/

Typeset by MPS Limited, Chennai, India
www.adi-mps.com

Printed and bound in USA

Contents

Contributors

Bhavik R. Bakshi
William G. Lowrie Department of Chemical & Biomolecular Engineering, The Ohio State University, Columbus, OH, USA

Umberto Berardi
Faculty of Engineering and Architectural Science, Ryerson University, Toronto, ON, Canada

Thomas Brecheisen
Institute for Environmental Science and Policy, University of Illinois at Chicago, Chicago, IL, USA

Heriberto Cabezas
US Environmental Protection Agency, Office of Research and Development, Cincinnati, OH, USA

Siwanat Chairakwongsa
The Petroleum and Petrochemical College, Chulalongkorn University, Bangkok, Thailand

Lidija Čuček
Centre for Process Integration and Intensification—CPI^2, Research Institute of Chemical and Process Engineering—MŰKKI, University of Pannonia, Veszprém, Hungary

Richard C. Darton
Department of Engineering Science, University of Oxford, Oxford, UK

Luca De Benedetto
Centre for Process Integration and Intensification—CPI^2, Research Institute of Chemical and Process Engineering—MÜKKI, Faculty of Information Technology, University of Pannonia, Veszprém, Hungary

Urmila Diwekar
Vishwamitra Research Institute, Crystal Lake, IL, USA

Tarsha Eason
US Environmental Protection Agency, Office of Research and Development, Cincinnati, OH, USA

Rafiqul Gani
CAPEC, Department of Chemical and Biochemical Engineering, Technical University of Denmark, Lyngby, Denmark

Carina L. Gargalo
CAPEC, Department of Chemical and Biochemical Engineering, Technical University of Denmark, Lyngby, Denmark

Erin L. Gibbemeyer
William G. Lowrie Department of Chemical & Biomolecular Engineering, The Ohio State University, Columbus, OH, USA

Alejandra González-Mejía
US Environmental Protection Agency, Office of Research and Development, Cincinnati, OH, USA

Arjen Y. Hoekstra
University of Twente, Twente Water Centre Enschede, The Netherlands

Boon Hooi Hong
Centre of Excellence for Green Technologies, The University of Nottingham Malaysia Campus, Selangor, Malaysia

Bing Shen How
Centre of Excellence for Green Technologies, The University of Nottingham Malaysia Campus, Selangor, Malaysia

Vikas Khanna
Department of Civil and Environmental Engineering, University of Pittsburgh, Pittsburgh, PA, USA

Jiří Jaromír Klemeš
Centre for Process Integration and Intensification—CPI^2, Research Institute of Chemical and Process Engineering—MÜKKI, Faculty of Information Technology, University of Pannonia, Veszprém, Hungary

Zdravko Kravanja
University of Maribor, Maribor, Slovenia

Hon Loong Lam
Centre of Excellence for Green Technologies, The University of Nottingham Malaysia Campus, Selangor, Malaysia

Zainuddin Abdul Manan
Process Systems Engineering Centre (PROSPECT), Faculty of Chemical Engineering, Universiti Teknologi Malaysia, Johor, Malaysia

Rajib Mukherjee
Oak Ridge Institute for Science and Education, Post Doctoral Research Fellow at National Risk Management Research Laboratory, US Environmental Protection Agency, Cincinnati, OH, USA

Michael Narodoslawsky
Graz University of Technology, Institut feur Prozess und Partikeltechnik, Graz, Austria

Alberto Quaglia
CAPEC, Department of Chemical and Biochemical Engineering, Technical University of Denmark, Lyngby, Denmark

Debalina Sengupta
Oak Ridge Institute for Science and Education, Post Doctoral Research Fellow at National Risk Management Research Laboratory, US Environmental Protection Agency, Cincinnati, OH, USA

Subhas K. Sikdar
Associate Director of Science National Risk Management Research Laboratory, US Environmental Protection Agency, Cincinnati, OH, USA

Gürkan Sin
CAPEC, Department of Chemical and Biochemical Engineering, Technical University of Denmark, Lyngby, Denmark

Shweta Singh
Department of Agricultural & Biological Engineering and Division of Environmental & Ecological Engineering, Purdue University, West Lafayette, IN, USA

Thomas Theis
Institute for Environmental Science and Policy, University of Illinois at Chicago, Chicago, IL, USA

Leisha Vance
US Environmental Protection Agency, Office of Research and Development, Cincinnati, OH, USA

Sharifah Rafidah Wan Alwi
Process Systems Engineering Centre (PROSPECT), Faculty of Chemical Engineering, Universiti Teknologi Malaysia, Johor, Malaysia

George G. Zaimes
Department of Civil and Environmental Engineering, University of Pittsburgh, Pittsburgh, PA, USA

Preface

How to measure our progress toward achieving sustainable development is one of the key issues of the sustainability agenda. In recent years, we have seen intensive efforts by a wide range of actors, such as international agencies, governments, nongovernmental organizations, communities, and businesses, to develop measurement systems and indicators that characterize progress toward sustainable development, or at least some of the dimensions of sustainability.

Measurement and indicators play an important part in introducing the concept of sustainable development into practice. They help decision-makers define sustainable development objectives and targets, and help assess progress toward meeting those targets.

It is fair to say that, despite some remarkable achievements in the development of sustainability indicators, there is still no broad consensus regarding the best way to measure progress toward sustainability. Thus, efforts to produce new and "better" indicators are continuing. We see constant evolvement of and improvement in the existing frameworks as well as emergence of new initiatives. They all attempt to answer that simple question (also posed by this book), how do we measure sustainability? Given the complexity of the concept of sustainable development and the inherent difficulty in measuring the "well-being" of both present and future generations, it should not be surprising that the search for the right answer is still ongoing.

Therefore, this book is timely as it provides broad insight regarding how to measure environmental impacts and other aspects of sustainability. It considers a range of methods and approaches, including, but not limited to, systems analysis and optimization, life-cycle assessment, carbon and water footprints, sustainable process design, and decision analysis regarding multiple criteria. I expect that this book will be of interest to a wide-ranging audience and that it will stimulate lively discussions about how to measure our progress toward sustainable development.

Adisa Azapagic
The University of Manchester, Manchester, UK

Acknowledgments

This project has been made possible due to dedicated contributions authored by teams of leading researchers in the field (listed here in the alphabetical order):

Prof. Bhavik R. Bakshi, The Ohio State University, Columbus, OH, USA
Prof. Umberto Berardi, Ryerson University, Toronto, Canada
Dr. Heriberto Cabezas, US Environmental Protection Agency, Cincinnati, OH, USA
Prof. Richard C. Darton, University of Oxford, Oxford, UK
Dr. Luca de Benedetto and Prof. Jiří J. Klemeš, University of Pannonia, Veszprém, Hungary
Prof. Urmila Diwekar, Vishwamitra Research Institute, Clarendon Hills, IL, USA
Prof. Rafiqul Gani, Technical University of Denmark, Lyngby, Denmark
Prof. Arjen Y. Hoekstra, University of Twente, Netherlands
Dr. Vikas Khanna, University of Pittsburgh, Pittsburgh, PA, USA
Prof. Zdravko Kravanja, University of Maribor, Maribor, Slovenia
DDr. Hon Loong Lam, University of Nottingham, Malaysia Campus Selangor, Malaysia
Prof. Dr. Zainuddin Abdul Manan and Dr. Sharifah Wan Alwi, UTM, Johor, Malaysia
Prof. Michael Narodoslawsky, Technical University Graz, Graz, Austria
Dr. Subhas K. Sikdar, US Environmental Protection Agency, Cincinnati, OH, USA
Prof. Thomas L. Theis, University of Illinois at Chicago, IL, USA

In many cases they have been supported by their research teams.

Introduction

Jiří Jaromír Klemeš and Petar Sabev Varbanov

Centre for Process Integration and Intensification—CPI[2], Research Institute of Chemical and Process Engineering—MÜKKI, Faculty of Information Technology, University of Pannonia, Veszprém, Hungary

INTRODUCTION

This book offers an overview of a project where a number of leading personalities in the research and development of the methodology in sustainability field agreed to join forces in the development of a comprehensive edited publication covering the major aspects and issues of assessing and measuring the environmental impact and sustainability. "Sustainability" is an issue that has attracted a lot of attention from researchers and, more recently, from policy-makers. A quick search on September 14, 2014, in SCOPUS (2014) showed 21,865 publications containing this word. However, looking for both "sustainability" and "assessment" in the title resulted in only 1,256 papers, and the search for "sustainability" and "measurement" revealed even less—just 76 papers—dealing with those issues. This means that from the overall pool of sustainability-related publications, only 5% deal with "sustainability and assessment" and only 0.3% deal with "sustainability and measurement."

That simple illustration has highlighted the importance of a project dealing with not solely "sustainability" but rather "sustainability" and "assessment" and "measurement," with the latter emphasizing the need for quantitative characterization of the issues being considered.

A number of relevant questions and issues have been formulated in relation to sustainability measurement and assessment. They include the following:

1. How do we define sustainability?
2. How do we assess sustainability?
3. How do we measure sustainability?
4. How do we set up a policy for sustainability?
5. How can Fisher Information be applied to measuring and assessing sustainability issues?
6. System Analysis Approach to sustainability
7. The Environmental Performance Strategy Map
8. Sustainable Process Index (SPI)
9. Environmental/Green House Gas Emissions (GHGE)/nitrogen/water footprints
10. Life-cycle sustainability aspects
11. What is a decision point in sustainability analysis?
12. How do we reach sustainable design?

A number of leading researchers in the field have been approached and the majority enthusiastically accepted the challenge to find, discuss, and present answers and potential solutions to these issues. The main topics used for developing the answers from various viewpoints have been compiled in this edited book. They (not solely) include:

- Systems analysis approach to sustainability
- The Application of Fisher Information to Environmental and Engineering Problems
- SPI
- How to reach a decision point in sustainability analyses
- Overview of environmental footprints
- Nitrogen footprint and the nexus between carbon and nitrogen footprints
- The analysis of the water footprint (WF) of industry
- Life-cycle sustainability aspects of microalgae biofuels
- Methods and tools for sustainable process design
- The built environment and embodied energy
- The Environmental Performance Strategy Map: An integrated life-cycle analysis (LCA) approach to support the strategic decision-making process
- Sustainable design through process integration
- Supply and demand planning and management tools driving development toward low carbon emissions
- Setting a policy for sustainability: the importance of measurement
- Sustainability assessment of buildings, communities, and cities.

The book has been structured into several main parts:

1. The introduction deals with suitability definitions, systems approach to sustainability, and ways to express and measure sustainability (Chapters 1–3).
2. The second part is devoted to an important part of sustainability assessment and quantification of environmental impacts—footprints. This topic has had turbulent development and the works studying and developing various footprints have been escalating quickly (Chapters 4–7).
3. The third part presents contributions to sustainable design, planning, and management. There are various aspects including planning and strategic tools, the design of sustainable processes, supply chains, which constitute an important (but not always sufficiently studied and optimized) part, and also management tools (Chapters 8–13).
4. The final part deals with policies steering industrial and economic development toward sustainability (Chapter 14) and a mostly chemical and mechanical engineering view of this problem regarding sustainability assessment of buildings, communities, and cities (Chapter 15). Communities and cities are important end-users of the several engineering disciplines and production processes.

A short overview of those main parts to provide an executive summary is presented.

SUITABILITY DEFINITIONS, SYSTEMS APPROACH TO SUSTAINABILITY, AND WAYS TO EXPRESS AND MEASURE SUSTAINABILITY

The first chapter, authored by Urmila Diwekar, deals with the title topic itself—Engineering Sustainability. The chapter provides the vision that sustainability analysis is, by definition, multidisciplinary. Further in the chapter, engineering sustainability is presented by using various systems analysis approaches from different disciplines. A hierarchical consideration of sustainability at several system levels is offered, starting with green manufacturing, extending to industrial networks, and then extending to the ecosystem level. All systems considered at those levels are complex and extend boundaries of current analysis frameworks. There have been uncertainties at every stage of analysis. To deal with the uncertainties in a systematic way has been a strong requirement. The presented analysis applies optimization and social science approaches.

What should be noted is one of most widely accepted definitions of sustainability or sustainable development that is attributed to the World Commission on Environment and Development (1987). This concept of sustainability, since the publication of the Brundtland report known as Our Common Future (1987), has been widely used as a rationale to focus on the long-term effects of anthropogenic activities on nature. That Commission stated in their final report that "Sustainable development is development that meets the needs of the present without compromising the ability of future generations to meet their own needs." There have been several other definitions of sustainability; however, this was one of the first and it still has its importance.

The Brundtland Commission's definition (1987) of sustainability originated from the concerns that current trends in population and economic developments are not sustainable. Already at the beginning of this century it has been obvious that a number of adverse environmental impacts are threatening the global ecosystem (Cabezas, 2013). Impacts that are considered include the following: global climate change; degradation of air, water, and land; depletion of natural resources, including freshwater minerals; and loss of agricultural land because of deforestation, soil erosion, and urbanization.

This first chapter further extends the presentation of the approach to engineering systems analysis for achieving sustainability. It has been based on the premise that sustainable manufacturing extends the boundaries of traditional design and modeling frameworks to include multiple objectives and should be started as early as possible. Uncertainties increase as the modeling is expanded. Industrial ecology broadens this framework to industrial networks level and accounting for time-dependent uncertainties becomes necessary. The author of the first chapter expressed strong opinion that the adaptive management is a key to sustainability at this level. Mental models for uncertainty analysis are considered potentially useful for solving tasks. Sustainability goes beyond industrial ecology and the boundary extends to our planet. The time scale for decision-making extends significantly and forecasting the consequences from proposed decisions is very important. As a result, the main sustainability challenges, such as modeling phenomena, modeling uncertainties, time-dependent decision-making, defining objectives, and forecasting, are still waiting for adequate answers.

The second chapter was authored by Alejandra González-Mejía, Leisha Vance, Tarsha Eason, and Heriberto Cabezas. It is based on the notion that assessing sustainability in human and natural systems is often hampered by complex dynamics, multiplicity of time scales, and inherent linkages among the observable properties. This has obviously been the case and makes sustainability assessment a complicated issue.

Although many indicators have been identified as useful in classifying and indicating trends of movement toward and away from sustainability, evaluating changes in system behavior requires simultaneously monitoring multiple variables over time. For this reason, great value can be found in methods that can capture the dynamics of multidimensional systems. Fisher Information (Fisher, 1922) offers a tool that can combine multiple variables into an index adequately representing its components that can be monitored over time to assess changes in the dynamic behavior of systems. The concept of Fisher Information has demonstrated promise in capturing dynamic order and detecting regime shifts in models and real systems, but challenges in handling data quality and interpreting results can limit its application. Overall, the second chapter summarizes ongoing activities that enhance the effectiveness of the approach and extends them to develop practical mechanisms for monitoring and managing highly complex and integrated systems toward sustainability.

The authors have demonstrated that the initial development and application of the approach involved using Fisher Information can derive fundamental equations of physics, thermodynamics, and population genetics (Frieden, 1998,) which have been further developed, including their own contributions. Fisher Information was later proposed as a sustainability metric by Cabezas and Fath (2002) and has been used to study changes in dynamic order in model systems, such as multispecies food webs (Pawlowski and Cabezas, 2008), pseudo-economies (Cabezas et al., 2005), and real ecosystems such as the Bering Strait, the western African savannah, the US Florida pine-oak, and global climate changes (Mayer et al., 2007).

The approaches described in Chapter 2 emphasize a number of points in relation to Fisher Information analysis:

- The selection of variables is critical for ensuring relevant analyses because inadequate representation of the system leads to results that are not of much use for monitoring and management.

- Natural systems often exhibit periodic behavior or cycles, by its own features or when interacting with human populations. Periodic behavior (e.g., wastewater treatment model) and scale of study impact the characterization of system dynamics and introduce uncertainty. However, Fischer Information can be used to distinguish patterns even when cycle periods are unknown.
- Data quality is a great challenge when reducing the uncertainties attributable to detection errors. In this regard, an analytical approach and mechanisms have been developed and demonstrated by the authors of Chapter 2 to help manage such difficulties.
- Interpretation of the results of Fischer Information analyses has been challenging and may have inhibited broader use of the index. Several approaches have been presented in this chapter that can increase the clarity and enhance the identification of important findings.

It can be concluded that Chapter 2 demonstrates the utility of Fischer Information as an effective tool for monitoring systems, assessing dynamic change, and informing actions to help guide complex human and natural systems toward sustainability. The authors found Fisher Information used in a number of studies assessing changes in city structure (Eason and Garmenstani, 2012), characterizing political instability in nation states (Karunanithi et al., 2011), and studying the dynamics of lead concentration levels in conjunction with regulatory actions (Gonzalez-Mejia et al., 2012). They stated that their future work should cover the following:

- Examining the usefulness of the continuous form method in capturing regime dynamics
- Using multivariate techniques to determine the drivers of system behavior
- Investigating the utility of the index studying the resilience of various system types (e.g., ecological, social, and supply chain systems)
- Using Fischer Information to assess spatial dynamics
- Further exploring signals in Fisher Information as a leading indicator in critical transitions.

Chapter 2 is also equipped with comprehensive annexes providing more detailed explanations of Fisher Information application.

Chapter 3 introduces a way to integrate sustainability assessment into engineering design on the same footing as economic considerations. Michael Narodoslawsky and his team introduced the SPI, which addresses this need and provides a comprehensive evaluation of sustainability from the viewpoint of ecology. It is based on the concept of strong sustainability (Narodoslawsky and Krotscheck 1995). It can be used even when only mass and energy balances are known to an engineer, as is usually the case during early stages of process design. The SPI calculates an ecological footprint over the whole life cycle of a product or service provided by technology. The main part of the evaluation is the calculation of the area necessary to bring to life the provision of the service or product sustainably into the ecosphere if neither global material cycles (e.g., the global carbon cycle) nor the quality of local environmental compartments (atmosphere, water systems, or soil) were disturbed. It takes into account all material flows exchanged with the environment over the life cycle, including raw material extraction, emissions, and waste generation.

The authors of Chapter 3 present SPI as a method that offers a way of assessing not only industrial processes but also various activities, products, services, and regional initiatives according to the normative concept of strong sustainability. Based on two principles that can be directly deduced from this normative fundament, the SPI method allows comprehensive assessment of the impact of exchanging material flows between the anthroposphere and the ecosphere. The method accounts for the global material cycles and the preservation of the ecological compartments defined as soil, water, and air as the basis for ecosystem functions. In terms of life-cycle assessment, the proposed procedure also includes the impacts of resource provision and emissions.

The SPI was developed with the intention of supporting engineers in industrial processes of design and operation. However, its conceptual rigor and generality allow it to be applied to a wide spectrum of activities including other business processes and activities undertaken by town and regional governments, provided that their actions and effects can be described in the terms of life cycle and impacts of the unit steps. The application of the SPI method on a large variety of processes and products has shown that this sustainability metric is extremely sensitive to the use of different resources, distinguishing sharply between renewable and fossil resources in particular. In addition, the metric is also very sensitive to the use and dissipation of substances that either occur naturally in very low concentrations or are alien to the ecosystem.

Several tools that have been developed on the basis of the SPI method, particularly SPIonWeb (2014) and Regional Optimizer – RegiOpt (Kettl et al., 2011), allow engineers to evaluate their designs or operations with acceptable effort when only basic knowledge of sustainability assessment is available. A comprehensive databank including the provision of most precursors to industrial processes as well as almost all forms of energy supply allows simple representation of complex life cycles, even for untrained users. The results of the assessment allow engineers to pinpoint sustainability "hot spots" that must be addressed to improve the ecological performance of their designs and/or operations.

SPI presented in Chapter 3 offers a quick and reliable method to estimate ecological impact, even of complex technological systems. Although it is no substitute for thorough legally prescribed environmental compliance assessment, as is also the case for all other single issue measures such as the carbon footprint or highly aggregated measures like energy, it provides consistent and comprehensive guidance for engineers who want to include sustainability in their decisions.

QUANTIFICATION OF THE ENVIRONMENTAL IMPACTS: FOOTPRINTS

Chapter 4 deals with a measurable "sustainability footprint" that can be constructed using chosen indicators. The combined indicator can be used for assessing the relative sustainability of a process or product. The authors—Debalina Sengupta, Rajib Mukherjee, and Subhas K. Sikdar—provide a step-by-step description using three case studies to illustrate how the sustainability footprint can be calculated based on Euclidean distance of a system from a chosen reference point. The authors determined the relative importance of the various indicators by multivariate statistical analysis. They claim that their analysis leads to the necessary and sufficient number of indicators for making unbiased decisions. The proposed analysis method leads to a decision point that, by their assurance, is perfectly general but of specific importance to process, product, or corporate sustainability.

Although the concept of sustainability, in the opinion of the authors of the chapter, is easy to grasp, its application needs to be made more straightforward and standardized. In terms of achieving simultaneous progress toward sustainability in the present with regard to economic, environmental, and societal aspects without jeopardizing the progress of future generations, the development of a standard method for analyzing sustainability is still in its beginnings. Sustainability of systems, such as agricultural (Conway and Barbier, 2013), industrial (Graedel and Allenby, 2010), and urban infrastructure, has been studied and considerable research (Čuček et al., 2012a,b) has provided direction regarding achieving stability of those systems. In other cases, countries have been compared regarding their sustainability status based on economic, environmental, and societal improvement efforts. Some examples of such methods are reviewed. Freudenberg (2003) gives a comprehensive assessment of the use of composite indicators for evaluating a country's sustainability. Esty et al. (2008) describes the development of the Environmental Performance Index of countries. Sengupta et al. (2014) demonstrates the use of an aggregate index in evaluating the relative sustainability of countries using the UN Millennium Development Goals indicators. The authors' review of previous work regarding decision-making concludes that none of the prescribed methods does an adequate job in this art.

BASF (2014) has formulated the approaches of eco-efficiency and socio-eco-efficiency. They use a graphical method (2D for eco-efficiency and 3D for socio-eco-efficiency) to arrive at satisfactory decisions. However, the details of this methodology have been kept proprietary and have not been disclosed. The authors also give an overview and utilize the AIChE (2014) suistainability definition and IChemE sustainable development process metrics (2014).

The conditions for sound decision-making are based on first choosing appropriate indicators. The chapter shows that cost is usually a missing element in case studies, but no sustainability can be validated without satisfying cost constraints.

The number of indicators would have to be sufficient as well. By including several closely related indicators, it is possible to make the decision tilted toward those indicators and introduce the risk of systematic bias in the decisions.

The point of Chapter 4 is to present a clear and sound methodology for performing sustainability analysis of any system with chosen indicators, leaving the quality of decisions to the experience and judgment of the decision-makers, because the methodology authors cannot guarantee that all applicable indicators are available to them. The authors of Chapter 4 claim that their method is algorithmic and is considered totally clear.

Their last point is that computation of the sustainability footprint usually makes it necessary to apply a weighting procedure using appropriate weighting factors. This is typically a societal choice and no scientific method can possibly help in identifying what would be the appropriate values to use. Therefore, the method authors have decided to use neutral position 1.0 as a default value for weight factors for all indicators. Such an assumption considers all indicators to be of equivalent importance. The results shown in the case studies are subject to that condition. Using practitioner experience, other more appropriate values can be incorporated into the calculation of the sustainability footprint.

The next chapter—Chapter 5 "Overview of Environmental Footprints"—was authored by Lidija Čuček, Jiří Jaromír Klemeš, and Zdravko Kravanja. The crucial question that they address is how to measure and reduce environmental burdens. This issue has researchers, organizations, policy-makers, and others putting considerable efforts into developing concepts and metrics for measuring environmental sustainability. Among developed concepts and metrics, environmental footprints are gaining increasing popularity and play an ever-increasing role in sustainability evaluations and research. Footprints have become ubiquitous for researchers, policy-makers, and the general public. Over the past years, carbon footprint has been almost the sole environmental protection indicator. More recently, it has been realized that other impacts are also very important and need to be accounted for; therefore, newer evaluation procedures include a variety of other footprints. However, there is no generally accepted footprint or footprint family that represents the overall impact on the environment. This chapter, therefore, provides an overview of environmental footprints as indicators defined as of June 2014 that can be used to measure sustainability for environmental decision-making.

Indicators are only part of the picture. To be useful, it is necessary to draw adequate boundaries of the systems. This is where LCA comes into play. LCA enables a holistic view, minimizing the risk of missing significant impacts. To avoid problem-shifting or limited sustainability evaluation, it is important to select boundaries as widely as possible. Several pathway options for defining the system boundaries have gained prominence:

- Cradle-to-cradle: from resource extraction (cradle) to recycling or producing new product (cradle) (100% utilization of waste)
- Cradle-to-grave: from resource extraction (cradle) to disposal (grave)

- Cradle-to-gate: from resource extraction (cradle) to factory gate, before being sent to consumers (gate), excluding the use and the disposal phases
- Gate-to-gate: includes only one process within the entire production chain (e.g., the processing within factory)
- Well-to-wheel: related to transport fuels and vehicles, from the energy sources to the powered wheels (Daimler, 2011); this comprises two parts: well-to-tank (energy supply) and tank-to-wheel (energy efficiency) (Fuel-Cell e-Mobility, 2014).

The preferred options among the listed ones are cradle-to-cradle and cradle-to-grave.

In many cases only global warming potential (GWP) or carbon footprint are evaluated as criteria for environmental sustainability, leading to a "castrated type of LCA" (Finkbeiner, 2009). Most of the effort and resources are spent on reducing carbon footprint. However, other aspects of environmental impacts may also be significant, leading to either synergies (non-trade-offs) or compromises (trade-offs) among environmental footprints. For example, photovoltaic systems exhibit synergies in carbon, water, nitrogen, and energy footprints. In contrast, biogas production reduces carbon, nitrogen, and energy footprints but increases the WF (Vujanović et al., 2014). Usually there are compromises in environmental footprints, and the possibilities for achieving the improvements within the system can be identified (Azapagic and Clift, 1999). For making proper decisions, it is usually necessary to adequately account for regional characteristics and climatic conditions. Therefore, more comprehensive analysis needs to be performed that should consider all aspects of the natural environment, human health, and resources (Finkbeiner, 2009).

There have been aggregate (scalar) measures of sustainability consisting of "one number," such as eco-efficiency and eco-profit and total profit, eco-NPV (Net Present Value) and total NPV, and other combined criteria. Their advantage is that a unique (optimal) sustainable solution could be obtained directly, which is especially important when solving large-scale problems when the models are huge and computational times are significant. However, this is not the case with indicators of potential environmental impacts and environmental footprints because of the difficulties in converting such impacts to monetary equivalents. When applying such metrics, a comprehensive list of objectives (indicators of potential impacts or footprints) should be taken into account and evaluated. To identify roadmaps toward more sustainable development, it is preferable to obtain optimal solutions accounting for these indicators directly without folding them into a single number. For this purpose multiobjective optimization (MOO) should be performed.

If several objectives are considered (e.g., all key environmental footprints), then there are some limitations. One of them is the increased time spent obtaining the entire solution space. Another is the difficulty of visualizing and interpreting the obtained solutions, in many cases making the choices unclear. It is possible to present, at most, four-dimensional Pareto projections (Čuček et al., 2014). For cases with more criteria, there is then the need to apply dimensionality reduction techniques for obtaining manageable sets of criteria while preserving the validity of the results. Such techniques greatly facilitate the comprehension of the solution space when many objectives are involved within MOO problems. Various dimension reduction approaches have been developed for this purpose and are generally supported computationally by linear and nonlinear models (Čuček et al., 2013). Among these are method minimizing the error of omitting objectives (Guillén-Gosálbez, 2011), principal component analysis (Sabio et al., 2012), factor analysis (Huang and Lu, 2014), representative objectives method (Čuček et al., 2012a,b), applied to direct (Čuček et al., 2012a,b) and total objectives (Čuček et al., 2014), and others.

A literature review in this chapter indicates that the major categories of footprints developed to date are carbon, ecological, water, and energy footprints, forming the so-called footprint family (see report authored by Galli et al., 2011 and another publication by Galli et al., 2012, and the more recent work of Fang et al., 2014). No single indicator *per se* is able to comprehensively monitor human impact on the environment, but indicators need to be used and interpreted jointly (Galli et al., 2012). Ecological, energy, carbon, and WFs rank as the most important footprint indicators in the existing literature. They are in close relation with the four worldwide concerns regarding threats to human society: food security; energy security; climate security; and water security (Fang et al., 2013).

Chapter 5 presents the metrics measuring environmental sustainability such as indicators of potential environmental impact, eco-efficiency, eco-profit and total profit, eco-NPV and total NPV (and other combined criteria), sustainability indexes, and environmental footprints. Further, key environmental footprints, such as carbon, water, ecological, energy, nitrogen, phosphorus, biodiversity, and land footprints, have been presented in detail. Finally, other environmental footprints have been accessed that are not widely known and used and are not recognized as key environmental footprints, but that may potentially be useful in further evaluations.

Chapter 6 deals with nitrogen footprint and the nexus between carbon and nitrogen footprints; it was prepared by Shweta Singh, Erin L. Gibbemeyer, and Bhavik R. Bakshi. The authors start from the premise that nitrogen is an indispensable nutrient for sustaining human existence and that a holistic approach for assessment and management of nitrogen flows is required. Footprint measures are one of the most widely used approaches for informing sustainable choices of products or processes. In this chapter, the methodology for nitrogen footprint calculation is reviewed and the techniques for footprint calculations at various scales are illustrated. The interaction between carbon and nitrogen footprint for decision-making is crucial to avoid the pitfalls of trade-offs in decision-making for sustainable choices. A case study is used as an illustration purpose of studying nitrogen footprint and the nexus between carbon and nitrogen footprints.

Nitrogen is the first element of group 15 in the periodic table and is classified as a nonmetal. The stable state of nitrogen is in the form of diatomic gas N_2, where the triple covalent bond makes the gas relatively inert. Even though N_2 is an inert gas (lifeless), the compounds of nitrogen are the key ingredients for supporting life. Nitrogen can exist in multiple oxidation states ranging from −3 in ammonia (NH_3) to +5 in nitrate (NO_3-), resulting in a wide range of reactive compounds formed by nitrogen. One of nitrogen's key roles is forming an essential part of all proteins. Approximately 78.09% by volume of air is N_2, nitrogen is a major component of fertilizers, and it exists in a plethora of chemical compounds such as plastics and polymers. The world today would be very different without nitrogen (Schlesinger, 1997).

Carbon footprint has been the focus of most analyses for product or process comparison. However, in recent years there has been an increased awareness that using multiple footprints for decision-making reflects the sustainability issue more correctly (Fang et al., 2014). To understand the relationships between carbon footprint and nitrogen footprint, it is important to understand the origin of the definition of carbon footprint in a similar fashion as nitrogen footprint. Similar to the effect of economic activities on the nitrogen cycle through reactive nitrogen mobilization and emissions, the carbon cycle is also significantly disrupted because of economic activities. Carbon (along with nitrogen) forms another basic building block for existence of life on Earth. Every living organism is composed of C, H, and O, along with N. Carbon also forms the basis of various products like food, fiber, and fuel, thus supporting economic activities. The natural biogeochemical cycle of carbon mainly relies on the process of photosynthesis to convert the oxidized form of carbon, CO_2, into a reduced form as organically bound carbon incorporated into various food products. This process of carbon fixation is the key link for closing the global carbon cycle.

Chapter 7, authored by Arjen Y. Hoekstra, discusses what new perspective the WF concept brings to the evaluation of sustainability compared with the traditional way of looking at water use. An important part of the problem concerns the water supply chains. The chapter provides a comparative discussion of three methods for tracing resource use and pollution over supply chains: environmental footprint assessment, life-cycle assessment, and environmentally extended input − output analysis. Some of the recent literature regarding direct and indirect WFs of various economy sectors is analyzed. The chapter also discusses the emerging concept of water stewardship for business and the challenge of creating greater product and business transparency.

One of the key issues in introducing the WF is the measure of freshwater appropriation underlying a certain product or consumption pattern. Three components are distinguished: blue, green, and gray WF (Hoekstra et al., 2011). These definitions are complemented by introducing methods to trace natural resource use and pollution over the supply chains.

The author classifies economic activities into three different sectors. The primary sector of the economy is considered the sector that extracts or harvests products from the earth. It has been classified as having the largest WF on Earth. This sector includes activities like agriculture, forestry, fishing, aquaculture, mining, and quarrying. The green WF of humanity is nearly entirely concentrated within the primary sector. It has been estimated that approximately 92% of the blue WF of humanity is in agriculture alone (Hoekstra and Mekonnen, 2012).

A very valuable part of Chapter 7 constitutes the provided case studies illustrating the influence of WFs in various human activities. The spatial patterns of water depletion and contamination are closely tied to the structure of the global economy. As currently organized, the economic system lacks incentives that would cause producers and consumers to move toward wiser use of our limited freshwater resources. For achieving sustainable, efficient and equitable water use worldwide, there is the need for greater product transparency, international cooperation, WF ceilings per river basin, WF benchmarks for water-intensive commodities, water pricing schemes that reflect local water scarcity, and some agreement about equitable sharing of the limited available global water resources among different communities and nations.

SUSTAINABLE DESIGN, PLANNING, AND MANAGEMENT

Chapter 8 deals with life-cycle sustainability aspects of microalgal biofuels; it was authored by George G. Zaimes and Vikas Khanna. Their core message is that the holistic systems analysis method of biofuel production has emerged that considers the environmental impacts over the entire fuel life cycle, from raw material extraction, to cultivation, to conversion, and to final use. The authors position the method as the leading one for quantifying the environmental sustainability of biomass-to-fuel systems. This chapter reviews the application of life-cycle assessment for environmental evaluation of emerging microalgal biofuels. Several prominent environmental performance indicators including carbon footprint, energy return on energy investment, and WF are reviewed in the context of microalgal systems and other leading biofuels/biofeedstocks. A case study of microalgal biodiesel production is provided in the text. The results are compared with previous microalgae LCAs and are discussed in a broader context of environmental sustainability.

Microalgae's promising characteristics such as high photosynthetic yields, high lipid content, and the ability to utilize waste resources such as wastewater (Zhou et al., 2014) or waste carbon dioxide (Bhatnagar et al., 2011) from industrial activities have the potential to make microalgae a favorable alternative to traditional terrestrial biofeedstocks for conversion to liquid fuel. Microalgae can be grown using different media, including brackish, saline, and wastewater, which may prove critical for reducing the high synthetic fertilizer and WFs associated with microalgae cultivation. Moreover, concerns regarding fossil fuel price volatility, petroleum supply constraints, and issues of global climate change are key drivers that motivate ongoing research in microalgal biofuel systems (Ferrell and Sarisky-Reed, 2010).

In this chapter, several energy metrics have been introduced. Some have been already used in the fuel/energy literature to quantify the viability of fuel production. These metrics quantify the amount of *primary energy* consumed throughout the fuel supply chain. Primary energy refers to any naturally occurring renewable or nonrenewable energy form or carrier, such as crude oil, coal, natural gas, uranium, solar, wind, tidal, biomass, and geothermal energy, that has not been subjected to any conversion or transformation process.

The important issue of WF of biofuels has been analyzed. Previous studies have shown that the production of biofuels/bioenergy is water-resource-intensive. Therefore, large-scale cultivation and commercial production of biofuels and bioproducts may deplete freshwater resources that would otherwise be used for food and crop production, thus becoming the bottleneck of this route for renewable energy supply. Understanding and quantifying the WF of emerging biofuels/bioproducts is critical for guiding the environmentally sustainable development of the biofuels/bioenergy industry. An overview of the WF for select biofeedstocks, biofuels, and petroleum fuels is provided in this chapter as well.

A comparative well-to-wheel LCA was conducted as a case study to investigate the effects of various combinations of coproduct scenarios and allocation schemes on the environmental sustainability of microalgal-derived biodiesel and renewable diesel produced via an integrated open raceway pond (ORP) biorefinery.

The authors conclude the chapter with a review of the application of life-cycle assessment for environmental evaluation of emerging microalgal biofuel systems. By comparing the impacts of emerging microalgal systems with regard to key LCA-based energy, water, and climate change performance indicators, the environmental sustainability of the systems can be benchmarked and compared with other leading commercial biofuels and petroleum fuels. The life-cycle methodology has been applied to US Federal Legislation such as the renewable fuels standard (RFS2) for setting specific sustainability thresholds for emerging renewable fuels.

Chapter 9 is entitled "Methods and Tools for Sustainable Chemical Process Design." It was authored by the team of Carina L. Gargalo, Siwanat Chairakwongsa, Alberto Quaglia, Gürkan Sin, and Rafiqul Gani. They made the observation that the pressure on chemical and biochemical processes to achieve a more sustainable performance increases as the need to define a systematic and holistic way for accomplishing it becomes more urgent. They presented a multi-level computer-aided framework for systematic design (CAD) of more sustainable chemical processes. The framework allows the use of appropriate methods and tools in a hierarchical manner for multi-level/criteria analysis that helps generate more sustainable process designs. The application of the framework, the methods, and the tools are illustrated through a case study of the production of bioethanol from various renewable raw materials.

An important part of the chapter is the overview of CAD tools for sustainable chemical process design. They stated that even though general guidelines for sustainable process design have been proposed, such as the framework for sustainable design by Hacking (2008), covering a wide and qualitatively broad spectrum of processes, they are not sufficient for effecting a complete design. They need to be complemented with tools incorporating specific indicators for ensuring sustainability of the chemical processes.

The chapter focuses on the following:

- The concept of waste reduction (Singh and Falkenburg, 1993)
- MOO techniques, simulation, and uncertainty analysis for chemical and material selection, management, and planning (Diwekar, 2003)
- Hierarchical approach to evaluate new or existing processes flow sheets, incorporating the stakeholders at each level
- Methodologies integrating the environmental constraints into the early stages of process design, along with more traditional economic and technical criteria.

The chapter authors have developed a sustainable process synthesis and design framework, which has been divided into four parts:

1. Problem definition
2. Superstructure generation—base case(s) identification
3. Process simulation, sustainability analysis, and bottleneck identification
4. Generation and screening of new design alternatives.

This has been implemented into a software tool *SustainPro* (Carvalho et al., 2013), which applies a retrofit methodology for proposing new sustainable design alternatives for a base case design considering a set of mass and energy indicators. *SustainPro* software needs input data regarding the mass and energy balances obtained from steady-state process simulation, connectivity, compound chemical properties, compounds, and utility costs.

The software categorized the given mass and energy balance information into open and closed paths. A path represents a compound that follows a certain route in terms of mass. A closed path means that a certain compound is being recycled, and an open path means that a compound is entering and leaving the system, meaning that it is not being recycled.

Chapter 10, "Life-Cycle Assessment as a Comparative Analysis Tool for Sustainable Brownfield Redevelopment Projects: Cumulative Energy Demand (CED) and Greenhouse Gas Emissions (GHE)," was written by Thomas Brecheisen and Thomas Theis. The sustainable redevelopment of brownfields reflects a fundamental, yet logical, shift in thinking and policy-making regarding pollution prevention. LCA is a tool that can be used to assist in determining the conformity of brownfield development projects to the sustainability paradigm. In this chapter, LCA was applied to the assessment of two brownfield redevelopment projects, one in Milwaukee, Wisconsin, and the other in Chicago, Illinois, to compare the cumulative energy required to complete the following redevelopment stages: (i) brownfield assessment and remediation; (ii) building construction/rehabilitation and site development; and (iii) long-term operation.

The results presented in the chapter indicate that the redeveloped site buildings consume energy at a significantly lower intensity than the median energy use intensity (EUI) of commercial buildings in the midwestern United States. Brownfield redevelopment offers a significant opportunity for the construction of more energy-efficient buildings and the associated long-term conservation of natural resources. The source of a building's energy supply plays a large role in the life-cycle impacts, for example, coal-fired power plants emit more greenhouse gases (GHGs) than nuclear power plants. Site location also plays a significant role in the redevelopment life cycle; secondary impacts can erode gains in building energy efficiency significantly, with as much as 44% of a site's annual operational energy being consumed by commuter transportation.

The scope of the LCAs was designed to determine the CEDs and the associated GHG (Green House Gas) emissions required to perform the brownfield remediation activities and building rehabilitation activities and to operate the redeveloped sites. Three primary life-cycle stages were analyzed:

1. Brownfield assessment and remediation
2. Building rehabilitation/construction and site redevelopment
3. Energy consumed during the operation of the sites.

The SimaPro software package, version 7.3.3 (Product Ecology Consultants, 2012), was used to perform the calculations for the LCA. The tool for the reduction and assessment of chemical and other environmental impacts (TRACI 2) (EPA, 2014), a model developed by the United States EPA, was used to calculate the GHG emissions (GWP) (Bare, 2002). TRACI 2 is a midpoint-based life-cycle impact assessment methodology in which the impact categories were characterized at the midpoint level for reasons including a higher level of societal consensus concerning the certainties of modeling at this point in the cause-and-effect chain (Goedkoop et al., 2010).

The operational energy is the dominant life-cycle stage for both sites after 10 years of operation and it is evident that the long-term operational energy consumption will be the most energy-intensive life-cycle stage over a building life cycle of 50–100 y. There is a need to replace commuter trips with alternative trips to reduce the energy consumption and the air pollution associated with the transportation sector. Transportation planning decisions present a major challenge in moving toward sustainability (Eckelman, 2013). It has been shown that cities with higher population densities have lower per capita transportation emissions (Kennedy et al., 2009).

Chapter 11, "The Environmental Performance Strategy Map: An Integrated LCA Approach to Support the Strategic Decision-Making Process," was prepared by Luca De Benedetto and Jiří Jaromír Klemeš. They address the need for a novel and consistent approach that complements environmental and financial considerations. In the process, they have introduced three new concepts: the environmental performance strategy map, the environmental technology routing, and the E^3 (environmental end-to-end) methodology. The first is particular graphical mapping that permits combining the main environmental indicators (footprints) with the additional dimension of cost. The sustainable environmental performance indicator is therefore introduced and defined as a single measure for sustainability. Comparison of different options for strategic decision-making purposes can be enhanced and facilitated by the use of this indicator.

Environmental technology routing is designed to evaluate the impacts of components and process on the environmental performance strategy map. The basic idea presented is that each component of a specific product has a certain environmental burden and consequently contributes to the environmental footprints. Those contributions are identified as environmental performance points (EPP). When an item is routed through different technology processes to build the final product, it keeps contributing to the overall environmental burden. These contributions are again identified as EPP. The sum of all EPPs provides the basis for the creation of the characteristic environmental performance strategy map. This new approach allows wide flexibility. The impact of changing a component, material, or production process will be reflected immediately on the map. Finally, these methods have been compiled in an environmental end-to-end methodology to provide the best combination of software and the aforementioned methods.

Considering the interrelations and the complexity of environmental issues, decision-making in this field is very difficult. This is particularly true if it is not supported by analytical tools and reliable metrics. The sustainable environmental performance indicator, presented in the chapter, does not aim at being the single metric that policy-makers should rely on; instead, it should be a tool that is used together with financial indicators to help in the decision-making process and that could be adopted to compare different options and their comparative impacts on societies and the ecosystems. With its "deviation-from-target approach," SEPI—index, defined as the Sustainable Environmental Performance Indicator (SEPI), provides a way to measure the effectiveness of environmental policies against performance targets. The proposed targets can be easily adjusted to reflect local communities, counties, or national reference targets.

SEPI can be successfully applied to provide an overall indicator of the environmental performance of existing applications or can be used as a supporting tool in comparing competing options in a strategic decision-making process. This new approach has been demonstrated with a specific case to illustrate the main steps needed to find the right balance between cost and environmental impacts. This offers potential balancing and minimizing of environmental impacts, such as energy/carbon footprint, water, emissions, and working environment, and quantifying them into one indicator.

The SEPI can be considered an important step in the debate regarding defining the appropriate metrics and methodologies for evaluating environmental performance. The authors point out the difficulties in converting some of the footprints to area requirements. In general, although it is relatively easy to express as area processes that are area-based, such as those in agriculture, converting to area processes that are not primarily area-based, such as a chemical process, can prove to be problematic. Another important development could be addressing the long-term human health and ecosystem degradation costs. Identifying best practices and costs associated with those, a target value could be set. This should be the basis for a comparison with current costs and practices.

Chapter 12, "Green Supply Chain towards Sustainable Industry Development," was written by Hon Loong Lam, Bing Shen How, and Boon Hooi Hong. Most of the latest developments in this field have been discussed in previous chapters of this book. To link-up the green technologies from pretreatment to process and delivery, green supply chain development is an essential activity.

A sustainable supply chain (or network) could be designed by appropriate operational management methods and optimization to reduce the environmental impact along the life cycle of a green product, from the raw material to the end product. The activities involved in the supply networks should lead to economic growth, environmental protection, and social progress for green technology utilization. To achieve sustainable development, the supply chain focuses on more than the transportation/logistic task. The special focus has to be given to the latest conservation of biomass (mass and energy) used in the process, the possibility of integrating green resources, the consideration of the industrial symbiosis relationship, and the network synthesis with multiple objectives of environmental, technical, economic, safety, and social factors. The chapter provides an overview of such approaches and the method of network synthesis to achieve this goal.

The process integration approach utilizes the developed regional energy clustering (REC) algorithm to partition biomass region into a number of clusters. Clusters are defined as geographic concentrations of interconnected suppliers, services providers, associated institutions, and customers in a region that compete but also cooperate (Porter, 1998). A cluster combines smaller zones to secure sufficient energy balance within the cluster (Lam et al., 2011). A zone can be a province/county, a community settlement/borough, an industrial park, or an agriculture compound from the studied region. The REC is used to manage the energy flow among the zones.

The P-graph approach was initially introduced by Friedler et al., (1992) and has been implemented in the systematic optimal design, including industrial processes synthesis and supply network synthesis (Friedler et al., 1996). P-graphs are bipartite graphs, with each comprising nodes for a set of materials, a set of operating units, and arcs linking them. There are two main algorithms—solution structure generator (SSG) and accelerated branch-and-bound (ABB)—used in the P-graph framework to determine all the feasible solutions and to obtain the optimal network designs. This is a very efficient graph theoretic method used to obtain the optimal (or near-optimal) green supply chain network (Lam et al., 2012).

This chapter provides a general overview of the model features and several methods of constructing the model. The model can be easily expanded and used with other strategies such LCA and game theory.

Chapter 13 provides a discussion of "Supply and Demand Planning and Management Tools Towards Low Carbon Emissions," authored by Zainuddin Abdul Manan and Sharifah Radifah Wan Alwi. The authors observed that a comprehensive and effective carbon emission reduction strategy for a region or a country entails addressing both energy *supply* and energy *demand* sides, covering electricity as well as thermal energy that affect carbon emissions. For stationary emission point sources, carbon is mainly emitted from power generation plants as well as from combustion of fuels. Three key categories of measures to reduce carbon emissions from stationary point sources include:

1. Supply management: Optimization of energy supply mix and loss reduction in energy supply
2. Demand management: Increasing energy efficiency at the demand side
3. End-of-pipe solution: Minimizing emissions through carbon reuse, demand, and source manipulations, regeneration, and carbon emission offsetting from point sources.

The first approach using the Pinch Analysis technique for carbon-constrained energy planning was developed by Tan and Foo (2007). They assumed that within a system there exists a set of energy sources, each with a specific carbon intensity characteristic of the fuel or the technology. At the same time, the system was also assumed to contain a set of demands, each with a specified carbon footprint limit. The aim of developing the approach was to determine the minimum amount of zero carbon energy resources needed to meet the specified emission limits and also to determine how the energy sources should be allocated to meet the different demands while complying with emission limits. This section, using the approach developed by Tan and Foo (2007), attempts to explain the method of generating the Energy Demand Composite Curve, the Energy Supply Composite Curve, targeting demand for zero carbon energy, and effects of reduced emission limits on zero carbon energy demand by using the hypothetical case.

The supply and demand planning and management tools as well as the end-of-pipe solution for carbon emission reduction from stationary Pinch point sources are further described. The techniques presented for carbon supply and demand planning include carbon Pinch Analysis design method, carbon footprint improvement with Pinch Analysis, and carbon emission Pinch Analysis (CEPA) following the methodology presented by Tan and Foo (2013).

POLICIES TOWARD THE SUSTAINABILITY

Chapter 14 was written by Richard C. Darton ("Setting a Policy for Sustainability: The Importance of Measurement"). The author's premise is that assessing the sustainability impact of any system—factory, business, institution, supply chain, industry, city, province, country—is essential for setting a policy for sustainability. The assessment can be performed by using a set of indicators. However, which indicator to select is not straightforward. This chapter describes the process analysis method (PAM), in which the system is considered as a group of processes causing triple bottom line impacts on the environment, the economy, and the human/social domain. The indicators are then related to these impacts. The PAM uses a set of sustainability indicators and metrics tailored to the particular system, but that should be similar for similar systems, facilitating comparison and benchmarking. The value of a particular indicator can be traced back through the analysis to a particular process activity, which is especially helpful in adjusting policy, because *cause* is linked to *effect* by the method.

A discussion is presented regarding a particular case in which none of the available options is completely sustainable; it may also be the case that one option is better in some regard (e.g., consumes less resources) and another is better in some other way (e.g., produces more human benefit). Although both of these difficulties will often be encountered, one cannot allow them to prevent choices being made using sustainability criteria. The ability to assess the sustainability of a development option is central to applying the concept in practice. The assessment allows the user to choose the "best" option based on the user-provided definition or to reject all options as insufficiently sustainable. Because sustainability is a holistic quality with many different aspects, much attention has been given to the criteria that should be used and formulated in terms of sets of indicators that address the important factors (Pinter et al., 2005; Bell and Morse, 2008).

The author discussed that in the commercial world, the Dow Jones Sustainability Index (DJSI) has been developed to monitor how companies treat economic, social, and environmental issues with regard to mitigating risk and exploiting opportunity (Searcy, 2009). The set of criteria (indicators) in the DJSI are designed to assess corporate sustainability, "a business approach that creates long-term shareholder value by embracing opportunities and managing risks deriving from economic, environmental, and social developments." This understanding of sustainability differs from that of the Brundtland Report (1987), illustrating the cardinal importance of being clear about the working definition of sustainability used in making any assessment. Dyllick and Hockerts (2002) pointed out that many companies have adopted eco-efficiency as a guiding principle, neglecting impacts on human and social affairs. Any assessment based on a restricted understanding of sustainability is bound to result in only a partial assessment of sustainability.

Another initiative overviewed in this chapter is the Global Reporting Initiative (GRI, 2002). This has developed a framework for facilitating the reporting of sustainability performance in a consistent and comprehensive way to give such reports a similar credibility to the financial reports that have long been standard practice in the world of audited accounts. The GRI guidelines include a set of indicators with broad coverage demonstrating *performance* against *goals* for the organization. Thus, policy is built into the assessment. The indicators are generic in nature and, for an assessment that is more focused either on business sector – related concerns or local conditions, the GRI is developing sector-specific supplements and national annexes. An example of sector-specific sustainability metrics is the set developed by The Institution of Chemical Engineers (2002).

These are consistent with the GRI approach and are intended for use by the chemical manufacturing industry.

The final method analyzed in this chapter is the PAM, which interprets and applies the concept of sustainable development to the practical problem of assessing the sustainability of a system in operation. The method is systemic, hierarchical, logical, and communicable, and it results in a set of indicators that aims to be comprehensive in its coverage of all three domains of sustainability. Because the selection of indicators for any particular application is based on an inventory of the system processes that give rise to sustainability impacts, the indictor set is tailored to that particular application. It is also based on a clearly stated definition of what constitutes a sustainable outcome. This contrasts with the situation sometimes encountered when indicators have been selected in an unsystematic fashion from published lists. The PAM bears some similarity to the pressure-state-response (PSR) approach, which also considers cause and impact (OECD, 1993).

Chapter 15, "Sustainability Assessments of Buildings, Communities, and Cities," was authored by Umberto Berardi. The author sees sustainability assessments in the built environment as an important part of addressing sustainability issues through multicriterion rating systems. Recent interpretations of the concepts of sustainability, assessment, and sustainability assessment in the built environment are discussed before reviewing existing assessment systems. A comparison between assessment systems at different scales is performed before focusing on the need of cross-scale evaluations. The tools reviewed include BREEAM, LEED, and CASBEE. The comparison of these tools shows that previous systems often accept weak sustainability, where natural resources may be substituted by other priorities. This chapter also shows that the dynamicity of the built environment suggests considering the sustainability assessment systems as tools to monitor the evolution of the built environment as well as the lives of citizens. Finally, the chapter shows the importance of adapting sustainability goals and indicators to each specific situation.

The dependence on scale has shown the importance of enlarging the spatial cross-boundaries in the evaluation of sustainability. When sustainability assessment at the community level was introduced, it seemed to represent the minimum unit of analysis for a complete evaluation, especially for the social and economic dimensions of sustainability. Later, assessors started considering criteria that could be solved at the level of a city. Among these criteria were:

- Adhere to ethical standards during development by ethical trading throughout the supply chain and by providing safe and healthy work environments and conditions
- Provide a mix of type zones
- Integrate development in the local context, conserve local heritage, and culture
- Guarantee access to local infrastructure and services to all citizens
- Involve all interested parties through a collaborative approach

- Provide social and cultural value over time and for all the people.

The author concludes that the chapter has shown several limits of the available systems because they lack appropriate assessment of social, economic, and environmental sustainability. The comparison among systems has revealed that they generally promote weak sustainability and accept that economic development can reduce natural capitals. This reduces their capability to measure sustainability in the long term.

CONCLUSION

This chapter provides an introductory summary of the core book chapters. The review of the presented material demonstrates the intensity of identifying and quantifying environmental and social impacts and stability. This indicates that the area has not yet reached full maturity and more intensive research is highly desirable. However, despite that, the presented contributions present a number of methodologies with demonstration applications that demonstrate high applicability in most cases. The collection of chapters provides readers with the opportunity to derive their own combinations of metrics and algorithms to use in assessing and measuring sustainability for their particular applications.

It is up to the readers of this book to assess how well the presented material fits their needs. Sustainability assessment and measurements has been developing quickly, and the authors are sure that more and more contributions are going to follow.

This book represents an attempt to bring together the leading researchers in the area of sustainability measurement and assessment. It has been most appreciated that the majority of them accepted the challenge and delivered high-quality contributions in a timely manner.

If this book encourages the readers to consider and think over those issues and, even better, to enter new research into the field, then the purpose of this book has been achieved.

REFERENCES

AIChE, 2014. Sustainability Index <www.aiche.org/ifs/resources/sustainability-index> (accessed 14.04.2014).

Azapagic, A., Clift, R., 1999. Life cycle assessment and multiobjective optimisation. J. Cleaner Prod. 7, 135–143.

Bakshi, B., Fiksel, J., 2003. The quest for sustainability: challenges for process systems engineering. AIChE J. 49, 1350.

Bare, J.C., 2002. Traci. J. Ind. Ecol. 6, 49–78.

BASF, 2014. How does the eco-efficiency analysis work and which criteria are used? <www.basf.com/group/corporate/en/function:rendering-service:/faqsearch/faqsearch-result/resultCat/faq-sd_eco-efficiency/functions/faqsearch/faqsearch#0900dea680518f46> (accessed 16.07.2014).

Bell, S., Morse, S., 2008. Sustainability Indicators: Measuring the Immeasurable? Earthscan, London, UK.

Bhatnagar, A., Chinnasamy, S., Singh, M., Das, K.C., 2011. Renewable biomass production by mixotrophic algae in the presence of various carbon sources and wastewaters. Appl. Energy 88, 3425–3431.

Brundtland, G. (Ed.), 1987. Our Common Future: The World Commission on Environment and Development. Oxford University Press, Oxford, UK.

Cabezas, H., 2013. In: Cabezas, H., Diwekar, U. (Eds.), Sustainability Indicators and Metrics in Sustainability: A Multi-Disciplinary Perspective, 197. Bentham e-books, Bejing, China.

Cabezas, H., Fath, B.T., 2002. Towards a theory of sustainable systems. Fluid Phase Equilib. 194–197, 3–14.

Cabezas, H., Pawlowski, C.W., Mayer, A.L., Hoagland, N.T., 2005. Simulated experiments with complex sustainable systems: ecology and technology. Resour. Conserv. Recy. 44, 279–291.

Carvalho, A., Matos, H.A., Gani, R., 2013. SustainPro—a tool for systematic process analysis, generation and evaluation of sustainable design alternatives. Comput. Chem. Eng. 50, 8–27.

Conway, G.R., Barbier, E.B., 2013. After the Green Revolution: Sustainable Agriculture for Development. Routledge, Taylor & Francis Group, London, UK, Sterling, VA, USA, 210 ps.

Čuček, L., Klemeš, J.J., Kravanja, Z., 2012a. A review of footprint analysis tools for monitoring impacts on sustainability. J. Cleaner Prod. 34, 9–20.

Čuček, L., Klemeš, J.J., Varbanov, P.S., Kravanja, Z., 2012b. Reducing the dimensionality of criteria in multi-objective optimisation of biomass energy supply chains. Chem. Eng. Trans. 29, 1231–1236.

Čuček, L., Klemeš, J.J., Varbanov, P.S., Kravanja, Z., 2013. Dealing with high-dimensionality of criteria in multiobjective optimization of biomass energy supply network. Ind. Eng. Chem. Res. 52, 7223–7239.

Čuček, L., Klemeš, J.J., Kravanja, Z., 2014. Objective dimensionality reduction method within multi-objective optimisation considering total footprints. J. Cleaner Prod. 71, 75–86.

Daimler, 2011. Well-to-Wheel. Available at: <www2.daimler.com/sustainability/optiresource/index.html> (accessed 21.5.2014).

Diwekar, U.M., 2003. Greener by design. Environ. Sci. Technol. 37 (23), 5432–5444.

Dyllick, T., Hockerts, K., 2002. Beyond the business case—for corporate sustainability. Bus. Strategy Environ. 11, 130–141.

Eason, T., Garmenstani, A.S., 2012. Cross-scale dynamics of a regional urban system through time. Region et Developpement 45, 55–77.

Eckelman, M.J., 2013. Life cycle assessment in support of sustainable transportation. Environ. Res. Lett. 8 (2013), 021004.

EPA, United States Environmental Protection Agency, 2014. Tool for the Reduction and Assessment of Chemical and Other Environmental Impacts (TRACI) <www.epa.gov/nrmrl/std/traci/traci.html> (accessed 09.10.2014).

Esty, D.C., Levy, M.A., Kim, C.H., De Sherbinin, A., Srebotnjak, T., Mara, V., 2008. Environmental Performance Index. Yale Center for Environmental Law and Policy, New Haven, USA, 382.

Fang, K., Heijungs, R., de Snoo, G., 2013. The footprint family: comparison and interaction of the ecological, energy, carbon and water footprints. Metallurgical Res. Technol. 110, 77–86.

Fang, K., Heijungs, R., de Snoo, G.R., 2014. Theoretical exploration for the combination of the ecological, energy, carbon and water footprints: overview of a footprint family. Ecol. Indic. 36, 508–518.

Ferrell, J., Sarisky-Reed, V., 2010. National Algal Biofuels Technology Roadmap. US Dept. Energy, Office of Energy Efficiency and Renewable Energy, Biomass Program, University of Maryland—College Park, USA.

Finkbeiner, M., 2009. Carbon footprinting—opportunities and threats. Int. J. Life Cycle Assess. 14, 91–94.

Fisher, R.A., 1922. On the mathematical foundations of theoretical statistics. Philos. Trans. R. Soc. Lond. 222, 309–368.

Freudenberg, M., 2003. Composite Indicators of Country Performance: A Critical Assessment. OECD (Organisation for Economic Co-operation and Development) Publishing, Paris, France. <http://dx.doi.org/doi:10.1787/405566708255> (accessed 27.11.2014).

Frieden, B.R., 1998. Physics from Fisher Information. A Unification. Cambridge University Press, New York, NY, pp. 318.

Friedler, F., Tarjan, K., Huang, Y.W., Fan, L.T., 1992. Graph-theoretic approach to process synthesis: axioms and theorems. Chem. Eng. Sci. 47 (8), 1973–1988.

Friedler, F., Varga, J.B., Feher, E., Fan, L.T., 1996. Combinatorially accelerated branch-and-bound method for solving the MIP model of Process Network SYnthesis. In: Floudas, C.A., Pardalos, P.M. (Eds.), State of the Art in Global Optimization. Kluwer Academic Publishers, Boston, MA, pp. 609–626.

Fuel-Cell e-Mobility, 2014. Well-to-Wheel—Integral Efficiency Analysis. Available at: <www.fuel-cell-e-mobility.com/h2-infrastructure/well-to-wheel-en/> (accessed 21.05.2014).

Galli, A., Wiedmann, T., Ercin, E., Knoblauch D., Ewing B., Giljum S., 2011. Integrating Ecological, Carbon and Water Footprint: Defining the "Footprint Family" and Its Application in Tracking Human Pressure on the Planet. Technical Document, Surrey, UK.

Galli, A., Wiedmann, T., Ercin, E., Knoblauch, D., Ewing, B., Giljum, S., 2012. Integrating Ecological, Carbon and Water footprint into a "Footprint Family" of indicators: definition and role in tracking human pressure on the planet. Ecol. Indic. 16, 100–112.

Global Reporting Initiative, 2002. Sustainability Reporting Guidelines. Global Reporting Initiative, Amsterdam, The Netherlands.

Goedkoop, M., Oele, M., An De Schryver, Vieria, M., Hegger, S., 2010. SimaPro 7 Database Manual. PRé Consutants. The Netherlands.

Gonzalez-Mejia, A.M., Eason, T.N., Cabezas, H., Suidan, M.T., 2012. Computing and interpreting Fisher Information as a metric of sustainability: regime changes in the United States air quality. Clean Technol. Environ. Policy 14 (5), 775–788.

Graedel, T.E., Allenby, B.R., 2010. Industrial Ecology and Sustainable Engineering. Prentice Hall, Upper Saddle River, New Jersey, USA, 320 ps.

Guillén-Gosálbez, G., 2011. A novel MILP-based objective reduction method for multi-objective optimization: application to environmental problems. Comput. Chem. Eng. 35, 1469–1477.

Hoekstra, A.Y., Mekonnen, M.M., 2012. The water footprint of humanity. Proc. Natl. Acad. Sci. 109 (9), 3232–3237.

Hoekstra, A.Y., Chapagain, A.K., Aldaya, M.M., Mekonnen, M.M., 2011. The Water Footprint Assessment Manual: Setting the Global Standard. Earthscan, London, UK.

Huang, H., Lu, J., 2014. Identification of river water pollution characteristics based on projection pursuit and factor analysis. Environ. Earth Sci.1–9.

Institution of Chemical Engineers, 2002. The Sustainability Metrics: Sustainable Development Progress Metrics Recommended for Use in the Process Industries. Institution of Chemical Engineers, Rugby, UK.

Karunanithi, A.T., Garmestani, A.S., Eason, T.N., Cabezas, H., 2011. The characterization of socio-political instability, development and sustainability with Fisher information. Glob. Environ. Change 21, 77–84.

Kennedy, C., Steinberger, J., Gasson, B., Hansen, Y., Hillman, T., Havranek, M., Pataki, D., Phdungsilp, A., Ramaswami, A., Mendez, G.V., 2009. Greenhouse gas emissions from global cities. Environ. Sci. Technol. 43, 7297–7302.

Kettl, K.-H., Niemetz, N., Sandor, N., Eder, M., Narodoslawsky, M., 2011. Regional Optimizer (RegiOpt)—sustainable energy technology network solutions for regions. Comput. Aided Chem. Eng. 29, 1895–1900.

Lam, H.L., Klemeš, J., Kravanja, Z., 2011. Model-size reduction techniques for large-scale biomass production and supply networks. Energy 36 (8), 4599–4608.

Lam, H.L., Klemeš, J., Varbanov, P., Kravanja, Z., 2012. P-graph synthesis of open-structure biomass networks. Ind. Eng. Chem. Res., 172–180.

Mayer, A.L., Pawlowski, C.W., Fath, B.D., Cabezas, H., 2007. Applications of Fisher information to the management of sustainable environmental systems. In: Frieden, B.R., Gatenby, R.A. (Eds.), Exploratory Data Analysis Using Fisher Information. Springer-Verlag, London, UK, pp. 217–244.

Narodoslawsky, M., Krotscheck, C., 1995. The sustainable process index (SPI): evaluating processes according to environmental compatibility. J. Hazard. Mater. 41 (2 + 3), 383–397.

OECD, 1993. OECD core set of indicators for environmental performance reviews, OECD Environment Monographs No. 83. OECD, Paris, France.

Pawlowski, C.W., Cabezas, H., 2008. Identification of regime shifts in time series using neighborhood statistics. Ecol. Complex. 5, 30–36.

Pinter, L., Hardi, P., Bartelmus, P., 2005. Sustainable Development Indicators: Proposals for a Way Forward; UNDSD Report. IISD, Winnipeg, MB, Canada.

Porter, M., 1998. Cluster and the new economics of competition. Harv. Bus. Rev. Cambridge, MA, USA, November–December, 76–90.

Product Ecology Consultants, 2012. SimaPro 7.3 Installation Manual. Amersfoort, The Netherlands.

Sabio, N., Kostin, A., Guillén-Gosálbez, G., Jiménez, L., 2012. Holistic minimization of the life cycle environmental impact of hydrogen infrastructures using multi-objective optimization and principal component analysis. Int. J. Hydrogen Energy 37, 5385–5405.

Schlesinger, W.H., 1997. The Global Cycles of Nitrogen and Phosphorus in Biogeochemistry: An Analysis of Global Change, second ed. Academic Press, Imprint of Elsevier, Amsterdam, the Netherlands.

SCOPUS, Elsevier, <www.scopus.com> (accessed 12.10.2014).

Searcy, C., 2009. The Role of Sustainable Development Indicators in Corporate Decision-Making. IISD, Winnipeg, MB, Canada.

Sengupta, D., Mukherjee, R., Sikdar, S.K., 2014. Environmental sustainability of countries using the UN MDG indicators by multivariate statistical methods. Environ. Prog. Sustainable Energy. Available at: http://dx.doi.org/doi:10.1002/ep.11963.

Singh, N., Falkenburg, D.R., 1993. A Green Engineering Framework for Concurrent Design of Products and Processes. Wayne State University, Detroit, MI, USA.

SPIonWeb, 2014, Ecological Footprint Calculator, <spionweb.tugraz.at/en/welcome> (accessed 11.03.2014).

Tan, R.R., Foo, D.C.Y., 2007. Pinch analysis approach to carbon-constrained energy sector planning. Energy 32 (8), 1422–1429.

Tan, R.R., Foo, D.C.Y., 2013. Chapter 17—Pinch Analysis for sustainable energy planning using diverse quality measures. In: Klemeš, J.J. (Ed.), Woodhead Publishing Series in Energy. Woodhead Publishing/Elsevier, Cambridge, UK, pp. 505–523. Handbook of Process Integration (PI), ISBN 9780857095930, http://dx.doi.org/doi:10.1533/9780857097255.4.505.

UN, United Nations, 2008. World urbanization prospects: the 2007 revision population database. <esa.un.org/unup/> (accessed 10.10.2014).

Vujanović, A., Čuček, L., Pahor, B., Kravanja, Z., 2014. Multi-objective synthesis of a company's supply-network by accounting for several environmental footprints. Process Saf. Environ. Prot. 92 (5), 456–466.

World Commission on Environment and Development, 1987. Our Common Future. Oxford University Press, Oxford, UK.

Zhou, W., Chen, P., Min, M., Ma, X., Wang, J., Griffith, R., Hussain, F., Peng, P., Xie, Q., Li, Y., Shi, J., Meng, J., Ruan, R., 2014. Environment-enhancing algal biofuel production using wastewaters. Renew. Sustainable Energy Rev. 36, 256–269.

CHAPTER

Engineering sustainability

1

Urmila Diwekar

Vishwamitra Research Institute, Crystal Lake, IL, USA

INTRODUCTION

The most widely accepted definition of sustainability or sustainable development is attributed to the World Commission on Environment and Development (1987). The Commission stated in its final report that "Sustainable development is development that meets the needs of the present without compromising the ability of future generations to meet their own needs." This definition clearly defines sustainability on a larger timescale by alluding to the future (Cabezas, 2013). Engineers are good at decision making when they can model the phenomenon correctly. However, sustainability that starts with green manufacturing involves beginning decision making at early stages of design and considering objectives that are not well quantified. This involves multi-criteria decision making in the face of uncertainties. Industrial ecology deals with complex industrial networks and dynamic decision making. Sustainability goes beyond industrial ecology when the system boundaries extend to our planet and long-term impacts need to be considered. Dynamic decision making and adaptive management are needed to address the complete problem of sustainability. Systems analysis approaches can provide tools for addressing these problems. However, there are challenges in applying current approaches to sustainable system analysis. Forecasting is an important part of decision making because sustainability is important for the future. Given that the future brings many uncertainties, addressing these uncertainties in both quantitative and qualitative forms is important in decision making. These uncertainties could be static or dynamic. It should be remembered that sustainability is a path, and sustainable decision making involves finding this path through time. There are no absolutes in sustainability; it is always relative (Sikdar, 2014). We present the problem of sustainability, starting with sustainable manufacturing as extending boundaries of traditional decision making, followed by systems analysis approaches and challenges to solve this problem.

Assessing and Measuring Environmental Impact and Sustainability.

EXTENDING BOUNDARIES

The Burndland commission definition (World Commission on Environment and Development, 1987) of sustainability originated from the concerns that the current trends in population and economic developments are not sustainable. Coupled with adverse environmental impacts like global climate change, degradation of air, water, and land, depletion of natural resources, including freshwater minerals, loss of agricultural land because of deforestation, soil erosion, and urbanization are threatening the ecosystem (Bakshi and Fiksel, 2003). To make the planet sustainable, the efforts should be made at all levels of ecosystem. At the industrial level, green engineering means green processes, green products, green energy, and eco-friendly management by considering environmental and other societal impacts as objectives and not constraints used in traditional design. Figure 1.1 shows the integrated framework recently developed (Diwekar, 2003) to include

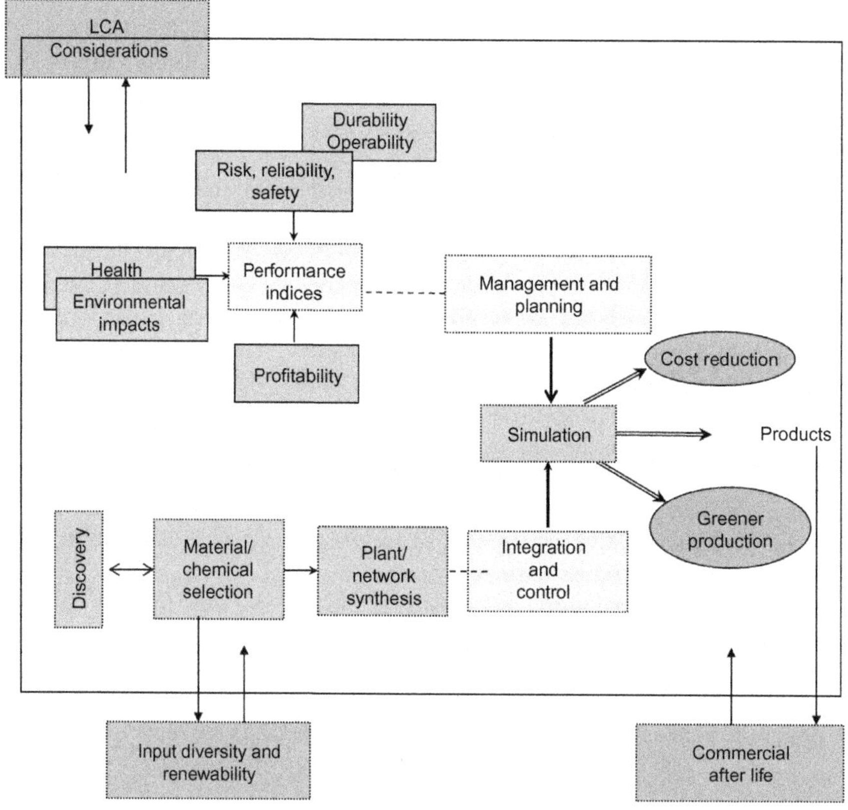

FIGURE 1.1

Integrated framework for green process design (Diwekar, 2003).

the green engineering principles at all stages. Unlike the traditional process design in which engineers are looking for low-cost options, environmental considerations include various objectives like the long-term and short-term environmental impacts. This framework includes decisions at all levels, starting from the chemical or material selection and the process synthesis stages, to the management and planning stage, which is linked to green objectives and goals (top left corner of the Figure 1.1). It is important to start as early as possible in the design process to enhance the impact of waste minimization (Figure 1.2) (Yang and Shi, 2000).

It should be remembered that as we extend the traditional boundaries of process and plant design to include early decisions like material selections, the models and data available for the analysis are fraught with uncertainties. Therefore, to address sustainability at the plant level, we need to use multi-objective methods in the face of uncertainties.

Life cycle considerations and commercial afterlife are also included in this decision making. They are shown beyond the boundaries of the framework, because life cycle analysis and afterlife analysis involve extending the boundary of the design further. Obviously, uncertainties increase as we extend the framework.

This framework shows that the first step toward green engineering is to extend the traditional boundaries of design. Because sustainability is a property of the entire system, it requires boundaries of the design to be greatly expanded beyond green design, green products, and green management to industrial ecology and to the ecosystem of the entire planet (Figure 1.3) (Diwekar and Shastri, 2010).

The next step toward sustainability in Figure 1.3 is industrial ecology. Industrial ecology is the study of the flows of materials and energy in industrial

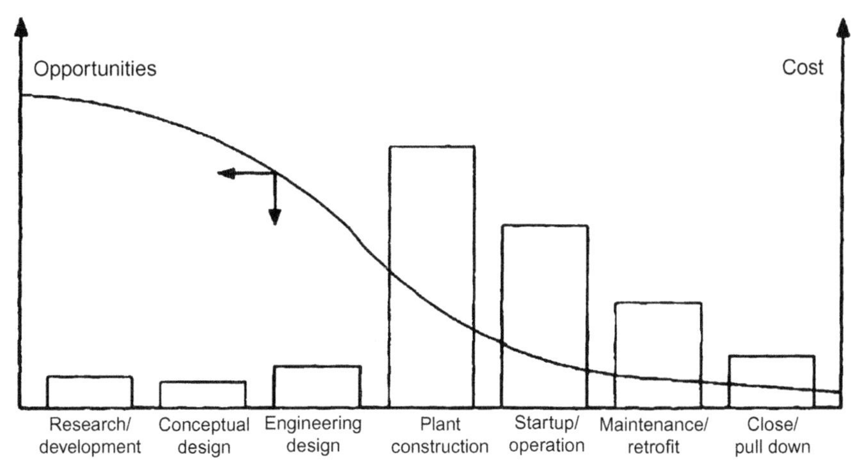

FIGURE 1.2

Opportunities of environmental impact minimization along the process life cycle.

Reproduced from Yang and Shi (2000).

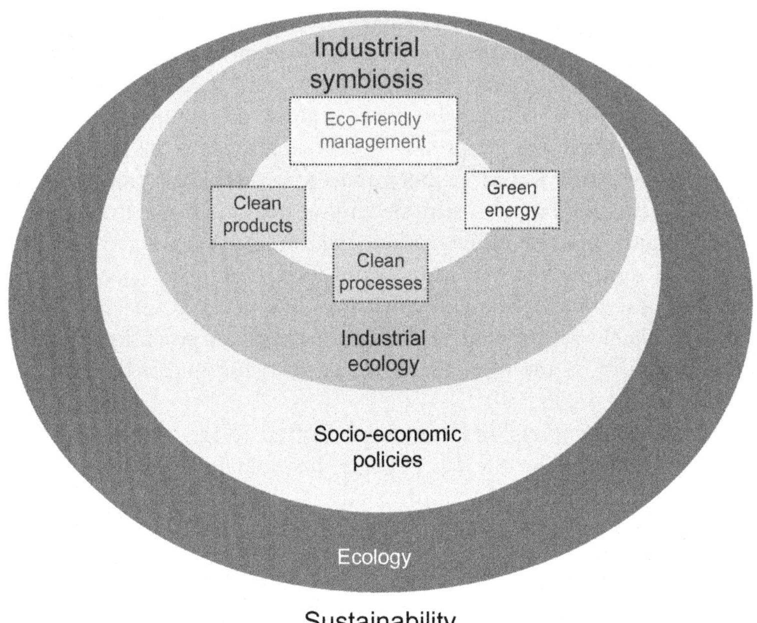

FIGURE 1.3

From green design to industrial ecology to sustainability (Diwekar and Shastri, 2010).

and consumer activities, of the effects of these flows on the environment, and of the influences of economic, political, regulatory, and social factors on the use, transformation, and disposition of resources (White, 1994). Industrial ecology applies the principles of material and energy balance traditionally used by scientists and engineers to analyze well-defined ecological systems or industrial unit operations to more complex systems involving natural and human interaction. These systems can involve activities and resource utilization over scales ranging from single industrial plants to entire sectors, regions, or economies. In so doing, the laws of conservation must consider a wide range of interacting economic, social, and environmental indicators. Figure 1.4 presents a conceptual framework for industrial ecology applied at different scales of spatial and economic organization evaluating alternative management options using different types of information, tools for analysis, and criteria for performance evaluation. As one moves from the small scale of a single-unit operation or industrial production plant to the larger scales of an integrated industrial park, community, firm, or sector, the available management options expand from simple changes in process operation and inputs to more complex resource management strategies, including integrated waste recycling and reuse options. The information changes from quantitative to order of magnitude to qualitative, and uncertainties increase. Special focus has been placed on implementing the latter via industrial symbiosis, for example,

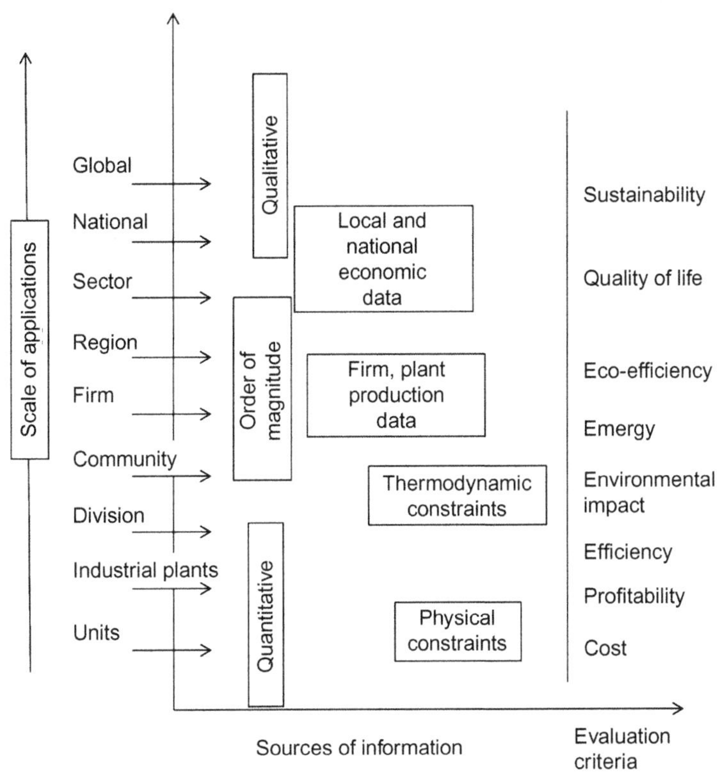

FIGURE 1.4

Industrial ecology sources of information and evaluation criteria.

through the pioneering work of integrating several industrial and municipal facilities in Kalundborg, Denmark (Ehrenfeld and Gertler, 1997). Business case of sustainability (Beloff, 2013) often describes the triple bottom line approach, namely, economic performance, environmental performance, and social performance of industry. However, industrial ecology advocates going beyond this triple bottom line. Industrial networks comprise any number of organizations that are linked to each other through the exchange of resources. A systems approach is a necessary prerequisite to understanding interaction between organizations and its consequences for the socio-economic and biophysical system in which organizations function. The function and structure of the industrial network are influenced by competition and cooperation between organizations trying to maintain or improve their position within the network (Petrie et al., 2013). It is this dynamic behavior that makes industrial networks inherently uncertain.

Sustainability goes beyond industrial ecology, where the natural boundary for sustainability science is the entire planet Earth. Earth can be considered a closed

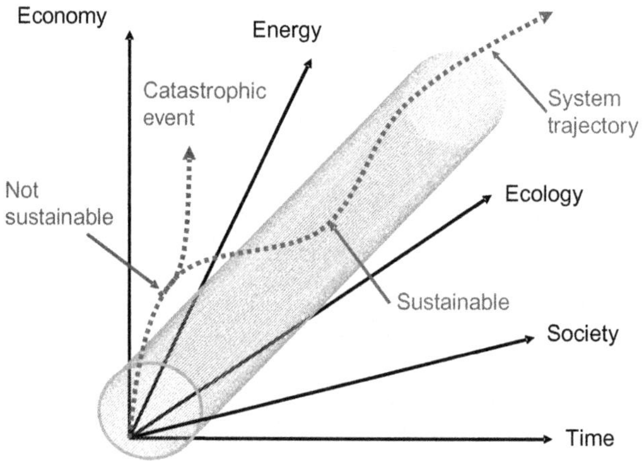

FIGURE 1.5

Sustainability as a path (Cabezas, 2013).

system, at least on the timescale of human history. The difficulty is that on the planetary scale when ecosystems, societies, and economies are considered, the system becomes very large and very complex. Uncertainties become time dependent and decision forecasting is essential to study the sustainability of the planet. Cabezas (2013) defines sustainability as a path or system trajectory moving in various dimensions of ecology, society, time, economy, and energy. This is depicted conceptually in Figure 1.5.

In summary, to achieve sustainability, green engineering principles need to be followed at the plant level, which extends the traditional boundary of design. Multi-objective analysis approaches that take into consideration uncertainties are necessary at this stage. However, sustainability considerations are not restricted to the plant level, but one has to consider large-scale industrial networks and society in general. The problem becomes complex and dynamic decision making enters the picture. Because sustainability involves our planet Earth and our ability to save the planet for future generations, a systems analysis approach involving dynamic decision making, forecasting, and considerations of time-dependent uncertainties is necessary. The following section describes some of these approaches and challenges.

SYSTEMS ANALYSIS APPROACH TO SUSTAINABILITY AND ENGINEERING CHALLENGES

The systems analysis approach to sustainability involves sustainable decision making at all levels of systems. An engineering systems analysis approach depends on optimization methods for decision making. This particularly involves providing attention to the multi-objective nature of the problem and uncertainties, decision forecasting, and the need for including adaptive learning instead of prescriptive decision making. The difficult part of this kind of decision making is dealing with uncertainties. The following section addresses uncertainties in these systems.

UNCERTAINTY ANALYSIS

Static uncertainties

At the plant level, green engineering demands green thinking at very early stages of design. The data and models at early stages are not well developed and, hence, involve significant uncertainties. Similarly, models of impact assessment are not accurate. In general, these uncertainties can be characterized as scalar uncertainties. These uncertainties can be characterized and quantified in terms of probabilistic distributions. Some of the representative distributions are shown in Figure 1.6. The type of distribution chosen for an uncertain variable reflects the amount of information that is available. For example, the uniform and log-uniform distributions represent an equal likelihood of a value lying anywhere

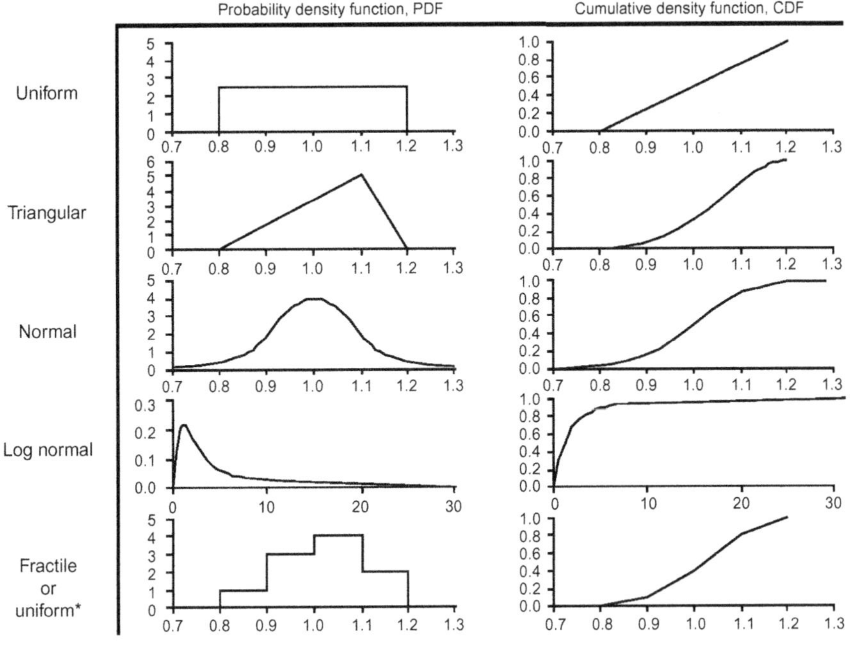

FIGURE 1.6

Examples of probability distribution functions.

within a specified range on either a linear or a logarithmic scale, respectively. Further, a normal (Gaussian) distribution reflects a symmetric but varying probability of a parameter value being above or below the mean value. In contrast, log-normal and some triangular distributions are skewed such that there is a higher probability of values lying on one side of the median than the other. A beta distribution provides a wide range of shapes and is a very flexible means of representing variability over a fixed range. Modified forms of these distributions, uniform* and log-uniform*, allow several intervals of the range to be distinguished. Finally, in some special cases, user-specified distributions can be used to represent any arbitrary characterization of uncertainty, including chance distribution (i.e., fixed probabilities of discrete values).

Once the uncertainties are modeled as probabilistic distributions, sampling methods (Diwekar and Ulas, 2007) can then be used to generate various scenarios for analysis.

Dynamic uncertainties

As stated, at the industrial ecology and ecosystem level, the issues of uncertainties become complex as: (i) the uncertainty in measuring the state of the system at a point in time so a projection can be initialized; (ii) the impossibility of developing a complete list of variables for complex natural systems, for example, ecosystems and economies; and (iii) the impossibility of modeling, representing, or understanding all of the relationships that exist between the variables of complex natural systems (Petrie et al., 2013). These issues call for consideration of dynamic uncertainties. In the literature, there are two ways to model these dynamic uncertainties. These are described in the following section.

Ito process representation

Diwekar et al. stated that because dynamic uncertainties are widely encountered in finance, the approach taken in finance in modeling dynamic uncertainties can be used for dynamic uncertainties encountered in dynamic systems in sustainability. They propose using Ito processes (Diwekar, 2008) for modeling dynamic uncertainties. Dynamic uncertainties in general result in stochastic dynamic models that require stochastic calculus for simulation and optimization. Ito (1951, 1974) developed Ito calculus to solve the stochastic differential equations resulting from inclusion of Ito processes. Ito processes have the ability to forecast future uncertainties from past data.

Ito processes are a large class of continuous time stochastic processes. One of the simplest examples of a stochastic process is the random walk process. The Wiener process, also called a Brownian motion, is a continuous limit of the random walk and serves as a building block for Ito processes through the use of proper transformations.

A Wiener process satisfies (Dixit and Pindyck, 1994) three important properties. First, it satisfies the Markov property (Markov, 1954). The probability distribution for all future values of the process depends only on its current value.

Second, it has independent increments. The probability distribution for the change in the process over any time interval is independent of any other time interval (non-overlapping). Third, changes in the process over any finite interval of time are normally distributed, with a variance that is linearly dependent on the length of time interval, dt. The general equation of an Ito process is as follows:

$$\mathrm{d}x = a(x,t)\mathrm{d}t + b(x,t)\mathrm{d}z \tag{1.1}$$

In this equation, dz is the increment of a Wiener process, and $a(x,t)$ and $b(x,t)$ are known functions. There are different forms of $a(x,t)$ and $b(x,t)$ for various Ito processes. In this equation, dz can be expressed as $\mathrm{d}z = \varepsilon_t\sqrt{\mathrm{d}t}$, where ε_t is a random number drawn from a unit of normal distribution.

The simplest generalization of Eq. (1.1) is the equation for Brownian motion with drift (Figure 1.7):

$$\mathrm{d}x = \alpha \mathrm{d}t + \sigma \mathrm{d}z \tag{1.2}$$

where α is called the drift parameter and σ is the variance parameter.

Other examples of Ito processes are the geometric Brownian motion with drift provided in Eq. (1.3) and the geometric mean reverting process given in Eq. (1.4).

$$\mathrm{d}x = \alpha x \mathrm{d}t + \sigma x \mathrm{d}z \tag{1.3}$$

$$\mathrm{d}x = \eta(\bar{x} - x)\mathrm{d}t + \sigma \mathrm{d}z \tag{1.4}$$

Where η is the speed of reversion and $\bar{x}$ is the nominal level that x reverts to. In geometric Brownian motion, the percentage changes in x and $\Delta x/x$ are

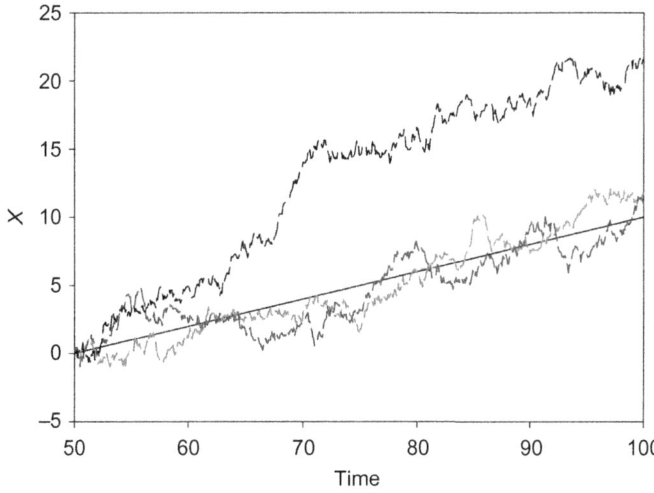

FIGURE 1.7

Sample paths of Ito process Brownian motion with drift (Diwekar, 2008).

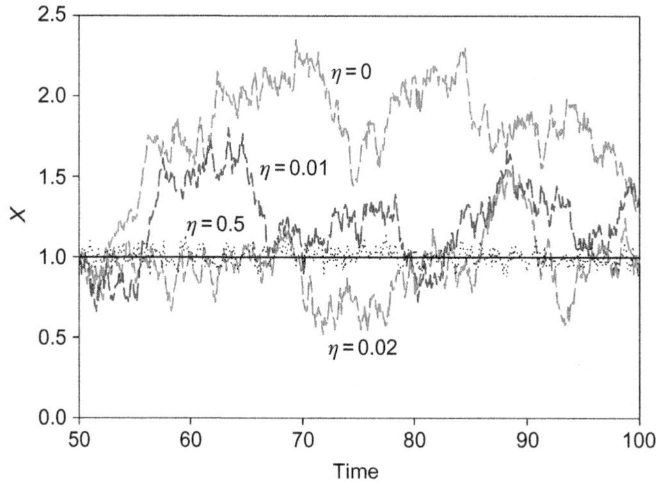

FIGURE 1.8

Sample paths of Ito process mean reverting process (Diwekar, 2008).

normally distributed (absolute changes are log-normally distributed). In mean reverting processes (Figure 1.8), the variable may fluctuate randomly in the short term, but in the longer term it will be drawn back toward the marginal value of the variable. In the simple mean reverting process the expected change in x depends on the difference between x and $\bar{x}$. If x is greater (less) than $\bar{x}$, then it is more likely to decrease (increase) in the next short interval of time.

Ito process fitting depends on availability of data and correctly assigning the proper Ito process. Further, it should be noted that Ito calculus can be a very arduous and challenging task.

Mental models

Petrie et al. (2013) argue that probabilistic representation uncertainty in the form of distributions or Ito processes are data intensive and do not take into consideration the qualitative nature of uncertainties and irrational behavior of organization. They proposed an approach of agent-based modeling (ABM) in which uncertainties can be modeled using "mental models," particularly for complex industrial networks. This methodology is less dependent on data and less sensitive to initial conditions.

Mental models have two components, a cognitive side and an ecological side (Gigerenzer and Goldstein, 2000). These two components are independent. An organization can operate in a fairly open environment and represent this environment accordingly using socially determined characteristics, such as social norms and values. Contrarily, an organization can be in a situation in which it also views the consequences of its choices as certain. Petrie et al. (2013) argue that by systematically exploring these two different extremes of mental models and how they deal with uncertainty, it is possible to develop a set of scenarios that can create a richer understanding of industrial network evolutions.

DECISION MAKING WITH UNCERTAINTY

The effect of uncertainty in decision making has been articulated around the question of whether the entity behaves as rational optimizers or whether seemingly irrational behavior exists. Rational behavior is generally based on a powerful analytical and data-processing apparatus. Optimization methods in the face of uncertainties fall under this category. However, simplifying strategies (Sterman, 2000) like use of routines (Nelson and Winter, 1982), heuristics (Tversky and Kahneman, 2000), social relationships (Festinger, 1954), and social norms (Giddens, 1984) all play an important role in strategic decision making. ABM and scenario analysis can include these strategies. These two approaches are described in the following sections.

Optimization approach

Optimization has been a valuable tool in designing and operating plants. In recent years, environmental considerations have been included in optimization of plants as constraints. In sustainable manufacturing, environmental and societal considerations are included not merely as constraints but also as part of the objective function. Thus, it is a multi-objective optimization problem. To include decisions starting from discovery stage, and to include various objectives at the other end, an optimization framework that includes discrete decision making for decisions, such as selection of material and selection of equipment, and continuous decisions, such as material flow, needs to be considered. A framework presented by Diwekar (2003) is shown in Figure 1.9.

Level 1 is the inner-most level. It corresponds with various models like material and energy balances, equipment design equations, and models for impact assessments. Obviously, these models come from various disciplines such as engineering, ecology, economics, social sciences, and policy analysis. The multi-disciplinary, multi-scale nature of modeling poses a number of challenges.

Level 2 is the sampling loop. The diverse nature of uncertainty, such as estimation errors and process variations, can be specified in terms of probability distributions. Once probability distributions are assigned to the uncertain parameters, the next step is to perform a sampling operation from the multi-variable uncertain parameter domain. Sampling is a major bottleneck in the computational efficiency of the stochastic programming (optimization under uncertainty) problems. One of the most widely used techniques for sampling from a probability distribution is the Monte Carlo sampling technique. Latin hypercube sampling (McKay et al., 1979) is one form of stratified sampling that can yield more precise estimates of the distribution functions than Monte Carlo. However, the main drawback of this stratification scheme is that it is uniform in one dimension and does not provide

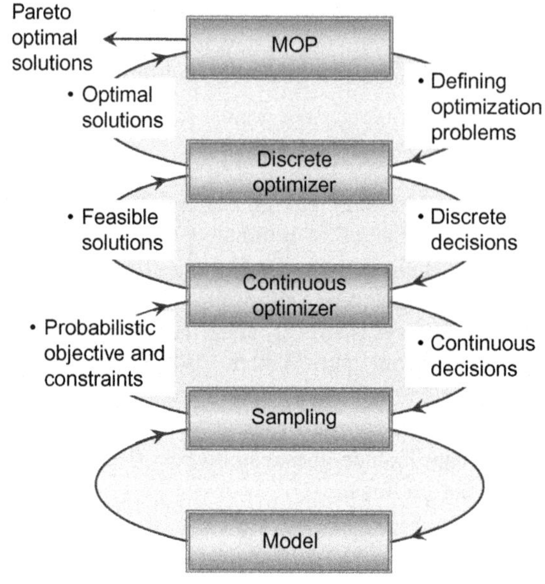

FIGURE 1.9

Optimization framework for sustainable manufacturing (Diwekar, 2003).

uniformity properties in k-dimensions. Efficient sampling techniques (Hammersley sequence sampling (HSS), Kalagnanam and Diwekar, 1997 and Latin hypercube Hammersley sampling (LHHS), Wang et al., 2004) based on Hammersley points have been proposed by Diwekar et al., who used optimal design schemes for placing the n points on a k-dimensional hypercube. These schemes ensure that the sample set is more representative of the population, showing uniformity properties in multi-dimensions, unlike Monte Carlo, Latin hypercube, and its variant, the Median Latin hypercube sampling technique (Diwekar and Ulas, 2007). It has been found that the HSS and LHHS techniques are an order of magnitude faster than LHS and Monte Carlo techniques; hence, they are the preferred techniques for uncertainty analysis and optimization under uncertainty.

Level 3 is the continuous optimizer. This step involves decisions regarding design and operating conditions of the plant. The computational efficiency of the stochastic optimization framework can be improved further based on the specialized structure of the problem.

Although the basic structure is that of the L-shaped method, BONUS helps reduce the computational burden. Traditional SNLP methods rely on improving the probabilistic objective function by repeated evaluation for each sample for every iteration; instead of running the model for given samples in every iteration, BONUS uses the concept of reweighting (similar to Bayesian updating) to obtain the probabilistic information about objective function, constraints, and derivatives needed for the non-linear programming optimization algorithm. This method is suitable for large-scale problems such as the sustainable manufacturing problem (Diwekar and David, 2014).

Level 4 is the discrete optimizer. Discrete decisions like selection of material, selection of equipment, and their order are decided at this level. There are a number of approaches for discrete optimization. Traditional mixed integer non-linear programming methods use the open equation system. For global optimization and black box models, probabilistic methods like simulated annealing, genetic algorithm, and colony optimization can be used. More details have been provided by Diwekar (2008).

Level 5 is multi-objective optimization. It is well known that mathematics cannot isolate a unique optimum when there are multiple competing objectives. Mathematics can, at most, aid designers in eliminating design alternatives dominated by others, leaving a number of alternatives called the Pareto set. For each of these decisions, it is impossible to improve one objective without sacrificing the value of another relative to some other alternative in the set. It is then the decision maker who decides which solutions of the Pareto set to consider. There are two types of generating methods available for this, the weighting method and the constraint method. In the weighting method, a single objective is formed by weighting various objective functions and the optimization problem is solved using sets of different weights to obtain the Pareto surface. In the constraint method, the basic strategy is also to transform the multi-objective optimization problem into a series of single objective optimization problems. The idea is to pick one of the objectives to minimize (i.e., Z_l) while each of the others (Z_i, $i = 1, \ldots, k, i \neq l$) is turned into an inequality constraint with parametric right sides (ε_i, $i = 1, \ldots, k, i \neq l$). Solving repeatedly for different values of $\varepsilon_i, \ldots, \varepsilon_{l-1}, {}_{-l+1}, \ldots,$ ε_k leads to the Pareto set. Alternative to the generating method is the goal programming method, which sets goals for different objectives and a single solution is then obtained (Diwekar, 2008). The challenge in sustainability is to define the proper objectives for this framework.

Although this framework is useful for obtaining greener designs and operating conditions at the plant level, as stated for industrial ecology and ecosystem sustainability, we need to consider forecasting and time-dependent decisions. Optimal control methods are useful for forecasting and time-dependent decision making. Time-dependent uncertainties are prominently featured in sustainable system dynamics. Diwekar et al. modified the framework shown in Figure 1.9. Figure 1.10 shows this new framework (Diwekar and Shastri, 2010). In this framework, time-dependent decisions (forecasting) and time-dependent uncertainties are handled by stochastic optimal control methods (Diwekar, 2008).

Again, it should be remembered that we will be dealing with large-scale complex systems for sustainability analysis and solutions. Therefore, any efficiency improvement in any level of these frameworks presented in Figures 1.9 and 1.10 will be of tremendous use.

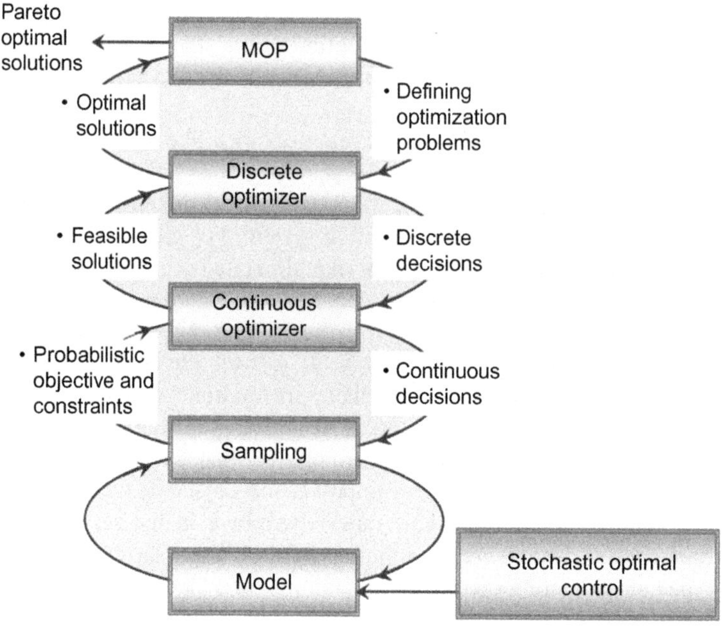

FIGURE 1.10

Optimization framework for sustainability (Diwekar and Shastri, 2010).

Agent-based scenario analysis

ABM is a simulation approach that has gained considerable attention in recent years for the modeling of industrial networks (Petrie et al., 2013). Such models depict the evolution of systems through a set of rules that describe the behavior and interactions of heterogeneous agents operating within a particular system. Any ABM is a multi-scale model with agent behavior "playing out" at one scale and consequential system evolution at a higher scale (both in space and time). The rules by which individual agents decide to act may vary, ranging from very simple to complex. The advantage of agent-based models over those characterized as general equilibrium, optimization, or econometric models is that they are able to reproduce and explain non-linear aggregate patterns without the need for regressions or simplifying assumptions. Further, ABM can include adaptive learning, which is a useful approach for sustainability in general and for industrial networks in particular. First, it disconnects the presumption of the analyst from the exploration of the system. In complex adaptive systems, this is an important advantage because it allows for an analysis of the system as it is, rather than as it is viewed by the analyst. In ABSA, ABM is coupled with a scenario generated based on mental models of uncertainties described previously to analyze the system.

ABSA involves four steps. First, different mental models used by individual organizations to describe uncertainty in their external environment (and informed by different "context scenarios") have to be identified. Second, these mental models have to be developed further into a set of explicit cognitive processes to support the decision making of individual organizations. Third, these cognitive processes need to be implemented within a defensible decision support framework that brings together the full spectrum of learning processes and dynamic behaviors that can take place within an industrial network—from simple adjustment of business practices for any individual organization to the development and growth of institutional behaviors across the network as a whole. Finally, the worth of scientific interventions needs to be assessed within a multi-criteria analysis (Petrie et al., 2013). This last step in ABSA can be performed using multi-objective optimization methods.

As stated, ABSA involves knowledge of qualitative as well as quantitative information and is very much problem dependent. Therefore, many ABMs limit their description of agent behavior to simple economic rules, but this approach can be too simplistic and has limitations in addressing future uncertainties.

CASE STUDIES

The first case study presented here corresponds to green manufacturing. This case study shows how extending the boundary to include early stage decisions like material selection can provide green engineering solutions. This is a case study from Eastman Chemicals (Kim and Diwekar, 2002).

Separation processes are not only vital for isolating and purifying valuable products but also crucial for removing toxic and hazardous substances from waste streams emitted to the environment. Figure 1.11 shows one example of separation processes from Eastman Chemicals plant using extraction and distillation for acetic acid (HOAc) separation and recycling. HOAc, an in-process solvent (IPS), is a valuable chemical, but it is also a pollutant when released to the environment. HOAc can be directly separated from water in a single distillation column; however, this requires a very large number of equilibrium stages (i.e., long column, high entropy system) because of close boiling temperatures of water and HOAc (100°C vs. 118°C), resulting in high capital and operating costs. Instead of using a single distillation, this separation, in practice, consists of an extraction column followed by a distillation column. An aqueous stream containing acetic acid enters the extraction column, in which a solvent extracts acetic acid from the water. The extract is then supplied to the azeotropic distillation column, where the bottom product is pure acetic acid and the top product is a heterogeneous water–new solvent azeotrope. The pure acetic acid product is recycled to upstream processes, whereas the azeotropic mixture is condensed and then decanted. The organic phase from the decanter is recycled, whereas the aqueous phase goes to the wastewater treatment facility. It has been observed that this process has few degrees of freedom. Because of this, a slight variation in feed not

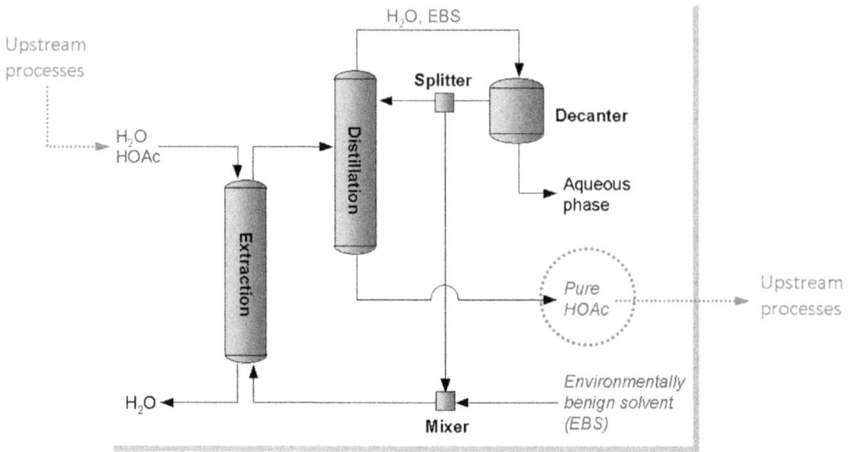

FIGURE 1.11

Coupled chemical and process synthesis (EASTMAN process) (Kim and Diwekar, 2002).

only results in loss of HOAc purity and yield but also results in a big environmental problem attributable to acetic acid concentrations in effluent. Eastman tried to change the process performance by changing operating conditions, but it did not work. As per green engineering principles, we decided to include early decisions in the process design and to select the extraction solvent (environmentally benign solvent), along with changing process design and synthesis. This is a coupled chemical and process synthesis approach. Obviously, this integration poses several challenges, such as combinatorial explosion of chemical and process design alternatives, and uncertainties in the group contribution models for chemical synthesis and multiple objectives. This case study utilizes all the algorithmic stages (loops) of Figure 1.9 to obtain greener and cost-effective designs.

The goals of this green process design are to achieve high HOAc recovery yield, high process flexibility, and low environmental impact. The coupled chemical and process synthesis involves integration of environmentally benign new solvent (EBS) selection and IPS recycling. EBS selection is an approach used to generate candidate solvent molecules that have desirable physical, chemical, and environmental properties. Computer-aided molecular design (CAMD) is commonly used for EBS selection. CAMD, based on the reverse use of group contribution methods, can automatically generate promising solvent molecules from their fundamental building blocks or groups (Joback and Reid, 1987). The group contribution method is a forward problem; if we know a molecule, then we can estimate its physical, chemical, biological, and health-effect properties based on its groups. In contrast, CAMD is a backward problem; if we know the desired properties or regulation standards, then we can find molecules that satisfy these properties or standards by combining groups. This case study was taken from Kim and Diwekar (2002).

The objectives of this multi-objective problem (MOP) under uncertainty are to maximize HOAc recovery, minimize environmental impacts (EI) based on LC_{50} and based on LD_{50}, and maximize the process flexibility. The continuous decision variable vector is [split fraction, distillation bottom rate, EBS makeup flow rate, heat duty]^T. To reduce the entropy of the system for recycling, solvent with no additional azeotrope needs to be selected. Therefore, the discrete decision design vector is [solvent type, distillation feed point, number of equilibrium stages]^T. As stated, one of the problems of this system is that the feed variation causes significant environmental problems; therefore, flexibility in the face of feed variation is explicitly included as an objective. However, even if flexibility in design is essential, design for worst-case scenarios is not advisable. Therefore, the problem selects two design surfaces flexible for uniformly varying (fractional variation) feed for percentage variations of 5% and 10%. To minimize the material diversity in the existing chemical plant, only a small number of solvents commonly available in such a plant from the 40 optimal solvents found previously (because this plant uses ASPEN Plus, ASPEN Plus databank is used as a representative for the plant) are included as candidate solvents. The candidate solvents are (i) methyl propyl ketone, (ii) methyl isopropyl ketone, (iii) diethyl ketone, (iv) ethyl acetate, (v) methyl propionate, (vi) isopropyl acetate, and (vii) propyl acetate. Apart from the solvent type, other discrete decisions include the total number of stages of the extractor and the distillation column (20 and 25), and the feed point to the distillation column. Continuous decisions include the distillation bottom flow, split ratio (i.e., EBS recycle ratio), EBS makeup flow rate, and heat duties of the distillation column. The feed rate is assumed to be 100 kmol/h with the molar feed composition of 0.7/0.3 in water/HOAc, and feed variation uncertainty is imposed on the feed flow rates. There is another important factor that should be considered in EBS selection and IPS recycling. In industrial practice, heat duty is critical in deciding alternatives for continuous distillation. Thus, before solving the MOP problem, the proposed EBS are also checked through the reboiler heat duty comparison to eliminate candidate EBS that require high heat duty.

Figures 1.12 and 1.13 show the Pareto optimal solutions for EBS selection and IPS recycling problems for the two feed variations. Because of a highly discretized solution surface and large infeasible regions (in terms of yield and feasibility), only six designs in the face of 5% feed variation are in the trade-off surface from the 50 optimization problems used to generate this surface. For 10% feed variation, this surface is reduced further to four designs. EBS ethyl acetate, which is the current extracting agent, has advantages in terms of two objectives: HOAc recovery and process flexibility for small feed variation. High HOAc recovery can be predicted because the distribution coefficient of ethyl acetate is the highest among the EBS present in the Pareto optimal solutions. Hence, the popular use of ethyl acetate can be understood from these two objectives. However, for 10% feed variation, ethyl acetate is no longer an optimal design. For the environmental impact based on LC_{50}, methyl propionate is the best solvent for this integrated problem. This is mainly because of the lowest value of

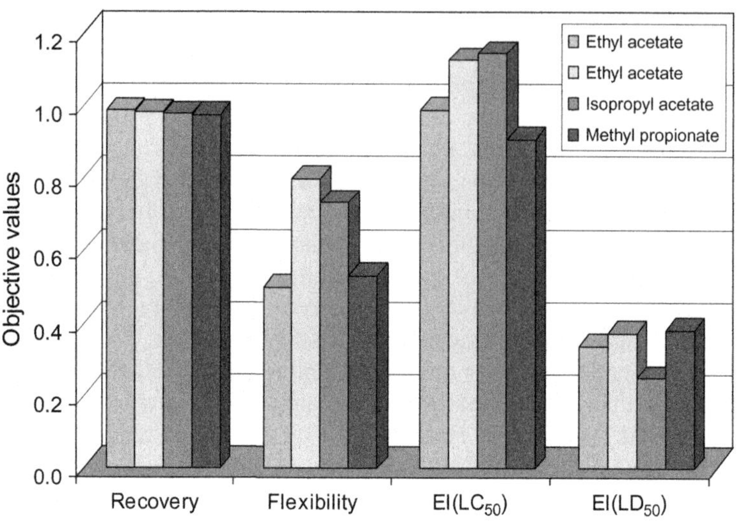

FIGURE 1.12

Pareto surface for 5% feed variation.

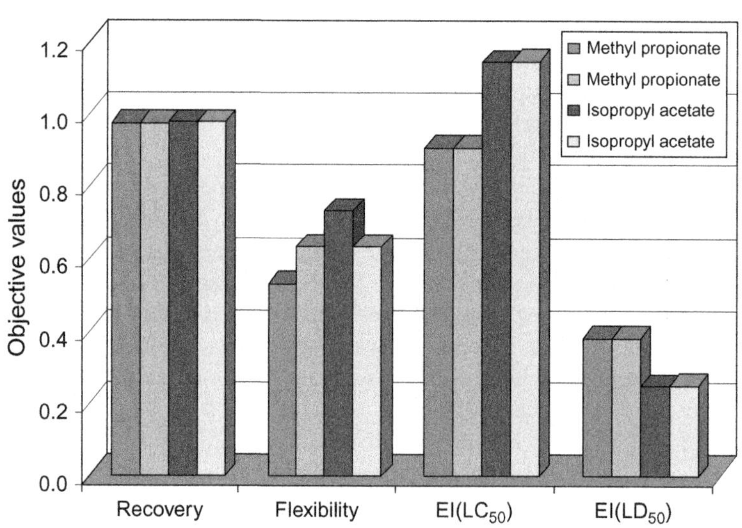

FIGURE 1.13

Pareto surface for 10% feed variation.

LC_{50}, 2,240 mg/L. For the environmental impact based on LD_{50}, isopropyl acetate is the best choice even though it does not have the lowest LC_{50} value. It has been found from simulation results that isopropyl acetate has the lowest fresh solvent makeup flow rate (e.g., 0.36 vs. 1.27 kmol/h of ethyl acetate), and this is the main reason for the lowest LD_{50} value. The process flexibility objective is highly dependent on the distillation feed point as well as EBS molecules.

From this case study, we can see that the integrated framework can provide different and better chemical and design alternatives to decision makers. In addition, uncertainties in CAMD models and operating conditions can play a significant role in the early design and synthesis stages.

The second case study presented here is from the sustainability literature. Coupled ecosystem economic models are often used to represent sustainability dynamics of the planet. One such model has been developed by my group (Kotecha et al., 2013) and is presented in Figure 1.14. This integrated model is

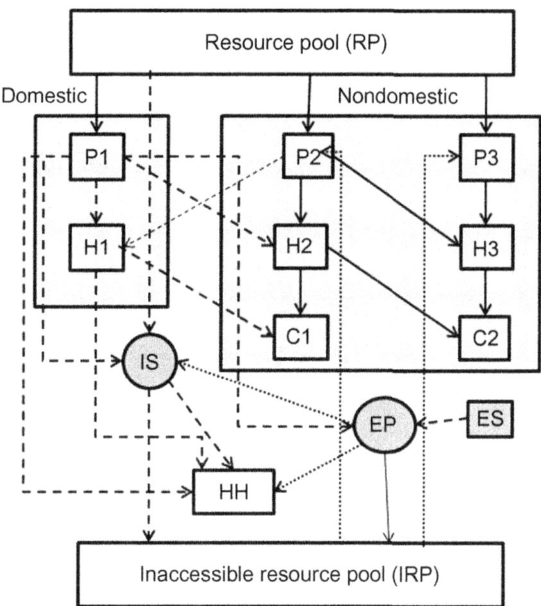

FIGURE 1.14

Compartment of the economic–ecological model. This figure depicts the compartment model that comprise plants (Pi), which are the primary producers, herbivores (Hi), carnivores (Ci), humans households (HH), a resource pool (RP), and an inaccessible resource pool (IRP). The arrows represent the mass flows from one compartment (origination) to another compartment (termination), and all living compartments have an implied flow back to the resource pool that represents death. IS is the industrial sector, whereas EP and ES are the energy producers and the energy source compartment, respectively.

based on food web dynamics of the ecosystem and the microeconomic model for the economy. It is a compartmental model comprising ecological and industrial compartments with energy producers in one compartment. Figure 1.14 shows the interactions of these compartments. The model has been simulated for two scenarios, namely, a base case scenario, which represents the current state of the system and is currently sustainable without any compartment being in the dying category, and the consumption increase scenario, which is projected to occur in the near future. The consumption increase scenario simulations are shown in Figure 1.15 (Doshi et al., 2014). It can be seen that a number of compartments start to become extinct as consumption increases. This is clearly an unsustainable system. We want to look at the future time span and forecast decisions for the future that can make this system sustainable for the near future. Optimal control methods, which deal with time-dependent decision making, are used for this purpose. It has been found that this simple system representing the planet is a complex network and requires at least six time-dependent decision variables to make the system sustainable for the near future (Figure 1.16) (Diwekar, 2005).

It should be remembered that the integrated model shown in Figure 1.14 has a number of uncertainties associated with it. Given the time span of decision

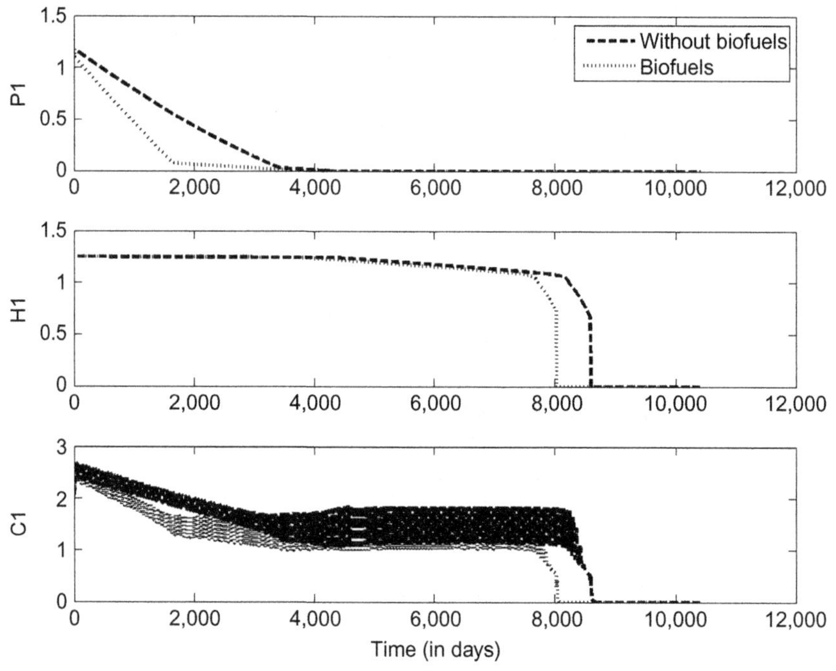

FIGURE 1.15

Consumption increase scenario for the three compartments: plants compartment, 1; herbivorous animals compartment, 1; and carnivorous animals compartment, 1.

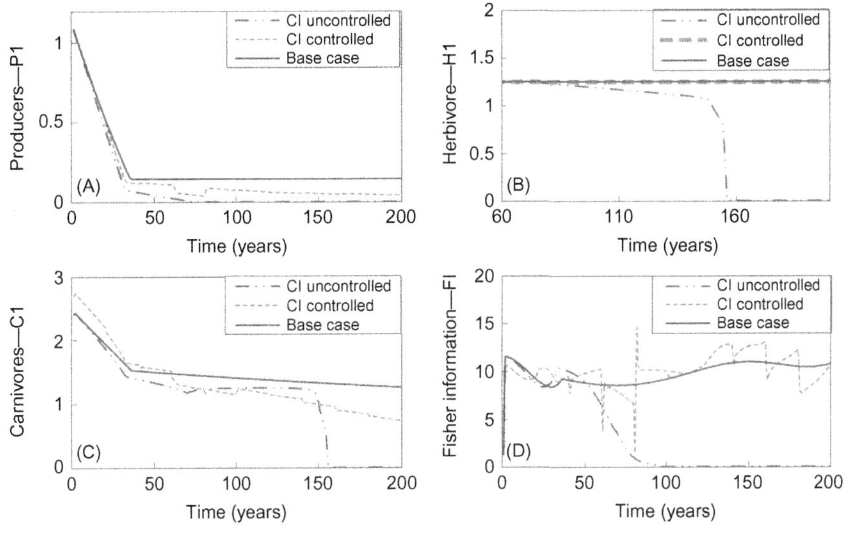

FIGURE 1.16

Six parameter multi-variable optimal control. Graphs (A)–(D) depict the effects of multi-variable optimal control on P1, H1, C1, and FI. In the legend, CI indicates consumption increase. The controlled scenario (dotted profile) is aiming to emulate the sustainable base case (solid line profile) and avoid extinction such as the uncontrolled profile (dashes and dots profile).

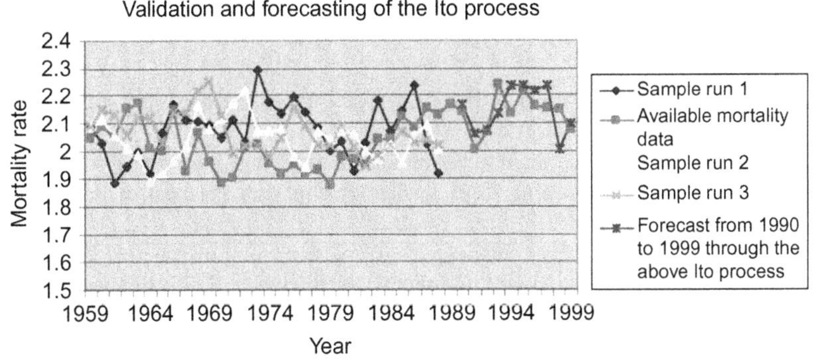

FIGURE 1.17

Ito process representation for mortality rate (Diwekar, 2005).

making, the uncertainties tend to be dynamic. One such uncertainty is shown in Figure 1.17. These uncertainties can be modeled using Ito processes, as stated previously. Figure 1.17 shows the mortality rate, a commonly used parameter in dynamic sustainability models for our planet, modeled as a geometric mean reverting process using data from 1959 to 1990 (Diwekar, 2005). The forecast from this Ito process from 1990 to 1999 is then compared with actual data (Figure 1.17). It can be seen that the forecasting ability of this Ito process model is good enough to study the impact of uncertainties. Stochastic optimal control methods can then be applied to time-dependent decision making instead of the deterministic optimal control profiles presented previously. These profiles will be a more robust attempt at time-dependent decision making because uncertainties are explicitly considered.

SUMMARY

This chapter presents an engineering systems analysis approach for sustainability. Sustainable manufacturing involves extending boundaries of traditional design and modeling framework to include multiple objectives and starting as early as possible. Uncertainties increase as the modeling envelop is expanded. Industrial ecology broadens this framework to industrial network levels and time-dependent uncertainties arise. Adaptive management is a key to sustainability at this level. Mental models for uncertainty analysis are useful. Sustainability goes beyond industrial ecology and the boundary extends to our planet. The timescale for decision making extends significantly and forecasting decisions is very important. This poses challenges such as modeling phenomena, modeling uncertainties, time-dependent decision making, defining objectives, and forecasting decisions.

REFERENCES

Bakshi, B., Fiksel, J., 2003. The quest for sustainability: challenges for process systems engineering. AIChE J. 49, 1350.

Beloff, B., 2013. The case and practice for sustainability in business, metrics. In: Cabezas, H., Diwekar, U. (Eds.), Sustainability: A Multi-Disciplinary Perspective, 310. Bentham e-books, Beijing, China.

Cabezas, H., 2013. Sustainability indicators and metrics. In: Cabezas, H., Diwekar, U. (Eds.), Sustainability: A Multi-Disciplinary Perspective, 197. Bentham e-books, Beijing, China.

Diwekar, U., 2003. Greener by design. Environ. Sci. Technol. 37, 5432.

Diwekar, U., 2005. From green process design to industrial ecology to sustainability. Resour. Conserv. Recycling 44, 215.

Diwekar, U., 2008. Introduction to Applied Optimization, second ed. Springer, New York, NY.

Diwekar, U., David, A., 2014. Better Optimizationof Nonlinear Uncertain Systems (BONUS) Algorithm with Real World Applications, in press, to be published as optimization briefs by Springer.

Diwekar, U., Shastri, Y., 2010. Green process design, green energy, and sustainability: a systems analysis perspective. Comput. Chem. Eng. 34, 1348.

Diwekar, U., Ulas, S., 2007. Sampling techniques. In: Kirk-Othmer Encyclopedia of Chemical Technology, Online Edition, vol. 26, p. 998.

Dixit, A.K., Pindyck, R.S., 1994. Investment under Uncertainty. Princeton University Press, Princeton, NJ.

Doshi, R., Diwekar, U., Benavides, P., Yenkie, K., Cabezas, H., 2014. Maximizing Sustainability of Ecosystem Model through Socio-Economic Policies Derived from Multivariable Optimal Control Theory, submitted to *Environmental Science and Technology*.

Ehrenfeld, J., Gertler, N., 1997. Industrial Ecology in Practice: the Evolution of Interdependence at Kalundborg. J. Ind. Ecol. 1 (1), 67.

Festinger, L., 1954. A theory of social comparison processes. Hum. Relat. 17 (2), 117.

Giddens, A., 1984. The Constitution of Society. University of California Press, Los Angeles, CA.

Gigerenzer, G., Goldstein, D.G., 2000. Reasoning the fast and frugal way: models of bounded rationality. In: Connolly, T., Arkes, H.R., Hammond, K.R. (Eds.), Judgement and Decision Making, 621. Cambridge University Press, Cambridge, UK.

Ito, K., 1951. On stochastic differential equations. Mem. Am. Math. Soc. 4, 1.

Ito, K., 1974. On stochastic differentials. Appl. Math. Optim. 4, 374.

Joback, R.C., Reid, K.G., 1987. Estimation of pure−component properties from group-contributions. Chem. Eng. Commun. 57, 233.

Kalagnanam, J., Diwekar, U., 1997. An efficient sampling technique for off-line quality control. Ann. Oper. Res. 38, 308.

Kim, K.-J., Diwekar, U.M., 2002. Solvent selection and recycling for continuous processes, invited paper. Ind. Eng. Chem. Res. 41, 4479.

Kotecha, P., Diwekar, U., Cabezas, H., 2013. Model based approach to study the impact of biofuels on the sustainability of an integrated system. Clean Technol. Environ. Policy 15, 21.

Markov, A.A., 1954. Theory of Algorithms [Translated by Jacques J. Schorr-Kon and PST staff] Imprint Moscow, Academy of Sciences of the USSR, 1954 [Jerusalem, Israel Program for Scientific Translations, 1961; available from Office of Technical Services, United States Department of Commerce] Added t.p. in Russian Translation of Works of the Mathematical Institute, Academy of Sciences of the USSR, v. 42. Original title: *Teoriya algorifmov*. [QA248.M2943 Dartmouth College library. U.S. Dept. of Commerce, Office of Technical Services, number OTS 60-51085.]

McKay, M.D., Beckman, R.J., Conover, W.J., 1979. A comparison of three methods of selecting values of input variables in the analysis of output from a computer code. Technometrics 21, 239.

Nelson, R.R., Winter, S.G., 1982. An Evolutionary Theory of Economic Change. Belknap Press, Cambridge, MA, USA.

Petrie, J., Kempener, R., Beck, J., 2013. Industrial ecology and sustainable development: dynamics, future uncertainty and distributed decision making. In: Cabezas, H., Diwekar, U. (Eds.), Sustainability: A Multi-Disciplinary Perspective, 243. Bentham e-books, Beijing, China.

Sikdar, S., 2014. Macro, Meso & Micro Aspects of Sustainability, Trans-Atlantic Research and Development Interchange on Sustainability, TARDIS 2014, Lecture 1. Estes Park, CO.

Sterman, J.D., 2000. Business Dynamics, Systems Thinking and Modeling for a Complex World. Irwin McGraw-Hill, Boston, MA, USA.

Tversky, A., Kahneman, D., 2000. Judgement under uncertainty: heuristics and biases. In: Connelly, L., Arkes, H.R., Hammond, K.R. (Eds.), Judgement and Decision Making, 35. Cambridge University Press, Cambridge, UK.

Wang, R., Diwekar, U., Gregoire-Padro, C., 2004. Latin hypercube Hammersley sampling for risk and uncertainty analysis. Environ. Prog. 23, 141.

White, A., 1994. Preface, National Academy of Engineers, 549 .The Greening of Industrial Ecosystems. National Academy Press, Washington, DC.

World Commission on Environment and Development, 1987. Our Common Future. Oxford University Press, Oxford, UK.

Yang, Y., Shi, L., 2000. Integrating environmental impact minimization into conceptual chemical process design—a process systems engineering review. Comput. Chem. Eng. 24, 1409.

CHAPTER

2 Recent developments in the application of Fisher information to sustainable environmental management

Alejandra González-Mejía, Leisha Vance, Tarsha Eason, and Heriberto Cabezas
US Environmental Protection Agency, Office of Research and Development, Cincinnati, OH, USA

INTRODUCTION

Developing and applying qualitative and quantitative metrics of sustainability has become a prominent focus in the sustainability literature. Although no consensus on universal measures exists, researchers continue to study trends in human and natural systems and the movement of indicators that affect the transition to a sustainable future (Kates and Parris, 2003). Further, researchers recognize that sustainable development of human systems involves evaluating social, environmental, and economic components, also known as people/planet/profit (Fisk, 2010) or equity/ecologic/economic areas (Flint and Houser, 2001). The common thread woven through these efforts is the focus on the interdisciplinary nature of sustainability with the goal of maintaining desirable conditions and avoiding catastrophic shifts in system condition. Regime shifts have been demonstrated in multiple systems such as ecosystems, often resulting in significant and sometimes irrecoverable damage (Brock and Carpenter, 2006). Hence, finding measures that can evaluate changes in the behavior of complex systems is of critical importance (Zellner et al., 2008). The dynamic reality of these linked systems underscores the importance of evaluating and managing patterns of growth and development (Karunanithi et al., 2011).

Information theory has shown great promise in this area because it affords the ability to assess ecosystem structure, complexity, stabililty, and diversity (Mayer et al., 2006). In particular, Fisher information (henceforth denoted as FI) is a measure of order in data (Fisher, 1922) and has been adapted into an index that provides a means of monitoring system variables to capture changes in dynamic order and, subsequently, its regimes and regime shifts (Fath et al., 2003). Initial development and application of the approach involved using FI to derive fundamental equations of physics, thermodynamics, and population genetics (Frieden, 1998).

Assessing and Measuring Environmental Impact and Sustainability.

From this foundation, new methods of estimating the index were developed. Further, FI was proposed as a sustainability metric by Cabezas and Fath (2002) and has been used to study changes in dynamic order in model systems, such as multispecies food webs (Pawlowski et al., 2005) and pseudoeconomies (Cabezas et al., 2005), and real ecosystems, such as the Bering Strait, the western African savannah, the US Florida pine-oak, and global climate changes (Mayer et al., 2006). The results from these studies demonstrated the power of the index and its ability to capture system dynamics. However, researchers noted key areas needed to enhance the approach, which include improved techniques for estimating FI from real data and advanced strategies for interpreting results and understanding drivers of complex system behavior. Such mechanisms would increase the efficiency of computations and expand the capacity for assessing transitions in real systems, and the techniques would also aid in interpreting results for application to management decisions. This chapter provides a summary of important recent activities that respond to these reseach needs and extends the utility and application of FI in sustainable environmental management.

INFORMATION THEORY

Information theory is a branch of science, mathematics, and engineering that studies information in a physical and mathematical context rather than a psychological framework. The first reference addressing information in this manner is attributed to Hartley (1928). However, the origin of information theory as we know it today is attributable to the work of Shannon (1948). Information theory generally addresses issues such as information content of observable phenomena, the creation and destruction of information, and the transmission of information. In this chapter, the discussion is oriented to concepts arising from statistics to include FI and Bayes theorem.

FISHER INFORMATION

FI was derived by the statistician R.A. Fisher (Fisher, 1922), who was interested in fitting model parameters from observation. FI is the information that is available from a set of measurements on a variable, s, that can be used to estimate the value of a parameter, θ. FI $I(\theta)$ is defined in Eq. (2.1), with $p(s|\theta)$ being the probability density of observing a particular value of s (i.e., independent variable) in the presence of θ.

$$I(\theta) \equiv \int \frac{1}{p(s|\theta)} \left[\frac{\partial p(s|\theta)}{\partial \theta} \right]^2 \mathrm{d}s \tag{2.1}$$

If the measurements of s contain no information about θ, then the partial derivative $(\partial p(s|\theta))/\partial \theta$ is zero, i.e., $p(s|\theta)$ does not depend on θ. In this case, a legitimate value of θ cannot be obtained from measuring s. Although this expression gives a rigorous definition of FI, it is not straightforward for use in practice because taking the partial derivative $\partial p(s|\theta)/\partial\theta$ from real system data is difficult.

To make Eq. (2.1) more accessible to practical calculation, we invoke the concept of shift invariance. Following the derivation by Mayer et al. (2007), we first let $\theta = \langle s \rangle$, where $\langle s \rangle$ is the mean value of s defined by $\langle s \rangle \equiv 1/(b-a)\int_a^b s(t)\mathrm{d}t$, and a and b represent the range of values of s. In this context, a shift invariant dynamic system is one where the shape of the probability density function (pdf) $p(s|\langle s \rangle)$ is independent of the value of $\langle s \rangle$, a constant. Because the FI depends on the value of $p(s|\langle s \rangle)$ and its derivative $\partial p(s|\langle s \rangle)/\partial\langle s \rangle$, it also becomes independent of the value of $\langle s \rangle$. This can be expressed formally by $P(s|\langle s \rangle) = p(s - \langle s \rangle|\langle s \rangle) = p(s - \langle s \rangle) = p(\bar{s})$ where, henceforth, s will be used in place of $\bar{s}$. After some manipulation, it can be shown that under shift invariance, the FI expression becomes Eq. (2.2),

$$I = \int \frac{1}{p(s)} \left[\frac{\mathrm{d}p(s)}{\mathrm{d}s}\right]^2 \mathrm{d}s \tag{2.2}$$

where $p(s)$ is the simple probability for observing a particular value of s. Now the FI I is simply the amount of information that is available from measurements of variable s. When time t dependence is important, as is the case with most dynamic systems [such that $s(t)$ is a function of time], Eq. (2.2) is simplified using the chain rule [i.e., $\mathrm{d}/\mathrm{d}s = (\mathrm{d}/\mathrm{d}t)(\mathrm{d}t/\mathrm{d}s)$ and $\mathrm{d}s = (\mathrm{d}s/\mathrm{d}t)\mathrm{d}t$], becoming Eq. (2.3),

$$I \equiv \int \frac{1}{p(s)} \left[\frac{\mathrm{d}p(s)}{\mathrm{d}t}\right]^2 \left(\frac{\mathrm{d}s}{\mathrm{d}t}\right)^{-1} \mathrm{d}t \tag{2.3}$$

A further and useful transformation of Eq. (2.3) is to express the FI in terms of the amplitude $q(s)$ of the probability density by use of its definition, $p(s) \equiv q(s)^2$. This has the benefit of avoiding division by $p(s)$, which can be problematic when $p(s)$ is a small number. By using this transformation, Eq. (2.4) is obtained.

$$I = 4\int \left[\frac{\mathrm{d}q(s)}{\mathrm{d}t}\right]^2 \left(\frac{\mathrm{d}s}{\mathrm{d}t}\right)^{-1} \mathrm{d}t \tag{2.4}$$

For a dynamic system that is continuous or approximately continuous, the variable s follows a trajectory through time so that each value of s is associated with a particular time. Hence, a one-to-one correspondence between values of s and time t exists. Consequently, the probability density $p(s)$ for seeing a particular value of s is associated with a time t, as shown in Eq. (2.5), where $p(t)$ is the probability of observing a particular time t.

$$p(s)\mathrm{d}s = p(t)\mathrm{d}t \tag{2.5}$$

Further, when sampling randomly along a time interval $[0 - T]$, all times are equally probable so that $p(t) = 1/T$ and Eq. (2.5) can be rewritten as Eq. (2.6).

$$p(s) = \frac{1}{T}\left(\frac{\mathrm{d}s}{\mathrm{d}t}\right)^{-1} \tag{2.6}$$

Differentiating the expression for $p(s)$ in Eq. (2.6) and inserting into Eq. (2.3) yields Eq. (2.7), where $\mathrm{d}s/\mathrm{d}t$ is the rate of change (velocity) of s and $\mathrm{d}^2s/\mathrm{d}t^2$ is the acceleration of s.

$$I = \frac{1}{T}\int_0^T \frac{(\mathrm{d}^2s/\mathrm{d}t^2)^2}{(\mathrm{d}s/\mathrm{d}t)^4}\mathrm{d}t \tag{2.7}$$

When s is a function of n variables, $s[x_1(t), x_2(t), \ldots, x_n(t)]$, each of them depending on time, the derivative $\mathrm{d}s/\mathrm{d}t$ is given by the Euclidean metric as shown in Eq. (2.8).

$$\frac{\mathrm{d}s}{\mathrm{d}t} = \sqrt{\sum_{i=1}^{n}\left(\frac{\mathrm{d}x_i}{\mathrm{d}t}\right)^2} \tag{2.8}$$

And the second derivative $\mathrm{d}^2s/\mathrm{d}t^2$ is obtained by differenting Eq. (2.8) to produce Eq. (2.9).

$$\frac{\mathrm{d}^2s}{\mathrm{d}t^2} = \left(\frac{\mathrm{d}s}{\mathrm{d}t}\right)^{-1}\sum_{i=1}^{n}\frac{\mathrm{d}x_i}{\mathrm{d}t}\frac{\mathrm{d}^2x_i}{\mathrm{d}t^2} \tag{2.9}$$

This framework, while rigorous, tacitly assumes that s is continuous and smooth enough to be twice differentiable. Athough this is true for many systems, there are also many situations in which differentiating s twice is not practical, often because of data scarcity or quality issues (Karunanithi et al., 2008).

Therefore, two methods have been developed for computing the index. Method 1 is a continuous approach in which FI is computed by evaluating the velocity and accelaration of the system trajectory using Eqs. (2.7)–(2.9). Method 2 involves "binning" points into discrete states of the system. These approaches are derived and thoroughly explained in Appendix 2.1. Regardless of computational method, we use FI as a measure of dynamic order by assessing the probability of observing states of a system, $p(s)$.

DYNAMIC ORDER, REGIME SHIFTS, AND THE SUSTAINABLE REGIMES HYPOTHESIS

To illustrate the concept of dynamic order, consider a system that has observable properties that follow regular patterns such that its condition does not change from one measurement to another. This system is said to be "perfectly ordered" because the condition is predictable. In contrast, a system constantly in flux has no recognizable patterns and is unpredictable. Given that FI measures the information that can be obtained from observation, it is evident that very little knowledge or information about the behavior of the system can be gathered by monitoring a disordered system (Mayer et al., 2006). From Eq. (2.2), note that FI is proportional to the slope of the probability of observing s, i.e., $\mathrm{d}p(s)/\mathrm{d}s$. Let us assume that in this case, s represents states (or particular conditions) of a system. This implies that a perfectly ordered system is always in the same state and has a high probability of exhibiting a particular condition (state); accordingly, $p(s)$ would have a high peak around that particular state (positive kurtosis), the slope $\mathrm{d}p(s)/\mathrm{d}s$ is high, and FI will approach infinity as the peak sharpens. Conversely, a perfectly disordered system is unpredictable and unbiased toward any state. As such, it has an equal likelihood of being in any state, i.e., $p(s) = p(1) = p(2) = \cdots p(n)$ for all states n. In this case, $\mathrm{d}p(s)/\mathrm{d}s = 0$ and FI tends toward zero. Hence, FI tracks order in complex dynamic systems. This is important because real dynamic systems must have some order to be able to function.

Real systems typically display complex dynamics that are between these two extreme cases (Mayer et al., 2007). Some systems (e.g., ecosystems) may display nonlinearity and traverse through multiple regimes (Garmestani et al., 2009), yet others may undergo changes and still exhibit stability because of underlying dynamics. In this context, sustainability is predicated on being able to maintan enough order for systems (ecosystems, economies, society) to function so that they can support social and economic development (Karunanithi et al., 2008) while ensuring persistent provisioning of ecosystem resources, goods, and services (Eason et al., 2014a).

The sustainable regimes hypothesis was developed (Cabezas and Fath, 2002) and adapted to guide the interpretation and use of FI as a measure of sustainability as a function of dynamic order (Karunanithi et al., 2008). Modifications to the hypothesis have been made to account for evaluating changes in trajectory (Gonzalez-Mejia et al., 2012b) and assessing resilience (Eason et al., 2014b). In summary, the hypothesis states that: (i) a system in an orderly dynamic regime has a characteristic set of natural patterns that may fluctuate within a normal and acceptable range of variation, but its overall condition does not change with time; hence, it has a steady nonzero (FI $>$ 0) FI over time (i.e., $\mathrm{d}\langle I\rangle/\mathrm{d}t \approx 0$); (ii) steadiliy decreasing FI (i.e., $\mathrm{d}\langle I\rangle/\mathrm{d}t < 0$), signifies a loss of dynamic order and denotes a system that is changing rapidly (increasing in variance) and losing stability; (iii) increasing FI indicates that the system is slowing and possibly becoming more ordered while maintaining function; and (iv) a significant and sometimes abrupt change in FI between two stable dynamic regimes denotes a regime shift. The sustainable regimes hypothesis tacitly implies that a sustainable dynamic regime is a necessary condition for sustainability that is beneficial to human existence.

To further illustrate the concept of regime shift, consider the flow of an incompressible fluid in a pipe. At low flow rate, the flow regime is laminar, which is well-characterized and orderly. Hence, the FI obtained from measuring variables is high. With a high flow rate, the flow is turbulent with eddies, and the flow is not as well-characterized. Thefore, the FI obtained from variable measurements is lower than is the case for laminar flow. In the transition between laminar and turbulent flow, the FI is lowest because neither the laminar nor the turbulent flow regimes are well-established. Thus, the flow regime and the measurable variables fluctuate most, leading to a low FI.

Increasing variance has often been shown to signal a regime shift and declining FI typically warns of impending transitions in empirical studies. Researchers studying system dynamics have noted an increase in variation in underlying variables preceding a transition to alternate system regimes (Carpenter and Brock, 2006). Accordingly, a system may undergo a period of instability (long or short) before experiencing a perturbation that causes a dramatic loss of stability in the regime and subsequent transition to another steady state. However, complex systems have distinct behavior and may not experience such signals as increasing variance before undergoing a transition (Seekell et al., 2011). Appendix 2.2 provides further discussion of the dynamics of FI as a system approaches a bifurcation point.

The true importance of assessing system dynamics is not only detecting a shift point but also being able to identify warning signals, monitor whether a system completely loses order without regaining stability, or monitor whether the system actually settles into a new stable regime characterized by distinct patterns. Regime shifts are characterized by significant and fundamental changes in system condition and typically have considerable economic, ecological, and social impacts. Often these shifts are caused by substantial changes in underlying drivers; however, not all are a function of large-scale disturbances (Biggs et al., 2009). Both subtle and dramatic patterns can be captured by FI (Eason et al., 2014b).

Note that FI affords the ability to evaluate the level of change in the system condition, but not the quality of its condition. Hence, shifting to a new regime does not ensure more or less humanly desirable conditions and requires assessment of observable variables to determine the desirability of the system's condition (Gonzalez-Mejia, 2011).

For systems that are constantly changing (i.e., do not have true steady states) or those with significant data quality issues, FI results computed using the velocity and acceleration approach (see Appendix 2.1 regarding method 1) tend to simply display peaks. In this case, FI does not capture regime shifts because stable regimes may not be present; rather, the peaks signify changes in the system trajectory (Gonzalez-Mejia et al., 2014). Hence, very useful results can still be obtained.

COMPARISON OF COMPUTATIONAL APPROACHES: ASSESSING SUSTAINABILITY IN MODEL SYSTEMS

A real dynamic system is typically characterized by multiple variables describing its condition over time. However, the study of simple model systems is useful for gaining insight into complex system dynamics. Therefore, this section compares and applies both methods for computing FI to model systems and serves as guide for the interpretation of FI when assessing system sustainability. Here, FI is used to evaluate: (i) elementary functions (i.e., linear, quadratic, exponential, and sinusoidal) that could represent prototypical system trajectories; (ii) a multivariate wastewater nitrification system model; and (iii) a shallow lake that illustrates a systemic shift from an oligotrophic to an eutrophic regime because of increasing phosporous levels. Further, approaches are illustrated for addressing the intrinsic challenges of assessing real systems, including handling noisy data and selecting the appropriate time window size for FI calculations. Unless otherwise specified, the FI result presented is a three-point mean displayed as I_1 for method 1 (hwin = winspace = 1 time step = 0.1 time units) and I_2 for method 2 (hwin = 8 time steps; winspace = 1 time step).

REPRESENTING THE TRAJECTORY OF A SYSTEM WITH ELEMENTARY FUNCTIONS

This section is intended as a tutorial to help provide a foundation and guidance for interpreting FI results. By using simple quadratic, exponential, linear, and sinusoidal functions, it is possible to illustrate some prototypical behaviors of FI (Gonzalez-Mejia et al., 2012b). Given these functions, FI is computed numerically and then the sustainable regimes hypothesis (see "Dynamic Order, Regime Shifts, and the Sustainable Regimes Hypothesis" section) is applied to assess the stability and sustainability of the system trajectories. In general, while stable systems are steady and predictable, unstable systems have trajectories that can be erratic and vary greatly over time.

Figure 2.1 is divided into two sections, where the top part provides examples of system trajectories described by elementary functions, that is, quadratic (x_1, x_2) (Figure 2.1A), exponential (x_3, x_4) (Figure 2.1B), linear (x_5, x_6, x_7) (Figure 2.1C), and sinusoid (x_8) (Figure 2.1D). The bottom part of Figure 2.1 displays the corresponding FI result computed using method 1 [e.g., $I_1(x_1)$] and method 2 [e.g., $I_2(x_i)$]. The analytical FI results for these functions were derived by Gonzalez-Mejia et al. (2012b) and correspond with the numerical results provided in Figure 2.1. The FI result is qualitatively the same for both methods, primarily differing in order of magnitude, but "Summary: Comparison of Methods 1 and 2" section compares and contrasts methods 1 and 2 in more detail.

Figure 2.1A provides a plot of a quadratic function and its mirror image (x_1, x_2), and the corresponding FI result. The FI result is qualitatively the same for both methods and captures the change in the system trajectory to include a peak at time 20 [$I_1(x_1) = I_1(x_2) \rightarrow \infty$, and $I_2(x_1) = I_2(x_2) = 8$]. These inflection points denote metastability defined by a slow rate of change, after which the trajectory shifts and picks up speed FI (rapidly declines). The FI of exponential (x_3, x_4) and linear functions with a slope (x_5, x_6) behave quite similarly (Figure 2.1B and C).

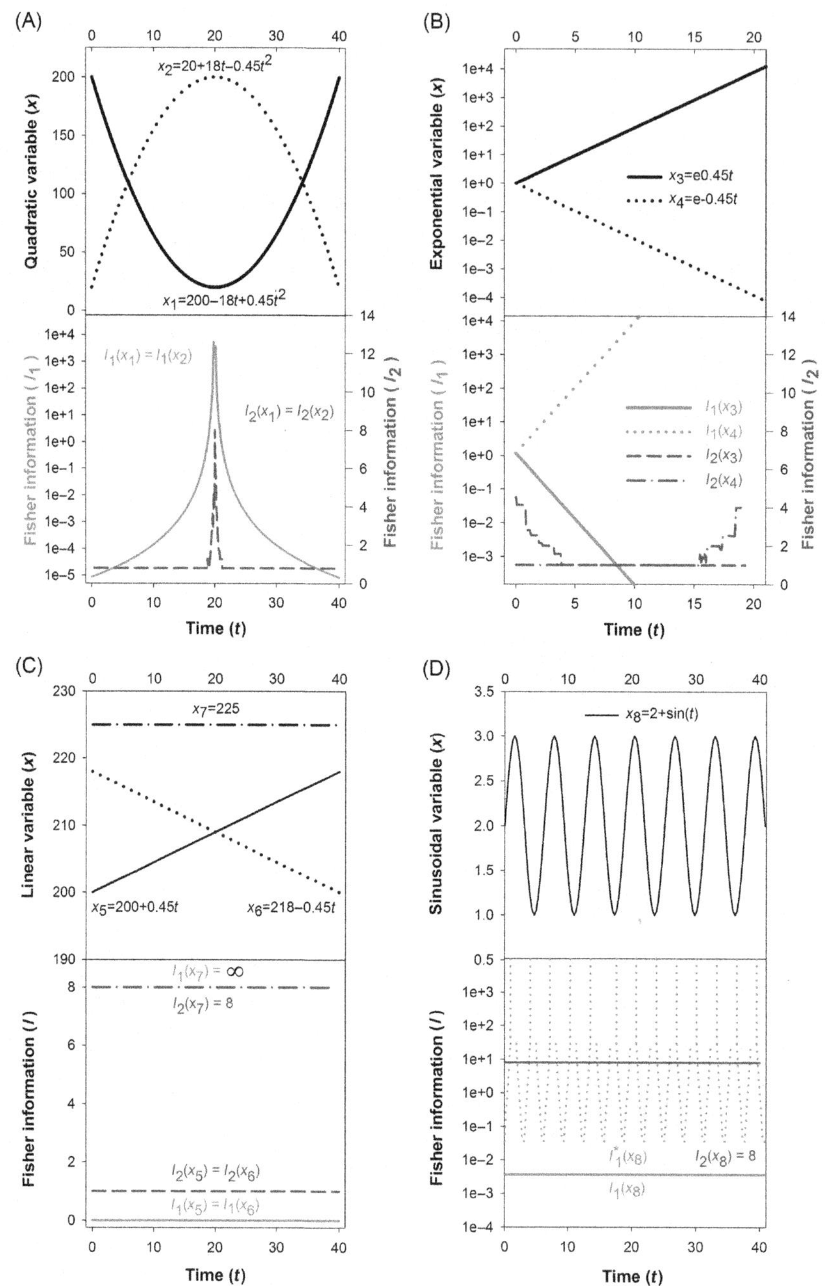

FIGURE 2.1

Sustainability analysis for system trajectories represented by elementary functions. The top part of the figure displays elementary functions (x_i) and the bottom part provides the corresponding FI result [method 1: $I_1(x_i)$; method 2: $I_2(x_i)$].

They represent cases in which a system is unpredictable, changing (for all or much of the system path) and has no stability. Consequently, no state of the system is significantly more probable than any other state, and little or no information can be obtained by observing the behavior of the system; hence, FI tends to zero. The difference in the FI result for the functions is attributable to the change being constant for the line with a slope [i.e., $x(t_i) = a + bt_i$ with $b \neq 0$; e.g., x_5] or varying for the exponential, which tends to "slow" at the beginning (e.g., x_3) or end of the trajectory (e.g., x_4). These system paths are unstable and unsustainable because they do not meet the sustainable regimes criteria (i.e., $\mathrm{d}\langle I\rangle/\mathrm{d}t \approx 0$).

In contrast, x_7 in Figure 2.1C illustrates a constant system trajectory [i.e., $x(t_i) = a + bt_i$ with $b = 0$]. Much like a quality-controlled production line, this represents a system that is predictable and generates products as desired; as such, the maximum information content can be obtained by observation. Hence, FI approaches infinity or its highest possible value at any given time [i.e., $I_1(x_7) \rightarrow \infty; I_2(x_7) = 8$], and the system path is stable [i.e., $\mathrm{d}\langle I(x_7)\rangle/\langle I\rangle/\mathrm{d}t \approx 0$]. A periodic function (i.e., x_8) is a very interesting case that represents cycles in system behavior (Figure 2.1D). These types of trajectories are common in natural systems, such as seasonal cycles of light, temperature, and predator–prey dynamics. Although a sinusoidal function is a simplified version of these systems, exploring its behavior is useful for understanding system dynamics and measuring sustainability. Note that Figure 2.1D has one function (x_8) and three FI results [i.e., $I_1^*(x_8)$, $I_1(x_8)$, and $I_2(x_8)$]. $I_1^*(x_8)$ is generated by setting the size of the computation window to $l = \pi/10$, which produces a cyclical result with peaks corresponding to periodic changes in the system path (the result is qualitatively the same using method 2). Since $\mathrm{d}\langle I_1^*(x_8)\rangle\langle I\rangle/\mathrm{d}t \neq 0$, this result indicates that the system is unstable. However, when $I_1(x_8)$ is computed using a window size of one period of the system (i.e., $l = 2\pi$), FI tends to infinity or a maximum [e.g., $I_2(x_8) = 8$] and does not change with time [i.e., $\mathrm{d}\langle I_2(x_8)\rangle/\mathrm{d}t = 0$]; thus, it meets the sustainable regimes criteria. Revisiting the $I_1^*(x_8)$ result, note that it displays a repeating pattern with equivalent peaks that approach infinity when the system exhibits metastability as it slows and changes direction (similar to the polynomial case). Although the FI is not steady, the periodic result indicates intrinsic organization and stability. This example shows the importance of the scale and time frames when computing FI. Because real systems have no exact cycles, "Wastewater Reactor Model for Nitrogen Removal" section proposes a systematic process to find the effective period of a system's trajectory. Further, the metastable behavior exhibited is an indication of a singularity that represents a change in trajectory and could be linked to bifurcation points (see Appendix 2.2).

Because multiple observable variables are needed to describe real systems, these elementary functions are not intended to present a true representation of complex system behavior. However, it is possible that stages of a system's trajectory may exhibit exponential, quadratic, linear, or periodic behavior. FI trends help to differentiate stable and unstable regimes, and changes in the system dynamics over time. Accordingly, the evaluation of elementary functions can serve as a guide to help deduce the behavior of FI in the presence of certain system conditions, including changes in the system path that may point to tipping points. Note that distinguishing the sustainability of regimes is hinged both on the stability of system dynamics and the desirability of the system condition. This idea is further explored in two models: a wastewater reactor for nitrogen removal and a lake shifting from an oligotrophic state (i.e., sustainable) to a eutrophic state (i.e., contaminated) because of phosphorus influx.

WASTEWATER REACTOR MODEL FOR NITROGEN REMOVAL

Evaluating complex systems is challenging because it involves tracking multiple variables with inherently distinct patterns. Hence, approaches are needed to make the evaluation straightforward. To illustrate this difficulty, we use a simple model to represent a wastewater treatment reactor exhibiting periodic behavior. The model is evaluated using FI and methods are presented that can be used to help determine periodicities for real systems and to examine the significance of cycles in sustainability assessment.

A food web model was used to represent the removal of nitrogen from municipal wastewater and the dynamics of nitrifying bacteria predation by snails (Gonzalez-Mejia, 2011). This model simulates a biological treatment process in an activated-sludge reactor that is completely mixed and aerated, with the microorganisms suspended and evenly distributed in the tank.

Nitrogen may exist in wastewater in several forms to include total Kjeldahl nitrogen, ammonia, nitrate, and nitrite. Although factors such as pH, temperature, and dissolved oxygen influence the form of nitrogen found in water (Wiesmann et al., 2007), these variables are not considered in this model. Thus, we simply refer to the substrate as nitrogen (N) measured in mg N/L.

Microorganisms are typically named according to the type of nitrogen transformation that they perform. For example, bacteria that oxidize ammonia are called *nitrosomonas*, and those that oxidize nitrites are named *nitrobacter* (Wiesmann et al., 2007). As in the substrate case, all microorganisms are considered *nitrifying bacteria* (X_{nitri}) in this model. Further, although there are several snail species that can predate on nitrifying bacteria, including *Physa gyrina* or *Physa integra* (Palsdottir and Bishop, 1997), these organisms are denoted as the snail population (X_{snail}). Denitrification (i.e., nitrate reduction) that may ultimately produce molecular nitrogen (N_2) is not considered in this model.

This model is described by a system of three differential equations with individual expressions for nitrogen [Eq. (2.10)], nitrifying bacteria [Eq. (2.11)], and snails [Eq. (2.12)], which are mass balances across a control volume ($V = 1/\theta$). The assumptions are that the system is a completely mixed continuous flow reactor at steady flow where N is consumed exclusively by X_{nitri} and the input nitrogen (N_0) is constant.

Equation (2.10) represents the rate of change of nitrogen concentration in a control volume (i.e., accumulation), which is equivalent to the input of nitrogen minus the nitrogen that is leaving the reactor and the substrate consumed by the nitrifying bacteria. As denoted in Eq. (2.10), the reaction rate of nitrogen utilization (i.e., $[r_N = (kX_{\text{nitri}}N)/(K_{sN} + N)]$) depends on the concentration of nitrogen and nitrifying bacteria, as well as the nitrogen consumption constant (k) or maximum specific nitrogen utilization rate. Another important element of the reaction rate is the saturation constant (K_{sN}). The K_{sN} is the nitrogen concentration at one-half of the maximum substrate utilization rate as in a Michaelis–Menten or Monod model (Tchobanoglous et al., 2003).

$$\frac{dN}{dt} = \frac{1}{\theta}(N_0 - N) - \frac{kX_{\text{nitri}}N}{K_{sN} + N} \tag{2.10}$$

Similarly, Eq. (2.11) describes the change in nitrifying bacteria concentration over time. In this case, the growth rate of X_{nitri} is modeled according to a yield of growth $\left(Y_{X_{\text{nitri}}}\right)$ with a constant death rate $\left(kd_{X_{\text{nitri}}}\right)$ (Tchobanoglous et al., 2003). In addition, predation of bacteria by snails depends on a consumption constant $\left(k_{X_{\text{nitri}}}\right)$ and a saturation constant $\left(K_{sX_{\text{nitri}}}\right)$; here, we assume the same saturation-type equation as the one used for nitrogen:

$$\frac{dX_{\text{nitri}}}{dt} = \frac{1}{\theta}(X_0 - X_{\text{nitri}}) + \left(\frac{Y_{X_{\text{nitri}}}k\,N\,X_{\text{nitri}}}{K_{sN} + N} - kd_{X_{\text{nitri}}}X_{\text{nitri}} - \frac{k_{X_{\text{nitri}}}X_{\text{nitri}}X_{\text{snail}}}{K_{sX_{\text{nitri}}} + X_{\text{nitri}}}\right) \tag{2.11}$$

Likewise, Eq. (2.12) describes the rate of change in snail concentration, where the growth rate of snails $dX_{\text{snail}/dt}$ depends on a yield of growth $\left(Y_{X_{\text{snail}}}\right)$ and is controlled by a death rate constant $\left(kd_{X_{\text{snail}}}\right)$.

$$\frac{dX_{\text{snail}}}{dt} = \frac{1}{\theta}(X_{0\ \text{snail}} - X_{\text{snail}}) + \left(\frac{Y_{X_{\text{snail}}}k_{X_{\text{nitri}}}X_{\text{nitri}}X_{\text{snail}}}{K_{sX_{\text{nitri}}} + X_{\text{nitri}}} - kd_{X_{\text{snail}}}X_{\text{snail}}\right) \tag{2.12}$$

Ideally, there are no bacteria or snails in the feedstock (i.e., $X_0 = 0$, $X_{0\ \text{snail}} = 0$). However, this wastewater treatment reactor describes an extraordinary event in which there are snails in the reactor. This model was simulated from 1 to 110 d (time step = 0.01 d) with a retention time of $\theta = 3$ d, nitrogen feed of 40 mgN/L, and parameter values set in Table 2.1 (Gonzalez-Mejia, 2011).

The differential expressions of Eqs. (2.10)–(2.12) were solved numerically using a fourth-order Runge–Kutta method (Chapra, 2008c). The simulation result for the system is shown in Figure 2.2A and reflects the periodicity of each variable attributable to nitrogen consumption and predation of bacteria by snails. Figure 2.2B is the phase plane plot for the trajectory of this periodic system, which illustrates one cycle of a stable trajectory. FI was calculated using a moving window equivalent to the system's cycle (Figure 2.2C). Because this system has multiple cycles, a simple procedure was developed to determine the effective cycle period of this system: FI method 1 is computed iteratively increasing the size of the calculation window until the minimum variation on FI is reached; in each trial the arithmetic mean and standard deviation (SD) of FI

Table 2.1 Parameters to Model a Wastewater Reactor for Nitrogen Removal

Variables (mg/L)		
Nitrogen: N	Nitrifying bacteria: X_{nitri}	Snail population: X_{snail}
Initial conditions		
$N_0 = 15.41$	$X_0 = 5.43$	$X_{0\ snail} = 3.29$
Constants		
$k = \frac{1.67\ mgN}{mgX_{snail}\ day}$ $K_{sN} = \frac{1.4\ mgN}{L}$	$Y_{X_{nitri}} = \frac{0.6\ mgX_{nitri}}{mgN}$ $kd_{X_{nitri}} = \frac{0.05}{day}$ $k_{X_{nitri}} = \frac{1.2\ mgX_{nitri}}{mg\ snail\ day}$ $K_{sX_{nitri}} = \frac{3\ mgX_{nitri}}{L}$	$Y_{X_{snail}} = \frac{0.4\ mg\ snail}{mgX_{nitri}}$ $kd_{X_{snail}} = \frac{0.01}{day}$

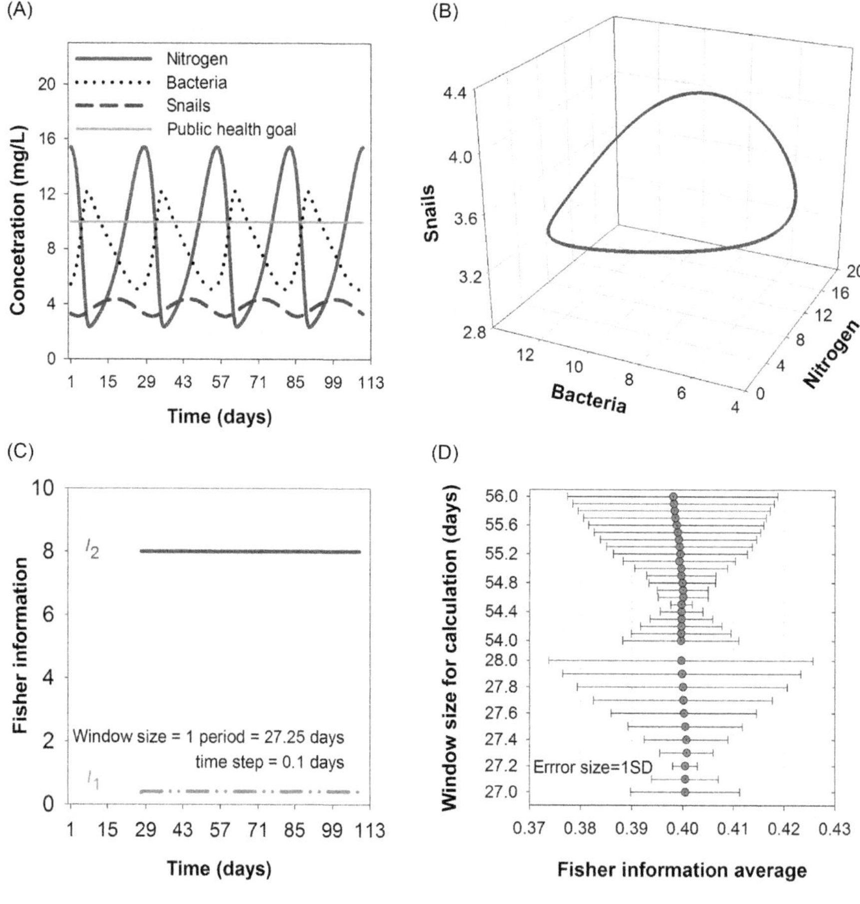

FIGURE 2.2

Wastewater treatment model simulation. (A) Time series. (B) Phase plane plot. (C) FI method 1 and method 2. (D) Optimal period of the system (minimum variation smallest SD on FI).

are computed for the complete time of study; and the minimum variation of FI is defined by the smallest SD (Gonzalez-Mejia, 2011). Figure 2.2D reports the optimal result with the window size set to one or two periods of the system cycle (i.e., 27.25 and 54.5 d). This model system serves to illustrate two useful results applicable to real systems: a system is stable if FI is the same over time (Figure 2.2D) or if FI displays repetitive patterns that indicate order and stability as in Figure 2.1D.

Besides the change in FI (i.e., $dI/dt = 0$), another important factor to consider for a sustainability analysis is the "desirability criteria." From a human perspective, nitrogen concentration must not exceed 10 mg/L in regulated US public water systems (US Environmental Protection Agency, 2009). Accordingly, although the system meets stability criteria as captured by FI, this system has approximately 10 d at the end of each month when the nitrogen threshold is exceeded (Figure 2.2A). Hence, it is critical to evaluate the dynamics of a system with FI along with the "desirability" of system conditions for a complete sustainability assessment.

SHALLOW LAKE MODEL: REGIME SHIFT

Some systems are steady and function with a great deal of stability, and others are periodic and exhibit orderly dynamic patterns. However, many systems have parameters that fluctuate with nominal variation until some aspect of the system reaches a threshold such that the behavior of the system shifts to a completely new regime with a different set of underlying conditions. Identiflying warnings of such behavior is critical to sustainable environmental management. Here, a set of differential equations are used to model a shallow lake experiencing increasing inflow of phosporous from runoff, resulting in a transition from an oligotrophic (clear and healthy) to an eutrophic regime (turbid and undesirable). This model was adapted from Carpenter's (2003) work on lake system regime shifts and was also explored by Pawlowski and Cabezas (2008) using neighborhood statistics. The stock of phosporous in the lake [$x(t+1)$] is represented by Eq. (2.13), with step changes in the input of phosporous (a) defined by Eqs. (2.14)–(2.19). Here, b is the rate of natural phosporous removal set to 0.58 (Figure 2.3A–C). To simulate the inherent noise in real data, Eq. (2.14) adds noise to the phosporous input [$Z(t+1)$] as a normally distributed error (ε) with mean of 0, variance of 0.0002, and a constant $p = 0.99$ (Karunanithi et al., 2008).

$$x(t+1) = ae^{Zt} - bx(t) + \frac{x^2(t)}{x^2(t)+1} \tag{2.13}$$

$$Z(t+1) = pZ(t) + \varepsilon(t) \tag{2.14}$$

$$0 < t < 200 \quad a = 0.02 \tag{2.15}$$

$$200 \le t < 210 \quad a = 0.02 + \left(\frac{t-200}{10}\right)0.06 \tag{2.16}$$

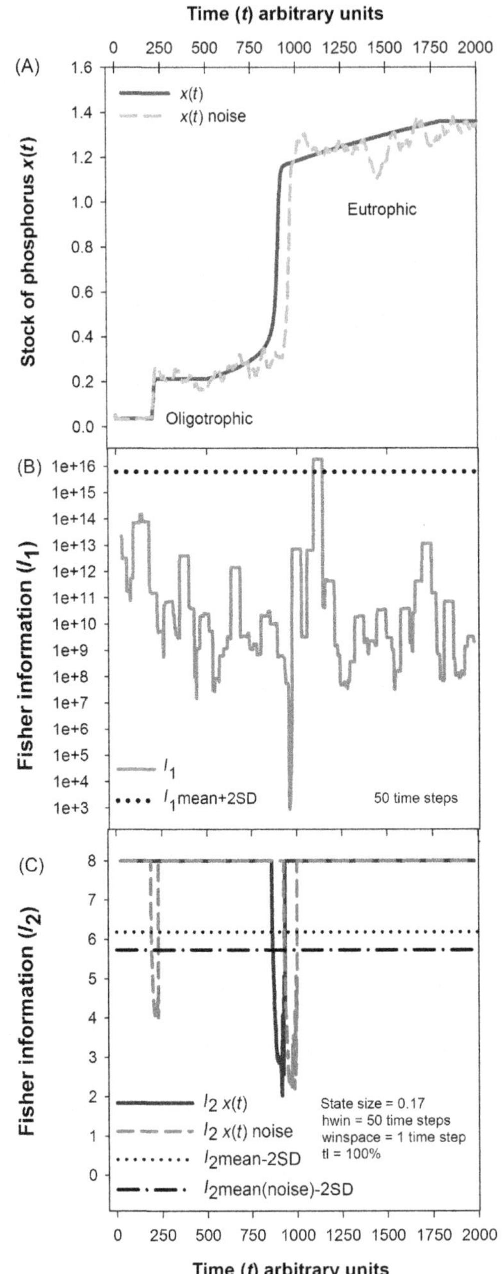

FIGURE 2.3

Simulated lake response to changes in phosporous input with noise. (A) Lake system equilibrium (with and without noise), (B) time series, (C) FI method 1:I_1, and (D) FI method 2:I_2.

$$210 \le t < 500 \quad a = 0.02 + 0.06 = 0.08 \tag{2.17}$$

$$500 \le t < 1800 \quad a = 0.08 + \left(\frac{t - 500}{1300}\right)0.06 \tag{2.18}$$

$$1,800 \le t \quad a = 0.08 + 0.06 = 0.14 \tag{2.19}$$

Figure 2.3A simulates the concentration of phosporus over time (with and without noise) to include step increases in line with Eqs. 2.13–2.19. FI was computed for both methods using an hwin of 50 time steps and winspace of one time step. For method 2, the size of states was set by computing the SD of the most stable region of the trajectory (first 500 points) and multiplying it by two (Karunanithi et al., 2008). Because the system was only defined by one variable, the tightening level (TL) was set to 100 (TL = 100%). The linear sections of the trajectory without noise [$x(t)$] resulted in discontinuous derivatives for method 1; hence, method 1 was only used to evaluate the noisy data set (mimicking real data) (Figure 2.3B). Method 2 displayed relatively consistent results when used to evaluate the dynamics of the system with and without noise [i.e., $x(t)$ and $x(t)$ noise] (Figure 2.3C). Note that the FI results reflect behavior indicative of a regime shift, that is, a significant decline (minimum) in FI as the stock of phosphorous quickly increases just before time of 1,000 (Figure 2.3B and C). The methods also detected a few nuances in the system trajectory. The FI result from method 1 ($I_1[x(t)\text{noise}]$) displayed a peak (more than 2 SD from mean FI) reflecting an inflection point where the noisy data settled, and then the trajectory changed from increasing to decreasing. A significant decrease (less than 2 SD from mean FI) was shown in the method 2 FI results ($I_2[x(t)\text{noise}]$) that corresponded with the initial step increases in phosphorus input.

In the context of sustainability, it is important to consider the overall results in conjunction with the human desirability of the system state. Note that although both the oligotrophic and eutrophic regimes are stable, the eutrophic condition is characterized by phosphorous concentration levels that promote algae growth and reduce the oxygen needed to support aquatic life. Accordingly, although the eutrophic condition is stable, there is no desire to sustain it.

SUMMARY: COMPARISON OF METHODS 1 AND 2

The methods presented for estimating FI are based on capturing changes in the condition of a system by evaluting the velocity and acceleration through the system trajectory (method 1) or assessing the probability of observing system states (method 2). Both methods can be used to collapse multiple variables into an index that can be tracked over time to assess changes in system behavior and, therefore, provide practical mechanisms for assessing stable and unstable periods in complex system dynamics.

Method 1 may be computed analytically or numerically when the trajectory of a system can be described by continuous functions that have smooth first and second derivatives. Normalization, interpolation (for sparse data), data filtering, and fitting may be used to preprocess empirical data in preparation for use with method 1; however, if there is no continuity in the derivatives, or if there is a very noisy data set, then computation is greatly hampered using this method. Because the binning approach is based on estimating the probability density of particular system states, method 2 is quite robust in handling noise and data quality issues inherent in real systems. Although there is no need for data preprocessing (other than interpolation for sparse data sets), TLs and estimations of measurement uncertainty have been strategically adopted to compensate for quality and uncertainty issues inherent in empirical data. Matlab code (Mathworks, Inc., 2012) has been written to automate these steps and compute FI using method 2. The advantages of method 1 are a stronger theoretical base and the ease of computation with minimal parameters to manage. However, if the system is defined by noisy data sets, then method 2 is usually the best option.

Through the elementary functions and wastewater treatment model, it is clear that the FI results are qualitatively the same for both methods. The core difference is the order of magnitude in the FI results ($I_1{:}0 \rightarrow \infty$; $I_2{:}0 \rightarrow 8$). However, the shallow lake model afforded the ability to explore additional distinctions. From the sustainable regimes hypothesis, regime shifts are typically characterized by significant declines in FI. In the lake model, a decrease in FI is evident for both methods, but if regime shift criteria is set to a standard measure of statistical difference (e.g., 2 SD below the mean FI), then the regime shift will not be identified for method 1 (Figure 2.3B). Further, method 2 produced quite a smooth result and even captured the dynamic change when the initial step increases in phosphorus occurred. As documented in "Information Theory" section, researchers such as Cabezas, Eason, Fath, Mayer, and Karunanithi, to name a few, have demonstrated the usefulness of both methods in identifying regime shifts. However, because of data quality issues in real systems, there has been great success with using method 2 to capture nuances to include distinct system dynamics (Mayer et al., 2006) transitional behavior and to detect regime shifts from empirical data (Eason et al., 2014b). Although additional work may be needed to use method 1 for studying regime dynamics in real systems, the detection of an inflection point (more than 2 SD from the mean FI) in the lake system provides evidence that method 1 can be used to assess changes in system trajectory. Capturing changes in trajectory is critical for evaluating systems like cities or metropolitan areas that are characterized by extremely sparse data (e.g., census data captured every 10 years) and are being assessed over a relatively short time frame (e.g., decades). Although these systems may not undergo drastic transitions, it is possible for underlying trends to change direction (e.g., increasing to decreasing) and impact system sustainability.

APPLICATION TO REAL URBAN, REGIONAL, AND NATIONAL SYSTEMS

Although much of the literature on metrics for detecting regime shifts is limited to ecosystems, the techniques developed in this arena may be useful on much higher spatial scales. For example, they can be used to help inform management and align policy in conjunction with local, regional, national, and global sustainability goals. This section provides a summary of some of the efforts of applying FI to support sustainable environmental management on various spatial scales. Further, it provides insight on how each of the methods for computing FI are applied to real systems.

URBAN: OHIO STATISTICAL METROPOLITAN AREAS

Given that 93.7% of the US population resides in urban areas (U.S. Census Bureau, 2011), long-term management and development practices are important for ensuring the preservation of human populations, socioeconomic structure, and natural environments. Thus, a comprehensive sustainability analysis is a powerful tool for studying and managing systems with the goal of encouraging the most favorable social, economic, and environmental trends toward a sustainable future. Using historical data, method 1 was applied to assess the impact of socioeconomic characteristics on the stability of six metropolitan statistical areas (MSAs) in Ohio from 1970 to 2009. These MSAs, specifically Cincinnati, Dayton, Cleveland, Akron, Columbus, and Toledo, were analyzed and compared to identify dynamic changes in these systems that highlight critical areas for policy and management. Further, Cincinnati, Ohio, was explored in greater detail to assess local dynamics (including the city and surrounding suburbs).

The area of study was delimited according the 2009 definitions for the city of Cincinnati and US MSAs (U.S. Census Bureau, 2010b). MSAs are defined by at least one urban core of no less than 50,000 inhabitants with a strong socioeconomic interconnection among the parts (U.S. Census Bureau, 2011). Time series data representing the suburbs (SUBS) were estimated by subtracting the city values from those of the MSA (i.e., SUBS = MSA − CITY). Forty variables were selected from US Census data and divided into two core components: 29 social variables and 11 economic variables, which were further classified into 7 subcomponents (i.e., social distribution, age distribution, household type, educational attainment, housing occupancy, employment status and poverty, and income). These social and economic variables were extracted from the decennial US Census in 1970, 1980, 1990, and 2000 (GeoLytics Inc., 2003) and the American Community Survey (ACS) for 2009 values (U.S. Census Bureau, 2010a). Because the variables are in different units (e.g., inhabitants, US dollars), each time series was normalized by the first datum to get unitless variables from a reference point in 1970. Further, because of the scarcity of data (i.e., only five data points per variable), interpolation and smoothing were required to generate enough data points to provide continuous first and second derivatives for calculating FI with method 1 (Gonzalez-Mejia et al., 2012a).

The dynamic behavior of the city and suburban area of Cincinnati, as well as the other five MSAs (i.e., Dayton, Cleveland, Akron, Columbus, and Toledo), were evaluated by computing FI separately for the 29 social variables (IS), the 11 economic variables (IE), and the overall system that included all 40 variables (ISE) (Figure 2.4A–C). FI was integrated numerically over 1 year [Eqs. (2.7)–(2.9)] with a winspace of one time step. Then, a moving average of 3 years was used as a high-frequency filter to focus on trends in the system's trajectory instead of on minor fluctuations. To detect significant changes in the behavior of these systems, local maxima were identified over the evaluation period (i.e., inflection points more than 2 SDs from the FI mean) and denoted as changes in trajectory. Further, Spearman rank order correlation (SROC), a nonparametric test, was used to explore the association strength between components and underling social and economic variables that may be possible drivers of change in the system dynamics. Statistically significant SROC results (ρ) are reported given a p-value $\ll 0.05$, considering 1 as the strongest correlation coefficient and [0.0–0.5] as no correlation; note that the pairs with positive correlation coefficients tend to increase together.

Figure 2.4A shows that the behavior of the city, suburban, and MSA of Cincinnati changed simultaneously between 2000 and 2005 (note FI peak). Similarly, an inflection point in the FI for the economic variables occurred (Figure 2.4B), which indicated that the most significant change in direction for economic variables also happened after 2000. In contrast, social variables changed direction years before the economic component (i.e., 1995–2000 in Figure 2.4C). Table 2.2 highlights strong correlation among all the areas undergoing study (i.e., MSA, suburbs, and city of Cincinnati) for social and economic variables (ISE). Note that while the economic component (IE) strongly correlated with the overall system (ISE) in each area, the social component (IS) did not. These results imply that the economic component was the main driver of significant changes in Cincinnati city, suburbs, and MSA (Table 2.2 and Figure 2.4).

Social and economic changes in the City, Suburban and Metropolitan area of Cincinnati, Ohio, from 1970 to 2009 are correlated (Table 2.2).

Although FI did not show the cause of significant change, it located the most significant deviation(s) of the system's trajectory. A further analysis is needed to determine whether the new path (i.e., after a FI maxima) is described by sustainable (e.g., desirable growth rate) or unsustainable trends. In the Cincinnati case, the economic component seemed to be driving changes in the overall system (Figure 2.4A and B and Table 2.2). Thus, by exploring the time series before and after the FI maxima (i.e., 2000–2005 in Figure 2.4A and B), two periods can be distinguished: one of socioeconomic growth between 1970 and 1999 and another of economic decline from 2000 to 2009. The next logical question to answer

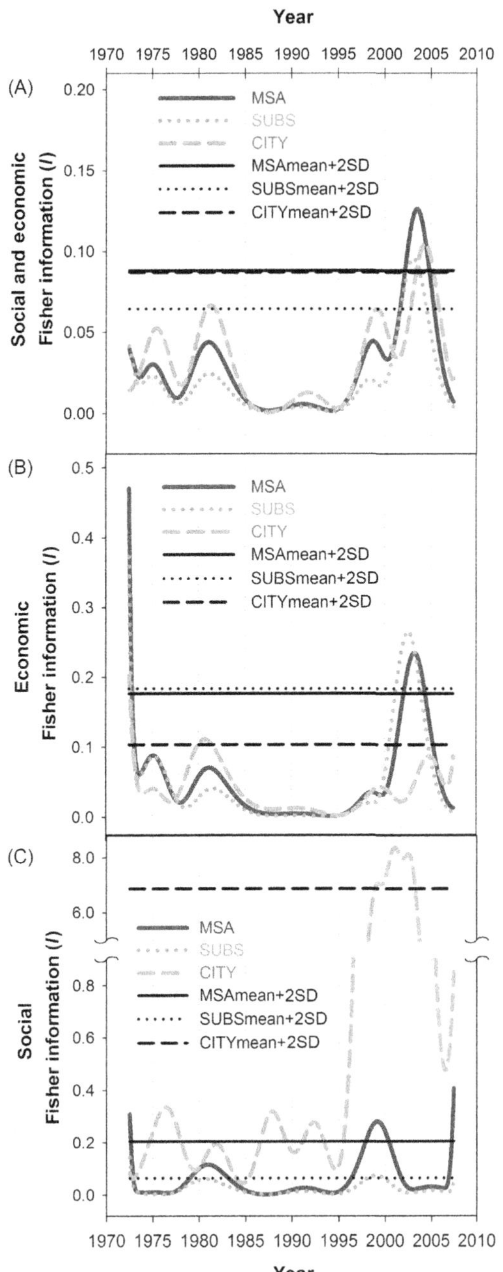

FIGURE 2.4

Sustainability assessment of Cincinnati city, suburbs, and MSA. Fisher information for (A) the social and economic components, (B) the economic component, and (C) the social component.

Table 2.2 SROC Results for Cincinnati City, Suburbs, and MSA OH-KY-IN

	ISE MSA	IE MSA	IS MSA	ISE SUBS	IE SUBS	IS SUBS	ISE CITY	IE CITY
IE MSA	**0.90**							
IS MSA	**0.63**	0.35						
ISE SUBS	**0.97**	**0.95**	0.54					
IE SUBS	**0.88**	**0.97**	0.35	**0.94**				
IS SUBS	0.53	0.26	**0.94**	0.43	0.24			
ISE CITY	**0.93**	**0.83**	0.56	**0.89**	**0.81**	0.48		
IE CITY	**0.75**	**0.77**	0.52	**0.73**	**0.68**	0.56	**0.77**	
IS CITY	0.45	0.29	0.28	0.41	0.37	0.10	0.42	−0.05

This table identifies social and economic components correlated to changes in the city, suburban, and metropolitan area of Cincinnati, OH, from 1970 to 2009. All results are at the significant level (p-value ≪0.05), considering 1 the strongest correlation coefficient and [0.0–0.5] no correlation. Note that the pair(s) with positive correlation coefficients tend to increase together. Strongest correlation is in bold.

was: which of the 40 variables have changed the most and may possibly describe the new path toward or away from a sustainable future?

The appearance of ISE maxima (Figure 2.5A) identified significant events that started in approximately 2002 for five MSAs (Akron, Cleveland, Cincinnati, Columbus, and Dayton) and 2000 (Toledo). Likewise, a change in the economic trend (identified by IE peaks) began in 2000 for Toledo and Columbus, in 2001 for Cleveland, Cincinnati, and Dayton, and in 2002 for Akron (Figure 2.5B). This demarcation of inflection points implies that the first 30 years of the study were significantly different than the last 9 years for urban systems in Ohio. As in the Cincinnati study, Cleveland ($\rho = 0.97$), Toledo ($\rho = 0.96$), Dayton ($\rho = 0.94$), Cincinnati ($\rho = 0.9$), and Akron ($\rho = 0.89$), Ohio MSAs displayed a high correlation between the economic component (Figure 2.5B) and the overall socio-economic system (Figure 2.5A). Therefore, the sustainability analysis will focus on the economic variables.

Table 2.3 describes the economic trend for each system from 2000 to 2009 (after the inflection point in FI). Results indicate that 73% of these variables presented no change, reflecting undesirable economic conditions that did not help counterbalance the increase in the number of vacant housing units, unemployed civilian labor force, and persons living below the poverty level reported in all Ohio MSAs. Because such conditions are not desirable for urban systems, they are indicative of movement away from sustainable socioeconomic tendencies in this region. Hence, Ohio MSAs were charaterized by socioeconomic growth (1970–1999) and an unsustainable period since 2000.

By using an arithmetic mean and SD of FI for the entire period of study (i.e., 1970–2009), these urban systems could be ranked to compare the stability of the MSAs. In Figure 2.6, each MSA is arranged by the average population size

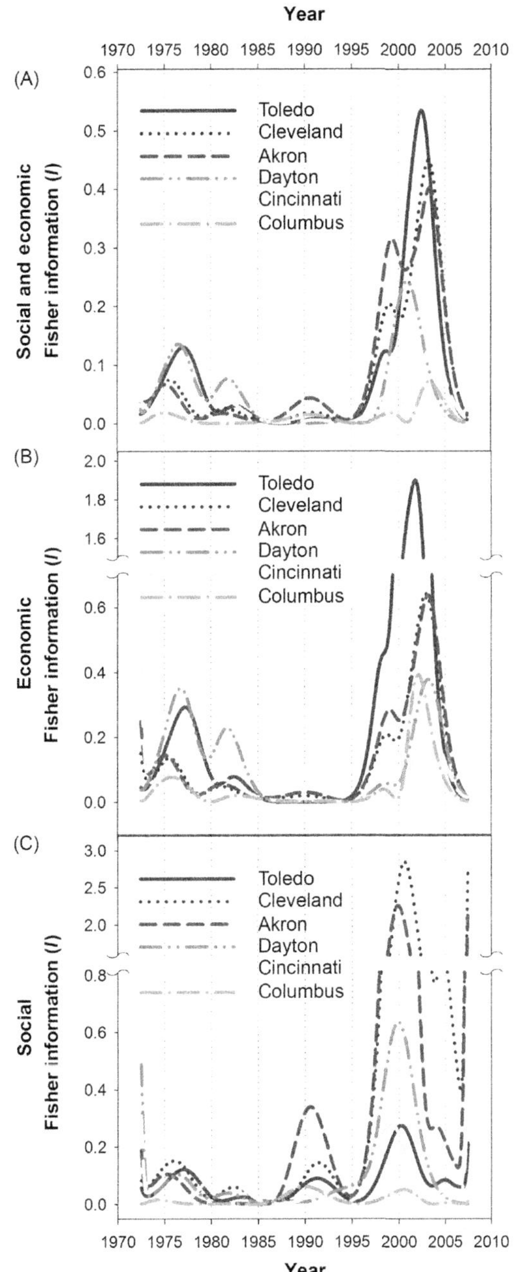

FIGURE 2.5

Changes in trajectory for Ohio MSAs. FI for Toledo, Cleveland, Akron, Dayton, Cincinnati, and Columbus. (A) Social and economic component, (B) economic component, (C) social component.

Table 2.3 Classification of Economic Trends (2000–2009) for MSAs in Ohio

Economic Component Change 2000–2009	Housing Tenure					Employment Status[a] and Poverty				Income	
	Total Housing Units	Total Occupied Housing Units	Total Vacant Housing Units	Total Owner-Occupied Housing Units	Total Renter-Occupied Housing Units	Employed Females	Employed	Unemployed	Total Persons Below the Poverty Level	Average Family Income	Average Household Income
Toledo	0.2	0.1	**3.5**	0.1	0.1	0.1	0.0	**3.5**	0.7	0.5	0.4
Akron	0.1	0.0	**1.5**	0.0	0.2	0.0	0.0	**2.2**	0.6	0.7	0.5
Cleveland	0.0	0.0	**1.8**	−0.1	0.0	0.0	−0.1	**2.2**	0.5	0.8	0.5
Dayton	0.1	0.0	**2.0**	0.0	0.1	−0.1	−0.1	**2.2**	0.6	0.7	0.5
Cincinnati	0.2	0.1	**2.0**	0.1	0.0	0.2	0.1	**2.9**	0.5	1.2	1.0
Columbus	0.3	0.2	**3.3**	0.2	0.2	0.2	0.1	**3.7**	1.3	1.0	0.9

The 2000 and 2009 values are normalized by the reference 1970 datum, and the table reports the difference between normalized 2009 and 2000 values.

In line with the change in the system trajectory identified by FI peaks in the economic component, underlying variables displayed unique characteristics in the last period of study.

[a]*Civilian labor force 16 years and older.*

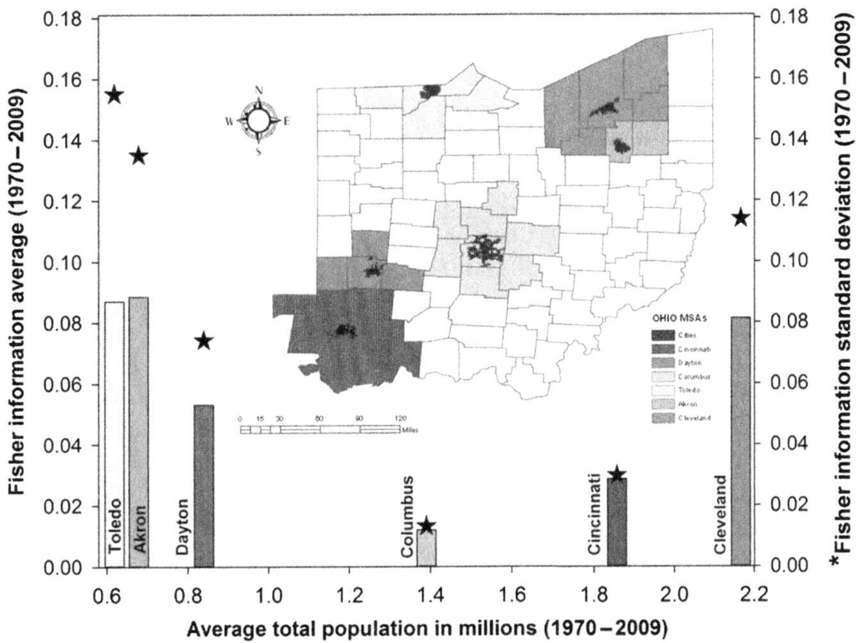

FIGURE 2.6

Sustainability assessment for Ohio MSAs (1970–2009). Ohio MSAs were arranged by population size, the mean, and SD in FI to rank stability. Using low SD in FI as the desired measure, that rank is: (1) Columbus, (2) Cincinnati, (3) Dayton, (4) Cleveland, (5) Akron, and (6) Toledo.

(x-axis), the mean FI is plotted as a bar (values on the left y-axis), and the SD in FI are shown as asterisks (values on the right y-axis). These socioeconomic systems can be divided in two groups: one characterized by relatively low SD and FI and another with both high SD and FI. This signifies that the organizational dynamics of these urban systems are driven by significant changes in trajectory. Because the sustainable regimes hypothesis points to the need for low variation in FI (*i.e.*, $dI/dt \approx 0$), the most stable MSAs are those with low SD in FI. Accordingly, Columbus was the most stable MSA (Figure 2.6), with a slight growth in economic trends toward a more sustainable future. In contrast, Toledo was the most unstable (i.e., high SD in FI) urban system in Ohio, with excessive increases in vacant housing units, unemployment, and inhabitants living below the poverty level (2000–2009).

Despite the similarities between Northern and Southern Ohio MSAs, many economic activities (e.g., educational services, health care and social assistance, and finance and insurance) grew in Toledo, Cleveland, and Akron MSAs between 2001 and 2009. Although employment increased in these same sectors for Columbus, Cincinnati, and Dayton, it also thrived in others (e.g., professional, scientific, and technical services, accommodation and food services, arts, entertainment, and recreation) (U.S. Bureau of Economic Analysis, 2012). Therefore, the private non-farm and government employment rates were more stable in the Southern Ohio MSAs compared with the northern ones. Unemployment was critical for all six MSAs, and housing vacancy may have also influenced the drastic change in the House Price Index from 2005 to 2009 ($HPI_{2005-2009}$), which was significantly greater in the southern MSAs: Toledo $HPI_{2005-2009} = -8.7$, Toledo $HPI_{1997-2001} = 31.1$, Cleveland $HPI_{2005-2009} = -7.0$, Cleveland $HPI_{1997-2001} = 26.3$, Akron $HPI_{2005-2009} = -3.8$, and Akron $HPI_{1997-2001} = 27.4$ (Federal Housing Finance Agency, 2010).

This study is not a cause-and-effect analysis, but it helps to find characteristic features of urban systems that are typically difficult to recognize by inspection of individual time series variables. This analysis provided interesting results useful for management and afforded the ability to test and debunk an assumption that larger areas such as Cleveland were not inherently more stable than smaller ones such as Columbus. Thus, the Ohio MSA results showed no link between population size and stable dynamics.

EARLY WARNING SIGNALS, REGIME CHANGE, AND LEADING INDICATORS

Ecological systems can be resilient and still experience large fluctuations or low stability (Holling, 1973). Although stability has been found to be related to the number of links between species in a trophic web (MacArthur, 1955) as the human population and its resource demands increase, systems typically move further away from equilibrium (Holling, 1973). As indicated previously, many complex systems have critical thresholds or tipping points that, when breached, cause the system to shift from one dynamic regime to another. This, in turn, has led to the idea that early warning signals may be derived from time series data gathered from systems on the verge of experiencing critical transitions (Scheffer et al., 2009). Wissel (1984) determined that when many systems are approaching a critical threshold, they undergo a "critical slowing down" in system activity. Other systems experience behavior such as enhanced flickering before the shift to a different regime occurs (Carpenter et al., 2008).

Indicators to include variance, skewness, kurtosis, and critical slowing have been proposed and studied by numerous researchers (Biggs et al., 2009) as indicators of impending regime shift. Eason et al. (2014b) provide a review of much of this literature and noted that these approaches have been extensively applied to assess models or simple systems with few variables. However, Scheffer et al. (2009) noted that work is still needed to determine if they are useful for evaluating real, complex, multivariate systems. This sentiment is echoed by Dakos et al. (2012), who explored multiple indicators that have been proposed as early warning signals and found difficulty in determining the best indicator for identifying impending transitions. Moreover, Biggs et al. (2009) found that these approaches typically do not signal a shift until it is underway and often too late for management intervention (Dakos et al., 2012).

Recently, FI has been examined as an indicator of pending regime change (RC). Eason et al. (2014b) used method 2 to explore the relationship between FI and traditional indicators of regime shift, and they comparatively evaluated their results when assessing simple and complex systems. From the study of simple model systems, FI was found to be negatively correlated with variance and positively correlated with kurtosis. The relationship between skewness and critical slowing (denoted by increasing autocorrelation defined by lag 1 autocorrelation coefficient, AR1) was inconclusive. However, the authors argue that because a more ordered system would have less deviation from mean behavior (less skewness), and because critical slowing is reflected by increasing autocorrelation (Scheffer et al., 2009), FI is expected to be negatively correlated with both skewness and AR1 as a system approaches a critical transition. Scheffer et al. (2012) indicated that a positive relationship exists between FI and critical slowing (increasing AR1). However, as noted by Seekell et al. (2012) and in Appendix 2.2, system dynamics may vary and, subsequently, so may indicator signals. For simple systems (two species of Lotka–Volterra and simulation of nitrogren release into a shallow lake), the indicators performed similarly with FI identifying shifts (local minimums) at approximately the same time or often before other indicators (Eason et al., 2014b).

Numerous researchers have studied the Pacific Ocean ecosystem and cited regional climate (McGowan et al., 1998) and biological changes (Grebmeier et al., 2006) resulting in two regime shifts in 1977 and 1989 (Hare and Mantua, 2000). Karunanithi et al. (2008) previously analyzed this system using FI and found dynamic changes in the system in line with expected shift periods. Here, we present results that extend the study to compare the performance of FI with traditional regime shift indicators. Sixty-five biological and climate variables were compiled from a Hare and Mantua (2000) study and, from this data, FI and variance were computed (hwin = 10, winspace = 1) to assess the dynamic behavior of the system. Method 2 was used to compute FI, and the size of states was determined by using Matlab code (Mathworks, Inc., 2012) developed to locate a period in the system with minimal variation to estimate the uncertainty for each dimension.

The results plotted in Figure 2.7A show that some variables displayed increasing variance but others did not. Such behavior is true of similar indicators (e.g., skewness, critical slowing) because using traditional indicators to explore system dynamics requires the indicator to be computed for each variable (Eason et al., 2014b). Hence, the traditional indicators provided unclear signals about the behavior of this complex ecosystem (and multivariate systems, in general).

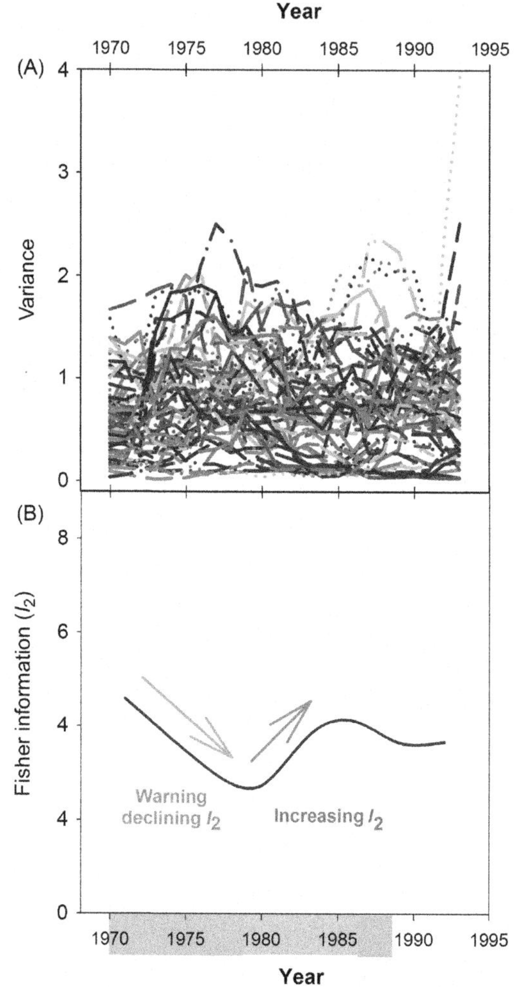

FIGURE 2.7

Comparing regime shift indicator performance for the Bering Strait ecosystem. (A) Variance and (B) FI (method 2).

Seekell et al. (2012) noted similar behavior and reported that there is evidence of conflicting patterns (increases or decreases) in autocorrelation, variance, and skewness as a system approaches a regime shift. When FI was computed from these data, there were local minimums in FI that corresponded with shift periods and decreases in FI prior to the shift (Figure 2.7B). Accordingly, Eason et al. (2014b) proposed that declines in FI should be explored as early warning signals of critical transitions because they provide evidence of loss of dynamic order and a “window of opportunity” for management intervention. Further expanding on this idea, Bayes theorem is used to augment the approach with classical statistical methods. Bayes theorem is not strictly part of information theory. However, we add it here because of its importance in interpreting and extending the results that can be obtained from FI analysis.

Bayes theorem

Bayes theorem was developed by the English Reverend Thomas Bayes (1702–1761) and first published in 1763 in the *Philosophical Transactions of the Royal Society of London*. The theorem deals with conditional probabilities, such as the likelihood of a particular event X occurring if another event Y has already occurred. For purposes of this work, the most important result from Bayes theorem in the simplest form is defined in Eq. (2.20), where $P(X|Y)$ is the probability of X being observed if Y has already occurred (the probability of X in the presence of Y), $P(X)$ is the probability of X ocurring, $P(Y|X)$ is the probability Y occurring if X has already happened, and $P(Y)$ is the probability of Y being observed.

$$P(X|Y) = \frac{P(X)P(\mathrm{Y}|\mathrm{X})}{\mathrm{P(Y)}} \tag{2.20}$$

Bayes theorem is most useful when there are reasonable estimates of $P(X)$ and $P(Y)$ and some information about the conditional probability $P(Y|X)$ exists. For assessing warning signals in FI, Bayes theorem is applied to estimate the likelihood that a decrease or a sequence of decreases in FI signals an impending RC. Hence, Eq. (2.21) denotes the appropriate expression using Bayes theorem, where $\Delta\mathrm{FI} < 0$ is a decline in FI, $P(\mathrm{RC}|\Delta\mathrm{FI} < 0)$ is the probability that there will be an RC if a decrease in FI has been observed, $P(\mathrm{RC})$ is the probability of observing an RC over the history of the system, $P(\Delta\mathrm{FI} < 0|\mathrm{RC})$ is the probability of seeing a decrease in FI if an RC has been observed, and $P(\Delta\mathrm{FI} < 0)$ is the probability of observing a decrease in FI.

$$P(\mathrm{RC}|\Delta\mathrm{FI} < 0) = \frac{P(\mathrm{RC})P(\Delta\mathrm{FI} < 0|\mathrm{RC})}{P(\Delta\mathrm{FI} < 0)} \tag{2.21}$$

Given that regime shifts have typically been preceded by declines in FI (Mayer et al., 2007), the probability of there being a decrease in FI ($\Delta\mathrm{FI} < 0$) if an RC has occurred is one, i.e., $P(\Delta\mathrm{FI} < 0|\mathrm{RC}) = 1$. $P(\mathrm{RC})$ can be estimated by counting the number of times an RC occurred and dividing it by the total number of FI results. Similarly, $P(\Delta\mathrm{FI} < 0)$ can be computed from the number of time steps where there was a decline in FI divided by the total number of FI time steps. The same general logic applies for using Bayes theorem to assess whether two or three sequential time steps of decreasing FI signals an RC (i.e., D2 and D3). Bayes theorem will be applied to the study of nation states.

REGIONAL: SAN LUIS BASIN, COLORADO

In an effort to explore methods of assessing the sustainability of a regional system, a group of US EPA researchers underwent a multiyear study to examine whether a system was moving toward or away from sustainability over time (U.S. Environmental Protection Agency, 2010). The goal was to develop an inexpensive methodology using readily available data to compute scientifically defensible measures that would provide results that are relevant, understandable, transferable, and able to inform decision-making. As a pilot project to determine the applicability and utility of such an approach, the team chose the San Luis Basin (SLB), a seven-county region (Alamosa, Conejos, Costilla, Hinsdale, Mineral, Rio Grande, and Saguache) in south-central Colorado (Figure 2.8). This study was sparked by interest within US EPA Region 8 to perform a sustainability assessment in this area, which is largely composed of public land that is managed/owned by multiple organizations (U.S. Environmental Protection Agency, 2010). Accordingly, it requires great collaboration and coordination to administer the land and its resources. The SLB contains multiple ecosystems and is primarily characterized as a rural agricultural region covering 21,000 km^2. Further, it is home to nearly 50,000 people and serves as a hinterland for exporting resources,

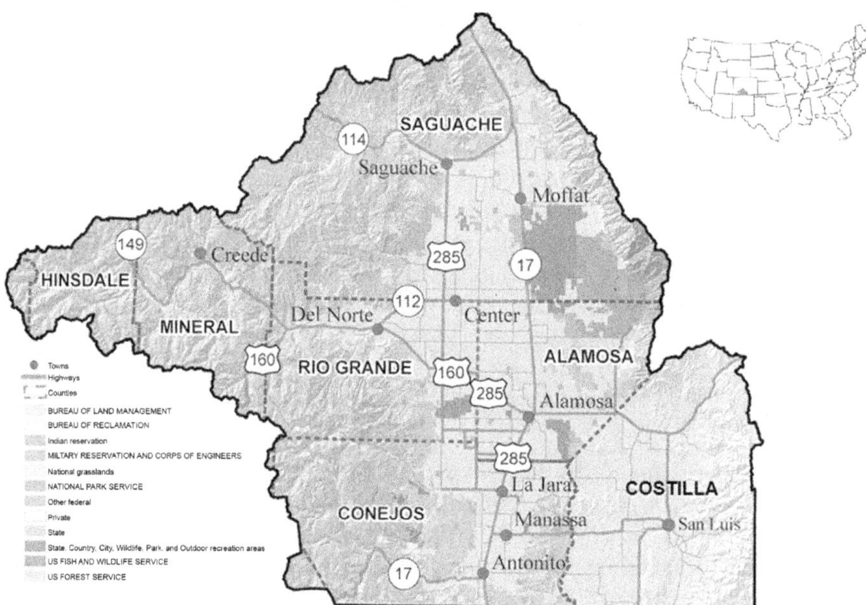

FIGURE 2.8

San Luis Basin Land Status Map. Slight modification of map provided by Bureau of Land Management–Colorado/San Luis Valley Field Office.

From Bureau of Land Management–Colorado/San Luis Valley Field Office (2014).

materials, and products to more developed areas within the United States, Mexico, and Canada (Campbell and Garmestani, 2012).

The initial step of the study was selecting a suite of indices or variables that would be used to evaluate the system as a function of the ecological impacts of human activity using ecological footprint analysis (EFA), economic well-being with green net regional product (GNRP), and available energy flow through the system using emergy analysis (EmA), and FI was used to examine system order and stability. Sustainability criterion was established for each index and used to evaluate trends from 1980 to 2005 (Figure 2.9).

Fifty-three variables were compiled from multiple publicly available data sources (e.g., Bureau of Economic Analysis, Energy Information Administration, US Department of Agriculture, and US Census Bureau) to characterize the consumption (energy and food), production, environmental, demographic, and land use aspects of the region for the period of study. FI was computed using method 2 and integration window parameters (i.e., hwin and winspace) were set to 8 years and 1 year. The level of uncertainty for the variables was unknown; hence, system data were used to estimate the level of uncertainty and to compute a size of states for each dimension. FI results were placed in the center of each window (e.g., data from 1980 through 1987 were used to compute the FI value that was plotted in 1984). To minimize noise in the computation, FI was "smoothed" using a three-point filter. Results revealed a slight increasing trend in FI until approximately 1995, after which there was a shallow decline for the remainder of the period. Because the FI over time stayed within ±2 SD of the mean FI, the system

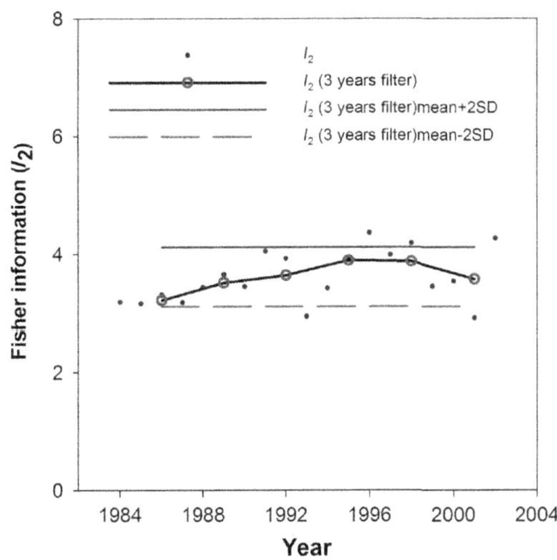

FIGURE 2.9

Sustainability analysis for the San Luis Basin, Colorado region. The plot provides the original FI computed from the data, smoothed result (3-year filter), and FI mean ±2 SD lines.

displayed great stability. However, the declining FI at the end of the study period is evidence of slight movement away from sustainability. In an effort to identify possible drivers of system behavior, SROC were used to compare the FI of the overall system with that of the system components (e.g., demographic, production, land use). From the SROC analysis, we were able to determine that the demographic component had a strong negative correlation ($\rho = -0.89$) with the FI of the overall system. Further exploration of underlying variables revealed that variation in population growth may have a notable influence on changes in system dynamics, suggesting that as population varies, there is a subsequent effect on the stability of the system. This result corresponds with the EFA of the SLB, which showed evidence of gradual movement away from sustainability as a result of population growth and subsequent declines in per capita biocapacity and ecological balance (Hopton and White, 2012). Details of the complete sustainability assessment of the SLB including the FI results (Eason and Cabezas, 2012) and the other metrics can be found in the US EPA report (U.S. Environmental Protection Agency, 2010). This work provided a test bed for the methodology and successfully demonstrated the abilility to assess patterns indicative of movement toward and away from sustainability. A follow-up study is being performed to test the utility and transferability of the approach to other systems with distinctively different characteristics.

NATIONAL: UNITED STATES

FI has been previously used to evaluate sociopolitical instability, development (Karunanithi et al. 2011), and sustainability in nation states (Mayer et al., 2007). Both studies provided interesting results and useful methods for assessing multiple systems and their path toward stability. Because many nations were being evaluated (e.g., one study examined nearly 70 nations), the authors opted to use mean FI values to aggreggate dynamic order over the entire period. The goal here was to examine system dynamics, including regimes and regime shifts. In this study, FI is used to evaluate dynamics in the United States, and Bayes theorem is used to examine the likelihood of a critical transition occurring.

Compared with many nations around the world, the United States has a documented history of stabililty. However, during the past century, the United States experienced national events, such as World Wars (WWs) I and II, the Great Depression, and the Dust Bowl (Figure 2.10). To assess whether the United States reflected fluctuations in its underlying dynamics that correspond with historical events in the twentieth century, this study used FI to examine the dynamic changes in key aspects of this well-established industrialized nation state.

Annual time series of 34 variables describing demographic, agricultural, labor, industry, trade, transport, finance, education, and prices were gathered from the *International Historical Statistics* series for the Americas (Mitchell, 2007) to characterize the United States from 1900 to 2000. Data were further categorized into "triple bottom line" sustainability classifications (economic, environmental,

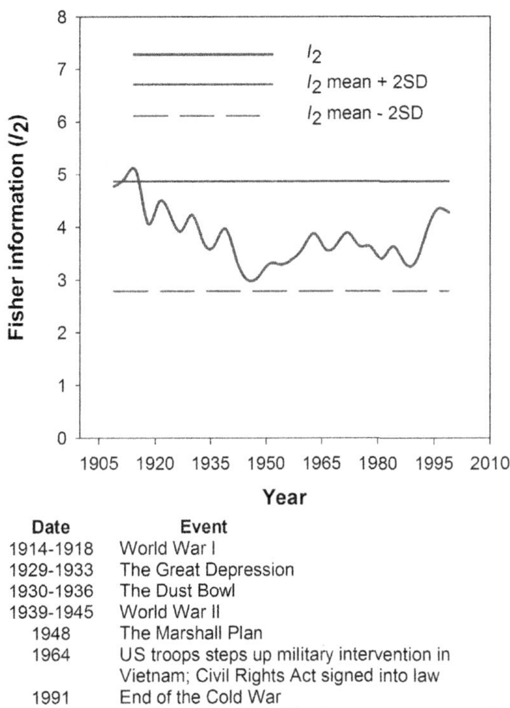

Date	Event
1914-1918	World War I
1929-1933	The Great Depression
1930-1936	The Dust Bowl
1939-1945	World War II
1948	The Marshall Plan
1964	US troops steps up military intervention in Vietnam; Civil Rights Act signed into law
1991	End of the Cold War
1992	North American Free Trade Agreement passed

FIGURE 2.10

Sustainability analysis for the United States. US Fisher Information with notable historical events from 1900 to 2001.

and social), and SROC analyses were also conducted to determine if FI calculated using economic, environmental, and social variables had significant correlation with overall FI. When necessary, an interpolation technique included in Matlab (Mathworks Inc., 2012) (e.g., piecewise cubic hermite interpolating polynomial, PCHIP) was again used to fill-in data gaps. FI was computed using method 2 (hwin = 8, winspace = 1) and uncertainity was estimated by finding a region with minimal variation in the time series, which was used to set the size of states.

FI results indicate that system order decreased until approximately 1950 and appeared to rebound and stabilize afterwards. Periods of decline coincided with calamitous events that included WWs I and II, the Great Depression, and the Dust Bowl (Figure 2.10). Actions such as the Marshall plan, end of the Cold War, and North American Free Trade Agreement characterized the more stable period. System order declined to a minimum in 1945 immediately after the end of WWII, but it never reached 2 SD below mean FI; hence, no regime shifts are considered to have occurred in the system during the twentieth century. SROC results indicated that although the overall system had positive correlations with demographic and agricultural FI ($\rho = 0.49$ and 0.51), the economic component had the strongest correlation with the overall system ($\rho = 0.76$) and appeared to drive the system dynamics.

EXPLORING DECLINES IN FI AS EARLY WARNING SIGNALS OF CRITICAL TRANSITION

In accordance with the proposal by Eason et al. (2014b) to explore declines in FI as an early warning signal of RCs, Bayes theorem was used to determine the frequency and significance of declines leading to critical transitions. Declines in FI were computed by calculating changes in the FI result over time (e.g., $FI_{1910} - FI_{1911}$). From this analysis, we found that the US FI had 14 single declines [$P(\text{D1}) = 14/30$], five double declines (two consequtive declines) [$P(\text{D2}) = 5/29$], and one triple decline [$P(\text{D3}) = 1/28$] during the study (Table 2.4). The sustainable regimes hypothesis denotes a regime shift as a significant and possibly abrupt change in FI between two regimes. Here, we explore criteria that may be used to classify the level of change indicative of a regime shift. To demonstrate the approach, we initally considered an RC as 1 SD below mean FI. Note that out of 30 total FI time steps, there was one time step in which FI was at least 1 SD below mean FI; therefore, the probability of an RC at 1 SD below the mean FI [i.e., $P(\text{RC})$] would be 1 in 30. Based on the occurrence of single declines during the time period, the probablility of observing an RC was 7.14% [i.e., $P(\text{RC}|\text{D1}) = 1 \times (1/30)/(14/30)$]. If double declines (D2) and triple declines (D3) are considered, then the probability of an RC increases to 19.3% and 93.3%. When the criteria for RC is made more stringent such that an RC requires a decline in FI of 2 SD below the mean FI, results indicate that no regime shift occurred in the United States during the twentieth century. Additional work is being performed to explore the utility of the approach and subsequent critieria for identifying regime shifts.

In summary, FI results tracked well with historical events, and declines in FI tended to correspond with periods of political and economic crises. Whereas the SROC analysis of categorical FI helped to determine which groups of variables

Table 2.4 Decline Analysis Using Bayes Theorem

Equation	Probability	Description
p(R|*D*1), std = 1	0.0714	[p(RG)p(*D*1|RG)]/p(*D*1) (RG = 1 SD below average FI)
p(R|*D*1), std = 2	0.0000	[p(RG)p(*D*1|RG)]/p(*D*1) (RG = 2 SDs below average FI)
p(R|*D*2), std = 1	0.1933	[p(RG)p(*D*2|RG)]/p(*D*2) (RG = 1 SD below average FI)
p(R|*D*2), std = 2	0.0000	[p(RG)p(*D*2|RG)]/p(*D*2) (RG = 2 SDs below average FI)
p(R|*D*3), std = 1	0.9333	[p(RG)p(*D*3|RG)]/p(*D*3) (RG = 1 SD below average FI)
p(R|*D*3), std = 2	0.0000	[p(RG)p(*D*3|RG)]/p(*D*3) (RG = 2 SDs below average FI)

Bayes' theorem calculation of the probabilities of an RC based on the number of single, double, and triple declines (D1, D2, and D3) at one and two SDs below FI.

might be influencing overall FI more than other variables, it is suggested that additional methods (such as principal component analyisis or Monte Carlo simulations) should be investigated to determine dominant variables of influence. Further, the use of the FI and Bayes theorem shows great promise as a leading indicator of critical transition. The combined application of these two approaches could be used as an ongoing management tool to aid in assessing whether a decline in system order warrants action or further investigation based on the historical behavior of the system.

CONCLUDING REMARKS

Human and natural systems display quite distinctive behavior, and managing these complex systems for sustainability requires methods that can capture both dramatic and subtle changes. FI shows great promise in this area and has been used to analyze a variety of real and model systems. However, challenges in handling data quality and difficulty in interpreting the results have hampered widespread use of the index (Rodionov, 2005). This chapter synthesizes recent developments that enhance the approach and extend the application of FI to sustainable environmental management. By summarizing ongoing activities (i.e., method development, interpretation guidance, data processing techniques, and case studies), efforts demonstrated the utility of the approach for managing highly complex and integrated systems toward sustainability.

Two approaches to computing FI were presented and model trajectories were examined to compare the methods. Although both are based on evaluating changes in the system trajectory, method 1 (continuous approach) has a stronger theoretical basis and is very effective at identifying changes in the trajectory in systems, particularly those with sparse data sets over short timescales. However, method 1 requires data preprocessing (e.g., normalization, smoothing, and model fitting) to ensure continuous first-order and second-order derivatives, and data quality issues (inherent in some real systems) may limit its utility. Further study is needed to evaluate its efficacy in assessing regime shifts and transitional dynamics in real systems. Method 2 (discrete approach) bins points in the trajectory of the system into distinct states and provides a robust approach to computing FI. Although techniques used to handle data quality (i.e., TLs and size of states) must be managed, these processes are automated and coded in Matlab (Mathworks, Inc., 2012). Method 2 is adept at characterizing regimes, regime shifts, and transitional dynamics that may warn of impending transitions. When used in conjunction with Bayes theorem, the method is enhanced to illustrate the significance of declines in FI and the likelihood of regime shifts.

As demonstrated by the study of Ohio MSAs, the San Luis Basin, and the United States, these approaches are effective for evaluating real systems at various scales. A variety of techniques were successfully used and demonstrated, which aided in interpreting results. Specifically, SROC tests (urban, regional, and national studies) classifying trends in significant descriptors of the system (Ohio MSA study) and comparing results with historical events (US study) and Bayes theorem (US study) were effective at helping to characterize and monitor system dynamics to support management of real complex systems. The approaches presented are not intended to provide a cause-and-effect analysis, but they offer some context to the results that may be important for understanding and administering these types of systems.

It is important to note that FI affords the ability to assess changes in system dynamics and not the quality of its condition (Eason et al., 2014b). Therefore, although steady FI values are associated with stable regimes, trends in underlying variables must be examined to determine the desirability of the conditions (e.g., shallow lake model). For example, Rico-Ramirez et al. (2010) used FI to evaluate the effectiveness of a cancer treatment technique by modeling conditions with and without immunotherapy. Because underlying conditions were known, they were able to determine that the therapeutic treatment used caused the system condition to shift from malignant to cancer-free (humanly desired).

When using the approaches described in this chapter, common considerations for system analysis still stand and their importance is emphasized for FI analysis:

- Variable selection is critical to ensuring relevant analyses because poor representation of the system leads to results that are useless for monitoring and management.
- Natural systems often exhibit periodic behavior or cycles, naturally or when interacting with human populations. Periodic behavior (e.g., wastewater treatment model) and scale of study impact the characterization of system dynamics; however, FI can be used to distinguish patterns even when cycle periods are unknown.
- Data quality is a great challenge for any analytical approach and mechanisms have been developed and demonstrated to help manage these difficulties.
- Interpretation of FI results has been challenging and may have inhibited broad use of the index. Approaches in this work that increase clarity and enhance identification of important findings have been presented.

This chapter demonstrated the utility of FI as an effective tool for monitoring systems, assessing dynamic change, and informing actions to help guide complex human and natural systems toward sustainability. FI has been used in a number of studies to include assessing changes in city structure (Eason and Garmenstani, 2012), characterizing political instability in nation states (Karunanithi et al., 2011), and studying the dynamics of lead concentration levels in conjunction with regulatory actions (Gonzalez-Mejia et al., 2012b). Future work includes examining the utility of method 1 (continuous form) in capturing regime dynamics, using multivariate techniques to determine drivers of system behavior, investigating the utility of the index in studying the resilience of various types of systems (e.g., ecological, social, and supply chain systems), using FI to assess spatial dynamics, and further exploring signals in FI as a leading indicator in critical transitions.

ACKNOWLEDGMENTS

The authors acknowledge the seminal work of Audrey Mayer, Christopher Pawlowski, Brian Fath, and Arunprakash Karunanithi. Alejandra González-Mejía served as a postdoctoral research associate with the Oak Ridge Institute for Science and Education. Leisha Vance was a postdoctoral research associate with the US Environmental Protection Agency's Federal Postdoctoral Research Program.

DISCLAIMER

The views expressed herein are strictly the opinions of the authors and in no manner represent or reflect current or planned policy by federal agencies. Mention of trade names or commercial products does not constitute endorsement or recommendation for use. The information and data presented in this product were obtained from sources that are believed to be reliable. However, in many cases the quality of the information or data was not documented by those sources; therefore, no claim is made regarding their quality.

APPENDIX 2.1 APPROACHES TO ESTIMATING FI

Because of data quality issues inherent in real systems, previous researchers noted challenges in computing and interpreting FI (Mayer et al., 2007). In these studies, behavior indicative of RC was detected; however, block averages were often used, making it difficult to determine when dynamic transitions actually occurred (Karunanithi et al., 2008). Further, several items for future work were specified that would enhance the ability to compute and interpret FI. In response, approaches have been developed and augmented to extend the capacity of the method in evaluating real, complex systems and informing management decisions. Here, two methods are presented that have unique mechanisms for handling the challenges posed.

METHOD 1: CONTINUOUS FORM OF FI AS A FUNCTION OF THE VELOCITY AND ACCELERATION OF THE SYSTEM'S TRAJECTORY

This section provides details on estimating FI based on the evaluation of changes in system condition as a function of the velocity and acceleration tangential to the system trajectory. Here, we use a statistical mechanics framework in which the system undergoing study is described by n-dimensions [i.e., variables, $x(t)$] that characterize the system's state (s) over time ($0 \le t \le T$). Therefore, $I(t)$ is assumed to be a continuous function defined by the phase space tangential velocity ($\mathrm{d}s/\mathrm{d}t$) and acceleration ($\mathrm{d}^2s/\mathrm{d}t^2$) of the system (Cabezas et al., 2003). Equation (2.7) may be used to calculate FI analytically in cases in which the system variables are defined by model equations where Euclidean metric is used to compute the first ($\mathrm{d}s/\mathrm{d}t$) and second ($\mathrm{d}^2s/\mathrm{d}t^2$) derivatives with Eqs. (2.8) and (2.9), respectively (Fath et al., 2003). For real systems with discrete data, these derivatives can be estimated numerically with high-accuracy finite differentiation formulas by using the Taylor series expansion that contains the most terms, thereby minimizing computational error (Chapra, 2008a). However, numerical differentiation (particularly second derivatives) is susceptible to noise and data artifacts and may produce indeterminate results, especially when the system speed ($\mathrm{d}s/\mathrm{d}t$) is close to or equal to zero. For that reason, two options are suggested to overcome this challenge. One is smoothing the time series $x(t)$ with dynamic filters and/or fitting the variable time series to a function (Gonzalez-Mejia et al., 2012a). This involves the use of at least three data points to compute $\mathrm{d}x/\mathrm{d}t$ and $\mathrm{d}^2x/\mathrm{d}t^2$. Another option is to estimate FI by integrating Eq. (2.7) for a period of time (T) wide enough to compensate for the inherent noise in the original variables (Mayer et al., 2007).

Because the assessment of movement toward and away from sustainability relies on trends rather than absolute values (Hopton et al., 2010), we recognize the importance of ensuring that data issues inherent in empirical studies do not inappropriately bias trends. Hence, methods that can aid in managing noisy, sparse data are needed when computing FI. Accordingly, when using the continuous form of FI, Gonzalez-Mejia et al. (2012a) have used several data preprocessing techniques including normalization, interpolation, filtering, smoothing, and fitting time series to a function.

In addition to data quality issues, the periodicity of the system undergoing study is an important factor to consider for computing and interpreting FI. Cabezas et al. (2003) indicated that for oscillatory systems, FI should be calculated with the integral spanning at least one period of the system; otherwise, it will appear noisy. However, it is typically difficult to find the cycle period of real systems because they have multiple periodicities. Mayer et al. (2006) suggested the use of system knowledge or the application of fast-Fourier transform methods to determine the system period. "Wastewater Reactor Model for Nitrogen Removal" section provides a summary of the systematic approach developed by Gonzalez-Mejia (2011) to aid in determining the effective period of a complex system.

Capturing dynamic order using $I(t_i)$ involves integrating the system parameters as a "block" or "moving" average using overlapping time windows of size l. Thus, the same value of FI can be assigned to the entire time interval (Mayer et al., 2007). If the period is unknown and/or the time series are very large, then an instantaneous $I(t_i)$ can be computed by moving the calculation one time step at a time (i) as expressed in Eq. (A.1.1).

$$I(t_i) = \frac{1}{l}\int_{t_i}^{t_{i+l}} \frac{(\mathrm{d}^2s/\mathrm{d}t^2)^2}{(\mathrm{d}s/\mathrm{d}t)^4}\,\mathrm{d}t \tag{A.1.1}$$

For numerical integration, the Newton–Cotes closed integration formulas (Chapra, 2008b) may be applied depending on the amount of available data for the $I(t_i)$ computation. There is no limit in the number of data points or time steps needed for each variable ($n \geq 1$); however, in practice, a minimum of five time steps of data per variable is required to obtain one value of $I(t_i)$. To focus on trends in the FI result and to further minimize the effect of biasing or short-term fluctuations, a moving filter (typically a three-point average) may be applied to compute a mean FI from the results (Eason and Cabezas, 2012).

METHOD 2: DISCRETE FORM AS A FUNCTION OF THE PROBABILITY DENSITY OF SYSTEM STATES

Because of challenges of data quality and issues related to obtaining suitable second-order derivatives from real system data using method 1 (Mayer et al., 2006), an alternative approach was developed based on the probability of observing different conditions (i.e., states) of the system. By assuming that the probability $p(s)$ of observing a particular system state s is proportional to the frequency at which state s is observed in the data. Based on this assumption, a method for computing the probability $p(s)$ has been developed (Karunanithi et al., 2008).

Building from the theoretical foundation described in the "Information Theory" section, we construct a mathematical framework of method 2 by using Eq. (2.2). Here, $p(s)$ is again the probability density of observing a state s of the system that is replaced with the amplitude $[q(s) \equiv \sqrt{p(s)}]$ to give Eq. (A.1.2).

$$I = 4\int \left[\frac{\mathrm{d}q(s)}{\mathrm{d}t}\right]^2 \mathrm{d}s \tag{A.1.2}$$

Equation (A.1.2) is then adapted for evaluating discrete data as shown in Eq. (A.1.3) by approximating the integral with a sum and replacing derivatives ($\mathrm{d}s$ and $\mathrm{d}q$) with finite differences $\Delta s = s_i - s_{i+1}$ and $\Delta q = q_i - q_{i+1}$ to give:

$$I \approx 4\sum_{i=1}^{n} \left[\frac{q_i - q_{i+1}}{s_i - s_{i+1}}\right]^2 (s_i - s_{i+1}) \tag{A.1.3}$$

where s_i is an index denoting a particular state of the system such that s_1 is state 1 and so on. Because $s_i - s_{i+1} = 1$, the final expression for computing FI using method 2 is expressed as:

$$I \approx 4\sum_{i=1}^{n} [q_i - q_{i+1}]^2 \tag{A.1.4}$$

Sections "Representing the Trajectory of a System with Elementary Functions" and "Shallow Lake Model: Regime Shift" provide details regarding how Eq. (A.1.4) is applied to computing FI. Key to the approach are grouping ("binning") points within the system trajectory into states as discussed here.

Binning approach

As in "Representing the Trajectory of a System with Elementary Functions" section, regarding method 1, a time-varying system is denoted by a phase space of n-dimensions and time that represents all possible states of the system. Accordingly, a point $y(t_i)$ in the system trajectory is defined by specifying a value for each variable at a particular time (t_i) (i.e., $y(t_i)$:$[x_1(t_i), x_2(t_i), x_3(t_i), \ldots, x_n(t_i)]$). Given that measurable variables have inherent uncertainty, it is reasonable to assume a "tolerance" (Δx_i), around each variable of the system. Hence, a particular system state is not a point but rather a region bounded by a distinct level of uncertainty for each variable. Hence, if two values of $x_i(t_i)$ and $x_i(t_j)$ of any system variable differ from each other by less than the uncertainty $\Delta x_i, |x_i(t_i) - x_i(t_j)| < \Delta x_i$, then the two points are deemed to be indistinguishable from each other. They then represent two separate observations of the same value and are "binned" together. When two different values for each of the variables in a point $y(t_i)$:$[x_1(t_i), x_2(t_i), x_3(t_i), \ldots, x_n(t_i)]$ differ from each other by less than their respective uncertainty, the two points are assumed to be repeat observations of the same point of the system and are "binned" together into the same state. The size of the volume over which data points can be "binned" into a state is termed the "size of state" (Karunanithi et al., 2008). The probability of observing a particular state is assumed to be proportional to the number of points inside the volume of the state. By using this approach, the probability of observing all possible states of the system is estimated, thereby affording the ability to compute FI and track changes in system order over time.

Now, establishing the level of uncertainty is necessary to distinguish normal variation from changes in behavior, and some approaches have been developed to determine this level of uncertainty (Karunanithi et al., 2008). Ideally, if the uncertainty measurement for the variables describing the system is known, then this information is used to bin points into states. Because such data often are unavailable, two basic strategies have been developed to address the challenge. One strategy is to locate a surrogate system (characterized by the same variables) that exhibits stability, and then to assume that the variation in the variables for this system is approximately equal to the uncertainty in the variables for the system being evaluated. When there are no suitable surrogates, the time series of the system undergoing study is sampled to find a period when there is minimal variation in each variable. SD is computed for each variable in this stable period, and it is assumed to represent the measurement uncertainty. Chebyshev theorem is then used (Karunanithi et al., 2008), which indicates that "the proportion of the observations falling within k SDs of the mean is at least $1-(1/k^2)$," (i.e. $P(|X - \mu|, < k\sigma) \geq [1 - (1/k^2)]$) (Lapin, 1975). This theorem applies to any arbitrary distribution; therefore, it is independent of the type or form of the pdf. Typically, $k = 2$ is selected for setting the size of states, indicating that at least 75% of the data are contained within 2 SDs of the mean (Eason and Cabezas, 2012). However, if the data are normally distributed, then 95% of the data would be found within 2 SDs of the mean (Clemen and Reilly, 2001).

As already discussed, the probability of observing a particular state of the system, $p(s)$, may be estimated by grouping ("binning") indistinguishable points in the trajectory into states of the system and counting the number of points inside each state (Karunanithi et al., 2008). The time evolution of the probability density is computed from period to period by dividing the time series describing the system into "time windows." The size of the time window (hwin) is determined based on the amount of data available and the behavior of the system. From empirical studies, hwin should be at least eight time steps (Eason and Cabezas, 2012). A sequence of overlapping windows is then created to process through the data and capture $p(s)$ changes during the length of the study. This is achieved by moving the time window forward by a time increment (winspace).

The mechanics of binning points are as follows. Within the initial window, the first point in the time series is taken as the center of the first state and all the points in the time window that fall within the uncertainty limits (i.e., $|x_i(t_i) - x_i(t_j)| < \Delta x_i$ for all i) are binned into the same state. The points located outside of the boundary ("unbinned") are then evaluated for inclusion in the next state. The process continues with the first *unbinned* point set as the center of the next state, and points within the boundary of this state are binned with it. The procedure is applied until all points within the first window are binned. Then, a new time window is created by moving by a time increment, and the process is repeated. This process is continued until all the points in the time series are binned (Eason and Cabezas, 2012).

Sensitivity analysis of the hwin and winspace parameters indicate that hwin is the key factor impacting the variability in FI and that winspace has a greater impact on the number of time windows established. Accordingly, for the best resolution of FI results (Cabezas and Eason, 2010), the moving window increment is typically set to one time step (winspace = 1).

After hwin, winspace and the size of states are determined, points are binned into states of the system, and $p(s)$ is calculated for each window ($p_i = \text{\#points in state/total\#points in time window}$). Next, the amplitude $[q(s) = \sqrt{p(s)}]$ is computed and FI for the time window is calculated using Eq. (A.1.4). The discrete form of FI is computed using a gradient to mimic the tails of the pdf. For example, if a system is in one state for the entire period, then $p(1) = 1 \therefore q(1) = \sqrt{p(1)} = 1$ and, as a convention, the tails of the pdf are defined as $q(0) = 0$ and $q(\text{end}) = 0$; hence, $I = 4 \times [(0-1)^2 + (1-0)^2] = 8$. Therefore, unlike the velocity and accelaration approach, which will range from 0 to ∞ (Mayer et al., 2007), the maximum value for the discrete form is 8.

As previously noted, the calculated FI value applies to the entire time window (Mayer et al., 2006). Previous work presented FI results as one steady value over each time window or placed it at the center of the time window. Because plotting FI as one point in each window reveals details of the system's dynamics that may be concealed by block averages (Karunanithi et al., 2008), we have continued this approach. However, recent studies involving regime shift detection, evaluating transitional dynamics, and assessing warning signals of impending RC have highlighted the need to account for the fact that for monitoring a system in real time, only historical data are available. Accordingly, we are exploring the utility of placing FI at the end of the time window for these types of studies.

One more step is added as an additional means of compensating for the effects of measurement error that may be present in data for system variables. A parameter, termed TL, is used to loosen the size-of-states constraint, such that a point can be binned into a state of the system when a certain percentage of the variables meet the binning criteria (Karunanithi et al., 2008). For example, let us consider the case of a system characterized by 100 variables. If 90 of the variables indicate that two points are indistinguishable, then the two points will be binned in a state at the 90% TL. In application, FI is calculated at multiple TLs between strict tightening (100%) and relaxed tightening (i.e., the lowest TL at which more than one state is present in the window). Note that that if TL is low enough, then all points fall into one state.

Now, the TL denotes the number of dimensions that meet the size-of-states criteria, which also affords the ability to study the pervasiveness of changes in dynamic order. A more pervasive change affects a wider number of system variables. From a practical standpoint, a smaller value for the lower bound of the TL is a possible indication that more variables (pervasive) are involved in the changes in system dynamics (Karunanathi et al., 2008).

After FI is computed at each TL, an average FI is calculated by taking the arithmetic mean of the values of FI at each TL between this upper boundary and lower boundary (Karunanithi et al., 2008). Further, as a convention, we typically report mean FI that is calculated by computing a three-point average from neighboring values. This "filtering" step helps to focus on trends in dynamic order rather than minor fluctuations (Eason and Cabezas, 2012). To make sure that the most recent time period is included in the computation, the $\langle \mathrm{FI}_j \rangle$ is computed in reverse order, such that w is the year corresponding to the last window of the FI computation [$\langle \mathrm{FI} \rangle \mathrm{I}_j = (1/3)(I_{j+1} + I_j + I_{j-1})$ with $j = w-$ odd number]. The procedure developed for computing and reporting FI in this method represents a process intended to conservatively capture important trends in dynamic order even in the presence of substantial data uncertainty (Eason and Cabezas, 2012).

Algorithm

Cabezas and Eason (2010) provide a simple example for calculating FI and the algorithm for estimating FI using the discrete form coded in Matlab (Mathworks, Inc., 2012). In summary, the basic steps of computing FI using the discrete approach (method 2) is as follows: (i) select variables that characterize the system and gather corresponding time series; (ii) determine the value of the computation parameters (hwin and winspace for developing the overlapping time windows) and the size of states for each variable from the uncertainty; (iii) divide the time series into a sequence of time windows; (iv) set the TL and bin each point into states of the system within each time window; (v) use the bned points to generate $p(s)$ for each time window; (vi) calculate FI from the $q(s)$ for each time window; (vii) repeat the steps for each time window; (viii) repeat the process (iv–vii) until FI is computed from strict to relaxed TL, from which an average FI is calculated; and (ix) compute a three-point mean Fisher information ($\langle FI \rangle$). To detect changes in dynamic order, FI is computed in overlapping windows and compared over time.

APPENDIX 2.2 FI AT TIPPING POINTS

A tipping point or bifurcation is defined as a qualitative change in the behavior of a dynamic system precipitated by a small, smooth change in a system parameter (Kuznetsov, 2004). We include it here for completeness and because extreme behavior in dynamic systems is often important. Although much of the bifurcation theory is related to the behavior of model systems represented by systems of equations, here we address the behavior of real systems similar to bifurcation-like behavior. In particular, we explore regime shifts that are special cases of bifurcations. Three particular scenarios are considered that are relevant to regime shifts. To explore these cases, consider the expression for FI in terms of the amplitude under the assumption of shift invariancc. By substituting Eqs. (2.8) and (2.9) into Eq. (2.7), Eq. (A.2.1) is produced where T is the total time of observations, so that $0 \leq t \leq T$. Equation (A.2.1) provides the working expresssion for FI that will be used to explore the aforementioned issue of tipping points.

$$I = \frac{1}{T}\int_0^T \frac{\left[\sum_{i=1}^{n} \frac{dx_i}{dt}\frac{d^2x_i}{dt^2}\right]^2}{\left[\sum_{i=1}^{n} \left(\frac{dx_i}{dt}\right)^2\right]^3} dt \tag{A.2.1}$$

The first case is one in which the observable variables that characterize the system fluctuate with increasing amplitude going to a singularity as the system approaches the point in time where the bifurcation is found. This is commonly the situation when ecosystems approach an RC (Brock and Carpenter, 2006) or when fluids approach a critical point (Sengers and Sengers, 1986). To explore this case, consider that each variable x_i has the following general functional form as in Eq. (A.2.2), where C_i is a smooth bound function that depends only on time and δ_i is defined by Eq. (A.2.3).

$$x_i = [1 + \delta(t - t_i^*)]C_i(t/T) \quad i = 1, 2, \ldots, n \tag{A.2.2}$$

$$\delta(t - t_i^*) \equiv \begin{cases} 0, t \neq t_i^* \\ +\infty, t = t_i^* \end{cases} \quad i = 1, 2, \ldots, n \tag{A.2.3}$$

Note that the expressions of Eqs. (A.2.2) and (A.2.3) are meant to represent a situation in which the system variables are well-behaved except at certain points in time represented by t_i^*, where they experience a singularity. Although many natural and human-made systems may or may not experience a "true" singularity, they are known to approach bifurcations with increasing variance or fluctuations of the variables as discussed. Hence, modeling these particular bifurcations as singularities [Eqs. (A.2.2) and (A.2.3)] is an adequate approximation to reality. The resulting expression for the FI is provided in Eq. (A.2.4).

$$I = \frac{1}{T}\int_0^T \frac{\sum_{i=1}^{n}\left[\left(\frac{d^2C_i}{dt^2}(1+\delta)+2\delta\frac{dC_i}{dt}+\delta C_i\right)\left(\frac{dC_i}{dt}(1+\delta)+\delta C_i\right)\right]^2}{\left[\sum_{i=1}^{n}\left(\frac{dC_i}{dt}(1+\delta)+\delta C_i\right)^2\right]^3}\,dt \tag{A.2.4}$$

If the bifurcation point (t_k^*) for any particular system variable (x_k) lies within the time bracket $0 \leq t \leq T$, then as the system approaches the bifurcation point ($t \to t_k^*$), consistent with Eqs. (A.2.2) and (A.2.3), the variable ($x_k \to +\infty$) approaches infinity. Its first ($dx_k/dt \to +\infty$) and second derivatives ($d^2x_k/dt^2 \to +\infty$) approach infinity as well. Then, the summations in Eq. (A.2.4) reduce to one dominant term k, where the singularity is present, giving Eq. (A.2.5):

$$I \to \frac{1}{T}\int_0^T \frac{\left[\frac{d^2C_k}{dt^2}+2\frac{dC_k}{dt}+C_k\right]^2}{\delta^2\left[\frac{dC_k}{dt}+C_k\right]^4}\,dt \quad t \to t_k^* \tag{A.2.5}$$

and in the limit where $t = t_k^*$, $I = 0$ with $\delta_k = +\infty$. Therefore, the FI approaches zero as any of the system variables approaches a bifurcation that has increasingly large amplitude.

$$I \to 0 \quad t \to t_k^* \tag{A.2.6}$$

In the second case, the same variables fluctuate with decreasing amplitude, culminating in no fluctuations as the system reaches the time of the bifurcation (i.e., the system arrives at the bifurcation as a steady state). This phenomena was noted by Dakos et al. (2012) in some model systems. Consider the example of a ball and cup system where there are two wells that denote distinct system states. This steady state may be represented by a relatively flat region between the two wells so that the ball slows greatly during the transition between the wells. Although such behavior can give the appearance of stability, it is short-lived and represents metastability, where the system is only stable in response to very minor perturbations. However, a small push either way can send the ball into one of the wells. For ecosystems, these dynamics may be present when a relatively stable transition exists, and an example from fluid systems is when a super-heated liquid continues to exist in the liquid phase without evaporating, provided it is not perturbed. Similar phenomena have been observed for solid and liquid transitions. An extensive treatment of this area exists for fluids (Debenedetti, 1996).

We start the analysis of the second case again using Eq. (A.2.1), where the expression for the measurable system variables (x_i) is now given by Eq. (A.2.7), where C_i is, as before, a smooth bound function depending only on time, K_i is a constant, and H is defined by Eq. (A.2.8).

$$x_i = [1 - H(t - t_i^*)]C_i(t/T) + K_i H(t - t_i^*) \quad i = 1, 2, \ldots, n \tag{A.2.7}$$

$$H(t - t_i^*) \equiv \begin{cases} 0, t \neq t_i^* \\ 1, t = t_i^* \end{cases} \tag{A.2.8}$$

Note that in the previous case each variable could have a different bifurcation point (t_k^*), but in the present case all of the variables are assumed to have the same bifurcation point (t^*). The reason is that having one or even a few of many variables go to a bifurcation point has negligible effect on FI when the trajectory is flat, and this makes for a very uninteresting case. Inserting the expressions from Eqs. (A.2.7) and (A.2.8) into the expression for FI from Eq. (A.2.1) gives Eq. (A.2.9).

$$I = \frac{1}{T}\int_0^T \frac{\left[\sum_{i=1}^{n} \frac{dC_i}{dt}\frac{d^2C_i}{dt^2}\right]^2}{(1-H)^2\left[\sum_{i=1}^{n}\left(\frac{dC_i}{dt}\right)^2\right]^3} dt \tag{A.2.9}$$

If the bifurcation point (t^*) for the system variables (x_i) is in the time bracket $0 \leq t \leq T$, then as the system approaches the bifurcation point t^*, consistent with Eqs. (A.2.7) and (A.2.8), $1 - H \to 0$ as $t \to t^*$ so that the integrant in Eq. (A.2.9) approaches infinity and I approaches infinity as well. In summary, the FI approaches infinity as the system approaches a bifurcation that has a "flat" trajectory during the transition.

$$I \to +\infty \quad t \to t^* \tag{A.2.10}$$

The third scenario explores the case where the system variables qualitatively change behavior at the bifurcation, but there is no extreme behavior such as a singularity or a flat trajectory near the bifurcation. This is probably the most common situation with dynamic systems. After all, complex dynamic systems such as ecosystems and economies spend the bulk of the time in reasonably stable regimes that do not have extreme bifurcations such as the aforementioned two cases. However, we still study the extreme cases because they are most likely to harbor unknown catastrophes or other undesirable phenomena. The present case is also one in which the fine details of system behavior matter most, and in which general statements are most difficult to formulate.

We start the analysis with the expression for I given by Eq. (A.2.1), and then define a bifurcation with Eq. (A.2.11), where C_i and D_i are two different but smooth bound functions depending only on time. This is the case where the system changes qualitative behavior as represented by the shift from function C_i to function D_i at the bifurcation point t^*.

$$x_i = \begin{cases} C_i(t/T), t < t^* \\ D_i(t/T), t > t^* \end{cases} \quad i = 1, 2, \ldots, n \tag{A.2.11}$$

Again, note that there are no singularities or flat spots. If we designate FI computed under C_i as I_C and FI computed under D_i as I_D, then by inspection of Eq. (A.2.1), the following three results become apparent. The first possibility is:

$$I_C = I_D \quad \text{if} \quad \sum_{i=1}^{n} \left(\frac{dC_i}{dt}\right)^2 = \sum_{i=1}^{n} \left(\frac{dD_i}{dt}\right)^2 \quad \text{and} \quad \sum_{i=1}^{n} \frac{dC_i}{dt}\frac{d^2C_i}{dt^2} = \sum_{i=1}^{n} \frac{dD_i}{dt}\frac{d^2D_i}{dt^2} \tag{A.2.12}$$

The expressions in Eq. (A.2.12) imply that the sum of the first and second derivatives of C_i and D_i must be equal, but it conveys nothing about the magnitude of C_i and D_i, which may or may not be equal. This approximately implies that the overall rate and pattern of change of the system under functions C_i is the same as that under functions D_i. Such could be a case in which, for example, there has been a change of magnitude ($C_i > D_i$ or $C_i < D_i$) but the rate of change is the same for both C_i and D_i. It could also represent a case in which the velocity (i.e., dC_i/dt and dD_i/dt) or the accelaration (i.e., d^2C_i/dt^2 and d^2D_i/dt^2) has increased for some variables and has decreased for others such that summations in Eq. (A.2.12) are equal.

A second possibility is stated in Eq. (A.2.13), which suggests that, on average, the velocity or rate of change dC_i/dt is smaller than the rate of change dD_i/dt, that is, the system is changing more rapidly after the bifurcation ($t > t^*$) than before the bifurcation ($t < t^*$).

$$I_C > I_D \quad \text{if} \quad \sum_{i=1}^{n} \frac{dC_i}{dt}\frac{d^2C_i}{dt^2} > \sum_{i=1}^{n} \frac{dD_i}{dt}\frac{d^2D_i}{dt^2} \quad \text{and/or} \quad \sum_{i=1}^{n} \left(\frac{dC_i}{dt}\right)^2 < \sum_{i=1}^{n} \left(\frac{dD_i}{dt}\right)^2 \tag{A.2.13}$$

Further, the acceleration component d^2C_i/dt^2 is larger than d^2D_i/dt^2, that is, the system is, on average, experiencing more acceleration before ($t < t^*$) the bifurcation than after it ($t > t^*$). This could be envisioned as a series of fast and slow segments along its trajectory such as a series of stops and starts (i.e., the rate of change increased at the bifurcation after which there is a smoother trajectory).

A third possibility is considered in Eq. (A.2.14), which implies that, on average, the velocity or rate of change dC_i/dt is larger than the rate of change dD_i/dt, that is, the system is changing more rapidly before the bifurcation ($t < t^*$) than after the bifurcation ($t > t^*$).

$$I_C < I_D \quad \text{if} \quad \sum_{i=1}^{n} \frac{dC_i}{dt}\frac{d^2C_i}{dt^2} < \sum_{i=1}^{n} \frac{dD_i}{dt}\frac{d^2D_i}{dt^2} \quad \text{and/or} \quad \sum_{i=1}^{n} \left(\frac{dC_i}{dt}\right)^2 > \sum_{i=1}^{n} \left(\frac{dD_i}{dt}\right)^2 \tag{A.2.14}$$

However, the acceleration component d^2C_i/dt^2 is smaller than d^2D_i/dt^2, that is, the system is, on average, experiencing less acceleration before ($t < t^*$) the bifurcation than after it ($t > t^*$). This phenomena may be represented as a series of fast and slow segments along its trajectory, such as a series of stops and starts (i.e., the rate of change slowed at the bifurcation but it is not as smooth of a trajectory). This third scenario illustrates that when there is subtle behavior in underlying variables, a bifurcation will occur when there is a difference in either the velocity or the acceleration in the system trajectory over time. Further, it demonstrates that as a system approaches a bifurcation, the relative value of the FI before and after the shift will depend on the particular behavior of the underlying variables.

REFERENCES

Biggs, R., Carpenter, S.R., Brock, W.A., 2009. Turning back from the brink: detecting an impending regime shift in time to avert it. Proc. Natl. Acad. Sci. USA 106 (3), 826–835.

Brock, W.A., Carpenter, S.R., 2006. Variance as a leading indicator of regime shift in ecosystem services. Ecol. Soc. 11 (2), 9–23.

Bureau of Land Management–Colorado/San Luis Valley Field Office, 2014. San Luis Valley Detailed Land Status Map. <www.blm.gov/pgdata/etc/medialib/blm/co/field_offices/slvplc/maps.Par.67051.Image.-1.-1.1.gif> (accessed 14.04.2014).

Cabezas, H., Eason, T., 2010. Fisher Information and Order. Cincinnati, OH, Report EPA/600/R-10/182, pp. 163–222.

Cabezas, H., Fath, B.D., 2002. Towards a theory of sustainable systems. Fluid Phase Equilibria 194, 3–14.

Cabezas, H., Pawlowski, C.W., Mayer, A.L., Hoagland, N.T., 2003. Sustainability: ecological, social, economic, technological, and systems perspectives. Clean Technol. Environ. Policy 5 (3–4), 167–180.

Cabezas, H., Pawlowski, C.W., Mayer, A.L., Hoagland, N.T., 2005. Simulated experiments with complex sustainable systems: ecology and technology. Resour. Conserv. Recycl. 44, 279–291.

Campbell, D.E., Garmestani, A.S., 2012. An energy systems view of sustainability: emergy evaluation of the San Luis Basin, Colorado. J. Environ. Manage. 95 (1), 72–97.

Carpenter, A., Brock, W.A., 2006. Rising variance: a leading indicator of ecological transition. Ecol. Lett. 9, 311–318.

Carpenter, S.R., 2003. Regime shifts in lake ecosystems: pattern and variation, Excellence in Ecology Series, vol. 15. Ecology Institute, Oldendorf/Luhe, Germany.

Carpenter, S.R., Brock, W.A., Cole, J.J., Kitchell, J.F., Pace, M.L., 2008. Leading indicators of trophic cascades. Ecol. Lett 11, 128–138.

Chapra, S.C., 2008a. Numerical differentiation, Applied Numerical Methods with MATLAB, for Engineers and Scientists, second ed. McGraw-Hill, New York, NY, pp. 448–471.

Chapra, S.C., 2008b. Numerical integration formulas. Applied Numerical Methods with MATLAB, for Engineers and Scientists. McGraw-Hill, New York, NY, USA, pp. 392–425.

Chapra, S.C., 2008c. Runge–Kutta methods. Applied Numerical Methods with MATLAB, for Engineers and Scientists. McGraw-Hill, New York, NY, USA, pp. 493–498.

Clemen, R.T., Reilly, T., 2001. Making Hard Decisions with Decision Tools. Southwestern Cengage Learning, Mason, OH, USA.

Dakos, V., Carpenter, S.R., Brock, W.A., Ellison, A.M., Guttal, V., Ives, A.R., Kéfi, S., Livina, V., Seekell, D.A., Van Nes, E.H., Scheffer, M., 2012. Methods for detecting early warnings of critical transitions in time series illustrated using simulated ecological data. PLoS One 7 (7), e41010.

Debenedetti, P.G., 1996. Metastable Liquids: Concepts and Principles. Princeton University Press, Princeton, NJ, USA.

Eason, T., Flournoy, A., Cabezas, H., Gonzalez, M., 2014a. Incorporating resilience into law and policy: a case for preserving a natural resource legacy and promoting a sustainable future. In: Garmenstani, A.S., Allen, C.R. (Eds.), Social-Ecological Resilience and Law. Columbia University Press, New York, NY, USA, p. 404.

Eason, T., Garmenstani, A.S., 2012. Cross-scale dynamics of a regional urban system through time. Region et Dev. 45, 55–77.

Eason, T., Garmenstani, A.S., Cabezas, H., 2014b. Managing for resilience: early detection of catastrophic shifts in complex systems. Clean Technol. Environ. Policy 16 (4), 773–783.

Eason, T.N., Cabezas, H., 2012. Evaluating the sustainability of a regional system using Fisher information in the San Luis Basin, Colorado. J. Environ. Manage. 94 (1), 41–49.

Fath, B.D., Cabezas, H., Pawlowski, C.W., 2003. Regime changes in ecological systems: an information theory approach. J Theor. Biol. 222 (4), 517–530.

Federal Housing Finance Agency, 2010. House Price Index Fourth Quarter 2009. Percent Change in House Prices with MSA Rankings. <www.fhfa.gov/AboutUs/Reports/ReportDocuments/2009-Q4-December_HPI_508.pdf> (accessed 07.03.2013).

Fisher, R.A., 1922. On the mathematical foundations of theoretical statistics. Philos. Trans. R. Soc. Lond. 222, 309–368.

Fisk, P., 2010. People Planet Profit: How to Embrace Sustainability for Innovation and Business Growth. Kogan Page Publishers, London, United Kingdom.

Flint, R.W., Houser, W.L., 2001. Living a Sustainable Lifestyle for Our Children's Children. iUniverse, Incorporated, Lincoln, NE, p. 324.

Frieden, B.R., 1998. Physics from Fisher Information. A Unification. Cambridge University Press, New York, NY, p. 318.

Garmestani, A.S., Allen, C.R., Gunderson, L., 2009. Panarchy: discontinuities reveal similarities in the dynamic system structure of ecological and social systems. Ecol. Soc. 14 (1), 15.

Geolytics Inc., 2003. Census CD neighborhood change database (NCDB): 1970–2000 tract data, selected variables for US Census tracts for 1970, 1980, 1990, 2000. E. Brunswick, NJ.

Gonzalez-Mejia, A.M., 2011. Fisher Information: Sustainability Analysis of Several US Metropolitan Statistical Areas. University of Cincinnati, Cincinnati, OH, p. 109.

Gonzalez-Mejia, A.M., Eason, T.N., Cabezas, H., Suidan, M.T., 2012a. Assessing sustainability in real urban systems: the greater Cincinnati metropolitan area in Ohio, Kentucky, and Indiana. Environ. Sci. Technol. 46 (17), 9620–9629.

Gonzalez-Mejia, A.M., Eason, T.N., Cabezas, H., Suidan, M.T., 2012b. Computing and interpreting Fisher Information as a metric of sustainability: regime changes in the United States air quality. Clean Technol. Environ. Policy 14 (5), 775–788.

Gonzalez-Mejia, A.M., Eason, T.N., Cabezas, H., Suidan, M., 2014. Social and economic sustainability of urban systems: comparative analysis of metropolitan statistical areas in Ohio, USA. Sustain. Sci. 9 (2), 217–228.

Grebmeier, J.M., Overland, J.E., Moore, S.E., Farley, E.V., Carmack, E.C., Cooper, L.W., Frey, K.E., Helle, J.H., Mclaughlin, F.A., Mcnutt, S.L., 2006. A major ecosystem shift in the northern Bering Sea. Science 311 (5766), 1461–1464.

Hare, S.R., Mantua, N.J., 2000. Empirical evidence for North Pacific regime shifts in 1977 and 1989. Prog. Phys. Geog. 47 (2–4), 103–145.

Hartley, R.V.L., 1928. Transmission of information. Bell Syst. Tech. J., 535–564.

Holling, C.S., 1973. Resilience and stability of ecological systems. Annu. Rev. Ecol. Syst. 4 (1–23), 23.

Hopton, M.E., White, D., 2012. A simplified ecological footprint at a regional scale. J. Environ. Manage. 111, 279–286.

Hopton, M.E., Cabezas, H., Campbell, D., Eason, T., Garmestani, A.S., Heberling, M.T., Karunanithi, A.T., Templeton, J.J., White, D., Zanowick, M., 2010. Development of a multidisciplinary approach to assess regional sustainability. Int. J. Sust. Dev. World 17 (1), 48–56.

Karunanithi, A.T., Cabezas, H., Frieden, B.R., Pawlowski, C.W., 2008. Detection and assessment of ecosystem regime shifts from Fisher information. Ecol. Soc. 13 (1), 22.

Karunanithi, A.T., Garmestani, A.S., Eason, T.N., Cabezas, H., 2011. The characterization of socio-political instability, development and sustainability with Fisher information. Glob. Environ. Change 21, 77–84.

Kates, R.W., Parris, T.M., 2003. Long term trends and a sustainability transition. Proc. Natl. Acad. Sci. USA 100 (14), 8062–8067.

Kuznetsov, Y.A., 2004. Elements of Applied Bifurcation Theory, third ed. Springer-Verlag, New York, NY, USA.

Lapin, L.L., 1975. Statistics: Meaning and Method. Harcourt Brace Jovanovich, New York, NY, USA.

Macarthur, R., 1955. Fluctuations of animal populations and a measure of community stability. Ecology 36 (3), 533–536.

Mathworks Inc., 2012. MATLAB Software: the language of technical computing version 7.8.0.347: Release R2011b. Natick, MA, USA.

Mayer, A.L., Pawlowski, C.W., Cabezas, H., 2006. Fisher information and dynamic regime changes in ecological systems. Ecol. Model. 195 (1–2), 72–82.

Mayer, A.L., Pawlowski, C.W., Fath, B.D., Cabezas, H., 2007. Applications of Fisher information to the management of sustainable environmental systems. In: Frieden, B. R., Gatenby, R.A. (Eds.), Exploratory Data Analysis Using Fisher Information. Springer-Verlag, London, UK, pp. 217–244.

Mcgowan, J.A., Cayan, D.R., Dorman, L.M., 1998. Climate–ocean variability and ecosystem response in the North Pacific Ocean. Science 281 (5374), 210–217.

Mitchell, B.R., 2007. International Historical Statistics: The Americas, 1750–2005. Palgrave Macmillan, New York, NY, USA.

Palsdottir, G., Bishop, P.L., 1997. Nitrifying biotower upsets due to snails and their control. Water Sci. Technol. 36 (1), 247–254.

Pawlowski, C.W., Fath, B.D., Mayer, A.L., Cabezas, H., 2005. Towards a sustainability index using information theory. Energy 30 (8), 1221–1231.

Pawlowski, C.W., Cabezas, H., 2008. Identification of regime shifts in time series using neighborhood statistics. Ecol. Complex. 5, 30–36.

Rico-Ramirez, V., Reyes-Mendoza, P.A., Ortiz-Cruz, J.A., 2010. Fisher information on the performance of dynamic systems. Ind. Eng. Chem. Res. 49 (4), 1812–1821.

Rodionov, S.N., 2005. A brief overview of the regime shift detection methods. In: Velikova, V., Chipev N. (Eds.), Proceedings of Large-Scale Disturbances (Regime Shifts) and Recovery in Aquatic Ecosystems: Challenges for Management Toward Sustainability, UNESCO-ROSTE/BAS Workshop on Regime Shifts, Varna, Bulgaria, pp. 17–24.

Scheffer, M., Bascompte, J., Brock, W.A., Brovkin, V., Carpenter, S.R., Dakos, V., Held, H., Van Nes, E.H., Rietkerk, M., Sugihara, G., 2009. Early-warning signals for critical transitions. Nature 461, 53–59.

Scheffer, M., Carpenter, S.R., Lenton, T.M., Bascompte, J., Brock, W., Dakos, V., Van De Koppel, J., Van De Leemput, I.A., Levin, S.A., Van Nes, E.H., Pascual, M., Vandermeer, J., 2012. Anticipating critical transitions. Science 338 (6105), 344–348.

Seekell, D.A., Carpenter, S.R., Pace, M.L., 2011. Conditional heteroscedasticity as a leading indicator of ecological regime shifts. Am. Nat. 178 (4), 442–451.

Seekell, D.A., Carpenter, S.R., Cline, T.J., Pace, M.L., 2012. Conditional heteroskedasticity forecasts regime shift in a whole-ecosystem experiment. Ecosystems 15 (5), 741–747.

Sengers, J.V., Sengers, J.M.H.L., 1986. Thermodynamic behavior of fluids near the critical point. Annu. Rev. Phys. Chem. 37, 189–222.

Shannon, C.E., 1948. A mathematical theory of communication. Bell Syst. Tech. J. 27, 379–423, pp. 623–656.

Tchobanoglous, G., Burton, F.L., Stensel, H.D., 2003. Microbial growth kinetics. Wastewater Engineering Treatment and Reuse. McGraw-Hill, New York, NY, USA, pp. 580–588.

U.S. Bureau of Economic Analysis, 2012. CA25N Total full-time and part-time employment by NAICS industry. <www.bea.gov/itable/iTable.cfm?ReqID = 70&step = 1#reqid = 70 &step = 1&isuri = 1> (accessed 06.07.2013).

U.S. Census Bureau, 2010a. 2009 American Community Survey 1-Year Estimates. <www.census.gov/acs/www/data_documentation/summary_file > (accessed 12.08.2012).

U.S. Census Bureau, 2010b. Metropolitan Statistical Areas and components, December 2009, with codes. Washington DC, USA, OMB Bulletin No. 10-02. p. 154.

U.S. Census Bureau, 2011. Population Distribution and Change: 2000 to 2010. Washington DC, USA, Report C2010BR-01, p. 12.

U.S. Environmental Protection Agency, 2009. National primary drinking water regulations. EPA 816-F-09-004. <www.epa.gov/safewater/consumer/pdf/mcl.pdf> (accessed 12.12.2010).

U.S. Environmental Protection Agency, 2010. San Luis Basin Sustainability Metrics Project: A Methodology for Evaluating Regional Sustainability. Cincinnati, OH, USA, Report EPA/600/R-10/182, pp. 231.

Wiesmann, U., Choi, I.S., Dombrewski, E.M., 2007. Biological nitrogen removal. Fundamentals of Biological Wastewater Treatment. Wiley-VCH Verlag GmbH & Co. KGaA, Weinheim, Germany.

Wissel, C., 1984. A universal law of the characteristic return time near thresholds. Oecologia 65 (1), 101–107.

Zellner, M.L., Theis, T.L., Karunanithi, A.T., Garmestani, A.S., Cabezas, H., 2008. A new framework for urban sustainability assessments: linking complexity, information and policy. Comput. Environ. Urban Syst. 32 (6), 474–488.

CHAPTER

Sustainable process index 3

Michael Narodoslawsky

Graz University of Technology, Institut feur Prozess und Partikeltechnik, Graz, Austria

MEASURING ECOLOGICAL IMPACT—THE NORMATIVE BASE OF THE SPI

The field of assessment methods for environmental impacts is wide and colorful, as is shown in the various chapters of this book. For engineers, this represents a particular challenge because they are used to methodological clarity of natural laws and technological design rules. They have learned to cope with economic assessment methods that provide them with guidance regarding how to increase the profit that may be generated by their design. These economic measures conveniently condense complex interactions between different economic factors into a single target function that engineers have learned to maximize (in the form of profit) or minimize (in the form of costs) with their engineering decisions. Assessment of environmental impacts, however, lacks both the clarity of design rules and the convenience of economic optimization. From an engineer's vantage point, the diversity of ecological assessment methods is not only unattractive but also outright confusing and alien to engineering thinking.

This brings us to the source of diversity in environmental evaluation. Environmental evaluation is, as the term indicates, based on normative value systems. This normative reference must not be seen as a liability but as a manifestation of the human way of making decisions. Human beings orient their decisions according to what is "better" or "worse." We assign "value" to the options that are open to us, which requires normative systems that distinguish between what is good and bad and that orient themselves to normative goals that we strive to achieve by our actions.

The diversity of environmental assessment methods that are currently overwhelming engineers that want to make environmentally conscious decisions is therefore a collateral effect of the still-raging societal discourse about the "right" relation between humans and nature. Because this discourse is far from decided, different normative systems, each representing a particular set of normative goals that will be achieved by human–nature interaction, compete for attention and dominance. The various evaluation methods are the translation of these different normative systems with the help of science and (mostly) quantitative measures that can guide decision making. The problem that engineers usually face is that these methods come as authoritative and (often) exclusive, without revealing their true normative content, and pretending that they are the result of pure scientific reasoning.

Environmental assessment methods range from single-issue indicators, such as carbon footprint (Wright et al., 2011) and water footprint (Hoekstra and Chapagain, 2008), to indirect (mostly thermodynamic) measures, such as exergy (Wall, 1988) and emergy (Ulgiati and Brown, 2001), to complex aggregated measures, such as material intensity per service unit (Schmidt-Bleek, 1998) and ecological footprints (Rees, 1992), to name just a few of the most common categories. A thorough analysis of the various assessment methods is still overdue. In short, single-issue indicators try to focus on particular environmental threats and, in some cases, such as the carbon footprint, on one particular factor in complex environmental changes, such as global warming. Therefore, they are epigones of the classical environmental protection attitude that characterized the last quarter of the twentieth century. Thermodynamic measures focus on the efficiency of utilizing energy or solar radiation. Complex aggregated measures (that also can be developed by weighing single-issue indicators) require an explicit formulation of their normative guiding principle because they usually comprise a broad range of human–nature interaction and use an overarching normative goal set to make them comparable.

The Sustainable Process Index (SPI) (Narodoslawsky and Krotscheck, 1995) is a member of the family of ecological footprints and, as such, a complex and highly aggregated environmental assessment method. The general normative base for these measures is the concept of *strong sustainability* that limits interchangeability of human and natural capital (Ekins et al., 2003). In essence, this concept requires a sustainable economy to live on natural income, which is solar radiation (and possibly geothermal energy). Because solar radiation requires area to be converted to useful goods and services, ecological footprints perceive human as well as natural processes to compete for area in sustainable development. This consequently leads to area as a measure of ecological sustainability in these assessment methods.

Contrary to other ecological footprint concepts, the SPI also takes natural material cycles as well as the quality of environmental compartments into account. The reason for this lays in the process engineering orientation of this measure because engineering cannot be reduced to handling renewable resources that result directly from the natural income of solar radiation. Fossil resources and also nonrenewable resources such as minerals and metals are also raw materials that engineers use to fulfill their societal tasks. Any sustainability assessment that engineers may apply to orient their work toward sustainability has to evaluate all materials used in their technological praxis.

To achieve engineering compatibility while at the same time not compromising the normative base of strong sustainability, the SPI uses a set of sustainability principles (EC, 1996):

1. Anthropogenic material flows must not alter the quality and quantity of global material cycles.
2. Anthropogenic material flows must not exceed local assimilation capacities and should be smaller than natural fluctuations in geogenic flows.

The SPI measures the area that is necessary to embed the metabolism of a certain human activity, meaning all material flows that this activity exchanges with the environment into the ecosphere. This area comprises the area necessary to close global material cycles (particularly the carbon cycle) in accordance with the first principle as well as the area necessary to dissipate all emissions and wastes resulting from this activity sustainably in accordance with the second principle. If the activity undergoing assessment comprises all flows exchanged with the environment along the life cycle of a product or service, then the SPI can be applied as an ecological life cycle assessment (LCA) that rates the ecological impact of the service or product according to strong sustainability.

Because of its normative foundation in the concept of strong sustainability, the SPI shares with all other measures from the ecological footprint family an *ecological budget* approach to the assessment: area (and thus the solar radiation received by our planet that can be converted to useful goods and services) is limited. This means that our natural income is limited (although it is granted practically for infinite time). From this argument follows that human society has to get along on a certain natural budget. If we assume that this natural budget is allocated equitably, then every human has to keep his/her ecological footprint for all goods and services he/she consumes within the statistical area, which is found by dividing the surface of our planet (when using the SPI method) by its inhabitants, currently approximately 70,000 m^2. Therefore, the SPI method can also be applied to define the fraction of the statistical natural income per person that is used to provide a particular good or service.

ASSIGNING FOOTPRINTS TO MATERIAL FLOWS

Using the normative foundation of strong sustainability for an ecological assessment applicable to LCA as well as to helping engineers in their decisions requires assigning ecological footprints to material flows caused by a technical activity. These footprints are the areas to embed the metabolism of the activity sustainably into the ecosphere. Because the environment is always dependent on the spatial context, the assessment of the ecological footprint has to take this aspect into account whenever appropriate. The SPI uses, according to the principles stated here, two different approaches for this conversion, depending on the substances involved. Flows triggered by a human activity as well as natural flows that form the reference for the assessment within the SPI concept are always considered on an annual base.

CALCULATING AREAS FOR MATERIALS SUBJECT TO GLOBAL MATERIAL CYCLES

The general strategy of the SPI method to evaluate flows of materials that are subject to global cycles is to link them to area requirements of natural processes critical in these cycles. In its current form, the SPI particularly addresses the water and carbon cycles.

In the water cycle, the key process is the transfer of water from atmosphere to land via precipitation. Water flows used as resources (regardless of their use) are therefore assigned the area that is necessary to catch the amount used via precipitation. Because part of the precipitated water will again evaporate, the actual amount of precipitated water that remains on the land is the reference for the SPI. These values are of course dependent on spatial context. If an activity requires 1,000 m^3 of water per year and the precipitation in the location where this activity takes place is 0.75 m^3/m^2y, of which two-thirds re-evaporate, then this resource flow will be assigned a footprint of 4,000 m^2.

Assessing the carbon cycle is trickier. In this case a distinction must be made between renewable and fossil resources. Renewable carbon resources (i.e., bioresources in all forms) are part of a short-term subcycle of the global carbon cycle. CO_2 from the atmosphere is fixed via photosynthesis to generate bioresources, requiring a certain area dependent on the yield of the particular bioresource in question. At the end of the life cycle of any product based on bioresources, carbon is again released to the atmosphere in the form of the same amount of CO_2 originally fixed by the growth of the bioresource, closing the short-term cycle of carbon exchange between atmosphere and vegetation. So the area required to fix CO_2 for the generation of the bioresource is the area necessary to close the carbon cycle for renewable carbon resources. If a certain activity requires 10,000 t/a of maize, and if the yield of maize in the region where this activity takes place is 1 kg/m^2y, then the footprint assigned to this resource flow is 10,000,000 m^2 (or 1,000 ha).

However, fossil resources are subject to a much slower subsystem of the global carbon cycle; they are retrieved from the long-term storage of carbon that formed over millions of years. The exchange between the more agile processes exchanging carbon between atmosphere, vegetation, and oceans is slow. The SPI concept focuses on sedimentation of carbon to the seabed as a main route of long-term storing of carbon, a process that is the origin of most crude oil and much of natural gas and that is still active today. For fossil carbon resources, the SPI method calculates the footprint as the area of seabed that is required to absorb the carbon used in a certain human activity.

The rate of sedimentation of approximately 0.002 kg carbon/y and m^2 of seabed is very slow (Bolin and Cook, 1983). This means that an activity requiring 20 t/y of fossil carbon will be assigned a footprint of 10,000,000 m^2, the same footprint as in the example for 10,000 t/y of maize.

CALCULATING AREAS FOR ALL OTHER MATERIALS

Human use of all materials that are not subject to global material cycles is inherently dissipative. This applies to all mineral materials such as metals; they are retrieved from point resources, converted to products, and at the end of their life cycle are emitted to the environment. For these materials the second sustainability principle will be used to calculate the ecological footprint. The ecological footprint is the area necessary to absorb the material flow emitted to the ecosphere without violating this principle. For these materials, the SPI assesses the ecological impact of dissipation on the environment.

The strategy for calculating the area to dissipate these material flows used in the SPI method consists of two concepts: the natural concentration of substances in environmental compartments (particularly soil and water) and the natural replenishment of these compartments. The calculation will be exemplified for dissipating a flow to the water compartment.

Suppose that a particular human activity emits 1 kg/y of arsenic to the water compartment. This compartment will be replenished by precipitation. Taking evapotranspiration into account and using the same data as in the example of the water cycle, the replenishment rate is 0.25 m^3/m^2y. This replenished water is assumed to be pure water that does not contain arsenic. The natural concentration of arsenic in (ground) water in this region is 0.00004 kg/m^3. If the arsenic from this particular activity is dissipated to this replenished water in a way so that the concentration is the same as in the natural compartment, then principle 2 is fulfilled and the quality of the natural water compartment is not changed. This requires 25,000 m^3/y of water that replenishes the groundwater. With the rate of 0.25 m^3/m^2y for replenishment of this compartment by precipitation and seepage to the groundwater, an area of 100,000 m^2 is necessary to provide for sustainable dissipation of the arsenic from this activity, which is the footprint assigned to this flow.

This approach also allows assessment of activities that release radioactive substances to water, as is the case in the life cycle of nuclear energy. In this case the natural concentration of radioactive substances such as deuterium in water serves as reference. Because this concentration is extremely low, small flows emitted to the environment lead to large ecological footprints even if these emissions are not necessarily surpassing legal threshold values.

The reasoning is similar for soil, except that the process of replenishing this compartment is the generation of compost (as a viable surrogate for the generation of top soil). Suppose that the concentration of cadmium in top soil in a certain region is 0.000001 kg/kg. Assume further that grass has a yield of 0.5 kg/m^2y in this region, and that 50% of this mass is lost in the process of composting.

This means that the "yield" of compost is 0.25 kg/m^{2}y, which is the replenishment rate of the compartment soil. Dissipating a flow of 1 kg of cadmium would then require 4,000,000 m^2 for sustainable dissipation into soil. This area is then assigned to this flow of cadmium.

However, the compartment air does not have a plausible replenishment rate. In this case, the SPI method compares the exchange between natural vegetation and air as a measure for sustainable dissipation, following the argument in principle 2 that anthropogenic flows must not exceed natural variation in geogenic flows. Because the medium natural exchange rate of methane between land and air is 0.0045 kg/m^{2}y (Bolin and Cook, 1983), an emission of 1,000 kg/y of methane will be assigned a footprint of 222,222 m^2.

In addition to materials that humans take from geogenic reserves and use dissipatively, technology generates materials that are alien to the ecosphere and that have long residence times in environmental compartments. For these materials, the SPI method applies the same logic as for other materials that are used dissipatively, because most of these substances already have measurable (if very small) concentrations in soil and water. Although this treatment somehow contradicts a rigid application of strong sustainability as the value base for the SPI, it allows engineers to estimate the impact of such problematic substances based on a pragmatic strategy that the existing state of the environment will at least not be degraded with persistent synthetic materials. The small measurable concentrations lead to large ecological footprints for their production and/or use, a fact that focuses the attention of SPI users on their environmentally detrimental impact.

For every material flow generated by a human activity, the SPI method assigns an ecological footprint. This footprint is equivalent to the largest area calculated for sustainable dissipation of any substance flow present in the material flow. This guarantees that for all other substances the ecological pressure for dissipation is below the limit defined by the second principle governing the preservation of the quality of local ecosystems.

LCA WITH THE SPI

The method described allows general assessment of the ecological footprint of material flows exchanged between the human society and the ecosphere. To utilize this general method for helping engineers to evaluate the ecological impact of their technologies, it must be further developed to assess technological processes. The SPI method follows the guidelines of LCA as described in the ISO standards 14040 (ISO, 2006). The method of assigning footprints to material flows and the requirements of engineering practice, however, lead to specific approaches within the SPI method.

Because the SPI method only assesses material flows exchanged with the environment, the system boundaries for any technological process have to be expanded until the extraction of resources from nature. Therefore, the footprint of any product is always the sum of the footprints of all processes in the value chain from extraction of resources to the provision of the product in question. If this value chain includes processes with multiple products, then appropriate allocation methods have to be used to assign ecological footprints accumulated so far to all products of this particular process step.

Substances subject to global material cycles (particularly carbon and water) are assigned a footprint representing the area necessary to close this cycle on extraction from nature and to accumulate further footprints for their production and handling along the value chain. Thus, the impacts of planting, tending, harvesting, and transporting of crops as well as extracting, transporting, and refining crude oil are accounted for by SPI LCA.

Substances that are not subject to a global material cycle (e.g., metals and minerals) are not assigned any ecological footprint at the point where they are extracted, but they accumulate ecological impact from mining, upgrading, conversion, and transport until they reach the factory gate of the process that converts them into final goods or services. However, they are assigned a (dissipation) footprint when they are discarded at the end of the life cycle. Likewise, materials that have already provided service to society and that are retrieved for re-use (e.g., used cooking oil or recycled glass) are not assigned footprints at the point of entry in a "new" life cycle.

Infrastructure and equipment are treated as products of life cycles and are assigned the ecological footprint incurred by the material flows exchanged between their production value chain and the environment. This total footprint covers their whole life span. For the assessment of the ecological footprint incurred by operating this infrastructure and equipment over the reference time of 1 y (which is the time base of the SPI calculation), this total footprint has to be (statically) depreciated according to the appropriate technical life span. In addition to infrastructure and equipment, technical processes also have direct area requirements. This area will be directly added to the ecological footprint of the process.

The footprint assigned to staff is equivalent to the statistical area available to a person. In most applications, however, this aspect will be disregarded because this area is small compared with other footprints.

Using these rules together with the methods to assign footprints to the material flows exchanged between the production chain that provides raw materials for a certain industrial process allows the calculation of the ecological footprint incurred by operating this process over the course of 1 y. Dividing this footprint by the number of products or service units provided by this process in 1 y gives the specific ecological footprint of the product or service unit. Finally, the SPI is the ecological footprint of a product or service unit divided by the statistical area per person and indicates how much of the natural income a person is entitled to and how much is consumed to provide this product or service via the technical process undergoing assessment.

APPLICATIONS OF THE SPI

Since its introduction in 1995, the SPI method has been widely used to assess the ecological impact of a broad range of technical processes, products, and even regions. The following survey lists a small collection of SPI-based studies that have been published in journals and books.

Fuel systems have been thoroughly assessed with the SPI method. Ofner et al. (1998) used the SPI to compare different fuel alternatives, with particular focus on dimethyl ether from natural gas, using a well-to-wheel life cycle. More recently, Niederl and Narodoslawsky (2004) presented an LCA of biodiesel from used vegetable oil and tallow, whereas Stoeglehner and Narodoslawsky (2012) used SPI assessment for a more general assessment of the advantages and disadvantages of biofuels.

Energy technologies have been another field of application for the SPI method. Krotscheck et al. (2000) investigated the ecological impact of bioenergy systems. Kettl et al. (2011a) expanded the field of evaluation to a wider field of renewable resource-based energy technologies and compared them with conventional energy technologies. Niemetz et al. (2012) used the SPI method to assess the ecological viability of biogas production systems based on intercrops.

In process industry applications, the SPI method has been particularly used to highlight ecological "hot spots" (i.e., process steps with large ecological impact) and to optimize life cycles from the view point of sustainability. Ku-Pineda and Tan (2006) have used the SPI to optimize water utilization in processes. Koller et al. (2012) investigated the ecological impact of polyhydroxic alcanoate (PHA) production from sugarcane, whereas Titz et al. (2012) used the SPI for assessing an integrated biorefinery process based on slaughterhouse waste to produce, among other products, PHA and biodiesel.

Agriculture is another field in which the SPI method has been successfully applied. In addition to a large number of research reports assessing ecological impacts of different crops, Bavec et al. (2012) compared conventional and organic cultivation methods for wheat and spelt production.

A relatively unconventional field for environmental assessment is the sustainability evaluation of local and regional development strategies. The SPI has been extensively applied to regional sustainable development issues. Narodoslawsky and Stoeglehner (2010) used the SPI for devising regional energy strategies, whereas Sandor et al. (2010) assessed the ecological impact of locally integrated energy systems linking industry and communities. Kettl et al. (2011b) presented a web-based tool that allows the optimization of regional technology systems utilizing endogenous resources and the assessment of their ecological impact in comparison with business as usual using the SPI method.

In addition to the applications cited herein, the SPI method has been used extensively in ecological product and process evaluations sponsored by business and/or public authorities. Aside from the aforementioned fields, the cases here range from assessing buildings, food, and products of daily consumption such as paper towels and even diapers.

CHARACTERISTICS OF THE SPI ASSESSMENT

Analyzing the results of the many case studies of SPI applications and taking into account comparative studies (EC, 1996), certain characteristics of the evaluation practice with the SPI method can be deduced. These main characteristics can be best discussed in a comparison of different technologies to provide electricity (Figure 3.1).

This figure shows that the SPI method distinguishes very sharply between nuclear and fossil-based technologies and those based on renewable resources.

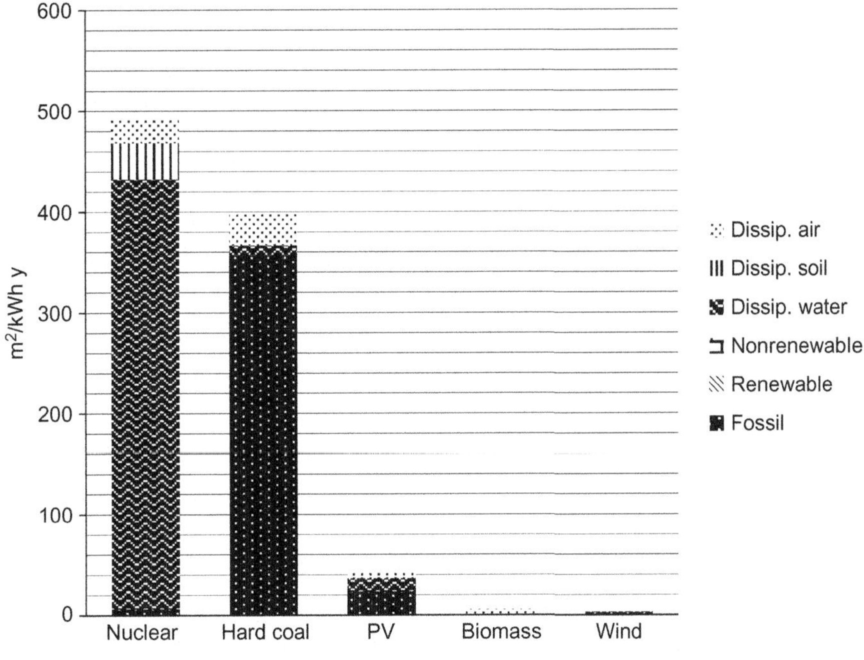

FIGURE 3.1

Ecological footprint for producing 1 kWh of electricity (reference time is 1 y).

There is a factor of 133 between the most unsustainable technology (nuclear power) and the technology rated best by the SPI method (wind power)! This, of course, is in accordance with the underlying normative concept of strong sustainability because technologies based on renewable resources use natural income, whereas nuclear and fossil resources are natural stocks and their utilization is rated less sustainable. The reason for the large footprint of nuclear electricity and electricity from hard coal-fired power plants differ considerably, which is interesting. In the case of hard coal, the reason is obviously the use of a resource that requires a large area to close its global cycle. In the case of nuclear power, it is the dissipation of radioactive material to the compartment water, caused mainly by deuterium emissions during processing of spent fuel rods within the life cycle of nuclear power. These emissions are, of course, below the legal threshold value; nevertheless, the concentration of this radioactive substance in the natural compartment is extremely small, which causes these emissions to considerably alter the local quality of this compartment. It may be noted that the SPI method clearly shows that fossil resources and, hence, global warming issues do not play a significant role in the ecological impact of nuclear power. Nevertheless, this form of electricity provision is assessed as being quite unsustainable compared with other electricity provision technologies.

The SPI method also distinguishes quite sharply between renewable energy technologies, with a factor of 11 between the worst (PV) and the best (wind power). Both technologies do not require any fuel, so their ecological impact is totally defined by the production and maintenance of equipment. In both cases the fossil nature of our current industrial system shows in the ecological footprint: 55% of the footprint of PV and even 72% of that of wind power (not seen in Figure 3.1 because of scale) are attributed to fossil resource use, mainly for energy and transport. The production process for PV infrastructure is much more energy intensive and material intensive than that for the wind power infrastructure as related to the energy produced in the technical life span. PV panel production produces considerable emissions, which are also reflected in the large dissipation areas for water, air, and (to a lesser extent) soil.

Electricity from biomass, in contrast to this pattern, has the biggest contribution to its footprint (82%) from the dissipation to air caused by nitrogen oxide emissions during incineration; 13% are caused by fossil resource use, mainly by fossil-powered harvesting and transport. The provision of the biomass itself as a renewable resource only accounts for 3% of the footprint, because the area for closing the short-term carbon cycle is small.

In general, the footprint calculated with the SPI method is strongly defined by fossil resource utilization. This usually leads to comparable results with carbon footprint or global warming potential (GWP) assessments. The SPI assessments, however, differ sharply from these methods when life cycles include steps that cause considerable noncarbon emissions. Compared with other ecological footprint calculations, particularly those according to Rees (1992), the results of the

SPI method show much larger footprints when fossil resources are used and tend to converge when renewable energy and raw materials are assumed in all steps of a life cycle.

TOOLS BASED ON THE SPI

Application of the SPI method is supported by a number of web-based tools that allow the user to assess ecological impacts of activities, services, products, and even regions. Most of these tools are accessible via the web page (TU Graz, 2013a,b) and provide a portfolio of environmental evaluation calculators that are geared to the requirements of different stakeholders, from engineers to farmers to regional development experts to the general public.

The basic tool for evaluating full life cycles of technical processes is SPIonWeb (TU Graz, 2013b). This web-based program calculates the ecological footprint according to the SPI methods as well as life cycle-wide carbon footprint and GWP based on a user-supplied ecological inventory of the life cycle. A database of almost 1,400 subprocesses helps the user to generate the life cycle for a particular process without detailed knowledge about the inventories of industrial (and agricultural!) processes providing key raw materials and process energy. This program evaluates the ecological impact of every process step in the life cycle and indicates which step and/or raw material contributes most to the impact and what particular substance is responsible for this impact. With this information, engineers can improve the ecological performance of their products by focusing on the step that defines the ecological impact of the whole life cycle.

Based on the results of this tool, other programs freely available on the Internet allow users to assess ecological impacts of various products and activities. The Environmental Long-term Assessment of Settlement (ELAS) calculator (IRUP, IPPE, and Studia, 2011) assesses ecological impacts (SPI—ecological footprint, life cycle-wide carbon emissions, cumulated energy demand) as well as socioeconomic implications of settlement. This calculator takes the life cycle of the whole infrastructure of settlements, their operation, as well as induced mobility of residents into account.

The program RegiOpt (TU Graz, 2013a) optimizes regional technology networks utilizing endogenous resources while providing energy as well as food to regional populations. It assesses the economic performance as well as ecological impact expressed as SPI footprint and life cycle carbon footprint of optimized systems versus business as usual and informs the user about the distribution between local and global economic and ecological impacts.

In addition to these tools, other target group-specific evaluation programs are available (TU Graz, 2013a). Most of these programs are only available in German language versions. They include calculators assessing ecological impacts ranging from individual lifestyles to the impact of energy use to farm operation and tourism as well as education institutions.

CONCLUSION

The SPI method offers a way to assess human activities, industrial processes, products, services, and regions according to the normative concept of strong sustainability. Based on two principles that can be directly deduced from this normative fundamental, the SPI method allows a comprehensive assessment of the metabolic impact of the exchange of material flows between anthroposphere and ecosphere. It takes into account the closing of global material cycles and the preservation of the ecological compartments of soil, water, and air as the basis of ecosystem functions. In terms of LCA, it includes the impacts of resource provision and emissions.

The SPI was developed with a clear focus on supporting engineers in their tasks of designing and operating industrial processes. The application of the SPI method to a large variety of processes and products has shown that this sustainability measure is extremely sensitive to the use of different resources, distinguishing sharply between renewable and fossil resources in particular. Also, the measure is very sensitive to the use and dissipation of substances that either are occurring naturally in very low concentrations or are alien to the ecosystem.

Tools that have been developed on the basis of the SPI method, particularly SPIonWeb, allow engineers who have only basic knowledge of sustainability assessment to evaluate their design or operation with acceptable effort. A comprehensive databank covering the provision of most precursors to industrial processes as well as almost all forms of energy provision allows easy representation of complex life cycles, even for untrained persons. The results of the assessment allow engineers to pinpoint sustainability "hot spots" that they must address to improve the ecological performance of their design and/or operation.

The SPI offers a quick and reliable method to estimate the ecological impact, even of complex technological systems. Although it is no substitute for thorough legally prescribed environmental compliance assessment (as is also the case for all other single-issue measures such as the carbon footprint or highly aggregated measures like emergy), it provides consistent and comprehensive guidance for engineers who want to include sustainability into their decisions.

REFERENCES

Bavec, M., Narodoslawsky, M., Bavec, F., Turinek, M., 2012. Ecological impact of wheat and spelt production under industrial and alternative farming systems. Renew. Agric. Food Syst. 27 (03), 242–250.

Bolin, B., Cook, R.B., 1983. The Major Biochemical Cycles and Their Interactions. SCOPE Series 21. John Wiley & Sons, Hoboken, NJ, USA.

EC, 1996. Operational Indicators for Progress towards Sustainability, End Report, Project Number: EV5V-CT94-0374, Brussels, Belgium.

Ekins, P., Simon, S., Deutsch, L., Folke, C., De Groot, R., 2003. A framework for the practical application of the concepts of critical natural capital and strong sustainability. Ecol. Econ. 44, 165–185.

Hoekstra, A.Y., Chapagain, A.K., 2008. Globalization of Water: Sharing the Planet's Freshwater Resources. Blackwell Publishing, Oxford, UK.

IRUPIPPE, Studia, 2011. <www.elas-calculator.euJuly/> (accessed 20.04.2014).

ISO, 2006. Environmental Management—Life Cycle Assessment—Principles and Framework ISO 14040, International Standardisation Organisation, Geneva, Switzerland.

Kettl, K.-H., Niemetz, N., Sandor, N., Narodoslawsky, M., 2011a. Ecological impact of renewable resource-based energy technologies. J. Fundam. Renew. Energy Appl. 1, 1–5, <http://dx.doi.org/10.4303/jfrea/R101101>.

Kettl, K.-H., Niemetz, N., Sandor, N., Eder, M., Narodoslawsky, M., 2011b. Regional Optimizer (RegiOpt)—sustainable energy technology network solutions for regions. Comput. Aided Chem. Eng. 29, 1895–1900.

Koller, M., Salerno, A., Reiterer, A., Malli, H., Malli, K., Kettl, K.-H., Narodoslawsky, M., Schnitzer, H., Chiellini, E., Braunegg, G. 2012. Sugarcane as feedstock for biomediated polymer production. In: J.F. Goncalves and K.D. Correia (Eds.), Sugarcane: Production, Cultivation and Uses, Nova Science Publishers, Haupauge, NY, USA. pp. 105–136.

Krotscheck, C., König, F., Obernberger, I., 2000. Ecological assessment of integrated bioenergy systems using the sustainable process index. Biomass Bioenergy 18 (4), 341–368.

Ku-Pineda, V., Tan, R.R., 2006. Environmental performance optimization using process water integration and Sustainable Process Index. J. Cleaner Prod. 14 (18), 1586–1592.

Narodoslawsky, M., Krotscheck, C., 1995. The Sustainable Process Index (SPI): evaluating processes according to environmental compatibility. J. Hazard. Mater. 41 (2 + 3), 383–397.

Narodoslawsky, M., Stoeglehner, G., 2010. Planning for local and regional energy strategies with the ecological footprint. J. Environ. Policy Planning 12 (4), 363–379.

Niederl A., Narodoslawsky M., 2004. Ecological evaluation of processes based on by-products or waste from agriculture—life cycle assessment of biodiesel from tallow and used vegetable oil, feedstocks for the future. In: Bozell, J., Patel, M. (Eds.), ACS symposium series, vol. 921.

Niemetz, N., Kettl, K.-H., Szerencsits, M., Narodoslawsky, M., 2012. Economic and ecological potential assessment for biogas production based on intercrops. In: Biogas., pp. 173–190, <http://dx.doi.org/10.5772/31870>.

Ofner, H., Gill, D.W., Krotscheck, C.H., 1998. Dimethyl Ether as Fuel for CI Engines—a New Technology and its Environmental Potential. Society of Automotive Engineers, Warrendale, PA, SAE Technical Paper Series Nr. 981158.

Rees, W.E., 1992. Ecological footprints and appropriated carrying capacity: what urban economics leaves out. Environ. Urbanisation 4 (2), 121–130.

Sandor, N., Eder, M., Niemetz, N., Kettl, K.-H., Narodoslawsky, M., Halasz, L., 2010. Optimizing the energy link between city and industry. Chem. Eng. Trans. 21, 295–300.

Schmidt-Bleek, F., 1998. The MIPS Concept: Less Consumption of Natural Resources—Better Quality of Life by a Factor of 10 (Das MIPS-Konzept: weniger Naturverbrauch—mehr Lebensqualität durch Faktor 10). Droemer, München, Germany (in German).

Stoeglehner, G., Narodoslawsky, M., 2012. Biofuels—the optimal second best solution? In: Machrafi, H. (Ed.), Green Energy & Technology. Bentham eBooks, Bussum, the Netherlands, SBN 978-1-605805-285-1.

Titz, M., Kettl, K.-H., Shazhad, K., Koller, M., Schnitzer, H., Narodoslawsky, M., 2012. Process Optimization for efficient biomediated PHA production from animal-based waste streams. J. Clean. Tech. Environ. Policy, http://dx.doi.org/10.1007/s10098-012-0464-7.

TU Graz, 2013a. <www.fussabdrucksrechner.at> (accessed 06.07.2014).

TU Graz, 2013b. <spionweb.tugraz.at> (accessed 06.07.2014).

Ulgiati, S., Brown, M.T., 2001. Emergy accounting of human-dominated, large-scale ecosystems. In: Jorgensen, S.E. (Ed.), Thermodynamics and Ecological Modelling. CRC Press LLC, Boca Raton, FL, pp. 63–113.

Wall, G., 1988. Exergy flows in industrial processes. Energy 13 (2), 197–208.

Wright, L., Kemp, S., Williams, I., 2011. Carbon footprinting: towards a universally accepted definition. Carbon Manage. 2 (1), 61–72.

CHAPTER

Moving to a decision point in sustainability analyses

4

Debalina Sengupta*,[1], Rajib Mukherjee*,[2], and Subhas K. Sikdar†

**Oak Ridge Institute for Science and Education, Post Doctoral Research Fellow at National Risk Management Research Laboratory, US Environmental Protection Agency, Cincinnati, OH, USA †Associate Director of Science National Risk Management Research Laboratory, US Environmental Protection Agency, Cincinnati, OH, USA*

Measure what is measurable, and make measurable what is not so.
—Galileo Galilei

INTRODUCTION

The concept of sustainability, since the publication of the Brundtland report known as Our Common Future (World Commission on Environment and Development, 1987), has been widely used as a rationale to focus on the long-term effects of anthropogenic activities on nature. The report advocated measures that would improve the condition of the environment, which is degraded, while continuing development that improves the economic and health conditions of humans. Although the concept is easy to grasp in terms of achieving simultaneous progress in the present with regard to economic, environmental, and societal aspects without jeopardizing the progress of future generations, the development of a standard method for analyzing sustainability is still in its nascent stage. Sustainability of systems such as agricultural (Conway and Barbier, 1990), industrial (Graedel and Allenby, 2010), urban infrastructures has been studied, and a considerable amount of research (Čuček et al., 2012) has provided directions to achieve stability of such systems. In other cases, countries have been compared regarding their sustainability status based on their performance in economic, environmental, and societal improvement efforts. In this context, Freudenberg (2003) provides a comprehensive assessment of the use of composite indicators for evaluating a county's sustainability, Esty et al. (2008) describe the development of the

[1]Post Doctoral Research Associate Artie McFerrin Department of Chemical Engineering, Texas A&M University, College Station, TX, USA
[2]Research Engineer Viswamitra Research Institute, Chicago, IL, USA

Environmental Performance Index of countries, and Sengupta et al. (2014) demonstrate the use of an aggregate index in evaluating the relative sustainability of countries using the UN Millennium Development Goals indicators.

In this chapter, the focus of the discussion is on analyzing the factors that determine sustainability and on moving to a decision point for the sustainability of engineered systems, such as manufacturing processes, products, end-use, and disposal strategies. We start this discussion with a background of some of the commonly used methods for sustainability analysis, especially the use of indicators or metrics. In what follows, we analyze how these individual methods help us draw conclusions for overall sustainability assessment for competing options of a product or a production system. Then, to fulfill the premise of this chapter of helping us move to decision points on sustainability, we introduce a new method for aggregating preselected indicators to compute a single index for comparing various technological options for a product. After this, we discuss the need for ranking the indicators on their relative importance in the sustainability performance and introduce a novel concept that helps with that determination. This new methodology helps us find only those indicators that are necessary and sufficient for adequate sustainability assessments. Four case studies are presented here as a step-by-step guide to arrive at a decision point in the sustainability analyses through the use of aggregation and ranking of indicators. Although the interest here is production systems, the methodology we introduce is applicable, in general, to all systems regardless of scale or nature.

DEFINING A SYSTEM IN THE CONTEXT OF SUSTAINABILITY

The aim of the World Commission of Environment and Development (WCED) was to create a united international community with shared sustainability goals. It proposed to achieve this by identifying sustainability problems worldwide, raising awareness about them, and suggesting the implementation of solutions. Thus, the organization's definition of sustainability centers on global conditions of ecology (i. e., environment), economic development (i.e., by technologies), and societal equity, where the envisioned system under consideration is the Earth. Since then, people have tried to subdivide this large system into its constituent systems to achieve a level of spatial and temporal granularity that can be observed, measured, and controlled by humans. This led us to discussions on the individual sustainability of these constituent systems, such as sustainability of communities, cities, businesses, and even of technologies. Sikdar (2003) identified four types of sustainable systems.

Type I: They typify global concerns or problems, such as global warming (GW) attributable to the emissions of carbon dioxide, methane, and nitrogen oxides; ozone depletion attributable to chlorofluorocarbons; or the use of genetically modified (GM) crops. A system of this type is defined as the Earth in relation to one of these global issues resulting from anthropogenic practices.

Because of the global nature of these issues, it is not surprising that global treaties and agreements are needed to address them. The Kyoto Protocol for greenhouse gases (Oberthür and Ott, 1999) and the Montreal Protocol (UNEP, 1987) for the phasing out of chlorofluorocarbons are examples of such treaties. Removing scientific uncertainties from environmental, developmental, and social impacts resulting from these global causes is the primary role of the science experts to enable the political leaders to craft robust agreements for the benefit of human health and the environment.

Type II: They are characterized by geographical boundaries. Examples are nations, provinces, cities, villages, or defined ecosystems. Technical disciplines that must be marshaled to consider these systems are civil engineering, environmental engineering, hydrology, ecology, urban planning, laws, regulations, and economics.

Type III: Businesses, either localized or distributed, constitute type III systems. Businesses strive to be sustainable by practicing cleaner technologies, recycling by-products, eliminating waste products, reducing emissions of greenhouse gases, eliminating the use of toxic substances, reducing water use, and reducing energy intensity of processes. For instance, by colocating manufacturing plants to minimize wastes (essentially achieving industrial ecology) or by establishing waste exchange networks, industries can achieve the threefold goal of economic development, environmental stewardship, and social good.

Type IV: They are the smallest of the systems and can be called "sustainable technologies." Any particular technology that is designed to provide economic value through clean chemistries would be an example of a type IV system. Clearly, type III and type IV systems are most suitable for the important role for technologists of all disciplines because the performance of these systems is dependent on process and product designs and manufacturing methods.

These four types of systems form the basic classification of the ways in which the boundaries and scope of a sustainability analysis can be determined. Each type of system will have specific attributes that can be examined through the use of indicators. The following section provides an overview of the indicator-based systems assessment for sustainability.

INDICATOR-BASED SYSTEM ASSESSMENT FOR SUSTAINABILITY

Indicators (or metrics) measure aspects of a system that allow us to characterize the conditions of the system. Indicators and metrics are used interchangeably in this chapter. The values of indicators assist in effective decision-making, such as business decisions to build or extend a new chemical plant, performance improvement decisions for a reactor, or government policy-making related to environmental pollution from a particular sector. Indicators are also essential for evaluating current conditions, tracking the outcomes of actions taken, and assessing progress toward overall goals. Indicators are also able to calibrate progress toward sustainable development goals. They can provide an early warning to prevent economic, social, and environmental setbacks.

The selection of indicators effectively determines the "looking glass" through which one views the technical or policy options within a system; therefore, it is extremely important in influencing decisions and judgments. Needless to say, the first hurdle in effective decision-making is the choice of the indicators that best represent a system. Once a set of specific indicators for a system assessment is chosen, data collection for those indicators may pose difficulties. The final step is decision-making with the indicator values. Combining the three pillars and finding a single value that encompasses the comprehensive set of indicators is a daunting, but necessary, task to incorporate physical and social science knowledge into decision-making. The choice of the set of indicators is determined *a priori*; in other words, the indicators are constructed first, followed by data collection and computation of these indicators. In most of the current practices, little or no analysis is performed for the indicators after the calculation of the values, but emphasis is given to tracking the values of the indicators themselves. To represent the indicators, a radar chart or star plot is used in most cases, and a unique decision point is never reached for overall performance of an option within the system. The radar plot, a visual diagram, can handle only a few indicators and fails when the system is defined by a large number of indicators. Also, because the data gathering and processing steps involved in the indicator calculation are resource-intensive, there is a significant need to identify whether an indicator is truly necessary for ascertaining the sustainability of a system. This is the rationale for the necessary and sufficient number of indicators.

A historical overview has been provided in this chapter covering the different methods that several eminent organizations and researchers have pioneered for making decisions with indicators. This review provides the background to introduce our methodology of sustainable decision-making with multivariate statistical methods. First, an aggregate index calculation method is established to determine overall sustainability from a large set of indicators. Then, a partial least-square variable importance in projection (PLS-VIP) method is applied to the dataset of indicators to determine the indicators contributing most to the aggregate index. The results of this methodology allow the user to provide a sound method to normalize and compute an aggregate index for comparing relative sustainability of competing systems and to rank the given set of indicators according to their contribution to the aggregate index, allowing a decision-maker to focus on the most needed areas of intervention for sustainable development.

HISTORY OF SUSTAINABILITY ANALYSIS THROUGH THE USE OF INDICATORS AND METRICS

The indicators measuring sustainability of a system vary according to the scale of that system. For example, the comparative progress of a country over another toward sustainable development can be partly measured by the percentage of the population with access to clean drinking water. The sustainability of a chemical process compared with another can be partly determined by the amount of material it uses or the amount of greenhouse gases it generates. Thus, sustainability analysis always involves a relative measure comparing one system to a reference system. The reference system can be another competing system or a synthetic one, or even the same system in a different time in the past. Over the years, various institutions have made progress in assessing the relative sustainability of a system of interest. Next, four such institutional sustainability evaluation efforts are explained.

AIChE SUSTAINABILITY METRICS SUITE/BRIDGES TO SUSTAINABILITY METRICS

One of the earliest studies in the development of sustainability metrics for decision-making was conducted by Canada's National Round Table on the Environment and the Economy (NRTEE) (Tanzil et al., 2003). This study by NTREE included eight companies from different sectors and recommended a set of core metrics that included material intensity, energy intensity, and dispersion of regulated toxics per unit of products or services. This study also suggested the use of complementary metrics such as greenhouse gas intensity. The World Business Council for Sustainable Development (WBCSD) recommended that, in addition to the material and energy consumptions, water consumption is another important sustainability metric. WBCSD also identified the emissions of greenhouse gases and ozone-depleting substances as the metrics having an environmental impact that can be calculated based on existing international consensus. The Center for Waste Reduction Technologies (CWRT) of the American Institute of Chemical Engineers (AIChE) (currently the Institute for Sustainability) with representatives from member companies was instrumental in developing sustainability metrics further (Beloff and Beaver, 2000). This effort concluded on a set of basic and complementary sustainability metrics expressed on a choice of denominators that include mass, revenue, or value added (Schwarz et al., 2002). BRIDGES to Sustainability™, a not-for-profit organization, compiled all of this information and developed an automated methodology and software known as Bridgesworks™ (Tanzil et al., 2003). BRIDGES conducted the research of the metrics development with significant funding from the US Department of Energy under a cooperative agreement with the AIChE CWRT.

Basic and complementary metrics of six impact categories were chosen to represent material, energy, water, solid wastes, toxic release, and pollutant effects. The five basic indicators corresponding to these categories were material intensity, energy intensity, water consumption, toxic emissions, and pollutant emissions (Table 4.1). The sustainability of the BRIDGES metrics are constructed as ratios with environmental impacts in the numerator and a physically or financially meaningful representation of output in the denominator, with the better process

Table 4.1 Basic Sustainability Metrics (Tanzil et al., 2003)

Output	
Mass of product	**Material intensity** $\frac{\text{Mass of raw materials—Mass of products}}{\text{Output}}$
or	**Water intensity** $\frac{\text{Volume of fresh water used}}{\text{Output}}$
Sales revenue	**Energy intensity** $\frac{\text{Net energy used as primary fuel equivalent}}{\text{Output}}$
or	**Solid waste to landfill** $\frac{\text{Total mass of solid waste disposed}}{\text{Output}}$
Value-added	**Toxic release** $\frac{\text{Total mass of recognized toxics released}}{\text{Output}}$

Table 4.2 Examples of Complementary Sustainability Metrics (Tanzil et al., 2003)

Material	**Solid Waste**
Packaging materials	Solid waste disposed relative to landfilling capacity
Nonrenewable materials	**Toxic release**
Toxics in product	Toxic release under each Toxics Release Inventory (TRI) category
Toxics in raw materials	Human toxicity (carcinogenic)
Water	Human toxicity (noncarcinogenic)
Rainwater sent to treatment	Ecosystem toxicity
Water from endangered ecosystem sources	**Pollutant effects**
Water use relative to water availability	Global warming potential
Energy	Tropospheric ozone depletion potential
Energy consumed in transportation	Photochemical ozone creation potential
Nonrenewable energy	Air acidification potential
	Eutrophication potential

being the one with a smaller value for the ratio (Table 4.1). These metrics have been used in various manufacturing facilities like Formosa Plastics (petrochemical), Interface Corporation (carpeting), and Caterpillar Inc. (tool manufacturing). Examples of complementary metrics related to the six impact categories are provided in Table 4.2.

IChemE SUSTAINABLE DEVELOPMENT PROCESS METRICS

The Institute of Chemical Engineers (IChemE) compiled a list of sustainable development process metrics for use by companies to report on their advancement toward sustainability (IChemE, 2014). These metrics allow for the comparison of different options within a company, or comparison between companies. This method was developed for assessing sustainability of corporations and individual technologies, but it is important that even for small operating units that wider implications and impacts are considered through this method. The metrics for the sustainable development include environmental, economic, and social indicators, as given in Table 4.3.

The IChemE sustainable development progress metrics are described in detail with equations required to compute the indicators in the report (IChemE, 2014).

Table 4.3 The IChemE Sustainable Development Process Metrics (IChemE, 2014)

Indicator Type	Category	Metrics
Environmental indicators	Resource usage	Energy
		Material (excluding fuel and water)
		Water
		Land
	Emissions, effluents, and waste	Atmospheric impacts
		Aquatic impacts
		Impacts to land
	Additional environmental items	Duty of care with respect to products and services produced for which environmental or health problem solutions are not yet known.
		Environmental impact of plant construction and decommissioning.
		Compliance
		Impacts on protected areas
		Impacts on local biodiversity or habitats
		Issues concerning long-term supply of raw materials from nonrenewable resources.
		Other possible relevant metrics.
Economic indicators	Profit, value, and tax	–
	Investments	–
	Additional economic items	–

(Continued)

Table 4.3 The IChemE Sustainable Development Process Metrics (IChemE, 2014) *Continued*

Indicator Type	Category	Metrics
Social indicators	Workplace	Employment situation
		Health and safety at work
	Society	–
	Other items	Issues concerning discrimination, concerning women and minorities or indigenous communities, the number in senior and middle management programs to improve employability, including focused education or training, and mentoring
		Incidents of child labor, forced labor, or violation of human rights, on the part of the company, its suppliers or contractors, and public protest concerning such issues. Report positive steps taken in this regard
		Performance of suppliers and contractors relative to criteria for their selection, incidents of noncompliance with sustainability requirements, e.g., responsible purchasing
		Other possible relevant metrics

A company wishing to disclose their sustainability performance is required to use these indicators, collect data for them, and compile the indicator values in a prescribed report format. Thus, the IChemE metrics help to collect information regarding the various aspects of process-related sustainability for a given company for external reporting. Internally, a particular company may make use of the indicators to compare between several competing processes. The report in itself is just a compilation of indicator data and two companies or processes are compared on a particular indicator on an *ad hoc* basis. No effort has been proposed by IChemE to combine this indicator information into a single index. This makes it a data collection process without significant analysis of assessing the path toward sustainability. However, with such high-quality numerical data involved with the indicators, significant progress can be made by identifying potential process improvements, the environmental impacts of the products and processes, or the health and safety improvements that have been achieved in a company, in regard to itself and also in regard to other companies that report data within a similar scope. With further availability of temporal data, the IChemE process metrics can be used to track sustainability performance over several years.

AIChE SUSTAINABILITY INDEX

The AIChE Sustainability Index (Cobb et al., 2009) was developed by engineers and scientific experts for benchmarking well-defined performance metrics or indicators. These include indicators such as environmental health and safety (EH&S) performance, innovation, and societal measures. These indicators were developed for assessing corporate sustainability in contrast to the other AIChE efforts through its CWRT for sustainability of technologies. Based on more than 30 sources of public data, these indicators allow companies to measure their efforts at both the company level and the sector level. The key indicators that constitute the AIChE Sustainability Index are provided in Table 4.4.

Table 4.4 Metrics for the AIChE Sustainability Index (AIChE, 2014)

Indicator Type	Category	Metrics
Strategic commitment to sustainability	Stated commitment	Public commitment to excellence in environmental and social performance throughout a company's value chain
	Commitment to voluntary codes	Public commitment to voluntary codes and standards, including responsible care, global compact, and others
	Sustainability reporting	Timely and comprehensive public reporting of sustainability performance
	Sustainability goals and programs	A comprehensive set of goals and programs that are specific and challenging
	Third-party ratings	Respected agencies' ratings on company-wide sustainability management and reporting
Sustainability innovation	General R&D commitment	Corporate commitment to research and development, as evident in the amount of R&D expenditure per net sales
	Sustainable products and processes	Development of products and processes with superior environmental, social, and economic performance
	Sustainability approaches in R&D	Use of sustainability considerations and decision-support tools in R&D and innovation processes
	R&D effectiveness	Results of the R&D investment, as reflected in the number of patents issued and commercialization of new products that enhance environmental and social sustainability
Environmental performance	Resource use	Intensity of energy, material, and water consumption, and use of renewable sources of energy and materials
	Greenhouse gas emissions	Intensity of greenhouse gas emissions

(Continued)

Table 4.4 Metrics for the AIChE Sustainability Index (AIChE, 2014) *Continued*

Indicator Type	Category	Metrics
	Other emissions	Air emissions, wastewater, and hazardous waste releases
	Compliance management	Environmental liability, fines, and penalties, and environmental capital investment
Safety performance	Employee safety	Recordable and days-away-from work injury rates
	Process safety	Number and trend of process safety incidents, normalized by number of employees, and occurrence of major safety incidents
	Plant security	Presence of an adequate plant security management system, represented by completion of a responsible care plant security audit
Product stewardship	Assurance system	Product stewardship policies and goals, incorporation of a responsible care product safety process, and engagement of value-chain partners to assure product safety
	Risk communication	Risk communication policies and goals, incorporation of a responsible care risk communication process, and preparation to meet REACH requirements
	Legal proceedings	Involvement in major legal proceedings related to product safety, risk, and toxicity
Social responsibility	Stakeholder partnerships	Extent of stakeholder engagement and partnership programs at the project, facility, and corporate levels
	Social investment	Contributions through employment, philanthropy, and community development projects
	Image in the community	Company image as indicated by reputable awards and recognition programs, including "most admired" and "best employer" ratings
Value-chain management	Environmental management systems (EMSs)	Presence of an EMS at the corporate and facility levels
	Supply chain management	Policies and procedures related to suppliers' sustainability, presence of sustainability evaluation, and audits for first-tier suppliers, and management of second-tier and higher-tier suppliers

The AIChE Sustainability Index project provides seven corporate measures of sustainability to enable a company to benchmark its performance among other companies in the same sector. Two important features of these indicators need to be pointed out. First, none of them are numerical, and thus expert judgment is required. Second, the indicators are composites of underlying factors that are measured and integrated to give a measure of that indicator. At the end, all data are numerically scaled to a range of 0–7, with the higher number indicating better performance in that composite indicator. For the company, these metrics allow measuring progress toward best practices at regular intervals and provide unbiased, expert interpretation of publicly available technical data. Thus, the AIChE Sustainability Index project focuses on measuring corporate sustainability through the use of various individually aggregated indicators.

An example of the results obtained from the AIChE Sustainability Index is shown in Figure 4.1. In this figure, we can see that the different lines on the radar plot show the performance of various companies in the seven indicator categories. The solid and dashed polygons represent performance of companies in size groups (big and small based on revenues). A tempting method to compare such performance data on a diagram like this could be the computation of the area occupied on the plot. Thus, for the scale of 0–7, with 0 being the worst and 7 being the best, the most area covered by a particular company will represent the highest sustainability achieved. The AIChE Sustainability Index project also gives the range of each indicator performance for the different options. However, the composite indicators used in this method have different units of measurement, unless they can all be made dimensionless, e.g., by monetizing. The validity of such integration will depend on the soundness of the scaling. However, AIChE did not suggest such an area computation, and the method of making inference gravitates to visual inspection or comparing individual data.

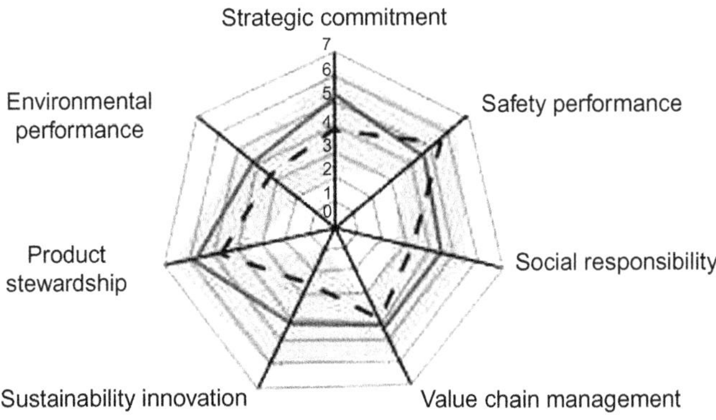

FIGURE 4.1

Representation of the AIChE Sustainability Index.

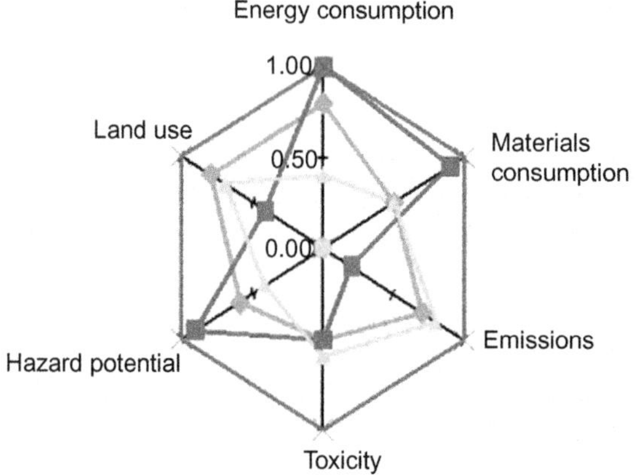

FIGURE 4.2

BASF ecological fingerprint (radar plot) (BASF, 2014a).

BASF: ECO-EFFICIENCY AND SEEBALANCE® ANALYSIS

Eco-efficiency analysis (EEA) (Saling et al., 2002) and more recently (Uhlman and Saling, 2010) was introduced and popularized by BASF to facilitate decision-making and enable a company to drive innovative product development toward bringing more sustainable products to the marketplace. The EEA harmonizes two of the pillars of sustainability by providing a relation between a product's or technology's economic benefits and its environmental impacts in six categories. These economic benefits and environmental impacts are calculated along the supply chain of the product in accordance with the ISO 14040 and 14044 standards for Life Cycle Assessment. Recently, BASF has extended the assessment to include social aspects in their SEEBALANCE method, which is a socio-EEA method for analyzing sustainability (Saling et al., 2005) and later (Kolsch et al., 2008).

In EEA, the environmental impact is described and calculated based on six categories: raw materials consumption, energy consumption, land use, air and water emissions and solid waste, potential toxicity, and potential risks. Combining these individual data gives the total environmental impact of a product or process. Combined environmental impact data are represented on a radar plot as shown in Figure 4.2. The economic data are compiled separately and include the various costs incurred in manufacturing or using a product. The economic analysis and the overall environmental impact are used to make eco-efficiency comparisons. Economic and ecological data are plotted on an *x*/*y* graph as shown in Figure 4.3. The costs are shown on the horizontal axis and the environmental impact is shown on the vertical axis. The graph reveals the eco-efficiency of a product or process compared with other products or processes; the left bottom corner denotes a zone for low eco-efficiency and the right top corner denotes

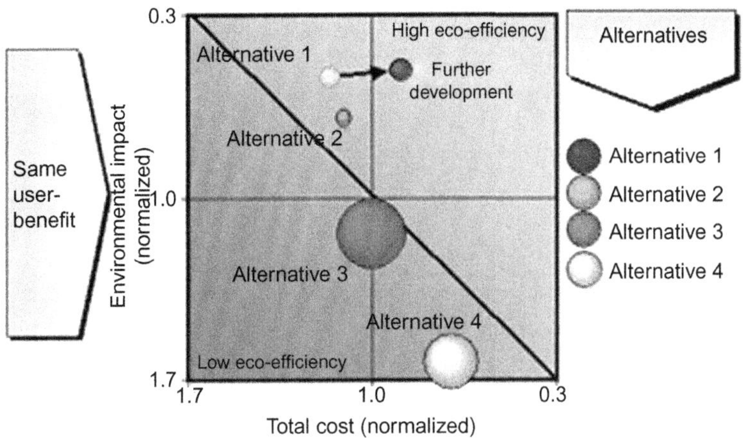

FIGURE 4.3

BASF eco-efficiency graph (BASF, 2014a).

high eco-efficiency. An option such as alternative 4 is the least preferred and alternative 1 is most preferred on this graph.

In SEEBALANCE, the social indicators are grouped into five stakeholder categories: employees, international community, future generations, consumers, and local and national community. Subcategories are measurable and include number of employees, occupational accidents occurring during production, and risks involved in the use of the product by the end consumer. These societal indicators are summarized in the so-called social fingerprint, similar to the environmental indicators. The three dimensions of sustainability are then represented on a three-dimensional graph called SEECube® (Figure 4.4).

The EEA thus utilizes two sets of information (economic and environmental) and the SEEBALANCE results combine the three pillars into the socio-eco-efficiency diagram. The indicators in individual categories are combined separately into composite indicators, with weights assigned to them. The weighting aspect and separately integrating the indicators into the respective categories of economic, environmental, and social suppress the original indicator values. Nevertheless, these graphical methods of BASF do represent an advancement assuming there is absolute transparency on two issues: the methods of creating the composite indicators for synthesizing the three measures and the method of weighting. Internally at BASF, this of course works for sound decision-making; however, for general applicability, more detailed information regarding the methodology must be made available. For instance, no information is available regarding the method of computing the environmental and societal impacts. A transparent and quantitative numerical aggregation method for one sustainability index for decision-making is needed.

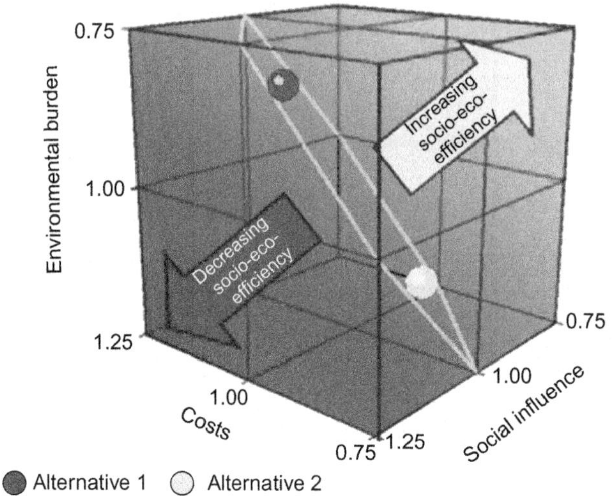

FIGURE 4.4

BASF SEECube (representing socio-eco-efficiency) graph (BASF, 2014b).

RESEARCH METHODOLOGY

NEED FOR AGGREGATE INDEX PLS-VIP METHOD

We see from the four efforts described that major emphasis has been given to the development and selection of indicators that track the progress of a business or technology system. However, unambiguous decisions regarding system performance, when the indicator datasets constitute numerical values of many indicators, are not easy to make. One common way of presenting the indicator data is by constructing radar plots, as shown in most of the cases here. The radar plot constitutes an n-dimensional polygon where n is the number of indicators. When several alternative options are represented on such a diagram, an easy decision is difficult to make from the clutter of the data, especially when the number of indicators is very high. BASF's methods do not suffer from this problem because they express the measures of the three pillars of sustainability in composite forms, thus allowing easy visual inspection for a decision. However, as pointed out, the methods of aggregation for creating the composite indices, especially representing environmental and societal impacts, are not publicly known. Another drawback of the radar plot is that it has to be standardized. This is related to the method of aggregation of the composite indices, the values of which are plotted, as we pointed out for AIChE sustainability index project. Various authors have argued in favor of a single aggregate sustainability index derived from these indicators (Nardo et al., 2005) and later (Hák et al., 2007). The aggregate sustainability index is a numerical composite of underlying indicators that reflects, in aggregate, the state of the system in question as measured by the values of the individual indicators. Aggregate indices are distinguished from composite indices in that composite indices are composed of more fundamental indicators and not necessarily a numerically related sum. Seminal work has been performed in this direction for composite index as captured in the Fisher information (Cabezas and Fath, 2002) followed by (Fath et al., 2003). Fisher information has been successfully applied to ecological systems to study regime shifts and resilience of systems. However, in their methodology, Fisher information is used as one indicator along with other indicators for sustainability assessment. Although the methodology of Cabezas and Fath (2002) has been applied before for regional systems (type II), it is now being applied to product systems (type IV) (Ingwersen et al., 2014).

Aggregate indices, on the contrary, are numerically combined indices of individual indicators that can be used to rank technology or policy options for a system. Decision-making with integrated information regarding sustainable development of a company has been demonstrated using the composite sustainable development index (I_{CSD}) to track data regarding economic, environmental, and social performance of a company with time (Krajnc and Glavič, 2005). For countries, aggregate indices have been used to compare their competitiveness, innovative abilities, degree of globalization, and environmental sustainability (Freudenberg, 2003).

For engineering applications, aggregate indices have not been widely used to assess the overall sustainability from environmental, economic, and social indicators. Some efforts that are currently being pursued tend to examine sustainability from different perspectives. For example, the Sustainable Process Index (SPI) described in Chapter 3 is based on the assumption that sustainable flow of solar exergy is required for a sustainable economy. The conversion of the solar exergy to services needs area that then becomes the limiting factor of a sustainable economy. A measure of areas added over the life cycle of a product gives the indicator of interest. The AIChE/CWRT total cost assessment (TCA) methodology (Beaver, 2000) effort calculates the overall sustainability by assigning an economic value to various aspects of sustainability and aggregating them to a single dollar amount, thus showing another method of aggregation. Sengupta and Pike (2012) demonstrated the use of the TCA method by maximizing a triple bottom-line profit for determining the sustainability of an industrial complex and quantified economic, environmental, and societal aspects into a single value measure for optimization. This methodology is useful when reliable costs are available for the three dimensions of sustainability assessment. Singh et al. (2007) also developed a sustainable performance for comparing steel industries. They have used Analytical Hierarchy Process (AHP) to calculate weighting factors of the indicators. The use of the composite sustainable development index (I_{CSD}) has been used to study the sustainability assessment of breweries (Tokos et al., 2012). The method has been extended with the use of geometric aggregation, linear aggregation, and sensitivity analysis by Zhou et al. (2012). Brandi et al. (2014) have used Canberra metrics to aggregate sustainability metrics and have shown the application of their method in different chemical processes. Olinto (2014) has used the vector space method for aggregating sustainability indices and has shown the results for different industrial processes. Thus, aggregate indices are useful for their ability to integrate large amounts of information into easily understood formats. However, one needs to be careful in the construction of aggregates because methodological difficulties often hinder the implementation of these indices.

The use of an aggregate index may trigger the curiosity to know how each indicator has contributed toward the aggregate indicator. This is helpful particularly when a large number of indicators are integrated and meaningful inferences are sought. A sustainability footprint D_e or D (called Euclidean distance and geometric mean distance, respectively) can be constructed by measuring a statistical distance between a multidimensional system and another similar and relevant multidimensional reference system. Given that the value of the sustainability footprint is available, the contribution of the indicators in the entire dataset of system options is analyzed using a multivariate statistical analysis known as PLS-VIP method. Thus, the aggregate index helps to rank the options in the system, and the PLS-VIP method is used to rank the indicators in the system. Together, they complete an analysis and reinforce decisions made regarding relative sustainability of a system.

STEPS IN SUSTAINABILITY ANALYSIS USING THE AGGREGATE INDEX PLS-VIP METHOD

The steps in constructing composite indicators are discussed in the following sections.

Step 1: Ensure quality and unidirectionality of indicator data

The first step in any indicator-based analysis starts with ensuring availability of good-quality data. Step 1 of the aggregate index PLS-VIP (AI-PLS-VIP) method starts after data have been collected for the system options using indicators that credibly characterize the particular system. It is absolutely necessary to have indicator data for all the system options. If some indicator data are missing for one of the options, then the following actions may be used (Freudenberg, 2003):

- **Data deletion.** Omitting entire records (for indicators or system options) when substantial data are missing. However, deleting data entirely for an option takes that option out of consideration.
- **Mean substitution.** This method takes the average of the remaining values of the same indicator, thus substituting a mean value computed from the other options available to fill in missing values. This can be used for temporal data.
- **Regression.** Using regressions based on other indicators to estimate the missing values. This is useful when temporal data for indicators are missing for a certain option.

- **Nearest neighbor.** Identifying and substituting the most similar case for the one with a missing value.
- **Ignore value.** This is essentially assuming a numerical value of zero for the indicator. Care must be taken to confirm that ignoring the value is physically consistent (that the indicator is actually zero). Otherwise, some of the other methods should be applied.

For making sustainability inferences, a system option with a smaller aggregate value of sustainability is considered more sustainable, in accordance with the convention that each indicator is fashioned in a way in which lower numerical values are more desirable than higher ones. To ensure this, an effective method for making all the indicators unidirectional is necessary. This can be accomplished in two ways. The first method is to design an indicator with attributes that make lower values better and higher values worse. The second method is to transform an existing indicator to make it conform to this convention. For instance, if an indicator value is given as a percentage, and if a higher percentage means better performance, then the indicator is changed by subtracting the value from 100. If an indicator value is not given in percentage, it can be transformed into a new indicator where an inverse of the original value can be used for complying with the convention. Thus, indicator data quality and unidirectionality are the prerequisites for assessing sustainability using the AI-PLS-VIP method.

Step 2: Compare relative sustainability of options: the aggregate index method for sustainability footprint

The second step in the AI-PLS-VIP method is the creation of the aggregate index for calculation of the sustainability footprint. Aggregate indices have been studied for determining sustainability of systems. For example, the aggregate index, D, proposed by Sikdar (2009) can successfully compare competing processes or products. However, D had limitations in the method when the indicator values could have a negative or zero numerical value such that the ratio will be negative. To overcome this, the D_e process is proposed (Sikdar et al., 2012). Later, however, this difficulty of D was overcome by synthesizing a reference point. Currently, both sustainability indices conform to the AI-PLS-VIP method.

The aggregate index computation involves two parts. The first part is the normalization of the indicators and the second part is the computation of the aggregate index.

Several techniques can be used to standardize or normalize the indicators. This step is important because the indicators characterizing a system are various, with different scales and units of measurement. Commonly used methods of normalization (Freudenberg, 2003), depending on the system being studied, include the following:

- *Standard deviation from the mean*, which imposes a standard normal distribution (i.e., a mean of 0 and a standard deviation of 1) on the data. Thus, positive (or negative) values for a given system option indicate above (or below) average performance: (actual value − mean value/standard deviation).

- *Distance from the group leader*, which assigns 100 to the leading option and other options are ranked as percentage points away from the leader: (actual value/maximum value) × 100.
- *Distance from the mean*, where the (weighted or unweighted) mean value is given 100 and data receive scores depending on their distance from the mean. Values higher than 100 indicate above-average performance: (actual value/mean value) × 100.
- *Distance from the best and worst performers*, where positioning is in relation to the maximum and minimum in the dataset and the index takes values between 0 and 100: (actual value − minimum value/maximum value − minimum value) × 100.
- *Categorical scale*, where each variable is assigned a score (either numerical, such as between [1...*k*], $k > 1$, or qualitative, i.e., high, medium, low) depending on whether its value is above or below a given threshold.

For the AI-PLS-VIP method, we used the distance from the best and worst performers as synthetically chosen by adopting a point in this space represented by the minimum values of the indicators encountered in the system. These indicators will represent an ideal case of a synthetic reference option X_0. Thus, with respect to X_0, any of the real options will always have a positive difference of the indicator values. This is essentially shifting the point of reference from one of the options to an imaginary option, and the exercise is to find how far a real option is from the reference option. In addition, if the indicator values are defined such that higher values are worse than smaller values, then a smaller difference of an option from the imaginary point would be more sustainable than one with a larger difference.

After normalization with respect to the X_0, the calculation of the aggregate index is computed by using the Euclidean distance method as shown in Eq. (4.1).

$$D_e \sqrt{\sum_{j}^{n} \left[c_j \frac{(y_j - x_{j0})}{(y_j - x_{j0})_{\max}} \right]^2} \tag{4.1}$$

where D_e, the measure of relative sustainability, is the Euclidean distance of a chosen option y_j at a point in time obtained after normalizing with regard to a synthetic reference option X_0. The idea of the arbitrary reference option in D_e calculation is to transform the dataset to avoid the occurrence of negative data points in the transformed data. The weighting factor c_j allows use of weighting preference (usually a societal choice) of any of the indicators in comparison with others. While considering time series data for options, the datasets can be represented by an $m \times n \times t$ matrix, where m is the number of options, n is the number of indicators, and t is the number of temporal points. The D_e value is computed for each system option each year.

Another method to compute the aggregate index is to use the geometric mean from a carefully normalized set of indicators. This normalization is achieved by a similar shifting of the reference point to ensure positive nonzero values

(Sikdar et al., 2012). Equation (4.2) represents the geometric mean of the ratios of the indicator values when any option is compared with only the reference option constituted with the minimum values assumed by the indicators for that dataset. In this case, the ratio of the lengths of the indicators from a fixed minimum is considered for the geometric mean approach.

$$\begin{aligned} D &= \left(\prod_{i}^{n} [c_i(y_i'/x_i')] \right)^{1/n} \\ y_i' &= y_i - (x_{i0} - C_{\text{offset}}) \\ x_i' &= x_i - (x_{i0} - C_{\text{offset}}) \end{aligned} \tag{4.2}$$

where x_{i0} is the minimum value assumed by the indicator i in the dataset and constitutes elements of the vector, $\mathbf{X}_0$. The distance will be zero where $x_i = x_{i0}$; hence, an offset from the minimum value is required for Eq. (4.2) to work. We can offset the point of reference arbitrarily with a constant, C_{offset}, to have all indicator distances greater than zero. Sikdar et al. (2012) discuss the selection criteria and effect of C_{offset} on the value of D.

Both D and D_e can be used to calculate the aggregate index in comparing the relative sustainability of options. We have used D_e in this chapter for demonstrating the applicability toward assessing sustainability of systems.

Step 3: Compare the ranking of indicators: PLS-VIP method

The third step in the AI-PLS-VIP method involves the ranking of the indicators in the order of their importance in their contribution toward the aggregate index (D_e). The importance of the indicator is measured by comparing the variability of the indicator with the aggregated index. PLS-VIP method is a multivariate regression method in which information from a data space of a larger number of variables is projected into that with a smaller number of variables. We start with the same data matrix, X, as used to calculate the aggregated index. The indicators are the variables in the data space. PLS-VIP is a supervised model for which an overall data behavior or pattern is required. This overall pattern can be represented by a response vector or a response matrix. In our modeling, we have used the aggregated index, D_e, as the response vector.

In the PLS-VIP analysis, the number of indicators is reduced in a way such that variations in the set of reduced indicators are most likely to be reflected in the response vector D_e. In other words, the overall data pattern would be unchanged with the reduced number of indicators. The easiest test of this expectation is if one can make the same conclusions regarding the overall data behavior with the original set and with this set with less indicators. The surplus indicators would thus be understood to be redundant or not important in contributing to the overall performance of the system in question. An application of PLS-VIP for sustainability assessment has been given by Mukherjee et al. (2013). In a multivariate method, there are variables that are inferred through a mathematical model of the original (observed) data and are known as latent variables. In contrast to the original variables, the latent variables are not explicit, nor can they be tweaked at will. In this chapter, the variables are the indicators; henceforth, we refer to the original variables as indicators and to the latent variables derived from the original indicators as latent indicators.

PLS regression model is used to deconstruct the original data matrix X into two orthogonal matrices, the loadings (L) and scores (T) of a number of latent indicators, and a residual matrix, E, as shown in Eq. (4.3) (Cinar et al., 2004).

$$X = TL^{\mathrm{T}} + E = \sum_{j=1}^{a} t_j l_j^{\mathrm{T}} + E \tag{4.3}$$

The score matrix T is related to the response vector D_{e} through a regression matrix b, as shown in Eq. (4.4) (Chong and Jun, 2005). F is the residual vector of D_{e}.

$$D_{\mathrm{e}} = Tb^{\mathrm{T}} + F \tag{4.4}$$

The values of the indicator for each option are represented as an option vector x_m. Each option vector x_m can be related to the score vector through weight vectors w_j, as given in Eq. (4.5).

$$t_j = w_j^{\mathrm{T}} x_m \tag{4.5}$$

The VIP for a particular indicator is calculated using the regression coefficient b, weight vector w_j, and score vector t_j, as given in Eq. (4.6).

$$\mathrm{VIP}_k = \sqrt{n \frac{\sum_{j=1}^{a} b_j^2 t_j^{\mathrm{T}} t_j \left(\frac{w_{kj}}{\|w_j\|} \right)^2}{\sum_{j=1}^{a} b_j^2 t_j^{\mathrm{T}} t_j}} \tag{4.6}$$

where w_{kj} is the kth element of the weight vector w_j.

PLS-VIP is used to identify the importance of each indicator in affecting the aggregate index D_{e}. To avoid the relative variability of the indicators to affect the result, normalized indicators are used for VIP calculation. This ensures variability within an indicator but avoids relative variability. Indicators with lower VIP scores have little influence on D_{e}, and those with the higher VIP scores contribute the most to D_{e}. The average of squared VIP score equals 1. VIP score greater than 1 is generally used as a criterion for detecting the relative importance of an indicator.

There are different algorithms available to solve PLS regression problems. In the present work, the regression for the PLS is based on the PLS1 algorithm (Martens, 1991). In the PLS1 algorithm, the correlation coefficient of X and D_{e} is used to obtain the first extracted score t_1 for the first latent indicator. After obtaining the first latent indicator, the regression follows by obtaining the second latent indicators from the residuals, and so on. This is most appropriate for our problem in which response matrix comprises one column constituting the aggregated indices, D_e. Regression coefficient and weight vectors from the first three latent indicator score vector are used for calculating VIP.

The code for calculation of D_e and PLS-VIP scores is written in MATLAB® and available on request from the authors.

CASE STUDIES FROM ENGINEERING SYSTEMS

The case studies selected in this section demonstrate the use of the AI-PLS-VIP method to calculate the sustainability footprint, followed by the ranking of indicators in their order of importance. These cases include detailed steps from data selection and making a decision with regard to sustainability, as outlined in the three-step method in the previous section.

CASE STUDY 1: COMPARISON OF BIOFUEL SYSTEMS

Pertaining to the renewable fuels standards (RFS) (Schnepf and Yacobucci, 2010), biofuels have gained much importance in the United States in recent years. In the rest of the world, Brazil has shown considerable progress toward implementation of bioethanol derived from sugarcane as a transportation fuel. Germany has taken approaches to establish biodiesel as a viable biofuel. Despite many advantages of biofuels, their widespread use has some limitations and raises some sustainability issues. First, the discussion of the use of food-grade feedstock for the production of fuels raises concern among various social groups around the world. These social issues may be apparently nonscientific in nature, but they form an important part in the policy and decision-making process for assessing the viability of biofuels in a particular region of concern. Second, the availability of agricultural land area for the cultivation of biofuels raises concern among agricultural and environmental scientists. The current arable land area is not sufficient to produce the biofuel volumes required to meet the transportation energy needs of today's world. One option is to convert nonagricultural land such as forests fallow lands into areas where crop cultivation can be made possible. This obviously raises concerns of deforestation, soil carbon dioxide release, and loss of biodiversity. Third, increasing the production of crops in a given land area results in rapid depletion of soil nutrition, increasing water consumption, and increased use of artificial chemical fertilizers, resulting in adverse environmental concerns. Finally, these technologies of converting crops to fuels are expensive and depend heavily on the availability of government subsidies for the viability of the biofuels industry. Thus, a thorough sustainability analysis incorporating the economic, environmental, and social aspects is of utmost importance when making any decision related to policies for biofuels.

In this case study, we analyze a two-part case of the sustainability of biodiesel feedstocks, followed by a comparison of biofuels in general. In both cases, five indicators, life cycle energy efficiency (LCEE), fossil energy ratio (FER), contribution to GW or carbon footprint (CF), land use intensity (LUI), and carbon stock change emissions (CSCE), are selected to represent the sustainability of the biodiesel and biofuel systems. From a qualitative point of view, the larger the values of LCEE and FER, the more sustainable the biofuel or feedstock. GW, CSCE, and LUI have the opposite meaning (i.e., the smaller the better).

1a: Comparison of various feedstocks for the sustainable production of biodiesel

Mata et al. (2011) considered the sustainability of several biodiesel options. These included the feedstocks soybean, rapeseed, sunflower, tallow, palm, jatropha, and microalgae. The numerals 1, 2, 3, and 4 denote various scenarios that Mata et al. (2011) considered in their work. Table 4.5 gives the raw data of the indicators obtained from the study by Mata et al. (2011). For this case, we have an $m \times n$ matrix representing the biodiesel system where the number of options is $m = 12$ and the number of indicators is $n = 5$.

Table 4.6 gives the various steps in the calculation of D_e for biodiesel options. The first step converts the LCEE and FER indicators by inverting them. This is performed to ensure unidirectionality of the indicators. Here, we see that data for

Table 4.5 Indicator Data on Biodiesel as Presented by Mata et al. (2011)

Indicator/Diesel Type	LCEE	FER	CF (kg CO_2-eq/MJ fuel)	LUI (m^2 y/MJ fuel)	CSCE (kg CO_2-eq/MJ fuel)
Fossil diesel (FD)	6.25	6.25	0.1		
Tallow biodiesel (TB)	1.66	1.6	0.13		
Palm biodiesel[123] (PB[123])	1.28	1.28	0.04	0.05	0.08
Sunflower biodiesel (SUB)	1.04	1.04	0.05	0.28	0.7
Rapeseed methyl ester[14] (RME[14])	1.5	1.5	0.12	0.31	0.78
Rapeseed methyl ester[2] (RME[2])	1.89	1.4	0.08	0.31	0.78
Rapeseed methyl ester[3] (RME[3])	2.9	1.15	0.04	0.31	0.78
Rapeseed ethyl ester[1] (REE[1])	0.81	0.81	0.07	0.31	0.78
Rapeseed ethyl ester[2] (REE[2])	2.97	1.32	0.07	0.31	0.78
Soybean biodiesel[12] (SB[12])	0.41	0.41	0.13	0.46	1.66
Jatropha biodiesel (JB)	21.31	1.74		0.41	1.04
Microalgae biodiesel[1] (MB[1])	1.84	0.56	0.014	0.01	0.01

[1,2,3,4]*Scenarios as presented by Mata et al. (2011).*
LCEE, life cycle energy efficiency, FER, fossil energy ratio; CF, carbon footprint; LUI, land use intensity; CSCE, carbon stock change emissions.

Table 4.6 Calculation of D_e for Comparing Options in the Biodiesel System

Indicator/Diesel Type	1/LCEE	1/FER	LUI	CF	CSCE	
		Y_i				
FD	0.16	0.16		0.10		
TB	0.60	0.63		0.13		
PB[123]	0.78	0.78	0.05	0.04	0.08	
SUB	0.96	0.96	0.28	0.05	0.70	
RME[14]	0.67	0.67	0.31	0.12	0.78	
RME[2]	0.53	0.71	0.31	0.08	0.78	
RME[3]	0.34	0.87	0.31	0.04	0.78	
REE[1]	1.23	1.23	0.31	0.07	0.78	
REE[2]	0.34	0.76	0.31	0.07	0.78	
SB[12]	2.44	2.44	0.46	0.13	1.66	
JB	0.05	0.57	0.41		1.04	
MB[1]	0.54	1.79	0.01	0.01	0.01	
X_0	0.05	0.16	0.01	0.01	0.01	
		$Y_i - X_0$				
FD − X_0	0.11	0.00	0.27	0.09	0.73	
TB − X_0	0.56	0.47	0.27	0.12	0.73	
PB[123] − X_0	0.73	0.62	0.04	0.03	0.07	
SUB − X_0	0.91	0.80	0.27	0.04	0.69	
RME[14] − X_0	0.62	0.51	0.30	0.11	0.77	
RME[2] − X_0	0.48	0.55	0.30	0.07	0.77	
RME[3] − X_0	0.30	0.71	0.30	0.03	0.77	
REE[1] − X_0	1.19	1.07	0.30	0.06	0.77	
REE[2] − X_0	0.29	0.60	0.30	0.06	0.77	
SB[12] − X_0	2.39	2.28	0.45	0.12	1.65	
JB − X_0	0.00	0.41	0.40	0.04	1.03	
MB1 − X_0	0.50	1.63	0.00	0.00	0.00	
$(Y_i - X_0)_{max}$	2.39	2.28	0.45	0.12	1.65	
Normalized indicator for biodiesel options $(Y_i - X_0)/(Y_i - X_0)_{max}$						D_e
FD	0.05	0.00	0.59	0.74	0.44	1.05
TB	0.23	0.20	0.59	1.00	0.44	1.28
PB[123]	0.31	0.27	0.09	0.22	0.04	0.48
SUB	0.38	0.35	0.60	0.31	0.42	0.95
RME[14]	0.26	0.22	0.67	0.91	0.47	1.27
RME[2]	0.20	0.24	0.67	0.57	0.47	1.04
RME[3]	0.12	0.31	0.67	0.22	0.47	0.91
REE[1]	0.50	0.47	0.67	0.48	0.47	1.17
REE[2]	0.12	0.26	0.67	0.48	0.47	0.99
SB[12]	1.00	1.00	1.00	1.00	1.00	2.24
JB	0.00	0.18	0.89	0.35	0.62	1.16
MB[1]	0.21	0.71	0.00	0.00	0.00	0.74

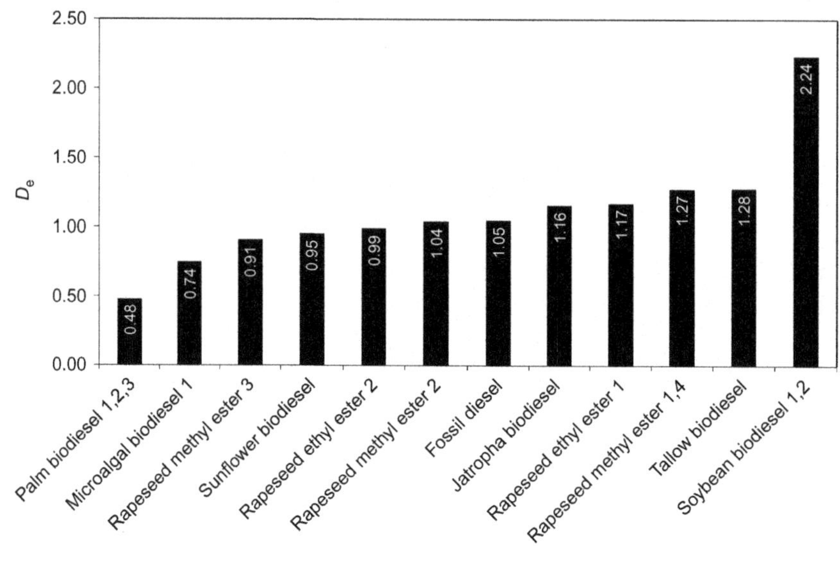

FIGURE 4.5

D_e for biodiesel options.

options FD (fossil diesel) and TB (tallow biodiesel) are missing in the LUI and CSCE indicator categories. For JB (jatropha biodiesel), the indicator value for CF is missing. In the second step in the calculation, we subtract the minimum value in each indicator category to obtain the distance from the minimum. The missing indicator data for an option are substituted with an average value computed from the indicator data for all of the other options. This is possible because the indicator data are fairly well populated for all of the options in this case study for the biodiesel system. If this was not the case (i.e., if sufficient data were not available for the indicators), then those indicators could not be considered in the calculation of D_e. Mata et al. (2011) applied the D method for aggregating indicators for a given option and came to the same conclusion that was attained with D_e.

In the third step, the normalized values of the indicators are obtained, from which D_e is computed using Eq. (4.1). Figure 4.5 shows the results for the D_e for the several biodiesel options. From this, we see that palm biodiesel is best in overall sustainability using the five indicators. The soybean biodiesel is the worst among all the biodiesel options.

Figure 4.6 shows the VIP scores of the five indicators. The CSCE indicator is the highest contributor among the five indicators chosen for analyzing the system and CF is the least contributor toward D_e. In general, a VIP score more than 1 is considered "important." Thus, from Figure 4.6, the CSCE and 1/LCEE are considered the most important indicators in the analysis of overall sustainability for the biodiesel options. However, sometimes this requirement can be reduced

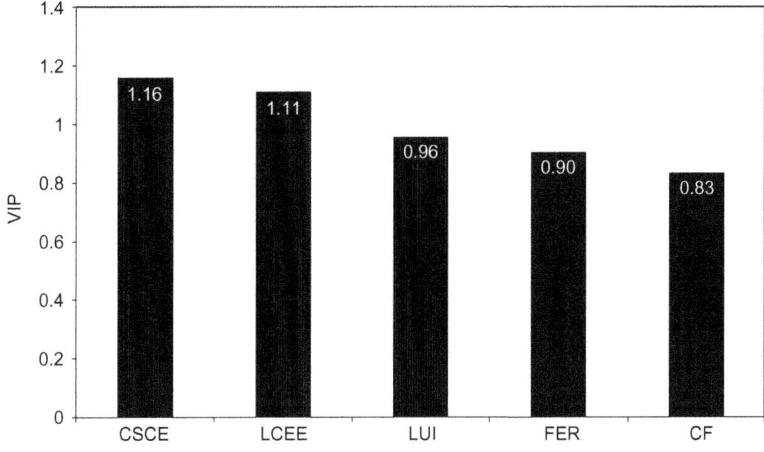

FIGURE 4.6

VIP scores for biodiesel indicators.

somewhat (for instance to 0.8), which would make all the indicators important in considering relative sustainability of the biodiesel options.

1b: Comparison of sustainability of biofuel options

The second part of this case study analyzes the sustainability of biofuel options in general with respect to gasoline and fossil diesel (Mata et al., 2013). The same five indicators used to study Case 1(a) for biodiesel were also used in this analysis. The data for the indicators were procured by Mata et al. (2013) from information available in the open literature and the availability of data to calculate them. These are given in Table 4.7. For this case, we have the $m \times n$ matrix representing the biofuel system where the number of options is $m = 11$ and the number of indicators is $n = 5$.

We calculated D_e for the biofuels system in Table 4.8 as we did for the biodiesel system in Case 1(a). For applying the AI-PLS-VIP method, we first need to reverse the direction of the LCEE and FER indicators to make higher values worse and lower values better for those indicators and follow the unidirectionality convention of D_e. However, in the second step for D_e calculation, we substitute the missing numerical data for gasoline, fossil diesel, and tallow biodiesel with zeroes. The assumption in this case was that there is minimal LUI and CSCE for these biofuels. Thus, following the method for D_e calculation, we arrive at the values obtained in Table 4.8.

The results for D_e are shown in Figure 4.7. The results have been organized in ascending order of D_e, from which we can infer that sugarcane bioethanol is the

Table 4.7 Indicator Data for Biofuels as Presented by Mata et al. (2013)

Indicator/Fuel Type	LCEE	FER	GW (kg CO_2-eq/MJ fuel)	LUI (m^2 y/MJ fuel)	CSCE (kg CO_2-eq/MJ fuel)
Gasoline (GS)	5.18	5.18	0.08	–	–
Fossil diesel (FD)	6.25	6.25	0.1	–	–
Sugarcane bioethanol (SBE)	7.63	7.63	0.06	0.03	0.07
Corn bioethanol (CBE)	1.53	1.84	0.06	0.11	0.29
Tallow biodiesel (TB)	1.66	1.66	0.13	–	–
Palm biodiesel (PB)	1.28	1.28	0.04	0.05	0.08
Sunflower biodiesel (SUB)	1.04	1.04	0.05	0.28	0.7
Rapeseed methyl ester (RME)	2.9	1.15	0.04	0.31	0.78
Rapeseed ethyl ester (REE)	2.97	1.32	0.07	0.31	0.78
Soybean biodiesel (SB)	0.41	0.41	0.13	0.46	1.66
Microalgae biodiesel (MB)	1.84	0.56	0.14	0.01	0.01

LCEE, life cycle energy efficiency; FER, fossil energy ratio; GW, global warming; LUI, land use intensity; CSCE, carbon stock change emissions.

Table 4.8 Calculation of D_e for Comparing Options in the Biofuel System

Indicator/Diesel Type	1/LCEE	1/FER	LUI	GW	CSCE
		Y_i			
GS	0.19	0.19	–	0.08	–
FD	0.16	0.16	–	0.1	–
SBE	0.13	0.13	0.03	0.06	0.07
CBE	0.65	0.54	0.11	0.06	0.29
TB	0.60	0.60	–	0.13	–
PB	0.78	0.78	0.05	0.04	0.08
SUB	0.96	0.96	0.28	0.05	0.7
RME	0.34	0.87	0.31	0.04	0.78
REE	0.34	0.76	0.31	0.07	0.78
SB	2.44	2.44	0.46	0.13	1.66
MB	0.54	1.79	0.01	0.14	0.01
X_0	0.13	0.13	0.01	0.04	0.01
		$Y_i - X_0$			
GS $- X_0$	0.06	0.06	0.00	0.04	0.00
FD $- X_0$	0.03	0.03	0.00	0.06	0.00
SBE $- X_0$	0.00	0.00	0.02	0.02	0.06
CBE $- X_0$	0.52	0.41	0.10	0.02	0.28
TB $- X_0$	0.47	0.47	0.00	0.09	0.00
PB $- X_0$	0.65	0.65	0.04	0.00	0.07
SUB $- X_0$	0.83	0.83	0.27	0.01	0.69

Table 4.8 Calculation of D_e for Comparing Options in the Biofuel System *Continued*

Indicator/Diesel Type	1/LCEE	1/FER	LUI	GW	CSCE	
RME $-X_0$	0.21	0.74	0.30	0.00	0.77	
REE $-X_0$	0.21	0.63	0.30	0.03	0.77	
SB $-X_0$	2.31	2.31	0.45	0.09	1.65	
MB $-X_0$	0.41	1.65	0.00	0.10	0.00	
$(Y_i-X_0)_{max}$	2.31	2.31	0.45	0.10	1.65	
Normalized indicator for biofuel options $(Y_i-X_0)/(Y_i-X_0)_{max}$						D_e
GS	0.03	0.03	0.00	0.40	0.00	0.40
FD	0.01	0.01	0.00	0.60	0.00	0.60
SBE	0.00	0.00	0.04	0.20	0.04	0.21
CBE	0.23	0.18	0.22	0.20	0.17	0.45
TB	0.20	0.20	0.00	0.90	0.00	0.95
PB	0.28	0.28	0.09	0.00	0.04	0.41
SUB	0.36	0.36	0.60	0.10	0.42	0.90
RME	0.09	0.32	0.67	0.00	0.47	0.88
REE	0.09	0.27	0.67	0.30	0.47	0.91
SB	1.00	1.00	1.00	0.90	1.00	2.19
MB	0.18	0.72	0.00	1.00	0.00	1.24

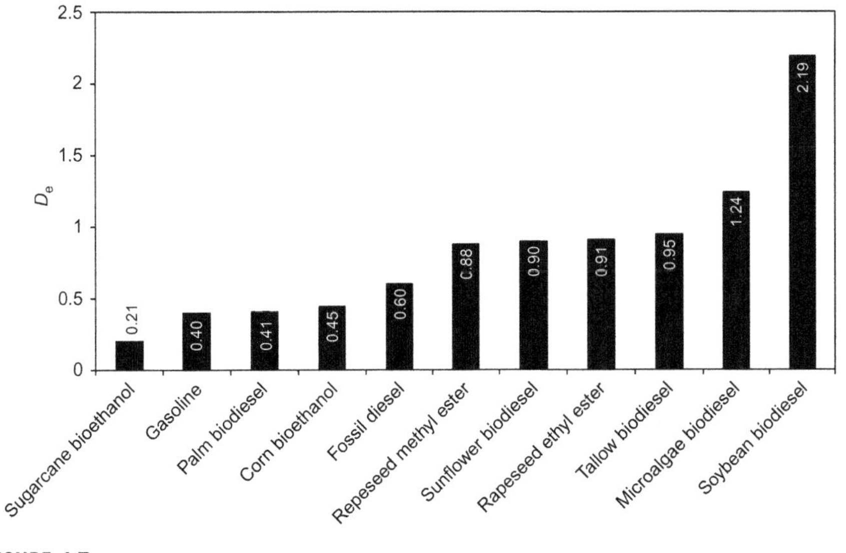

FIGURE 4.7

D_e for biofuel options.

best in sustainability performance and soybean biodiesel is significantly worse in its performance.

Figure 4.8 shows the VIP scores for the indicators: 1/FER has contributed most toward D_e, which suggests that it has been most influential in the sustainability index. This also shows that there had been maximum variability in the data for 1/FER compared to GW, which had the minimal variability among the options for the biofuels in this study.

Results from Mata et al. (2011) also agree with this inference that ethanol from sugarcane is the most sustainable option, even when compared with dominant fossil fuels, whereas no feedstock for biodiesel production emerges as a better option.

Two parts of the case in which two similar systems are compared with the use of indicators are described. For the biodiesel system, there were 12 options. For the biofuels system, there were 11 options. Comparing the values for D_e in the two systems for similar options shows that the numerical values change in the two systems, implying that the absolute numerical value of D_e should not be used to compare the same options in two different systems. The choice of the data substitution using an aggregate value in the biodiesel system and data deletion in the biofuel system obviously affected the relative position of the fossil diesel in the two systems. With respect to the VIP scores for the indicators, we see that the order of importance of the indicators is different despite the two systems being similar to each other. The numerical values of the indicator data play a role again in this inference.

It needs to be pointed out that although our analysis applies in general, inferences in each case depend critically on whether all applicable indicators were chosen. For instance, an important economic indicator is cost. This indicator was

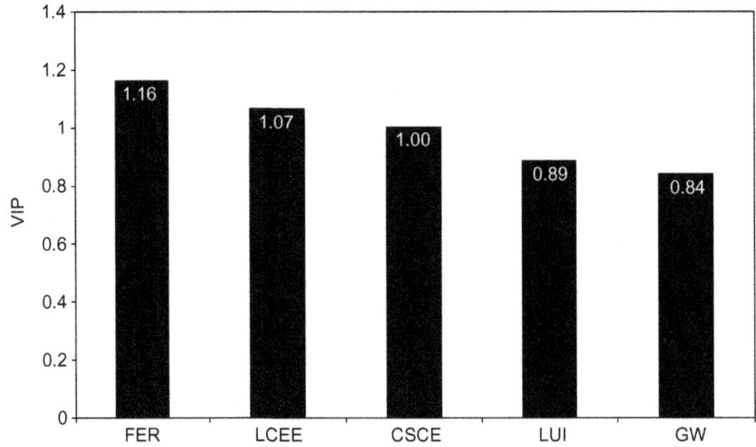

FIGURE 4.8

VIP scores for biofuel indicators.

not available for the two cases of biofuels discussed. Perhaps because of pricing complexity arising from varying subsidy regimes around the world, Mata et al. (2013) did not have access to such data. Had the cost data been available, the inferences certainly would have been different. Therefore, one would be advised to be careful about initial indicator selection to reach a decision point in sustainability analysis.

CASE STUDY 2: ENVIRONMENTAL IMPACT COMPARISON OF INDOOR AND OUTDOOR GROWTH OF ALGAL SPECIES FOR BIODIESEL PRODUCTION

Itoiz et al. (2012) studied the environmental impact of marine microalgal biomass species for the production of biodiesel. For this case, there are three marine algal species and two options for cultivating them, either indoors or outdoors. Itoiz et al. (2012) also calculated the environmental indicators abiotic depletion (AD), acidification (AC), eutrophication (E), global warming potential (GWP), ozone layer depletion potential (ODP), human toxicity (HT), freshwater aquatic ecotoxicity (FWAE), marine aquatic ecotoxicity (MAE), terrestrial ecotoxicity (TE), and photochemical oxidation (PO) for comparing the six options. For such a system, the overall environmental impact can be calculated using the AI-PLS-VIP method, and the environmental indicators can be ranked according to their influence on the overall environmental impact. Thus, we are studying the $m \times n$ matrix of algal cultivation system where $m = 6$ and the number of indicators is $n = 10$. The data reported in this chapter are given in Table 4.9.

Table 4.10 gives the calculation of aggregate index, D_e. In this case, indicator data for all the algal strains were available. This was possible because Itoiz et al. (2012) calculated the indicator data from their experimental results. Thus, indicators directly calculated from process data are better sources of information for calculating D_e.

Figure 4.9 shows the overall environmental impact of the three algal strains and two growth options. It is clear from the values of D_e that growth outdoors is preferable to growth indoors, which is in agreement with the conclusions of Itoiz et al. (2012). However, Itoiz et al. (2012) made the following inference from the data given in Table 4.9: "Specifically, *Alexandrium minutum* outdoor production had the lowest environmental impact in all categories. By contrast, *Alexandrium minutum* indoor production had the highest impact for all categories." While making this inference, the authors overlooked that the value for ODP in the *Alexandrium minutum* (outdoors) is higher than that of the other two strains. When we look at the comparative D_e values, we see that the strain *Alexandrium minutum* (outdoors) is the third best case, solely because of the higher value of ODP. Thus, the inference of the authors is contradicted, and a better result for relative overall impact of the options for algal species and cultivation method is made possible by the use of the aggregate index method.

Table 4.9 Environmental Indicator Data on Algal Strains and Cultivation Method as Presented by Itoiz et al. (2012)

Indicator/Algal Strain	AD	AC	E	GWP	ODP	HT	FWAE	MAE	TE	PO
Heterosigma akashiwo (indoors) (HA_i)	1.06	1.36	0.07	144.00	0.0	42.90	9.57	24,200.0	2.41	0.05
Heterosigma akashiwo (outdoors) (HA_o)	0.18	0.20	0.01	23.80	0.0	5.82	1.35	3,190.0	0.31	0.01
Alexandrium minutum (indoors) (AM_i)	1.12	1.44	0.07	153.00	0.0	45.60	10.20	25,700.0	2.56	0.05
Alexandrium minutum (outdoors) (AM_o)	0.17	0.19	0.01	22.90	0.0	5.64	1.30	3110.0	0.30	0.01
Karlodinium veneficum (indoors) (KV_i)	1.10	1.42	0.07	151.00	0.0	44.70	9.97	25,200.0	2.51	0.05
Karlodinium veneficum (outdoors) (KV_o)	0.17	0.20	0.01	23.50	0.0	5.77	1.33	3,160.0	0.31	0.01

AD, abiotic depletion; AC, acidification; E, eutrophication; GWP, global warming potential; ODP, ozone layer depletion; HT, human toxicity; FWAE, freshwater aquatic ecotoxicity; MAE, marine aquatic ecotoxicity; TE, terrestrial ecotoxicity; PO, photochemical oxidation.

Table 4.10 Calculation of D_e for Comparing Strains and Growth Options in the Algal System

Algae Strain	AD	AC	E	GWP	ODP	HT	FWAE	MAE	TE	PO	
					Y_i						
HA_i	1.06	1.36	0.07	144.00	0.0	42.90	9.57	24,200	2.41	0.05	
HA_o	0.18	0.20	0.01	23.80	0.0	5.82	1.35	3,190	0.31	0.01	
AM_i	1.12	1.44	0.07	153.00	0.0	45.60	10.20	25,700	2.56	0.05	
AM_o	0.17	0.19	0.01	22.90	0.0	5.64	1.30	3,110	0.30	0.01	
KV_i	1.10	1.42	0.07	151.00	0.0	44.70	9.97	25,200	2.51	0.05	
KV_o	0.17	0.20	0.01	23.50	0.0	5.77	1.33	3,160	0.31	0.01	
X_0	0.17	0.19	0.01	22.90	0.0	5.64	1.30	3,110	0.30	0.01	
					$Y_i - X_0$						
$HA_i - X_0$	0.89	1.17	0.06	121.10	0.0	37.26	8.27	21,090	2.11	0.04	
$HA_o - X_0$	0.01	0.01	0.00	0.90	0.0	0.18	0.05	80	0.01	0.00	
$AM_i - X_0$	0.95	1.25	0.06	130.10	0.0	39.96	8.90	22,590	2.26	0.05	
$AM_o - X_0$	0.00	0.00	0.00	0.00	0.0	0.00	0.00	0	0.00	0.00	
$KV_i - X_0$	0.93	1.23	0.06	128.10	0.0	39.06	8.67	22,090	2.21	0.05	
$KV_o - X_0$	0.00	0.01	0.00	0.60	0.0	0.13	0.03	50	0.00	0.00	
$(Y_i - X_0)_{max}$	0.95	1.25	0.06	130.10	0.0	39.96	8.90	22,590	2.26	0.05	
	Normalized indicator for algal cultivation options $(Y_i - X_0)/(Y_i - X_0)_{max}$										D_e
HA_i	0.94	0.94	0.93	0.93	0.86	0.93	0.93	0.93	0.93	0.93	2.93
HA_o	0.01	0.01	0.01	0.01	0.00	0.00	0.01	0.00	0.00	0.01	0.02
AM_i	1.00	1.00	1.00	1.00	1.00	1.00	1.00	1.00	1.00	1.00	3.16
AM_o	0.00	0.00	0.00	0.00	0.09	0.00	0.00	0.00	0.00	0.00	0.09
KV_i	0.98	0.98	0.98	0.98	0.91	0.98	0.97	0.98	0.98	0.98	3.08
KV_o	0.00	0.00	0.01	0.00	0.00	0.00	0.00	0.00	0.00	0.00	0.01

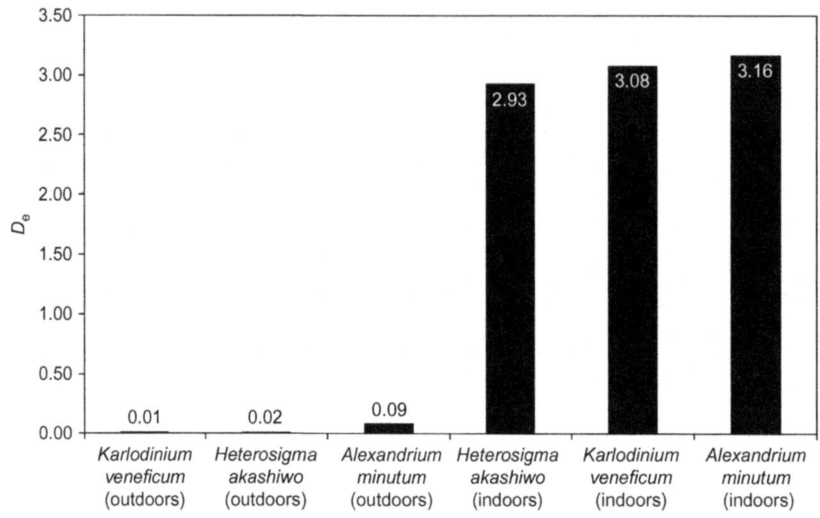

FIGURE 4.9

D_e for algal strain and cultivation options.

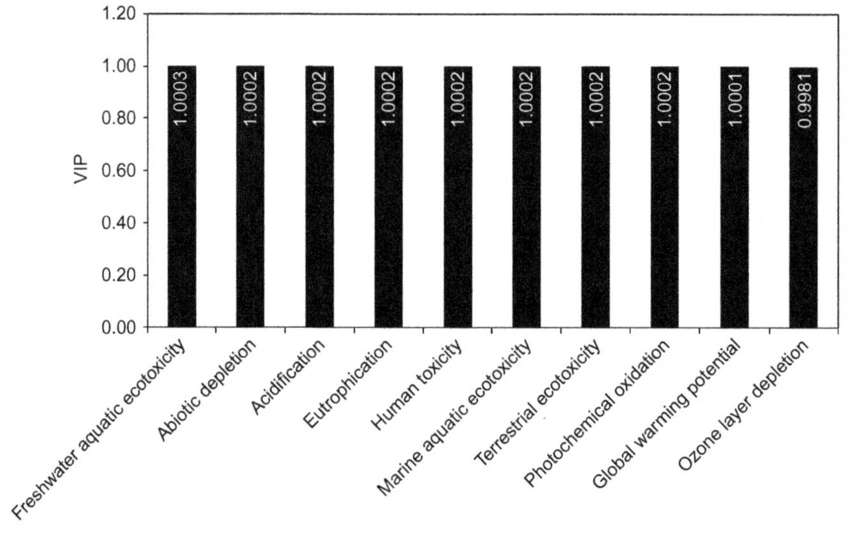

FIGURE 4.10

VIP scores for algal indicators.

The relative importance of the indicators using the PLS-VIP method as shown in Figure 4.10 is even more fascinating. Here, we see that all the indicators are almost equally important in the calculation of the overall environmental impact. The values of the indicator VIP scores vary by an insignificant amount, which proves that none of the indicators should be discarded for any future analysis of the algal system. The only exception to this is the ODP indicator, which had a value less than 1, although not significantly. Excluding this indicator would bring the same inference as that of Itoiz et al. (2012), affecting a change in the ranking of the options. However, this would lead to an incorrect inference when the aim is to find the species with the minimum overall environmental impact for cultivation.

From this case study, we see how the AI-PLS-VIP method helps to arrive at a correct decision when simple observation of various indicator data in isolation can lead to wrong inferences being made.

CASE STUDY 3: COMPARISON OF ENVIRONMENTAL IMPACTS OF POLYMERS

Tabone et al. (2010) evaluated the efficacy of green design principles with respect to environmental impacts, as used in life cycle assessment (LCA) methodology. They presented a case study of 12 polymers, seven derived from petroleum, four derived from biological sources, and one derived from both. They evaluated the environmental impacts of each polymer's production using standard databases for LCA methodology. The authors also used the metrics of atom economy, mass from renewable sources, biodegradability, percent recycled, distance of furthest feedstock, price, life cycle health hazards, and life cycle energy use for assessing adherence to green design principles. From this point, they used a decision matrix to generate single value metrics for each polymer evaluating either adherence to green design principles or life cycle environmental impacts. Table 4.11 gives the values of the indicators for the polymer options.

Results from the authors show a qualified positive correlation between adherence to green design principles and a reduction in the environmental impacts of production. The qualification results from a disparity between biopolymers and petroleum polymers. While biopolymers rank highly in terms of green design, they exhibit relatively large environmental impacts from production. Biopolymers rank 1, 2, 3, and 4 based on green design metrics; however, they rank in the middle of the LCA rankings. Polyolefins rank 1, 2, and 3 in the LCA rankings, whereas complex polymers, such as PET, PVC, and PC, are place at the bottom of both ranking systems. These ranks are given in Table 4.12.

We applied the AI-PLS-VIP method to find the rank of the various polymer options. The D_e gives the overall rank of the various polymer options, thus allowing a combined rank instead of separately ranking according to green design or LCA. For this, we start by analyzing the data given in Table 4.11. The data contain an $m \times n$ matrix of polymers system where $m = 12$ and the number of indicators is $n = 11$. However, from Table 4.11, we see that two of the indicators are qualitative in nature, which reduces the number of indicators for D_e computation to $n = 9$. This limitation could be overcome by finding numerical values of the

Table 4.11 Indicator Data for Polymer Materials as Presented by Tabone et al. (2010)

Indicator/Polymer Material	OAE	C	NC	RE	E	CED	RM	DOF	R	B	P
Polyethylene terephthalate (PET)	80	1.1E − 02	62.9	4.90E − 03	5.72	123.8	0	Intern.	18	N/A	4.13
Bio-polyethylene terephthalate (B-PET)	62	1.3E − 02	72.7	5.70E − 03	6.98	146.2	15	Intern.	18	N/A	4.13
Polyvinyl chloride (PVC)	55	1.1E − 02	31.7	7.30E − 03	0.4	82.9	0	Intern.	0	N/A	4.02
Polylactic acid (Nature Works) (PLA-NW)	80	6.1E − 02	22.5	1.20E − 03	1.21	79.4	100	Region.	0	Indus.	4.66
Polylactic acid (General) (PLA-G)	80	8.4E − 03	37.5	3.10E − 03	4.31	98.3	100	Region.	0	Indus.	4.66
Polyhydroxy alkanoate (general) (PHA-G)	48	7.2E − 03	30	3.10E − 03	2.76	91.5	100	Region.	0	Backyard	6.2
Polyhydroxy alkanoate (utilizing stover) (PHA-S)	48	1.1E − 02	30	2.10E − 03	2.76	91.5	100	Region.	0	Backyard	6.2
High-density polyethylene (HDPE)	100	6.5E − 04	18.7	1.30E − 03	0.65	73.4	0	Intern.	10	N/A	1.52
Low-density polyethylene (LDPE)	100	6.9E − 04	19.6	1.50E − 03	0.82	72.3	0	Intern.	5	N/A	1.58
General purpose polystyrene (GPPS)	98	3.2E − 03	92.7	2.50E − 03	1.79	92.2	0	Intern.	1	N/A	2.35
Polycarbonate (PC)	59	3.0E − 03	85.6	9.50E − 03	3.13	128.9	0	Intern.	0	N/A	5.25
Polypropylene (PP)	100	5.8E − 04	16.8	1.20E − 03	0.54	67.6	0	Intern.	0	N/A	1.78

OAE, overall atom economy %; C, carcinogens (kg benz. eq/L); NC, noncarcinogens (kg tolu. eq/L); RE, respiratory effects (kg PM2.5 eq/L); E, ecotoxicity (kg benz. eq/L); CED, cumulative energy demand (MJ eq/L); RM, % renewable material; DOF, distance of feedstocks; R, % recovery; B, biodegradable; P, price (USD/L).

Table 4.12 Green Design and LCA Ranks as Given by Tabone et al. (2010)

Polymer Material	Green Design Rank	LCA Rank
Polylactic acid (nature works) (PLA-NW)	1	6
Polyhydroxy alkanoate (utilizing stover) (PHA-S)	2	4
Polyhydroxy alkanoate (general) (PHA-G)	2	8
Polylactic acid (general) (PLA-G)	4	9
High-density polyethylene (HDPE)	5	2
Polyethylene terephthalate (PET)	6	10
Low-density polyethylene (LDPE)	7	3
Bio-polyethylene terephthalate (B-PET)	8	12
Polypropylene (PP)	9	1
General purpose polystyrene (GPPS)	10	5
Polyvinyl chloride (PVC)	11	7
Polycarbonate (PC)	12	11

distance (e.g., km of transit in whole life cycle) and the biodegradability of polymers (e.g., hours or years required to decompose). The indicators overall atom economy (OAE), renewable material, and recovery, all expressed as percentages, do not follow the unidirectionality requirement for D_e calculation. Because all three indicators are percentages, constructing a new indicator by subtracting the original indicator value from 100 would reverse the direction of the data. Thus, for example, 100-OAE represents cases in which the lower values (such as zero in the case of high-density polyethylene (HDPE), low-density polyethylene (LDPE), or polypropylene (PP)) are perceived as good and higher values (such as 52% for polyhydroxy alkanoate (general) (PHA-G) and polyhydroxy alkanoate (utilizing stover) (PHA-S)) are considered bad for sustainability. Table 4.13 gives the steps for calculation of D_e.

Figure 4.11 shows the relative D_e values for the 12 polymer options. The HDPE ranks best among the polymers options, whereas polycarbonate has the worst sustainability performance. The biopolymers polylactic acid (general) (PLA-G), Polylactic acid (Nature Works) (PLA-NW), PHA-S, and PHA-G are ranked 4, 5, 8, and 9, respectively.

Figure 4.12 shows the comparison in the ranking of the options calculated using the aggregate index method to those reported by Tabone et al. (2010) using the green design and LCA ranks. The D_e ranks of most of the options fall between the green design rank and the LCA rank. The exceptions to this observation are HDPE, LDPE, PHA-S, and PHA-G. The D_e ranks of HDPE and LDPE are lower than both the green design and LCA ranks, whereas those of PHA-S and PHA-G are significantly higher. One possible reason for this discrepancy may be attributable to the exclusion of the indicators for distance of feedstocks and biodegradability. If the numerical values of these indicators could be included in the ranking of the options using the aggregate index method, then we will

Table 4.13 Calculation of D_e for Comparing Options in the Polymer System

Indicator/Polymer Material	100-OAE	C	NC	RE	E	CED	100-RM	100-R	P	
				Y_i						
PET	20.00	0.01	62.90	0.00	5.72	123.80	100	82	4.13	
B-PET	38.00	0.01	72.70	0.01	6.98	146.20	85	82	4.13	
PVC	45.00	0.01	31.70	0.01	0.40	82.90	100	100	4.02	
PLA-NW	20.00	0.06	22.50	0.00	1.21	79.40	0	100	4.66	
PLA-G	20.00	0.01	37.50	0.00	4.31	98.30	0	100	4.66	
PHA-G	52.00	0.01	30.00	0.00	2.76	91.50	0	100	6.20	
PHA-S	52.00	0.01	30.00	0.00	2.76	91.50	0	100	6.20	
HDPE	0.00	0.00	18.70	0.00	0.65	73.40	100	90	1.52	
LDPE	0.00	0.00	19.60	0.00	0.82	72.30	100	95	1.58	
GPPS	2.00	0.00	92.70	0.00	1.79	92.20	100	99	2.35	
PC	41.00	0.00	85.60	0.01	3.13	128.90	100	100	5.25	
PP	0.00	0.00	16.80	0.00	0.54	67.60	100	100	1.78	
X_0	0.00	0.00	16.80	0.00	0.40	67.60	0	82	1.52	
				$Y_i - X_0$						
PET − X_0	20.00	0.01	46.10	0.00	5.32	56.20	100	0	2.61	
B-PET − X_0	38.00	0.01	55.90	0.00	6.58	78.60	85	0	2.61	
PVC − X_0	45.00	0.01	14.90	0.01	0.00	15.30	100	18	2.50	
PLA-NW − X_0	20.00	0.06	5.70	0.00	0.81	11.80	0	18	3.14	
PLA-G − X_0	20.00	0.01	20.70	0.00	3.91	30.70	0	18	3.14	
PHA-G − X_0	52.00	0.01	13.20	0.00	2.36	23.90	0	18	4.68	
PHA-S − X_0	52.00	0.01	13.20	0.00	2.36	23.90	0	18	4.68	
HDPE − X_0	0.00	0.00	1.90	0.00	0.25	5.80	100	8	0.00	
LDPE − X_0	0.00	0.00	2.80	0.00	0.42	4.70	100	13	0.06	

GPPS – X_0	2.00	0.00	75.90	0.00	1.39	24.60	100	17	0.83	
PC – X_0	41.00	0.00	68.80	0.01	2.73	61.30	100	18	3.73	
PP – X_0	0.00	0.00	0.00	0.00	0.14	0.00	100	18	0.26	
$(Y_i - X_0)_{max}$	52.00	0.06	75.90	0.01	6.58	78.60	100	18	4.68	
	Normalized indicator for polymer options $(Y_i - X_0)/(Y_i - X_0)_{max}$									D_e
PET	0.38	0.17	0.61	0.45	0.81	0.72	1.00	0.00	0.56	1.79
B-PET	0.73	0.21	0.74	0.54	1.00	1.00	0.85	0.00	0.56	2.11
PVC	0.87	0.17	0.20	0.73	0.00	0.19	1.00	1.00	0.53	1.92
PLA-NW	0.38	1.00	0.08	0.00	0.12	0.15	0.00	1.00	0.67	1.63
PLA-G	0.38	0.13	0.27	0.23	0.59	0.39	0.00	1.00	0.67	1.50
PHA-G	1.00	0.11	0.17	0.23	0.36	0.30	0.00	1.00	1.00	1.82
PHA-S	1.00	0.17	0.17	0.11	0.36	0.30	0.00	1.00	1.00	1.81
HDPE	0.00	0.00	0.03	0.01	0.04	0.07	1.00	0.44	0.00	1.10
LDPE	0.00	0.00	0.04	0.04	0.06	0.06	1.00	0.72	0.01	1.24
GPPS	0.04	0.04	1.00	0.16	0.21	0.31	1.00	0.94	0.18	1.76
PC	0.79	0.04	0.91	1.00	0.41	0.78	1.00	1.00	0.80	2.42
PP	0.00	0.00	0.00	0.00	0.02	0.00	1.00	1.00	0.06	1.42

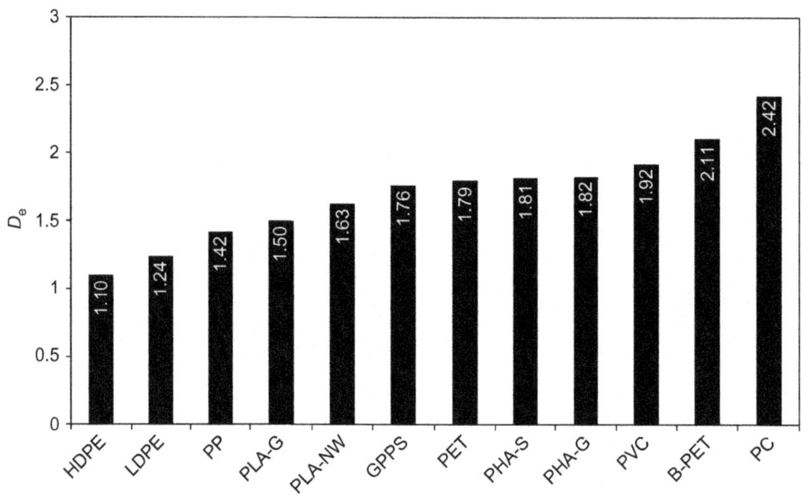

FIGURE 4.11

D_e for polymer options.

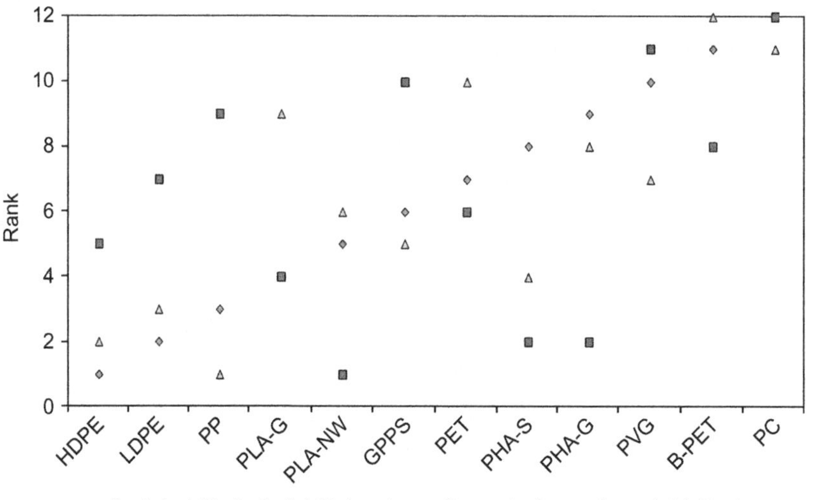

FIGURE 4.12

Comparison of ranking polymer options by different methods.

probably see different ranks. Nevertheless, the aggregate index allows a transparent and quantitative method of arriving at a decision point with regard to sustainability of the polymer options, compared with the decision matrix method used for green design and LCA ranks.

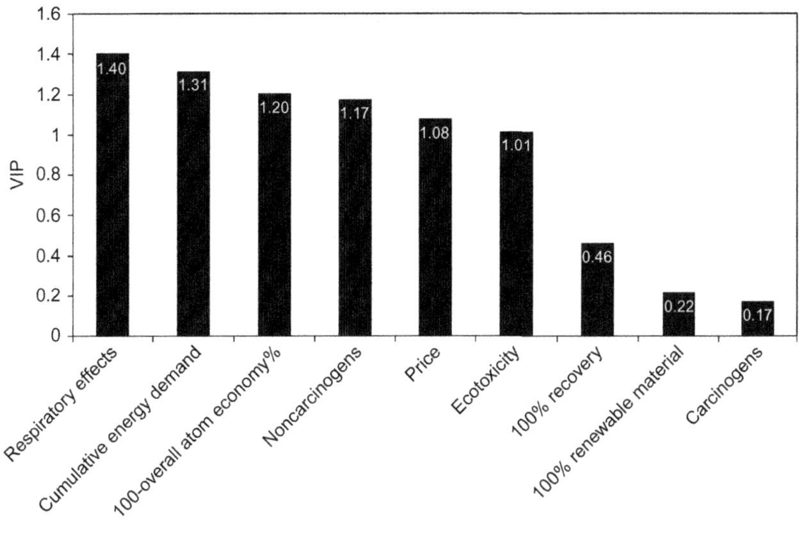

FIGURE 4.13

VIP scores for polymer indicators.

Figure 4.13 shows the relative ranks of the indicators in contributing to the values of D_e for the polymer system. The maximum variation in the original data is for respiratory effects followed by cumulative energy demand, whereas there is little variation in the data for 100% renewable material and carcinogens. This is reflected in the VIP scores, where very low scores are obtained for 100% renewable material and carcinogens.

The VIP data indicate that only six indicators are important in making decisions regarding the green designs. Three indicators, recovery, renewable material, and carcinogens, do not contribute to the overall D_e and, hence, can be considered not significant in green designs of this ensemble of alternatives. This inference, as in other cases, is subject to the caveat that the chosen indicator must be able to represent the totality of the system. The methodology is perfectly general; however, if not all applicable indicators are chosen, then errors in inference may occur.

CONCLUSIONS

In this chapter we emphasize the point that it is not sufficient to collect indicator data about various options of a chosen system and it is not possible to move to a decision point regarding the supremacy of one option relative to other options in a transparent quantitative manner. In the large majority of the cases, not all indicators will be favorable in any of the competing options, so a judgment will have to be made by the decision-maker. This judgment is not likely to be clear and accepted by all. Our construction of a sustainability footprint, D_e, is based on Euclidean distances of various system options that are characterized by an indicator set of multiple dimensions from a compatible reference system. The presented hypothesis was that system performance is completely described by the chosen indicator set. There is no scientific methodology for choosing indicators. Satisfactory selection of indicators comes from the practitioners' knowledge of the system undergoing study. There is no guarantee that a chosen indicator set covers all possible considerations affecting a system.

A review of previous work on decision-making shows that none of the prescribed methods do an adequate job in this art. The BASF approach of both eco-efficiency and socio-eco-efficiency uses a graphical method (two-dimensional for eco-efficiency and three-dimensional for socio-eco-efficiency) to arrive at satisfactory decisions. However, the details of their methodology are not publicly disclosed. Our method is algorithmic and considered totally clear.

The conditions for sound decision-making start with choosing necessary indicators. One can fail in this endeavor, because we show that cost is a missing element in the case studies, yet no sustainability can be valid without satisfying cost constraints. Second, the number of indicators would have to be sufficient. By including several closely related indicators, it is possible to make the decision tilted toward those indicators.

The suggested step-by-step procedure provided here leads to the calculation of the sustainability indices of the competing system options. An Excel plot for the various alternatives will clearly show the positions of the competing systems from which a clear decision can be made regarding the superiority of one of the contenders. We illustrated that with four case studies with datasets obtained from the published literature. The question regarding the relative importance of the indicators in contributing sufficiently to the sustainability index remained. This is the point about checking the sufficiency condition for the indicators. Suggested use of the multivariate statistical analysis based on the PLS-VIP leads to the rank order of the indicators in contributing to the sustainability footprint (i.e., to the aggregate performance of the competing systems). Thus, having made a decision regarding superiority of an option, PLS-VIP helps us check the sufficiency of the indicator set. When it is discovered that some indicators we thought would be important to include are unnecessary or redundant, the analysis can be repeated without using the unneeded indicators. We usually found that the decision is the same. Because data collection is expensive, this technique is likely to save money.

The point of the chapter was to present clear and sound methodology for performing sustainability analysis of any system with chosen indicators, not to dwell on the quality of decisions, because we cannot guarantee that all applicable indicators were available to us. One last point is that computation of the sustainability footprint may require the use of weighting factors for the indicators. This is typically a societal choice, and no scientific method can possibly help us acquire what would be appropriate to use. We decided to use the null position of 1.0 as a weight factor for all indicators. In other words, they are all of equivalent importance. All the results shown in the case studies are subject to that condition. If a practitioner has that sense, then it can be easily incorporated into the calculation of the sustainability footprint.

ACKNOWLEDGMENT

This work was supported by the Environmental Protection Agency. The conclusions presented here, however, are those of the authors and cannot be construed as representing the Agency. This project was supported in part by an appointment to the Research Participation Program at the Office of Research and Development, National Risk Management Research Laboratory, US Environmental Protection Agency, administered by the Oak Ridge Institute for Science and Education through an interagency agreement between the US Department of Energy and EPA.

REFERENCES

AIChE, 2014. Sustainability Index. <www.aiche.org/ifs/resources/sustainability-index> (accessed 14.04.2014).

BASF, 2014a. How does the eco-efficiency analysis work and which criteria are used? <www.basf.com/group/corporate/en/function:rendering-service:/faqsearch/faqsearch-result/resultCat/faq-sd_eco-efficiency/functions/faqsearch/faqsearch#0900dea680518f46> (accessed 16.07.2014).

BASF, 2014b. SEEBALANCE®. <www.basf.com/group/corporate/en/sustainability/eco-efficiency-analysis/seebalance> (accessed 16.07.2014).

Beaver, E., 2000. LCA and total cost assessment. Environ. Progr. 19, 130–139.

Beloff, B., Beaver, E., 2000. Sustainability Indicators and Metrics Project of CWRT. Bridges to Sustainability, Houston, TX.

Brandi, H., Daroda, R.J., Olinto, A.C., 2014. The use of the Canberra metrics to aggregate metrics to sustainability. Clean Technol. Environ. Policy 16, 911–920.

Cabezas, H., Fath, B.D., 2002. Towards a theory of sustainable systems. Fluid Phase Equilib. 194, 3–14.

Chong, I.-G., Jun, C.-H., 2005. Performance of some variable selection methods when multicollinearity is present. Chemom. Intell. Lab. Syst. 78, 103–112.

Cinar, A., Palazoglu, A., Kayihan, F., 2004. Chemical Process Performance Evaluation. CRC Press, Boca Raton, FL.

Cobb, C., Schuster, D., Beloff, B., Tanzil, D., 2009. The AIChE sustainability index: the factors in detail. Chem. Eng. Prog. 105, 60.

Conway, G.R., Barbier, E.B., 1990. After the Green Revolution: Sustainable Agriculture for Development. Earthscan Publications Ltd., London, UK and New York, USA.

Čuček, L., Klemeš, J.J., Kravanja, Z., 2012. A review of footprint analysis tools for monitoring impacts on sustainability. J. Clean. Prod. 34, 9–20.

Esty, D.C., Levy, M.A., Kim, C.H., De Sherbinin, A., Srebotnjak, T., Mara, V., 2008. Environmental Performance Index. Yale Center for Environmental Law and Policy, New Haven, CT, p. 382.

Fath, B.D., Cabezas, H., Pawlowski, C.W., 2003. Regime changes in ecological systems: an information theory approach. J. Theor. Biol. 222, 517–530.

Freudenberg, M., 2003. Composite Indicators of Country Performance: A Critical Assessment. OECD Publishing, Paris.

Graedel, T.E., Allenby, B.R., 2010. Industrial Ecology and Sustainable Engineering. Prentice Hall, Upper Saddle River, NJ.

Hák, T., Moldan, B., Dahl, A.L., 2007. Sustainability Indicators: A Scientific Assessment. Island Press, Washington, DC.

IChemE, 2014. The Sustainability Metrics—Sustainable Development Progress Metrics Recommended for Use in the Process Industries. Institution of Chemical Engineers, Rugby, UK.

Ingwersen, W., Cabezas, H., Weisbrod, A.V., Eason, T., Demeke, B., Ma, C., Lee, S.J., Hawkins, T., Bare, J.C., Ceja, M., 2014. Integrated metrics for improving the life cycle approach to assessing product system sustainability. Sustainability 6, 1386–1413.

Itoiz, E.S., Fuentes-Grünewald, C., Gasol, C., Garcés, E., Alacid, E., Rossi, S., Rieradevall, J., 2012. Energy balance and environmental impact analysis of marine microalgal biomass production for biodiesel generation in a photobioreactor pilot plant. Biomass Bioenergy 39, 324–335.

Kolsch, D., Saling, P., Kicherer, A., Grosse-Sommer, A., 2008. How to measure social impacts? A socio-eco-efficiency analysis by the SEEBALANCE® method. Int. J. Sustain. Dev. 11, 1–23.

Krajnc, D., Glavič, P., 2005. A model for integrated assessment of sustainable development. Resour. Conserv. Recycl. 43, 189–208.

Martens, H., 1991. Multivariate Calibration. John Wiley & Sons, NJ, USA.

Mata, T.M., Martins, A., Sikdar, S., Costa, C.V., 2011. Sustainability considerations of biodiesel based on supply chain analysis. Clean Technol. Environ. Policy 13, 655–671.

Mata, T.M., Caetano, N.S., Costa, C.A., Sikdar, S.K., Martins, A.A., 2013. Sustainability analysis of biofuels through the supply chain using indicators. Sustain. Energy Technol. Assess. 3, 53–60.

Mukherjee, R., Sengupta, D., Sikdar, S.K., 2013. Parsimonious use of indicators for evaluating sustainability systems with multivariate statistical analyses. Clean Technol. Environ. Policy 15, 699–706.

Nardo, M., Saisana, M., Saltelli, A., Tarantola, S., 2005. Tools for Composite Indicators Building. Institute for the Protection and Security of the Citizen Econometrics and Statistical Support to Antifraud Unit, European Commission, Ispra, Italy.

Oberthür, S., Ott, H.E., 1999. The Kyoto Protocol: International Climate Policy for the 21st Century. Springer, Berlin, Germany.

Olinto, A., 2014. Vector space theory of sustainability assessment of industrial processes. Clean Technol. Environ. Policy, 1–6.

Saling, P., Kicherer, A., Dittrich-Krämer, B., Wittlinger, R., Zombik, W., Schmidt, I., Schrott, W., Schmidt, S., 2002. Eco-efficiency analysis by BASF: the method. Int. J. Life Cycle Assess. 7, 203–218.

Saling, P., Maisch, R., Silvani, M., König, N., 2005. Assessing the environmental-hazard potential for life cycle assessment, Eco-Efficiency and SEEBALANCE. Int. J. Life Cycle Assess. 10, 364–371.

Schnepf, R., Yacobucci, B.D., 2010. Renewable Fuel Standard (RFS): Overview and Issues. Congressional Research Service, Washington, DC, USA.

Schwarz, J., Beloff, B., Beaver, E., 2002. Use sustainability metrics to guide decision-making. Chem. Eng. Prog. 98, 58–63.

Sengupta, D., Mukherjee, R., Sikdar, S.K., 2014. Environmental Sustainability of Countries Using the UN MDG Indicators by Multivariate Statistical Methods. Environmental Progress & Sustainable Energy, <http://dx.doi.org/10.1002/ep.11963>.

Sengupta, D., Pike, R.W., 2012. Chemicals from Biomass: Integrating Bioprocesses into Chemical Production Complexes for Sustainable Development. CRC Press, Boca Raton, FL, USA.

Sikdar, S.K., 2003. Sustainable development and sustainability metrics. AIChE J. 49, 1928–1932.

Sikdar, S.K., 2009. On aggregating multiple indicators into a single metric for sustainability. Clean Technol. Environ. Policy 11, 157–161.

Sikdar, S.K., Sengupta, D., Harten, P., 2012. More on aggregating multiple indicators into a single index for sustainability analyses. Clean Technol. Environ. Policy 14, 765–773.

Singh, R.K., Murty, H.R., Gupta, S.K., Dikshit, A.K., 2007. Development of composite sustainability performance index for steel industry. Ecol. Indicators Vol. 7 (Issue 3), 565–588.

Tabone, M.D., Cregg, J.J., Beckman, E.J., Landis, A.E., 2010. Sustainability metrics: life cycle assessment and green design in polymers. Environ. Sci. Technol. 44, 8264–8269.

Tanzil, D., Ma, G., Beloff, B., 2003. Sustainability metrics. Innovating for sustainability. The 11th International Conference of Greening of Industry Network, San Francisco, CA, USA, pp. 12–15.

Tokos, H., Pintarič, Z., Krajnc, D., 2012. An integrated sustainability performance assessment and benchmarking of breweries. Clean Technol. Environ. Policy 14, 173–193.

Uhlman, B.W., Saling, P., 2010. Measuring and communicating sustainability through eco-efficiency analysis. Chem. Eng. Prog. 106, 17–26.

UNEP, 1987. Montreal Protocol on Substances that Deplete the Ozone Layer. US Government Printing Office, Washington, DC, p. 26.

World Commission on Environment and Development, 1987. Our Common Future, Vol. 383. Oxford University Press, Oxford, UK.

Zhou, L., Tokos, H., Krajnc, D., Yang, Y., 2012. Sustainability performance evaluation in industry by composite sustainability index. Clean Technol. Environ. Policy 14, 789–803.

CHAPTER

Overview of environmental footprints

5

Lidija Čuček*, Jiří Jaromír Klemeš* and Zdravko Kravanja†

**Centre for Process Integration and Intensification - CPI², Research Institute of Chemical and Process Engineering - MŰKKI, University of Pannonia, Veszprém, Hungary*

†University of Maribor, Maribor, Slovenia

"The greatest shortcoming of the human race is our inability to understand the exponential function"
– Dr. Albert A. Bartlett, Arithmetic, Population and Energy (1969)

"We never know the worth of water, till the well is dry."
– Thomas Fuller, Gnomologia (1732)

"The road we have long been traveling is deceptively easy, a smooth superhighway on which we progress with great speed, but at its end lays disaster."
– Rachel Carson, Silent Spring (1962)

GLOSSARY

Carbon footprint (CF) is the more popular environmental protection indicator that commonly signifies a certain amount of gaseous emissions that are relevant to climate change. It is associated with human production or consumption activities (Wiedmann and Minx, 2008). It usually represents the total amount of CO_2 and other greenhouse gases (GHGs) emitted over the full life cycle of a system, and it is expressed in mass units of CO_2 equivalents per functional unit (UK Parliamentary Office of Science and Technology (POST), 2011).

Ecological footprint (EF) is a measure of human demand on the environment and represents the amount of biologically productive land and sea areas necessary for supplying the resources a human population consumes and to assimilate the associated waste (Wackernagel et al., 2002).

Environmental footprint is a quantitative measurement describing the appropriations of natural resources by humans (Hoekstra, 2008). It describes how human activities can impose different burdens and impacts on the global environment (UNEP/SETAC, 2009). The larger the footprint, the more resources that are needed to support human lifestyles.

Environmental indicator is a numerical value that helps to provide information and insight into the state of the environment. Indicators are developed based on the quantitative measurements or statistics of environmental conditions that are tracked over time (Attorre, 2014).

Footprint tools are tools for footprint calculations and suggested reduction paths. The more common tools are calculators, especially for CF (Padgett et al., 2008) but also for EF, water footprint (WF), nitrogen footprint (NF), and other footprints (Čuček et al., 2012c). Other tools are optimization frameworks, mathematical programming tools such as GAMS (GAMS Development Corporation, 2013), MIPSYN (Kravanja, 2010), and others, and structural optimization tools based on process graphs such as PNS solution (Bertok et al., 2012). Furthermore, there are specific tools available and include Environmental Performance Strategy Map (EPSM) (De Benedetto and Klemeš, 2009), which combines footprints and cost; SPIonExcel (Sandholzer and Narodoslawsky, 2007), which calculates Sustainable Process Index (SPI) (Krotscheck and Narodoslawsky, 1996); regional optimizer (RegiOpt) (Niemetz et al., 2012), which is a tool for sustainability evaluation of regional renewable energy networks; software package Bottomline[3] (Integrated Sustainability Analysis (ISA), 2014) for creating sustainability reports for companies and organizations; and many others.

Footprint family is a set of indicators able to track human pressures on the planet and from different angles (Galli et al., 2012). It represents the major categories of footprints developed to date, which are CF, EF, WF, and energy footprint (ENF), and is related to climate, food, water, and energy security (Fang et al., 2014).

Life cycle analysis (LCA) is a structured, comprehensive, internationally standardized tool (ISO 14040 (International Organisation for Standardisation (ISO), 2006a) and ISO 14044 (International Organisation for Standardisation (ISO), 2006b)) for quantifying those emissions, resource consumptions, and environmental and health impacts associated with processes, products, or activities. It is commonly referred as a "cradle-to-grave" analysis or as an open loop. It takes into account the system's whole life cycle from the extraction and processing of resources through manufacturing, usage, and maintenance to recycling or disposal, including all transportation and distribution steps (Guinée et al., 2002). The LCA principle and framework are divided into four phases: goal and scope definition; life cycle inventory (LCI); life cycle impact assessment (LCIA); and interpretation.

Sustainable development is "a process for improving the range of opportunities that would enable individual humans and communities to achieve their aspirations and full potential over a sustained period of time while maintaining the resilience of economic, social, and environmental systems" (Munasinghe, 2004). It represents the "development that meets the needs of the present without compromising the abilities of future generations to meet their own needs" (World Commission on Environment and Sustainable Development (WCED), 1987).

Water Footprint (WF) is closely linked to the virtual water concept (Hoekstra and Chapagain, 2007). It accounts for the appropriation of natural capital in terms of the water volumes required for human consumption (Hoekstra, 2009). It consists of blue, green, and gray WFs. The blue WF refers to the consumption of surface and groundwater, the green WF refers to the consumption of rainwater stored within the soil as soil moisture, and the gray WF refers to pollution and is defined as the volume of freshwater required for assimilating the load of pollutants based on existing ambient water quality standards (Mekonnen and Hoekstra, 2010).

INTRODUCTION

Environmental and social issues, such as global warming, water pollution, food supply, exponential population growth, security of energy supply, and others are attracting greater awareness when addressing sustainability issues preferably leading toward more sustainable development. Many different methods and tools have been developed over recent decades for measuring and monitoring sustainability and sustainable development to assess and evaluate progress toward more sustainable systems (De Benedetto and Klemeš, 2008). Environmental sustainability has especially emerged as a key issue among the three sustainability pillars, social ("People"), economic ("Prosperity" or "Profit"), and environmental ("Planet").

Environmental impacts are usually defined through a life cycle assessment (LCA). LCA is a set of tools and ideas for evaluating the sustainability of a system (products, processes, or services) throughout the full life cycle of the system (whole supply chain). LCA is usually associated with environmental components (von Blottnitz and Curran, 2007).

Since 1962, when Silent Spring written by Carson (Carson, 1962) was published, and especially from 1987 when the Brundtland report (World Commission on Environment and Sustainable Development (WCED), 1987) was published, the concepts and directions toward quantifying sustainability and sustainable development started to develop. The concept of environmental sustainability was developed in 1995 by Goodland (1995). It "seeks to improve human welfare by protecting the sources of raw materials used for human needs and ensuring that the sinks for human wastes are not exceeded, in order to prevent harm for humans". During last decades environmental sustainability has become widely recognized and accepted, and has led to several policies and treaties, such as The Clean Air Act (Martineau and Novello, 2004), Kyoto protocol (United Nations Framework Convention on Climate Change, 1998), European Union policies (European Commission, 2014), policies in China (China.org.cn, 2014), and many other policies around the world.

Sustainable development and environmental sustainability have been interpreted in many different ways because they are inexactly defined and thus allow certain leeway (Zaccai, 2012). In the 1980s, a variety of definitions had appeared regarding sustainability concepts (Pezzey, 1989). The most quoted definition is that it represents "development that meets the needs of the present without compromising the ability of future generations to meet their own needs." This definition was published in the Brundtland Commission report "Our Common Future" in 1987 (World Commission on Environment and Sustainable Development (WCED), 1987).

In addition, several different concepts and methods have been developed for the sustainability evaluations of particular processes, products, or activities (Jeswani et al., 2010). Still, the actual measurements of environmental sustainability and sustainable development remain an open question, and novel concepts and measures are being developed. To achieve sustainable development, the following are needed (Krajnc and Glavič, 2003):

- Changes in industrial processes
- Types and quantities of resources used
- Proper treatment of waste
- Controlling the emissions
- Controlling the produced products

The main goals of environmental protection are (Juhász and Szőllősi, 2008):

- To reduce world consumption of fossil fuels
- To reduce and clean up all sorts of pollution with the future goal of zero pollution
- Emphasis on clean, alternative energy sources with low carbon emissions
- Sustainable use of water, land, and other scarce resources
- Preservation of existing endangered species
- Protection on biodiversity

In this chapter the first part briefly reviews the following: the approach mostly used in environmental assessment, such as life cycle thinking and the LCA framework; the concepts of direct, indirect, and total effects on the environment; and measurements of environmental sustainability. In the second part, special attention is given to key and other environmental footprints, their importance, and overview.

LIFE CYCLE THINKING AND LCA FRAMEWORK

When moving toward (more) sustainable processes, products, or activities, it is essential that the entire life cycle is considered (Allen, 2008); therefore, environmental indicators should be and are usually defined on the basis of LCA (Pozo et al., 2012). LCA is a structured, comprehensive, internationally standardized tool (environmental management standards ISO 14040 (International Organisation for Standardisation (ISO), 2006a) and 14044 (International Organisation for Standardisation (ISO), 2006b)) for quantifying those emissions, resource consumptions, and environmental and health impacts associated with processes, products, or activities. LCA is a

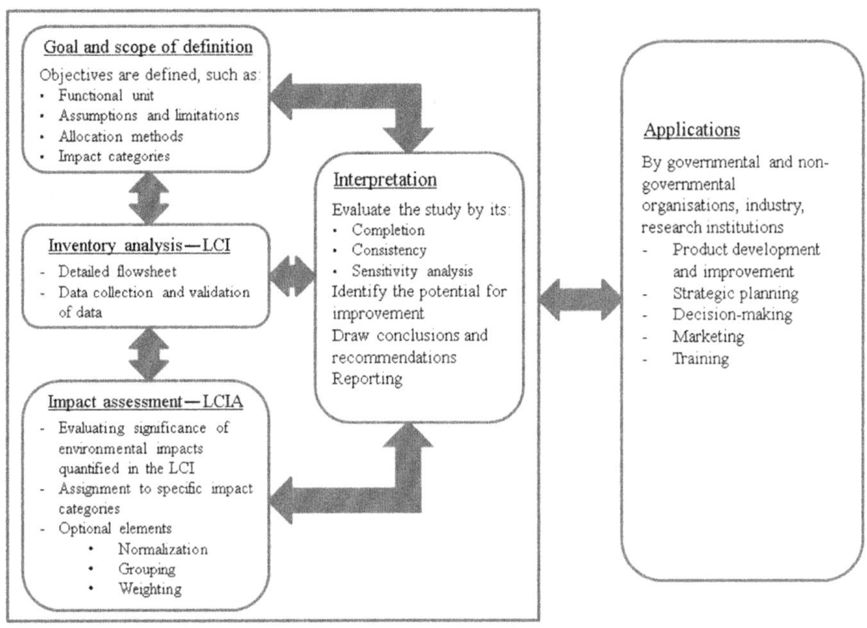

FIGURE 5.1

Four phases and direct applications of life cycle assessment.

Modified from Rebitzer et al. (2004).

relatively young approach that started in the late 1960s and early 1970s; however, before 1990 there was little public concern about LCA (Hunt et al., 1996).

LCA is commonly referred to as a "cradle-to-grave" analysis (Glavič and Lukman, 2007) or as an open loop. It takes into account a system's full life cycle from the extraction and processing of resources through manufacturing, usage, and maintenance to recycling or disposal, including all transportation and distribution steps (Bojarski et al., 2009). Over the years, a "cradle-to-cradle," or closed-loop, perspective has been introduced, which attempts to reach 100% utilization of all types of waste (Haggar, 2007).

The comprehensive scope of LCA is useful to avoid problem-shifting, for example, from one phase of the life cycle to another, from one region to another, or from one environmental problem to another (Finnveden et al., 2009). LCA can help reduce environmental pollution and resource usage, and it often improves profitability (McManus, 2010). LCA is an adequate instrument for environmental decision support and has gained wider acceptance over recent years within both academia and industry (von Blottnitz and Curran, 2007).

An LCA principle and framework is divided into four phases: goal and scope definition; Life Cycle Inventory (LCI) LCI; Life Cycle Impact Assessment (LCIA) LCIA; and interpretation. Those phases and direct applications of the LCA framework are shown in Figure 5.1.

1. First phase—Goal and scope definition: During the first phase, the objectives of the analysis and the system's boundaries should be defined, such as functional unit, assumptions and limitations, life cycle stages, allocation methods (when there are several products or functions of the system), and the chosen impact categories. The goals and scope can be adjusted during the iterative process of the analysis.
2. Second phase—Inventory analysis (LCI): The second phase involves data collection relating to inputs of materials and energy and outputs, including releases into air, soil, and water. All data should be related to the functional unit defined during the first phase.
3. Third phase—Impact assessment (LCIA): The third phase of LCA is aimed at evaluating the significances of those environmental impacts quantified in the LCI. The relative contribution of each environmental impact should be assigned to specifically selected impact categories (global warming potential (GWP), acidification potential, CF, NF, land usage, etc.). Other optional LCIA elements such as normalization (e.g., comparing the results to a population or area of Europe), grouping (sorting and ranking of impact categories), and weighting may also be conducted. Weighting, however, brings a high degree of subjectivity into LCA analyses.
4. Fourth phase—Interpretation: This is the last phase of the LCA analysis. It should evaluate the study in a systematic way by considering its completion, consistency, and sensitivity analysis. Interpretation should also identify areas that have the potential for improvement within a system and draw conclusions and recommendations.

LCA enables a holistic view and, therefore, a view from the life cycle perspective. To avoid problem-shifting or limited sustainability evaluation, it is important to select boundaries as wide as possible. The preferred options among the ones listed here are "cradle-to-cradle" and "cradle-to-grave":

- **"Cradle-to-cradle"**: From resource extraction (cradle) to recycling or producing new product (cradle) (100% utilization of waste)
- **"Cradle-to-grave"**: From resource extraction (cradle) to disposal (grave)
- **"Cradle-to-gate"**: From resource extraction (cradle) to factory gate before being sent to consumers (gate); it excludes the use and the disposal phases
- **"Gate-to-gate"**: Includes only one process within the entire production chain (e.g., the processing within a factory)
- **"Well-to-wheel"**: Related to transport fuels and vehicles, from the energy sources to the powered wheels (Daimler, 2011). It comprises two parts: "well-to-tank" (energy supply) and "tank-to-wheel" (energy efficiency) (Fuel-Cell e-Mobility, 2014)

These selected boundaries, except "well-to-wheel," are shown in Figure 5.2 (modified from Procter & Gamble Science-in-the-Box, 2014).

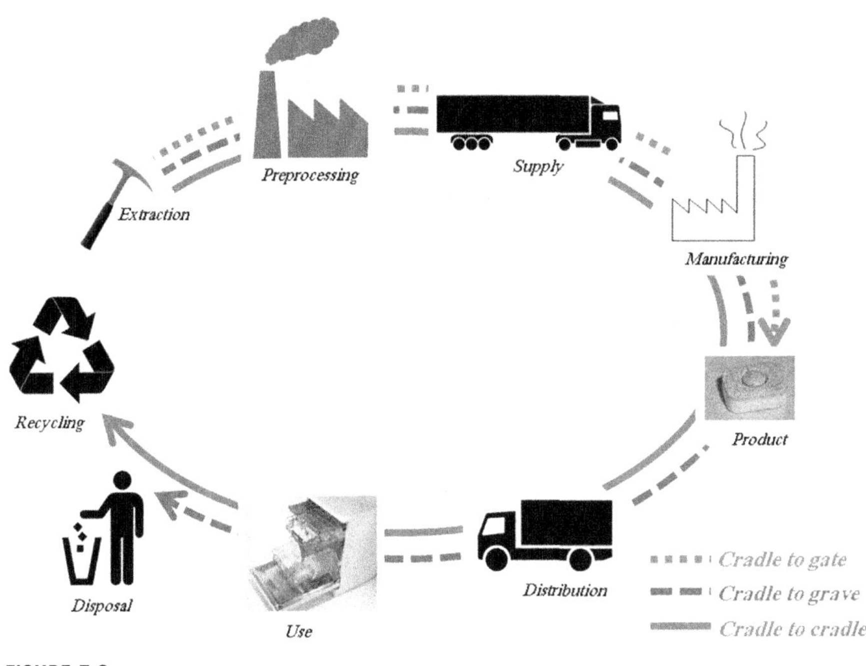

FIGURE 5.2

Mostly used boundaries of LCA.

Modified from Procter & Gamble Science-in-the-Box (2014).

LCA methodology and sustainability assessment, in general, still have some limitations that need to be overcome (Čuček, 2013). The main limitation is the high degree of uncertainty arising from the LCI, which gives rise to results with high variability. Data quality and quantity are often insufficient for comprehensive LCA. A possible consequence of discrepancies in data is that two independent studies analyzing the same products may generate very different results, and different LCIA methodologies can yield different results (McManus, 2010). The results also have low spatial and temporal resolution (de Haes et al., 2004), and LCA only assesses potential impacts and not real impacts (Quantis Sustainability counts, 2009). Another limitation is the lack of a systematic method for generating and identifying sustainable solutions (Grossmann and Guillén-Gosálbez, 2010). There is no single method that is universally acceptable (Hendrickson et al., 1997). It is very challenging to define indicators that are not too broad or too specific (De Benedetto and Klemeš, 2010). Performing the LCA analyses can be costly regarding data and resources, and they can be time-intensive. Table 5.1 illustrates the advantages and drawbacks of LCA approaches.

Table 5.1 Advantages and Drawbacks of LCA Approaches in General

Advantage	Drawback
Comprehensive scope of analysis	Quality and availability of data (uncertainty)
Cradle-to-grave or even cradle-to cradle	Base assumptions and system boundaries
Avoids problem-shifting	Model of the process is needed
Widely accepted	Quantification and normalization
Supports decision-making	Health and safety issues are difficult to assess
Identification of "hot spots"	Limited inclusion of economic and social aspects
Identification of sustainable options	Different studies analyzing the same products may generate very different results
Market advantage (purchasing decisions)	Different LCIA methodologies can yield different results
Enables progress toward more sustainable lifestyles	Costly (money, time, and resources)
	Lack of systematic method for sustainable solutions (economic, environmental, and social)
	May be subjective
	High dimensionality of criteria could be obtained
	Conventional LCAs are generally based on average technology
	For emerging technologies, LCAs are based on an average technology at lab scale or pilot scale extrapolated to large scale
	Improving some criteria may cause worsening of another ones
	In many cases there are several simplifications, because no detailed data are available
	Low spatial and temporal resolution
	LCA only assesses potential impacts and not real impacts

DIRECT, INDIRECT, AND TOTAL EFFECTS

Environmental impacts of raw materials and energy usage, processes, products, and services are usually measured only by their burdening effects on the environment (Hundal, 2000). However, a broader view should also incorporate any possible unburdening effects of an activity (Čuček et al., 2012e). An activity can then be considered environmentally sustainable if its unburdening exceeds its burdening. This may happen, for example, when waste is transformed into a precious green product (Kravanja and Čuček, 2013). In such a case, unburdening occurs twice, first because of the waste not being deposited and second because of fossil-based products being supplemented by the green product. A concept of direct (burdening) and unburdening (indirect) effects, which together form total effects (Kravanja and Čuček, 2013), is described in the following paragraphs.

Direct environmental footprints and any sustainability metric conventionally only measure the directly harmful effects (burdens) on the environment. Direct

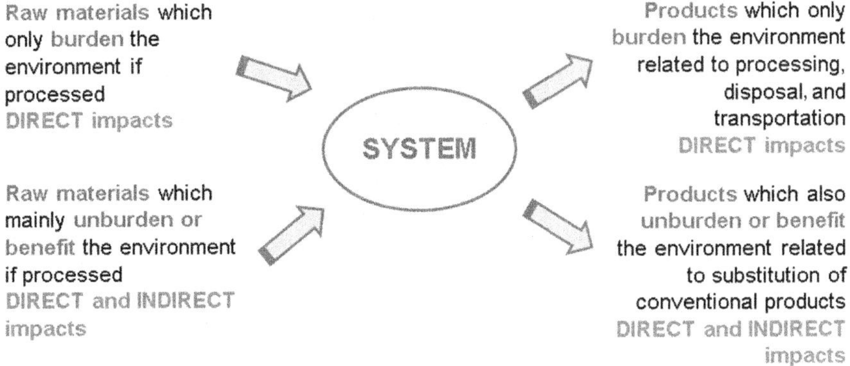

FIGURE 5.3

Representation of direct, indirect, and total effects.

environmental metrics, e.g. footprints, are related to the extraction of resources, production of materials, usage, maintenance, and recycling and/or disposal, including all transportation and distribution steps. However, when a system, in addition to its direct burdening effects on the environment, exhibits a significant unburdening effect on the environment by considering only direct effects, misleading solutions may result (Čuček et al., 2013c).

Indirect footprints represent an unburdening of the environment by substituting harmful products with benign products and the utilization of harmful products rather than discarding them. Several examples include when waste is utilized instead of being discarded, when environmentally benign raw materials, products, and services are used instead of harmful ones, or when conventional energy is replaced by renewable energy. The indirect effect is a reduction in the footprint. Indirect footprints indirectly unburden or benefit the environment (Čuček, 2013).

Total footprints are defined as a sum of the direct and indirect footprints (Kravanja and Čuček, 2013). Considering total effects enables the ability to obtain more realistic solutions than in those cases when only direct effects are considered. An appropriate sustainable synthesis should identify those solutions that unburden the environment the most, rather than only proposing the least-burdening solutions (Čuček, 2013). The concept of total effects is shown in Figure 5.3.

When the sustainability aspect is improved, and therefore minimized, in terms of the direct effects only, there the system does not operate, e.g. there is no production (Kravanja and Čuček, 2013). However, when the total aspects are considered, there are several metrics that are improved and several that are worse. It should be noted that trade-off or compromise solutions are obtained when only direct effects are considered; trade-off and non-trade-off solutions are obtained when total effects are considered. Such a situation could also occur when comparing two systems. Figure 5.4 presents an example using

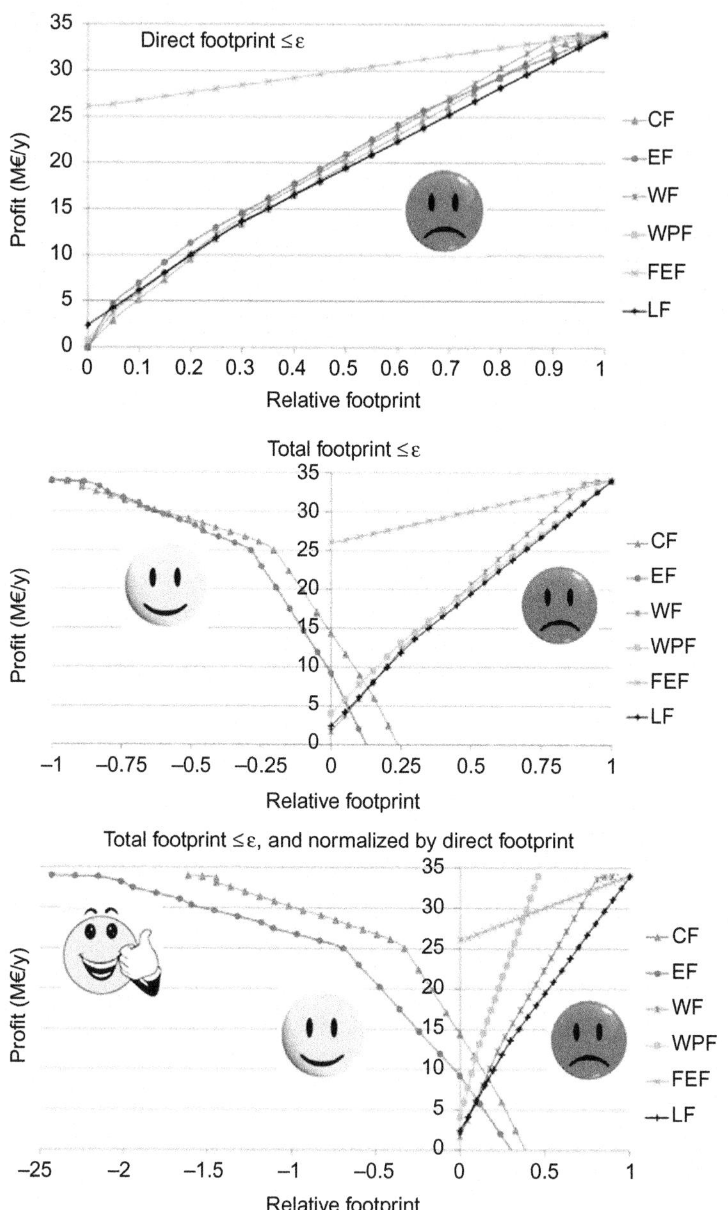

FIGURE 5.4

Difference between solutions obtained by considering only direct, indirect, and total/indirect effects.

Modified from Čuček et al. (2012c).

multi-objective optimization (MOO) of economic vs. environmental criteria - annual profit vs. environmental footprints.

MEASURING ENVIRONMENTAL SUSTAINABILITY

Measuring environmental sustainability requires methods and tools that are to be used for defining the environmental impacts of human activities. Indicators and metrics that can be used to measure and quantify environmental sustainability need to be developed to provide a basis for decision-making. Among the developed metrics measuring environmental sustainability are:

- Indicators of potential environmental impacts
- Eco-efficiency
- Environmental footprints (Čuček et al., 2012c)
- Sustainability indexes (Kravanja and Čuček, 2013)
- Eco and total profit (Čuček et al., 2012a), eco and total net present value (NPV) (Čuček et al., 2013a), and other combined criteria (Kravanja and Čuček, 2013)

INDICATORS OF POTENTIAL ENVIRONMENTAL IMPACTS

Indicators of potential environmental impacts deal with the potential effects and impacts on humans, environmental health, and resources from the LCI (Saur, 1997). They are divided into midpoint-oriented and end-point-oriented or damage impact categories (Figure 5.5).

The following midpoint categories are usually considered:

- *Ozone depletion potential* is the potential for the reduction in the protective stratospheric ozone layer. The ozone-depleting substances are freons, chlorofluorocarbons, carbon tetrachloride, and methyl chloroform. It is expressed as CFC-11 equivalents.
- *Global warming potential* represents the potential change in climate attributable to increased concentrations of CO_2, CH_4, and other GHG emissions that trap heat. It leads to increased droughts, floods, losses of polar ice caps, sea-level rising, soil moisture losses, forest losses, changes in wind and ocean patterns, and changes in agricultural production. It is expressed in CO_2 equivalents usually for time horizon 100 y.
- *Acidification potential* is based on the potential of acidifying pollutants (SO_2, NO_x, HCl, NH_3, HF) to form H^+ ions. It leads to damage to plants, animals, and structures. It is expressed in SO_2 equivalents.
- *Eutrophication potential* leads to an increase in aquatic plant growth attributable of nutrients left by over-fertilization of water and soil, such as nitrogen and phosphorus. Nutrient enrichment may cause fish death, declining

water quality, decreased biodiversity, and foul odors and tastes. It is expressed in PO_4^{3-} equivalents.
- *Photochemical ozone creation potential* is also known as ground-level smog, photochemical smog, or summer smog. It is formed within the troposphere from a variety of chemicals including NO_x, CO, CH_4, and other volatile organic compounds (VOCs) in the presence of high temperatures and sunlight. It has negative impacts on human health and the environment and is expressed as C_2H_4 equivalents.
- *Ecotoxicity (freshwater, marine, terrestrial) potential* focuses on the emissions of toxic substances into the air, water, and soil. It includes the fates, exposures, and effects of toxic substances and is expressed as 2,4-dichlorophenoxyacetic acid equivalents.
- *Human toxicity potential* deals with the effects of toxic substances on human health. It enables relative comparisons between a larger number of chemicals that may contribute to cancer or other negative human effects for the infinite time horizon. It is expressed as 1,4-dichlorobenzene equivalents.
- *Abiotic depletion potential* is concerned with the protection of human welfare, human health, and ecosystems, and represents the depletion of non-renewable resources (abiotic, non-living (fossil fuels, metals, minerals)). It is based on concentration reserves and the rate of de-accumulation and is expressed in kg antimony equivalents.
- Land use
- Water use

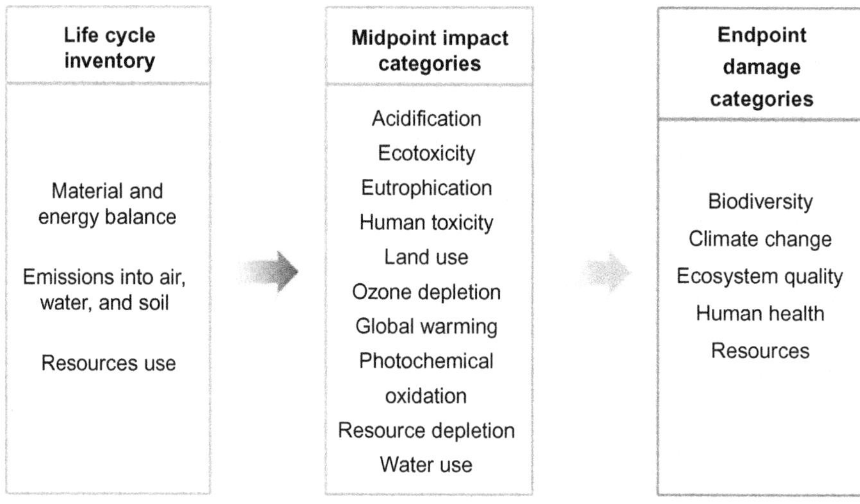

FIGURE 5.5

From LCI to endpoint damage categories.

ECO-EFFICIENCY

Eco-efficiency is a management strategy of doing more with less (Glavič et al., 2012). It is based on the concept of creating more goods and services while using fewer resources and creating less waste and pollution (Glavič et al., 2012). Eco-efficiency is a sustainability measure combining environmental and economic performances. However, a commonly agreed definition does not exist yet (Huppes and Ishikawa, 2005). Eco-efficiency can be seen either as an indicator of environmental performance or as a business strategy for sustainable development (Koskela and Vehmas, 2012). The most common eco-efficiency is defined as (Koskela and Vehmas, 2012):

- A ratio between environmental impact and economic performance
- A ratio between economic performance and environmental impact

Eco-efficiency is achieved through three objectives:

- Increasing product or service values
- Optimizing the usages of resources
- Reducing environmental impacts (Government of Canada, 2013)

A wide variety of terminology referring to eco-efficiency has been developed, with some divergence (Huppes and Ishikawa, 2005). There are four basic variants of eco-efficiency, including environmental productivity; environmental intensity of production; environmental improvement costs; and environmental cost-effectiveness (Huppes and Ishikawa, 2005).

Eco-efficiency offers a number of practical benefits, such as (Government of Canada, 2013):

- Reduced costs through more efficient usages of energy and materials
- Reduced risk and liability by "designing out" the need for toxic substances
- Increased revenue by developing innovative products and increasing market share
- Enhanced brand image through marketing and communicating improvement efforts
- Increased productivity and employee morale through closer alignment of company values with the personal values of the employees
- Improved environmental performance by reducing toxic emissions and increasing the recovery and reuse of "waste" material

ENVIRONMENTAL FOOTPRINTS

Environmental footprints are quantitative measures showing the appropriation of natural resources by humans (Hoekstra, 2008). Footprints are divided into environmental, economic, and social footprints, and combined environmental, social, and/or economic footprints (Čuček et al., 2012c). The concept of "footprint" originates from the idea of ecological footprint - EF (Fang et al., 2014) introduced by Rees (1992). After its introduction, several other environmental footprints have been introduced. Each footprint indicates particular classes of pressures associated with process, product, or activity from the life cycle perspective (Galli et al., 2013).

Several footprints are identified as key footprints because they are essential for sustainability and sustainable development. These recognized footprints are CF, WF, ENF, NF, phosphorus (PF), biodiversity (BF), and land (LF), and they are presented comprehensively. Besides these, various footprints have recently been introduced and are presently being reviewed. For these footprints there is a lack of applications; therefore, more comprehensive reviews cannot be made at this early stage of their development. Environmental footprints are presented in detail in the following sections.

ECO AND TOTAL PROFIT, ECO AND TOTAL NPV, AND OTHER COMBINED CRITERIA FOR MEASURING DIRECT, INDIRECT, AND TOTAL EFFECTS

The eco cost, eco benefit, and eco profit coefficients have the advantage that they can be directly incorporated within the objective function, together with a given economic objective (Kravanja and Čuček, 2013). Therefore, subjective weighting between sustainability objectives is avoided. The advantages of eco cost and eco benefit coefficients are as follows: they are expressed as monetary values; there is no need to compare two products, processes, or services (as is often the aim of LCA); calculations are based on current European price levels; and updated eco cost coefficients can be used (Delft University of Technology, 2012). Economic and eco profits can be merged together, and the preferred solutions are those with maximal total profit (Kravanja and Čuček, 2013). These approaches enable obtaining the better unique solutions and represent fair prices by properly accounting for environmental problems. However, the burden-prevention step is still subjective to some extent. When considering environmental or economic and environmental dimensions within one value, these criteria are:

- Eco cost (Vogtländer et al., 2010)
- Eco benefit or eco revenue (Čuček et al., 2012a)
- Eco profit (Čuček et al., 2012a)
- Net profit (Kravanja and Čuček, 2013)
- Total profit (Čuček et al., 2012a)
- Net NPV (Čuček et al., 2013a)
- Eco NPV (Čuček et al., 2013a)
- Total NPV (Čuček et al., 2013a).

All these criteria use eco cost coefficients (Delft University of Technology, 2012) based on LCA.

To define these criteria, different sets are used, such as R_{B}, R_{UNB}, P_{B}, and P_{UNB} (Čuček et al., 2012a), as defined and used throughout the model for raw materials and products (Figure 5.3):

- R_{B}—set of those raw materials that only burden the environment if processed
- R_{UNB}—set of those raw materials that also unburden or benefit the environment when used
- P_{B}—set of those products that only burden the environment in relation to processing, disposal, and transportation
- P_{UNB}—set of those products that also unburden or benefit the environment (Čuček et al., 2012a)

Eco cost (burdening the environment)

Eco cost (c^{Eco}) is an indicator based on LCA and describes environmental burden on the basis of preventing that burden. It is the sum of the marginal prevention costs during the life cycle, sum of eco costs of material depletion, eco costs of energy and transport, and eco costs of emissions. Eco costs are those costs that should be made to reduce the environmental pollution and material depletion to a level that is in line with the carrying capacity of the Earth (Vogtländer et al., 2010). They are virtual costs and are not yet integrated within current prices ("what if" basis) (Vogtländer et al., 2002). Eco costs include those burdens that originate from the extraction of raw materials, pre-processing, processing, and disposal of harmful products or from purification of the polluted products, from the transportation of raw materials to the plants, from distribution of products to consumers, and from the transportation of products to the locations of disposal.

Eco cost is defined as the sum of all the negative impacts from burdens on the environment, where eco cost coefficients (c_p^{s} and $c_p^{s,\mathrm{tr}}$) are used (Čuček, 2013):

$$c^{\mathrm{Eco}} = \sum_{p \in R_{\mathrm{B}} \cup R_{\mathrm{UNB}} \cup P_{\mathrm{B}} \cup P_{\mathrm{UNB}}} q_p^m \cdot c_p^{\mathrm{s}} + \sum_{p \in R_{\mathrm{B}} \cup R_{\mathrm{UNB}} \cup P_{\mathrm{B}} \cup P_{\mathrm{UNB}}} q_p^m \cdot l_p \cdot D_p \cdot c_p^{s,\mathrm{tr}} \tag{5.1}$$

This summation is performed over all raw materials and products, because all of them contribute to the burdening. The second term represents burdening due to transportation, where the pth flow rate is multiplied by an inverse of the load factor l_p, the distance (D_p/km), and a specific eco cost coefficient for transportation $c_p^{s,\mathrm{tr}}$/(kg($t \cdot$ km), kg/(m$^3 \cdot$ km), . . .). Note that the inverse of the load factor is set at 2 when the transport is fully loaded in one direction and empty in the other (Čuček, 2013).

Eco benefit (unburdening the environment)

Eco benefit or eco revenue (R^{Eco}) is defined as the sum of all the positive impacts of unburdens on the environment. Positive impacts are related to raw materials that mainly benefit the environment when used (e.g., the utilization of waste such as industrial wastewater, manure, sludge, etc.); their direct harmful impact on the environment is thus avoided. Positive impacts are also related to products that benefit the environment (e.g., if they are substitutes for harmful products). They are defined using a substitution factor as the ratio between the quantity of conventional product ("currently" used product) and the quantity of the produced product (Čuček et al., 2012a).

Eco benefit or eco revenue (R^{Eco}) represents the unburdening of the environment and is defined as the sum of all the positive impacts of unburdening the environment. Here, the eco benefit coefficients (c_p^{s} and $c_p^{\mathrm{s,tr}}$) are used (Čuček et al., 2012a):

$$R^{\mathrm{Eco}} = \sum_{p \in R_{\mathrm{UNB}}} q_p^m \cdot c_p^{\mathrm{s}} + \sum_{p \in P_{\mathrm{UNB}}} q_p^m \cdot f_p^{S/P_{\mathrm{UNB}}} \cdot c_p^{\mathrm{s}} \tag{5.2}$$

Eco profit (= eco benefit − eco cost)

Eco profit (P^{Eco}) is defined as an analogy with economic profit as the difference between unburdening and burdening the environment expressed as eco benefits and eco costs (Čuček et al., 2012a):

$$P^{\mathrm{Eco}} = R^{\mathrm{Eco}} - c^{\mathrm{Eco}} \tag{5.3}$$

Net profit

Net profit (P^{N}) is defined as the economic profit (P^{Econ}) reduced by the eco cost (c^{Eco}). The synthesis problem, where economic profit and eco cost are optimized simultaneously, takes the following form (Kravanja and Čuček, 2013):

$$\begin{aligned} &P^{\mathrm{N}}(x,y) = \max(P^{\mathrm{Econ}}(\mathbf{x},\mathbf{y}) - c^{\mathrm{Eco}}(\mathbf{x},\mathbf{y})) \\ &\text{s.t.} \quad \mathbf{h}(\mathbf{x},\mathbf{y}) = 0 \\ &\quad \mathbf{g}(\mathbf{x},\mathbf{y}) \le 0 \\ &\quad (\mathbf{x}^{\mathrm{LO}}\mathbf{x} \le \mathbf{x}^{\mathrm{UP}}) \in \mathbf{X} \subset \mathbf{R}^n, \mathbf{y} \in \mathbf{Y} = \{0,1\}^m \end{aligned} \tag{5.4}$$

Total profit

Total profit (P^{T}) is the summation of the economic profit (P^{Econ}) and eco profit (P^{Eco}). For the synthesis problem, where economic profit and eco profit are optimized simultaneously, the solutions obtained are those with maximal total profit (*TP*) (Kravanja and Čuček, 2013):

$$\begin{aligned} &P^{\mathrm{T}}(x,y) = \max(P^{\mathrm{Econ}}(\mathbf{x},\mathbf{y}) + P^{\mathrm{Eco}}(\mathbf{x},\mathbf{y})) \\ &\text{s.t.} \quad \mathbf{h}(\mathbf{x},\mathbf{y}) = 0 \\ &\quad \mathbf{g}(\mathbf{x},\mathbf{y}) \le 0 \\ &\quad (\mathbf{x}^{\mathrm{LO}}\mathbf{x} \le \mathbf{x}^{\mathrm{UP}}) \in \mathbf{X} \subset \mathbf{R}^n, \mathbf{y} \in \mathbf{Y} = \{0,1\}^m \end{aligned} \tag{5.5}$$

Eco NPV

The eco NPV is an analogy of the economic NPV, which takes into account the complete economics of the project throughout the project's life cycle. Income within the NPV's yearly cash flow is represented by unburdening (eco benefit) and outcome by burdening (eco cost) of the environment (Čuček et al., 2013a). The following simplified formula can be used when fixed yearly cash flows are assumed:

$$W_{\mathrm{NP}}^{\mathrm{Eco}} = \left[(1-r_t)(R^{\mathrm{Eco}} - c^{\mathrm{Eco}})\right]\left[\frac{(1+r_d)^{t_D}-1}{r_d(1+r_d)^{t_D}}\right] \tag{5.6}$$

Total NPV

Similar to economic and eco profit being merged into total profit, economic NPV and eco NPV form total NPV, and the preferred solutions are those with maximal total NPV (Čuček et al., 2013a).

$$\begin{aligned} & W_{\mathrm{NP}}^{\mathrm{T}}(x,y) = \max(W_{\mathrm{NP}}^{\mathrm{Econ}}(\mathbf{x},\mathbf{y}) + W_{\mathrm{NP}}^{\mathrm{Eco}}(\mathbf{x},\mathbf{y})) \\ & \text{s.t.}\quad \mathbf{h}(\mathbf{x},\mathbf{y}) = 0 \\ & \quad \mathbf{g}(\mathbf{x},\mathbf{y}) \le 0 \\ & \quad (\mathbf{x}^{\mathrm{LO}} \le \mathbf{x} \le \mathbf{x}^{\mathrm{UP}}) \in \mathbf{X} \subset \mathbf{R}^n, \mathbf{y} \in \mathbf{Y} = \{0,1\}^m \end{aligned} \tag{5.7}$$

SUSTAINABILITY INDEXES

Sustainability usually has environmental, economic, and social dimensions (Jørgensen et al., 2008). Therefore, the relative sustainability index (RSI) is composed of economic, environmental, and social indicators (Tallis et al., 2002). Here only environmental indicators are briefly summarized.

Environmental indicators

Environmental metrics should provide a balanced view of the environmental impact of inputs (resource usage) and outputs (products, services, activities produced, emissions, effluents, and waste). Environmental indicators are typically grouped into resource usage indicators (material, energy, water, and land) and pollution indicators (global warming, atmospheric acidification, photochemical smog formation, human health effect) (Tallis et al., 2002). The environmental indicators by the optimal solution at the first level (base case solution) often yield the reference point for the second level, which then yields a sustainable solution. For example, the relative environmental index (REI) can be calculated as follows (Kravanja et al., 2005):

$$\begin{aligned} & \mathrm{REI} = \frac{1}{N}\sum_{j=1}^{N}\frac{\mathrm{EI}_j}{\mathrm{EI}_j^0} \quad j \in J \\ & \mathrm{REI} = \frac{1}{N}\left[\underbrace{\sum_{k\in K}\frac{q_{m,k}}{q_{m,k}^0}}_{\text{material usage}} + \underbrace{\sum_{e\in E}\frac{\phi_e}{\phi_e^0}}_{\text{energy usage}} + \underbrace{\sum_{w\in W}\frac{q_{m,w}}{q_{m,w}^0}}_{\text{water usage}} + \sum_{c\in C, j\in J,(c,j)\in CJ}\ \sum_{o\in O}\frac{q_{m,c,o}}{q_{m,c,o}^0}\mathrm{PF}_{j,c}\right] \end{aligned} \tag{5.8}$$

where

j—the index of environmental indicators from the set of environmental indicators J
k—the index of raw material input streams from the set of input streams K
e—the index of energy sources consumed or generated from the set of energy sources E
w—the index of water streams consumed from the set of water streams W
c—index of substances contributing to the jth pollution indicator from the set of substances C
CJ—set of pairs of substances contributing to the jth pollution indicator, and jth pollution indicator
o—the index of output streams from the set of output streams O
N—the number of environmental indicators
EI_j—the environmental indicator
EI_j^0—the environmental indicator for the base case
$q_{m,k}$—mass flow rate of input stream k (mass unit/time unit)
ϕ_e—heat or energy flow rate of heat or energy source consumed e (energy unit)
$q_{m,w}$—mass flow rate of water stream consumed or generated w (mass unit/time unit)
$q_{m,c,o}$—mass flow rate of substance c in output stream o (mass unit/time unit)
$\mathrm{PF}_{j,c}$—the potency factor of substance c that contributes to environmental indicators $j, j \in J$

Note that in these definitions all the indicators are normalized by their reference values as obtained at the first level.

Relative sustainability index

Different indicators are expressed in different units. Environmental indicators are usually expressed as a burden per a certain functional unit. Because their units are different, they cannot be composed unless they are normalized. When the indicators ($I_{i,f}$) of a studied alternative represent "cradle-to-grave" impacts and are compared with those of the selected base case (I_f^0), relative indicators are obtained, which can then be composed into an RSI_i by suitable weighting factors (w_f, $\sum_f w_f = 1$, $\forall f \in F$):

$$\mathrm{RSI}_i = \sum_f w_f \cdot \frac{I_{i,f}}{I_f^0}, \quad f \in F, \quad i \in I = \{1,\ldots,N+1\} \tag{5.9}$$

where $f \in F$ is a sustainability (economic and/or environmental and/or social) indicator. If only direct effects on the environment and society are considered, then the relative direct sustainability index (RDSI_i) is obtained (Eq. (5.10)). If indirect effects are also considered, then the relative total sustainability index ($RTSI_i$) is obtained (Eq. (5.11)), which shows how much better $RTSI_i$ is than $RDSI_i$:

$$RDSI_i = \sum_{f=1}^{N} w_i \cdot \frac{I_{i,f}^{d}}{I_f^{d,0}}, \quad f \in F, \quad i \in I = \{1, ..., N+1\} \tag{5.10}$$

$$RTSI_i = \sum_{f=1}^{N} w_i \cdot \frac{I_{i,f}^{\mathrm{d}} + I_{i,f}^{\mathrm{ind}}}{I_f^{\mathrm{d},0}} = \sum_{f=1}^{N} w_i \cdot \frac{I_{i,f}^{\mathrm{t}}}{I_f^{\mathrm{d},0}}, \quad f \in F, \quad i \in I = \{1, ..., N+1\} \tag{5.11}$$

$I_{i,f}^{\mathrm{d}}$ represents the direct indicator, $I_{i,f}^{\mathrm{ind}}$ represents the indirect indicator, and $I_{i,f}^{\mathrm{t}}$ represents the total indicator. The resulting $RDSI_i$ usually ranges from 1 (worst possible value) to 0 (best possible value), and i is the index of the selected points between 1 and 0. However, the $RTSI_i$ can range from -1, or from more negative values (positive for the environment), to 1. The higher the quotient between the total and direct effects (in the case of alternatives that significantly unburden the environment), the more this indicator changes the $RTSI_i$ and the lower the value of $RTSI_i$ (Kravanja and Čuček, 2013).

MULTI-OBJECTIVE OPTIMIZATION

MOO, also known as multi-criteria, vector, multi-attribute, Pareto optimization, and multi-objective programming (Chakrabortty and Hasin, 2013), simultaneously integrates two or more objectives or goals that are subject to certain constraints. Usually, compromise solutions (trade-offs) that reveal the possibilities for achieving improvements within the system are obtained (Azapagic and Clift, 1999). In general, no single solution exists when all the objectives are optimized simultaneously, but a number of Pareto optimal solutions (a feasible region of optimal solutions, Pareto front) can be obtained (Rao, 2009).

MOO is the use of specific methods for determining the best solutions to the problem subject to given constrains. It is the key methodology used for sustainable system design and synthesis (Klemeš et al., 2010). However, from simultaneous MOO, a number of Pareto optimal solutions (a feasible region of optimal solutions) can be obtained (Rao, 2009) (Figure 5.4).

Several methods have been developed for solving MOO problems. The simplest method is to transform the MOO problem into a single-objective optimization (SOO) problem by applying weights to different criteria (the weighted objective method). Other widely used optimization methods are: the ε-constraint method (Haimes et al., 1971), in which a sequence of constrained single-objective problems are solved; the goal-programming method, in which the solution is obtained by minimizing a weighted average deviation of the objective functions from the goal set by the decision-maker and evolutionary algorithms that involve random search techniques (Bhaskar et al., 2000). The solution for such problems is a set of "non-inferior" or Pareto points (Čuček et al., 2012c).

The ε-constraint method is usually applied (Pieragostini et al., 2012), and a set of non-inferior Pareto optimal solutions are thus generated. A two-level system synthesis is performed when e.g. an economically effective synthesis is performed at the first level to obtain a solution that is then considered as a base case or reference solution for the multi-objective synthesis performed at the second level. The synthesis problem in which a sequence of N problems (MP-E)$_e$ is performed as a maximization or minimization of one objective (e.g., maximization of profit) (O_1) subjected to environmental metric $\mathbf{e} \in E$ takes the following form:

$$
\begin{aligned}
& O_1(x,y) = \max f(\mathbf{x},\mathbf{y}) \;\text{ or }\; \min f(\mathbf{x},\mathbf{y}) \\
& \text{s.t.}\quad \mathbf{h}(\mathbf{x},\mathbf{y}) = 0 \\
& \quad \mathbf{g}(\mathbf{x},\mathbf{y}) \le 0 \qquad (\text{MP-}E)_e \\
& \quad O_e(\mathbf{x},\mathbf{y}) \le \varepsilon_i, \mathbf{e} \in E \\
& \quad (\mathbf{x}^{\text{LO}}\mathbf{x} \le \mathbf{x}^{\text{UP}}) \in \mathbf{X} \subset \mathbf{R}^n, \mathbf{y} \in \mathbf{Y} = \{0,1\}^m \\
& \quad \varepsilon_i = \varepsilon_{i-1} - \Delta\varepsilon \\
& \quad \Delta\varepsilon = \frac{1}{N},\; \varepsilon_i = \frac{1}{N}\cdot(i-1),\; i \in I = \{1,\ldots,N+1\},\; i_0 = N+1
\end{aligned}
$$

where x denotes the vector of continuous variables (mass flow rates, temperatures, design variables, etc.), y denotes vector of binary variables (discrete decisions), and $f(x)$, $h(x,y)$, and $g(x,y)$ are (non)linear constraint functions. Objectives O_e decrease sequentially by a suitable step-size until there is no feasible solution. Pareto inferior solutions are usually obtained in this way. However, non-trade-off solutions can also be obtained when objectives have synergistic effects. Inequality ($O_e(\mathbf{x},\mathbf{y}) \le \varepsilon_i$) should be modified into an equality constraint in such a case.

A more detailed discussion about the advantages and weaknesses of different sustainability measurements, such as footprints, LCA indexes, eco cost, eco profit, and total profit, in relation to MOO synthesis of sustainable systems can be found elsewhere (Kravanja, 2012).

AGGREGATE MEASURE OF ENVIRONMENTAL SUSTAINABILITY

In many cases, GWP or CF are evaluated as criteria for environmental sustainability (castrated type of LCA) (Finkbeiner, 2009). Most of the effort and resources are spent to reduce the CF. In general, there may be either synergies (non-trade-offs) or compromises (trade-offs) within environmental metrics. For example, photovoltaic shows synergies in CF, WF, NF, and ENF. Biogas production, however, reduces CF, NF, and ENF but increases the WF (Vujanović et al., 2014). Usually, there are compromises in environmental footprints and the possibilities for achieving the improvements within the system are revealed (Azapagic and Clift, 1999). Proper decisions are strongly dependent on regional characteristics and climatic conditions. Therefore, more comprehensive analyses that consider all aspects of the natural environment, human health, and resources should be performed (Finkbeiner, 2009).

There are aggregate measures of sustainability consisting of "one number," such as eco-efficiency, eco profit, and total profit, eco NPV and total NPV, and other combined criteria. Their advantage is that a unique (optimal) sustainable solution could be obtained directly, which is especially important when solving large-scale problems when the models are huge and computational times are significant. However, this is not the case with indicators of potential environmental impacts and environmental footprints.

When applying these metrics, a comprehensive list of objectives (indicators of potential impacts or footprints) should be taken into account and evaluated. To "create a pathway" toward more sustainable development, preferred optimal solutions need to be obtained. For this purpose, MOO should be performed.

If several objectives are considered (e.g., all key environmental footprints), then there are limitations. Increased time is spent obtaining all the solutions (the entire solution space) in some cases. Visualization and interpretation of solutions are difficult and, in many cases, unclear. Also, it is possible to present, at most, four-dimensional Pareto projections (Čuček et al., 2014). It is necessary to apply a criteria dimensionality reduction technique to facilitate the comprehension of the solution space when many objectives are involved within MOO problems. Various dimension reduction approaches have been developed for this purpose and are generally categorized into linear and non-linear methods (Lygoe, 2010). Among the methods developed are a method for minimizing the error of omitting objectives (Guillén-Gosálbez, 2011), principal component analysis (Sabio et al., 2012), factor analysis (Huang and Lu, 2014), representative objectives method (Čuček et al., 2012d) applied to direct (Čuček et al., 2013b) and total objectives (Čuček et al., 2014), and others.

KEY ENVIRONMENTAL FOOTPRINTS

A literature review indicates that the major categories of footprints developed to date are CF, EF, WF, and ENF, forming the so-called footprint family (Galli et al., 2011). No single indicator has been able to comprehensively monitor human impact on the environment, but indicators need to be used and interpreted jointly (Galli et al., 2012). EF, ENF, CF, and WF rank as the more important footprint indicators in the existing literature. They are in close relationship with the four worldwide concerns regarding threats to human society: food security; energy security; climate security; and water security (Fang et al., 2013).

Besides these footprints, NF, PF, BF, and LF are recognized as key environmental footprints because they are among those essential for health security, food and water security, and land and species security. The importance of these footprints can also be seen in Figure 5.6, which presents the planet boundaries. These boundaries are associated with the planet's biophysical subsystems or processes (Rockström

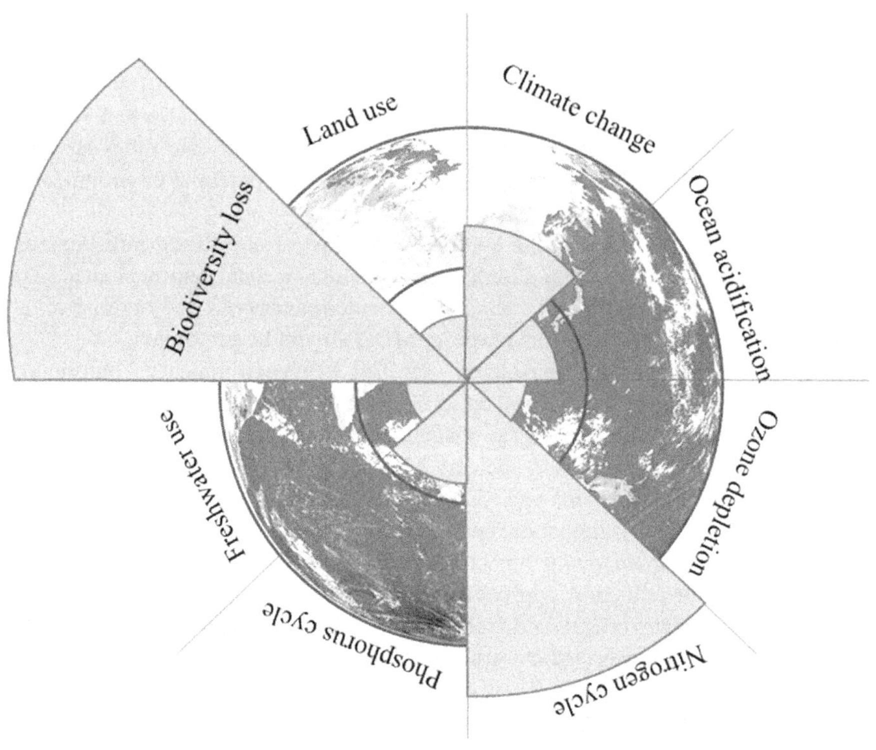

FIGURE 5.6

Planetary boundaries.

Modified from Rockström et al.(2009).

et al., 2009) and define human-determined acceptable levels of a key global variable (Carpenter and Bennett, 2011). It can be seen that climate change (CF), rate of biodiversity loss (BF), and interference with the nitrogen cycle (NF) have already transgressed the boundaries. In addition, other boundaries such as freshwater use (WF), land use (LF), ocean acidification, ozone depletion, and interference with global phosphorus cycle (PF) are fast approaching. The more quoted definitions, policies, and ways of reducing key environmental footprints are presented in Table 5.2.

The footprint family of indicators are frequently used indicators in scientific literature (Figure 5.7). Also, the remaining indicators recognized as key are becoming more and more popular. For those indicators it is foreseen that they will also become "hot footprints" over the next few years. Figure 5.7 shows the number of publications addressing the "footprint family" of indicators in scientific search engines Science Direct and Scopus. Figure 5.7 shows an increased interest in key environmental footprint evaluation in Science Direct and Scopus can be seen over the years to 2014 (June 2014).

Table 5.2 Key Environmental Footprints: Definitions, Policies, and Ways for Footprint Reduction or Prevention

Footprint	Definition	Policy	How to Reduce It
Carbon footprint (CF)	The more quoted definition: CF stands for the amount of CO_2 and other GHGs, emitted over the full life cycle of a process or product. It is expressed in mass of $CO_2{}^{-}eq$ (UK Parliamentary Office of Science and Technology (POST), 2011)	Policies include targets for emissions reduction, increased use of renewable energy, and increased energy efficiency. The main current international agreement on reducing emissions of GHGs is the Kyoto protocol (United Nations Framework Convention on Climate Change, 1998)	(Čuček et al., 2012b): – Technological development – Carbon capture and storage (CCS) – Carbon offsetting (compensation for emissions made elsewhere) – Reduced energy consumption – Improved energy efficiency (Perry et al., 2008): – Integration of waste and renewable energy during energy production (Carbonfund.org, 2014): – Reducing, reusing, and recycling of products – Consuming less meat – Reducing food waste – Buying local products if possible

(Continued)

Table 5.2 Key Environmental Footprints: Definitions, Policies, and Ways for Footprint Reduction or Prevention *Continued*

Footprint	Definition	Policy	How to Reduce It
Water footprint (WF)	Indicator of direct and indirect water use by a consumer or producer (Hoekstra et al., 2011). It is divided into blue (consumption of freshwater), green (consumption of rainwater), and gray WFs (indicator of pollution)	Due to increasing pressures on water resources, policies should help to secure water resources for future generations. The Water Framework Directive of the EU aims to protect all water bodies by preventing pollution at the source and set a target date of 2015 to achieve "good status" for all waters (European Commission, 2000). Clean Water Act is the primary law in the United States that addresses surface water quality protection and water pollution control (Clean Water Act, 2002). Also, in other countries there are water protection legislations, such as Water Law of the People's Republic of China (The Central People's Government of The People's Republic of China, 2002), Water Act of Australia (Australian Government, 2013), etc.	– Responsible and efficient water use at all levels – Better water conservation and management – Avoid wasting of blue water (for irrigation, etc.) – Recycling water as much as possible (in industries) – Make better use of green water (planting, use in toiletries, etc.) – Minimize gray WF – Substitute the products with smaller WF (Gruener, 2010)
Ecological footprint (EF), biocapacity, and ecological overshoot	EF is a resource accounting indicator that measures how many bioproductive land and water areas are available on Earth, and how much of this area is appropriated for human use (Kitzes et al., 2007). EF is related to Earth global biocapacity and overshooting Biocapacity is the capacity of an area to provide the resources and absorb waste Overshoot or ecological deficit stands for usage of more resources that Earth can renew, and can persist only for some	Very limited impact in terms of policy lessons because EF is an aggregated indicator (van den Bergh and Grazi, 2014)	– Carbon capture and storage (CCS) – Carbon offsetting – Reduced energy and water consumption – Improved energy efficiency – Renewable energy utilization – Responsible and efficient use or products – Reducing, reusing, and recycling of products

	time. The human economy can deplete stocks and fill waste sinks (Ewing et al., 2010). It is a difference between biocapacity and EF (Global Footprint Network, 2012)		– Reducing waste – Using more sustainable transportation
Energy footprint (ENF)	ENF is currently superficially defined. It is suggested that it stands for the specific energy usage per functional unit when considering fossil-based and renewable-based energy (Sobhani et al., 2012)	Efficient energy use and renewable energy are the main pillars of sustainable energy policies. Various countries around the world have adopted different energy policies. The European Union set binding targets by 2020 to reduce GHG emissions by 20% from 1990 levels, raising the share of energy production from renewable resources to 20%, improvement in energy efficiency, and a minimum target of 10% of the renewable energy in the transportation sector (European Commission, 2009a). The United States established several energy management goals and requirements due to economic and energy security considerations. Among the acts, the more important are Energy Policy Act of 2005 (Energy Policy Act, 2005) and Energy Independence and Security Act of 2007 (Energy independence and Security Act, 2007). China's twelfth 5-year plan (2011–2015) set a compulsory target of non-fossil fuels proportion of 11.4% by 2015 and 15% by 2020, energy intensity reduction by 16%, and carbon intensity reduction by 17% (Li and Wang, 2012). Renewable Energy Target scheme in Australia requires 20% of Australia's electricity coming from renewable sources by 2020 (Australian Government and Department of the Environment, 2014), etc.	– Technological developments – Reduced energy consumption – Improved energy efficiency as one of the larger and least costly opportunities – Using more sustainable transportation – Renewable energy production and consumption

(Continued)

Table 5.2 Key Environmental Footprints: Definitions, Policies, and Ways for Footprint Reduction or Prevention *Continued*

Footprint	Definition	Policy	How to Reduce It
Nitrogen footprint (NF)	NF is the total amount of reactive nitrogen released into the environment as a result of an entity's resource consumption, expressed in mass units of reactive nitrogen (Leach et al., 2012). It represents disruption of the regional to global nitrogen cycle and its consequences due to human activities (Čuček et al., 2012b)	Several directives are related to reactive nitrogen and phosphorus emissions in air and water bodies to address the problem of nutrient build-up in ecosystems. The European Union's main directives on nitrogen and phosphorus in the atmosphere are: directive 2001/81/EC, National Emission Ceilings; 2008/1/EC, Integrated Pollution, Prevention, and Control (IPPC); and 2008/50/EC, Ambient Air Quality. The European Union's main directives related to nutrient emissions in water bodies are: 2006/118/EC, Groundwater Directive; 208/56/EC, Marine Strategy Framework Directive; 91/676/EEC, Nitrates Directive; 91/271/EEC, Urban Waste Water Treatment Directive; and 2000/60/EC, Water Framework Directive (Oenema et al., 2011). The main legislations in the United States related to nutrient emissions are Clean Water Act. (2002) and The Clean Air Act (2004)	(Čuček et al., 2012b): – Reducing energy consumption – Changing diets to more sustainably prepared food and fish – Consuming less meat, particularly beef – Reducing food waste – Crop genetic engineering to create symbiotic Rhizobium bacteria – Traditional or enhanced breeding techniques – Precision farming based on soil analysis – Cover crops (Smil, 1997): – Stabilization of population – Recycling of organic waste – Crop rotation – Legume cultivation (European Commission DG Environment News Alert Service, 2012): – Using more sustainable transportation – Improving wastewater treatment

Phosphorus footprint (PF)	PF represents disruption of phosphorus cycle. It is expressed in mass units of phosphorus per functional unit	See policies for NF reduction	– Reducing food waste – Crop genetic engineering – Search for new phosphorus sources (Smil, 2000): – Stabilization of population – Consuming less meat and dairy products – More efficient fertilization (e.g., terracing, no-till agriculture, cover crops) – Precision farming – Removal of phosphorus from sewage and industrial waste Reduce, reuse, and recycle (Vaccari, 2009): – Improving wastewater treatment by removing pollutants, especially heavy metals – Using agricultural residues and animal waste, including bones as fertilizer – Halt excreted phosphorus
Biodiversity footprint (BF)	BF is defined as "the summation of all the pressures that have potential consequences for biodiversity" (de Bie and van Dessel, 2011)	Several biodiversity-related legislations have been developed, and approximately 170 countries now have national biodiversity strategies and action plans (Secretariat of the Convention on Biological Diversity, 2010). Among them are the EU Biodiversity Strategy to 2020 (European Commission, 2011b), Australia's Biodiversity Conservation Strategy 2010–2030 (Natural Resource Management Ministerial Council, 2010), Canada Species at Risk Act (Minister of	– CF, NF, WF, and PF mitigation (Secretariat of the Convention on Biological Diversity, 2010): – Careful land-use planning – Sustainably produced and harvested food (Hooper): – Reduce (avoid), reuse, and recycle

(*Continued*)

Table 5.2 Key Environmental Footprints: Definitions, Policies, and Ways for Footprint Reduction or Prevention *Continued*

Footprint	Definition	Policy	How to Reduce It
Biodiversity footprint (BF)		Justice, 2014), Biological Diversity Act of India (National Biodiversity Authority, 2014), National Policy of Biological Diversity of Malaysia (Ministry of Science Technology and The Environment Malaysia, 1998), etc.	– Opt for renewable energy and energy efficiency – Reducing the use of pesticides and fertilizers – More resource-efficient economy (European Commission, 2011a): – Protecting areas (both in land and in coastal waters) to conserve the native habitats and biodiversity – Conservation of particular species – Restoration of terrestrial, inland water and marine ecosystems and their services, and conserve biodiversity – Combating invasive alien species – Ensuring the sustainable use of fisheries resources
Land footprint (LF)	LF, also termed actual land demand, is related to land requirement and is defined as the summation of all the areas directly and indirectly required to satisfy the consumption (Giljum et al., 2013a)	Agricultural policies should support the implementation of sustainable agricultural practices. Policies related to CF, ENF, and BF also affect LF. Biofuel policies (Sorda et al., 2010) could have significantly negative impact on LF	– Careful land-use planning (Giljum et al., 2013b): – Reduce consumption of meat and other animal products – Opt for renewable energy that has low LF and, therefore, avoid the use of biofuel

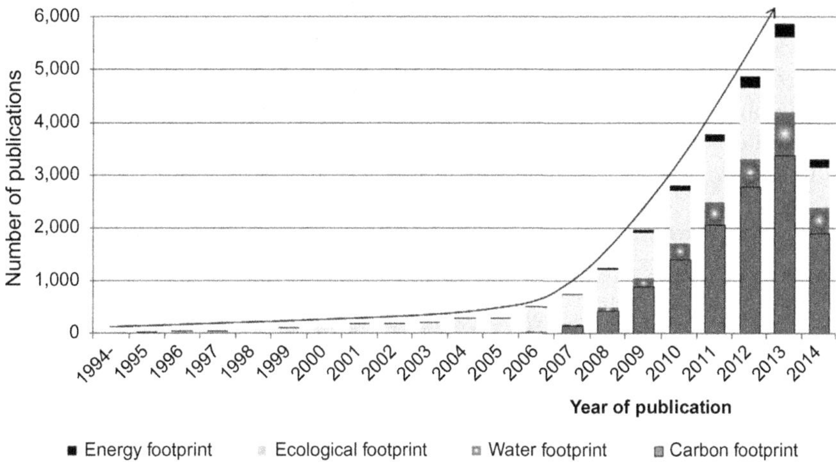

FIGURE 5.7

Increasing number of scientific publications related to evaluation of the footprint family of indicators in Science Direct and Scopus (to June 2014).

CARBON FOOTPRINT

Over the past few years, CF, has become one of the more important environmental protection indicators (Wiedmann and Minx, 2008). It has been widely used in scientific literature. Several research studies have analyzed the CF, its impact, minimization, mitigation, and even sequestration (Klemeš et al., 2006). Five-thousand fifty articles in Science Direct and 7,500 articles in Scopus (May 2014) address the CF. The more quoted definitions, policies, and methods regarding CF reductions are presented in Table 5.2.

CF is related to human-induced climate change and global warming, which are recognized by many as the greatest environmental threat for the twenty-first century (Abbott, 2008). The term "global warming" considers the continuing increase in the average temperature of the Earth's atmosphere and oceans. Many researchers assume that it is caused by increased concentrations of GHGs in the atmosphere, resulting (most likely) from human activities such as deforestation, other land use changes, and the burning of fossil fuels. The main GHGs, the concentrations of which are increasing, are CO_2, CH_4, N_2O, hydrochlorofluorocarbons (HCFCs), hydro-fluorocarbons (HFCs), and ozone in the lower atmosphere (World Meteorological Organization, 2014). CO_2 is the single most important human-emitted GHG in the atmosphere, contributing approximately 63.5% to the overall global radiative forcing, not considering water vapor, which is continuously within the Earth's climatic circulation (World Meteorological Organization, 2009). The burning of fossil fuels releases additional (not released by natural processes) CO_2 into the atmosphere. Approximately half of this excess CO_2 is absorbed by the land and oceans, but the remainder accumulates within the atmosphere and enhances the natural greenhouse effect (Dilling et al., 2003). Global emissions from the burning of fossil fuels, the clearing of forests, and agricultural activities have led to the release of more than 1,100 Gt of CO_2 into the atmosphere since the mid nineteenth century (Sims et al., 2007).

It has been declared that the global average near-surface atmospheric temperature had already increased by 0.78°C during the twentieth century, with much of this warming (0.61°C) occurring over the past 30 y (Staudt et al., 2008). Temperatures are predicted to increase by at least an additional 1.1–6.4°C over the next 100 y. This warming would cause significant changes in sea levels (an increase of 0.18–0.59 m), ecosystems, the melting of glaciers, the extent of ice and snow, precipitation, water availability, and the probable expansions of subtropical deserts. Other consequences would include higher maximum temperatures, fewer colder days, changes in agricultural yields, and an increase in infectious diseases. Global warming and climate change may also be associated with deterioration in the health of humans, more intense hurricanes, tropical cyclone activity, flooding, drought, wildfires, the insect populations, and ocean acidification (Staudt et al., 2008).

CF is, in general, a consumption-based indicator (Gleick et al., 2014). It usually represents the amount of CO_2 and other GHGs emitted over the full life cycle of a process or product (UK Parliamentary Office of Science and Technology (POST), 2011). CF is quantified using indicators such as the GWP (European Commission, 2007), which represents the quantities of GHGs that contribute to global warming and climate change. A specific time horizon is considered, usually 100 y (Intergovernmental Panel on Climate Change, 2009). A similar definition indicates that the CF is a result of life cycle thinking applied to global warming (climate change) (European Commission, 2007).

The land-based definition of the CF is that it represents the land area required for the sequestration of fossil fuel CO_2 emissions from the atmosphere through afforestation (De Benedetto and Klemeš, 2009). Wiedmann and Minx (2008) proposed that the CF is a measurement of the exclusive direct (on-site, internal) and indirect (off-site, external, embodied, upstream, and downstream) CO_2 emissions of an activity or over the life cycle of a product, measured in mass units. Wright et al. (2011). suggested that only two carbon-based gases, CO_2 and CH_4, the data collection of which is relatively straightforward, should be used when determining a CF. It includes the activities of individuals, populations, governments, companies, organizations, processes, and industrial sectors (Galli et al., 2012).

Other different terms relating to GHG emissions have been suggested and/or used, such as climate footprint (Wiedmann and Minx, 2008), CO_2 footprint (Huijbregts et al., 2008), GHG footprint (Downie and Stubbs, 2013), methane footprint (Wiedmann and Barrett, 2011), and GWP footprint (Meisterling et al., 2009).

CF is calculated using LCA or input–output analysis (Hertwich and Peters, 2009). It can be calculated using various tools. The simplest are CF calculators that are aimed at reducing emissions related to individual lifestyles (Jones and Kammen, 2011). They have become relatively common on the Internet and can

Table 5.3 List of Selected CF Calculators

Application	Unit	Developer	Website
Calculates the CF for different countries from mobility and from options for living with carbon offsets	t/y and €/y	Carbon Footprint Ltd	www.carbonfootprint.com
For the United States, calculates the CF from transportation, housing, and shopping with actions to reduce the CF, possibly to zero	$t_{CO2eq.}$/y	University of California, Berkeley	coolclimate.berkeley.edu
For the United States, calculates the CF based on the state average and compares it with the United States and global average CF per person	$t_{CO2eq.}$/y	The Nature Conservancy	www.nature.org
For the United States, calculates the household CF and compares it to the average CF	lb_{CO2}/y	US EPA	www.epa.gov

promote public awareness of carbon emissions from the behavior of individuals. However, these calculators can generate varying results, even for the same individual activity (Čuček et al., 2012c). Overall, these calculators lack consistency and furnish insufficient information about their methods and estimates. Several of the CF calculators also promote methods for mitigating CO_2 emissions through offsets or investments in renewable energy technology (Padgett et al., 2008). The similarities and differences among 10 US-based calculators were reviewed by Padgett et al. (2008). CF calculators are intended primarily for individuals and households but can also be applied to businesses. Table 5.3 contains a list of CF calculators selected from several currently available CF calculators (Chait, 2009).

WATER FOOTPRINT

In recent decades, the need for measuring the WF has appeared to be attributable to freshwater security and pollution. The qualities and quantities of surface and groundwater resources are being affected by the impact of population growth, migrations to cities, increasing resource consumption, and climate change (United Nations Environment Programme, 2007). Over the next decade (by 2025), it is estimated that two-thirds of the world's population will be subjected to a shortage of water (United Nations Environment Programme, 2007), mainly attributable to water pollution (Conserve Energy Future, 2014). It is seen by many as a water crisis (Holden, 2014). Therefore, WF evaluation and minimization are priorities for sustainability. The good news is that reducing the WF to a sustainable level is possible with consumption pattern change, even with an increasing population (Ercin and Hoekstra, 2014).

WF was introduced in 2002 by Hoekstra and Hung (2002). It has also attracted significant and increased attention within the scientific community, but much lower than that of the CF (Figure 5.7).

WF is an indicator of direct and indirect water usage by an individual, community, business, or nation (Hoekstra, 2008). It is considered a consumption-based metric because it attributes the water usage to the consumer rather than the producer (Gleick et al., 2014). It is closely linked to the concept of virtual water (Hoekstra and Chapagain, 2007) and represents the total volume of direct and indirect freshwater used, consumed, and/or polluted. Table 5.2 presents the definition, policies, and possibilities regarding how to decrease WFs.

WF consists of (Hoekstra et al., 2011):

- Blue WF: Indicator of consumption of freshwater (surface and groundwater) to produce the goods and services by the individual or community. It includes evaporated water, water incorporated within products, water not returned to the same catchment area, and water not returned during the same period.
- Green WF: Consumption of rainwater that does not run-off or recharge the groundwater but is stored within the soil as soil moisture. For crops and wood, it is the sum of production-related evapotranspiration and incorporation of water into the harvested products.
- Gray WF: Indicator of pollution. It is defined as the volume of freshwater that is required to dilute pollutants to such an extent that the quality of the water remains above agreed water quality standards.

WF is a method for quantifying water usage for a particular product for any well-defined group of consumers (e.g., an individual, city, province, state, or nation) or producer (e.g., a public organization, private enterprise, or economic sector). It can also reflect the embodied or virtual water in imports and exports (Dong et al., 2013). A WF is measured in terms of water volumes consumed (evaporated or incorporated within the product) and polluted per unit of time or per functional unit (Galli et al., 2012).

The strength of the WF concept is that it provides a broad perspective on the water management of the system and allows for a deeper understanding of water usage. The WF integrates water usage and pollution over the complete supply chain (Galli et al., 2011). It measures water usage of different types, such as blue, green, and gray waters. It also connects water uses to specific places and times, and evaluates the hydrological sustainability of a system (Dourte and Fraisse, 2012).

The weaknesses of the WF are that it represents just the quantity of water used without an estimation of the related environmental impacts, the lack of required data, the estimation of gray WF is subjective (Jeswani and Azapagic, 2011), and no uncertainty studies are available even though uncertainty can be significant (Galli et al., 2011). There is still greater potential for improvement regarding accounting standards as well as data coverage, disaggregation, and quality (Giljum et al., 2013a).

There is a huge WF associated with some products. Table 5.4 shows the WF of some selected products and of primary energy carriers.

Table 5.4 Average WF of Some Products and Primary Energy Carriers

Product (Hoekstra and Chapagain, 2007)	Virtual Water (L)	Primary Energy Carrier (1 GJ[a]) (Gerbens-Leenes et al., 2008)	Virtual Water (m^3)
1 glass of beer (250 mL)	75	Natural gas	0.11
1 cup of coffee (125 mL)	140	Coal	0.16
1 slice of bread (30 g)	40	Crude oil	1.06
1 cotton t-shirt (250 g)	2,000	Uranium	0.09
1 egg (140 g)	135	Wind energy	≈0
1 hamburger (150 g)	2,400	Solar thermal energy	0.30
1 kg of rice	2,500	Hydropower	22.3
1 kg of beef (Water Footprint Network, 2014)	15,400	Biomass energy	10–250

[a]*277.8 kWh.*

Table 5.5 List of Selected WF Calculators

Application	Unit	Developer	Website
For different countries, it calculates the water required to produce the goods and services consumed by humans	m^3/y	Hoekstra, Chapagain, Mekonnen (Water Footprint Network, 2014)	www.waterfootprint.org
For different countries, it calculates the water required to produce the goods and services consumed by humans	L/d, gal/d	Kemira (2014)	www.waterfootprintkemira.com
For the United States, it individually calculates the WF relating to water used in homes and yards, food consumption, transportation and energy, and purchased products	L/d, gal/d	National Geographic (2014a)	environment.nationalgeographic.com
For US individuals and households, it calculates the WF, compares it with that of the average American user, and provides water-saving tips	L/d, gal/d	GRACE Communications Foundation (2014)	www.gracelinks.org

WF can also be calculated using various tools, such as WF calculators, graph-based tools, mathematical programming tools, and input–output analysis. The simplest are WF calculators, which have become common tools for evaluating WF. Table 5.5 contains a list of a few selected WF calculators.

ECOLOGICAL FOOTPRINT

The EF was developed in 1992 by Rees (1992). It answers the question: "How much of the biological capacity of the planet is required by a given human activity or population?" (Global Footprint Network, 2011). The average world citizen has an EF of approximately 2.7 global average hectares, but there are only 2.1 global average hectares of bioproductive land and water per capita on Earth (Rees, 2010). According to Sadik (Motavalli, 1999), many environmentalists think that the carrying capacity of the Earth is a maximum of 4 billion. Currently, the world population exceeds 7.2 billion people (Worldometers—Real Time World Statistics, 2014), and it is estimated that it will reach and stabilize at 10 billion (Worldometers—Real time World Statistics, 2014).

Humanity had already overshot global biocapacity by 50% during approximately 2007 (Ewing et al., 2010), which means that humanity used approximately 1.5 Earths to support its consumption (Galli et al., 2014). The increase in global demands is mainly attributable to CF, fishing ground footprint, and agricultural footprint (Galli et al., 2014). This overshoot, if it continues, will put global ecosystems at serious risk for degradation or collapse (Kitzes et al., 2007). For these reasons, EF evaluation is also attracting increased interest within the scientific community (Figure 5.7). Furthermore, Table 5.2 offers definitions of EFs, biocapacity, and ecological overshooting. In terms of policy, there are only limited measures; however, there are several possibilities for decreasing EFs.

EF has emerged as the world's primary measurement of humanity's demands on nature (Wackernagel and Rees, 1996) and is now widely used as an indicator for measuring environmental sustainability. EF is defined as a measurement of the human demand for land and water areas and compares the human consumption of resources and absorption of waste with the Earth's ecological capacity to regenerate (Global Footprint Network, 2010). EF provides an aggregated assessment of multiple anthropogenic pressures (Galli et al., 2012). It is a composite indicator that combines (Toderoiu, 2010):

- Built-up area (built-up land footprint)—Area covered by human infrastructure for housing, transportation, and industrial production (settlements, roads, hydroelectric dams, etc.). By best estimates approximately 0.2×10^9 ha or 1.3% of the total land is built-up land.
- Land for carbon absorption or uptake (CF)—Biologically productive area required to absorb the CO_2 not sequestered by the oceans (Kitzes et al., 2007). It is calculated using the carbon absorption potential of the world

average of forests. The footprint is primarily attributable to fossil fuel burning, international trade, and land use practices (Borucke et al., 2013).

- Fishing grounds (fishing ground footprint)—Area of inland and marine waters necessary for primary production of fish, seafood, and other marine products. Approximately 2.3×10^9 ha (6.4% of total water surface) of fishing ground exist (only continental shelves and inland waters are included).
- Forest area (forest land footprint)—Area of forest required to support the annual harvest of timber products, fuel wood, and pulp. Approximately 3.9×10^9 ha (26.2% of total land) of forest are available worldwide.
- Grazing land (grazing land footprint)—Area of grassland used in addition to crop feeds required to raise livestock for meat, hides, wool products, milk, and dairy. It comprises all grassland used to provide feed for animals, including cultivated pastures as well as wild grasslands and prairies. There are approximately 3.5×10^9 ha (23.5% of total land) of grassland and pasture worldwide.
- Cropland (cropland footprint)—Area required to grow all crop products required for human consumption (food and fiber) and to grow livestock feeds, fish meals, oil crops, and rubber. There are approximately 1.5×10^9 ha (10.1% of total land) of cropland worldwide.

The main strength of the EF concept is that it is attractive and intuitive (Schaefer et al., 2006), and that its methodology is continuously improving. It is especially useful in raising the awareness of environmental loading imposed by humanity (Ferng, 2002). The EF helps in the understanding of the complex relationships between the many environmental problems by exposing humanity to a "peak everything" situation (Galli et al., 2011). However, it should be noted that the EF measures only one major aspect of sustainability, namely, the environmental aspect, and not all environmental concerns (Galli et al., 2012). EF excludes economic or social indicators.

EF is, in general, a consumption-based indicator and includes all biologically productive areas worldwide for satisfying consumption at specific local levels, including those embodied in import and export (Giljum et al., 2013a). EF's balance for local consumption (e.g., within a country), EF_C is calculated as follows (Galli et al., 2014):

$$EF_C = EF_P + EF_I - EF_E \tag{5.12}$$

where

EF_P—ecological footprint of production
EF_I—footprint embodied in imported commodity flows
EF_E—footprint embodied in exported commodity flows

The EF of each single product i, EF_i, locally produced, imported, or exported is defined as (Galli et al., 2014)

$$EF_i = \frac{P_i}{Y_{W,i}} \cdot EQF_i \tag{5.13}$$

where

P_i—amount of each primary product i harvested, amount of CO_2 emitted
$Y_{W,i}$—annual world average yield for the production of commodity i, carbon uptake capacity
EQF_i—equivalence factor for the land use type producing product i

Local biocapacity, BC, is defined in Eq. (5.14) and provides an assessment of the capacities of assets to produce renewable resources and ecological services (Galli et al., 2014):

$$BC = \sum_i A_{N,i} \cdot YF_{N,i} \cdot EQF_i \tag{5.14}$$

where

$A_{N,i}$—bioproductive area that is available for the production of each product i at the local level
$Y_{W,i}$—specific yield factor for the land producing product i

EF and biocapacity can be applied at scales ranging from single products to households, cities, regions, and countries or to humanity as a whole; however, it is most effective, meaningful, and robust at aggregate levels (Wackernagel et al., 2006). National footprint accounts (nation-level EF assessments) are regarded as the most complete (Kitzes et al., 2009).

EF is usually measured in global area units as the amount of bioproductive space (Hoekstra, 2008) and in global area units per person (Ewing et al., 2010). Each global hectare represents the same fraction of the Earth's total bioproductivity and is defined as one hectare of land or water normalized to the world average productivity from all of the biologically productive land and water within a given year. The total biologically productive area available on the Earth is approximately 12 billion hectares (Galli et al., 2011). Biologically productive areas include cropland, forests, and fishing grounds but exclude deserts, glaciers, or the open ocean (Kitzes and Wackernagel, 2009). However, converting the data to area units can be problematic. It also has limited data availability, uncertainty of data, and geographic specificity.

EF can also be calculated using various tools. The simplest are EF calculators, which have become common tools. Table 5.6 contains a list of a few selected EF calculators.

ENERGY FOOTPRINT

Population and income growth are the two more powerful driving forces behind the demand for energy (British Petroleum, 2011). Consequently, insecure energy supply, high energy prices, and ever-increasing energy demand are among the more important issues in today's society (Brandi et al., 2011). Most likely, the global energy usage will continue to increase by using predominantly fossil fuels (British Petroleum, 2011). In 2012, the global primary energy consumption was

Table 5.6 List of Selected EF Calculators

Application	Unit	Developer	Website
For 17 countries, it calculates how many planets and how much land area is required to support a human's lifestyle, including food, shelter, mobility, goods, and services	$N_{\text{Planet Earths}}$ and global hectares	Global Footprint Network	www.footprintnetwork.org
It calculates how many planets are required to support a human's lifestyle, including food, travel, home, and staff	$N_{\text{Planet Earths}}$	World Wide Fund for Nature	footprint.wwf.org.uk
For various countries, it calculates EF and provides suggestions for reducing carbon, food, housing, and goods and services footprints	$N_{\text{Planet Earths}}$ and global hectares	Center for Sustainable Economy	myfootprint.org

522 EJ (12.47 Gtoe) (British Petroleum, 2013) and grew by 1.8% per year. Oil remained the world's leading fuel with 33.1% of global energy consumption, natural gas's share of global primary energy consumption was 23.9%, global coal's consumption was 29.9%, and nuclear global energy consumption was 4.5%. The renewable's share in global energy consumption was 8.6%, of which 6.7% was attributable to hydroelectricity. It should be noted that those numbers include only commercially traded fuels (British Petroleum, 2013).

Humankind is rapidly exhausting fossil fuels and, as a consequence, people are going to depend on non-fossil energy sources in future. The timing of a global peak regarding oil production is less certain, although there is a growing view that maximum production will occur between 2010 and 2020 (Maggio and Cacciola, 2012). The global peak for coal extraction from the existing coalfields was predicted by Patzek and Croft (2010) to occur approximately during 2011. Some other estimates predicted peak coal later, generally between 2010 (Mohr and Evans, 2009) and 2062 (Maggio and Cacciola, 2012). Natural gas is performing slightly better, with the world's natural gas production peak being predicted between 2025 (Maggio and Cacciola, 2012) and 2066 (Mohr and Evans, 2011). An energy crisis will be an urgent challenge handled within the twenty-first century. Developing clean and renewable energy resources ranks as one of the greatest challenges facing humankind in the medium-term to long-term (Mata et al., 2011). No single energy technology or combination of technologies exists that can address all challenges in a sustainable manner (Ma et al., 2011). ENFs are therefore of significant importance. The suggested definition, various policies, and possibilities for reducing ENFs are shown in Table 5.2.

In general, various definitions of an ENF have been offered. The global footprint network (Global Footprint Network, 2012) defined it as the sum of all areas used to provide non-food and non-feed energy. ENF is the sum of the areas of carbon uptake land, hydropower land, forested land for wood fuel, and cropland for fuel crops. Palmer (1998) defined an ENF as a measurement of the land required to absorb those CO_2 emissions originating from energy usage. Another definition of an ENF is that it represents the area required to sustain energy consumption and is measured as the area of forest that would be required to absorb the resulting CO_2 emissions, excluding the proportion absorbed by the oceans and the area occupied by hydroelectric dams and reservoirs for hydropower (World Wide Fund for Nature, 2002). De Benedetto and Klemeš (2009) calculated an ENF by multiplying the final energy usage of different energy carriers by their land indices and adding the results to the footprint of the whole supply chain. Yet another definition of an ENF is that it corresponds to the demand for non-renewable energy resources (Schindler, 2013), such as cumulative fossil and nuclear energy demand in terms of primary non-renewable energy use, such as energy content of a material as well as the non-renewable energy spent on its extraction (Hermann et al., 2011). It is also defined as the specific energy usage per functional unit (Sobhani et al., 2012) considering fossil-based and renewable-based energy (Vujanović et al., 2014). Because of differences in the qualities of heat and electrical energies, ENF can be divided into electricity–transportation footprint and heat footprint (Vujanović et al., 2014).

ENF can be measured in local (the surface area of a specified region's average biologically productive land and sea over a given year) (Global Footprint Network, 2012) or global (the surface area of the Earth's average biologically productive land and sea over a given year) (Wiedmann and Lenzen, 2007) area units or in units of energy (higher or lower heating value (Hermann et al., 2011) per functional unit.

ENF includes sub-footprints, such as the fossil or fossil ENF (Stoeglehner and Narodoslawsky, 2009), nuclear ENF (Stoeglehner et al., 2005), renewable ENF (Chen and Lin, 2008), wind footprint (Magoha, 2002), solar footprint (Denholm and Margolis, 2008), heat footprint (Vujanović et al., 2014), and others.

NITROGEN FOOTPRINT

Nitrogen is essential for life and it is a critical limiting element for food production (Smil, 1997). It is the most common element in the Earth's atmosphere and a primary component of crucial biological molecules, including proteins and nucleic acids such as DNA (deoxyribonucleic acid) and RNA (ribonucleic acid). It plays a key role in helping to feed the growing populations. Smil (2001) estimated that in the mid 1990s, 40% of the global population was dependent on crops fertilized with reactive nitrogen. In 2008, an updated estimate was that 48% of the world's population was fed by Haber–Bosch nitrogen (Erisman et al., 2008). Since 1970, the world's population has increased by 78% and production of reactive nitrogen, N_r (all N species except N_2), increased by 120% (Galloway et al., 2008). Crops need large amounts of N to grow, but only N_r can be readily used by most organisms, including crops.

However, the world's nitrogen cycle has been dramatically altered because of human activities (Galloway et al., 2008). The amount of human-caused reactive or biologically available N_r within the global environment has increased by a factor of 12.5 (187 Mt N/y in 2005) since the nineteenth century (15 Mt N/y in 1860), in association with the increased use of fertilizers (Galloway et al., 2008). Agriculture is responsible for approximately 80% of the N_r produced worldwide (Union of Concerned Scientists, 2009). More than half of the synthetic N fertilizer ever used on the planet has been used since 1985 (Millennium Ecosystem Assessment, 2005). N_r is also created by the burning of fossil fuel and biomass, by manure run-off, and by the planting of legumes. The primary nitrogen emission sources are transportation, agriculture, power plants, and industry.

Much anthropogenic nitrogen is lost in the air, water, and soil, and has a destructive effect on the ecosystem and human health (Galloway et al., 2008). Approximately 80% of nitrogen used in food production is lost before the consumption and the remainder is lost after consumption as human waste. In biofuel production, all of the nitrogen used is lost to the environment (N-Print Team, 2014). Nitrogen deposition significantly alters the global nitrogen cycle, reduces the biodiversity, pollutes and degrades waters, impacts the human health (shortness of breath, blue baby disease (U.S. Environmental Protection Agency (EPA), 2014), and some cancers (Smil, 1997)), and affects the global environment (Vitousek et al., 1997). It contributes to the eutrophication of coastal rivers and bays (non-potable water, blooms of cyanobacteria and algae, fish kills, and consequently dead zones), formation of smog, soil acidification (Smil, 1997), depletion of ozone in the upper atmosphere, and global warming (European Commission DG Environment News Alert Service, 2012).

NF is a measurement of the amount of reactive nitrogen released into the environment as a result of human activities, expressed in total units of N_r (Čuček et al., 2012b) (see also the NF website) (N-Print Team, 2014). The definition, policies, and NF reduction suggestion paths are represented in Table 5.2. NF mainly covers the N_r emissions NO_x, N_2O, NO_3^-, and NH_3, and they can be rapidly interconverted from one N_r form to another (Galloway et al., 2003). An N_r atom that starts out as part of NH_3 in the Haber–Bosch process is used to produce fertilizer. N_r is then partly incorporated within the crops and then is partly released as NH_3, NO, N_2O, N_2, or NO_3^-. The N_r species can be rapidly interconverted from one N_r form to another (Galloway et al., 2003). NF represents disruption of the regional to global nitrogen cycle and its consequences. The weakness of the NF is the lack of data, its uncertainty (Leach et al., 2012), and high data demand (Gu et al., 2013).

NF can be calculated using various tools. The better-known tools are NF calculators. Table 5.7 contains a list of available NF calculators.

Table 5.7 List of NF Calculators

Application	Unit	Developer	Website
For the United States, United Kingdom, Germany, and the Netherlands, it determines an individual's contribution to nitrogen losses to the environment from resource uses. It focuses on four main areas of consumption: food, housing, transportation, and goods and services (Leach et al., 2012)	mass unit/y	N-PRINT Team	n-print.org (N-Print Team, 2014)
For Chesapeake watershed, it calculates average household NF and provides tips to reduce the footprint	lb/y	Chesapeake Bay Foundation	www.cbf.org (Chesapeake Bay Foundation, 2014)

PHOSPHORUS FOOTPRINT

Phosphorus is a fossil mineral found in rocks that accumulates as a result of geological processes. In addition to nitrogen, phosphorus is a critical element in food security. Today, food could not be produced at current global levels without the use of processed mineral fertilizers; therefore, humanity has become addicted to phosphate rock (Cordell et al., 2009).

Phosphorus is a finite resource that might become scarce and might peak within the next decades (Cordell, 2010). It is predicted that current global reserves may be depleted during the next decades (Cordell et al., 2009). However, depletions and peaks in mineral resources are difficult to forecast (Carpenter and Bennett, 2011) because of changes in production costs, data reliability, changes in demand and supply (Cordell et al., 2009), and possible future extraction technologies. Nonetheless, peak phosphorus is worthy of careful consideration because declines in phosphorus availability could significantly decrease agricultural yields (Carpenter and Bennett, 2011). The PF is therefore of significance. It may even be more important in the future because of a growing global population and biofuels. The main definition, policies, and potential ways to reduce PF are shown in Table 5.2.

The PF addresses the phosphorus imbalance within crops (Lott et al., 2009). Leakage of phosphorus leads to eutrophication of surface freshwater and some coastal waters (Carpenter and Bennett, 2011), which spurs blooms of cyanobacteria depleting fish and creating dead zones. More than 400 dead zones have been reported worldwide, affecting an area of more than 245,000 km^2 (Diaz and Rosenberg, 2008).

BIODIVERSITY FOOTPRINT

Humans have transformed 40–50% of the ice-free land surface, changing prairies, forests, and wetlands into agricultural and urban systems (Chapin et al., 2000).

They are destroying biological diversity (biodiversity) at an alarming rate (Purvis and Hector, 2000) and causing changes in the global distribution of organisms (Chapin et al., 2000).

The main drivers of biodiversity loss are (Sodhi et al., 2009):

- Landscape modification, habitat loss or change, fragmentation, human infrastructure, and deforestation, especially tropical humid forests that are the more species-rich
- Spread of invasive non-native species or genes outcompeting endogenous species and the spread of disease (Galli et al., 2014)
- Overharvesting and overexploitation of specific species, especially attributable to fishing and hunting. The global trading system for food, cosmetics, and pharmaceuticals is related to it
- Pollution that affects the health of species, particularly in aquatic ecosystems, such as excess nutrient (nitrogen and phosphorus) loading in waters, pesticide use in farming, and dumping of garbage at sea
- Climate change and the greenhouse effect

Ultimately, the main drivers are human activities (European Commission, 2011a) and the size, growth, and resource demands of the human population (Nag, 2008). As the world population and economy grow, so do the pressures on biodiversity (Galli et al., 2014).

Extinction is the most common way biodiversity is lost or reduced (Treat and Callahan, 2013). When a species becomes endangered or extinct, usually more than just that species is affected. According to some studies, humanity has already caused the extinction of 5–20% of species (Chapin et al., 2000) in many groups of organisms, and current rates of extinction are estimated to be 100 times to 1,000 times faster than natural extinction rates (Pimm et al., 1995). Some estimates are that the extinction rate is up to 140,000 species per year, and a much larger number of species are endangered (Nag, 2008).

BF is defined as "the summation of all the pressures that have potential consequences for biodiversity" (de Bie and van Dessel, 2011). It is related to reduction in biodiversity and, therefore, to the rate of biodiversity loss (Hanafiah et al., 2012). BF is also measured as a global area required to compensate for the mean species abundance loss caused by direct land use and by fossil-based CO_2 emissions for the life cycle of a product (Hanafiah et al., 2012). Definition, policy, and BF reduction options are summarized in Table 5.2.

Biodiversity loss threatens the well-being of humans (Díaz et al., 2006). It is estimated that between 50,000 and 70,000 plant species are used in medicine worldwide (Schülke, 2008), and approximately 1×10^8 t of aquatic life, including fish, molluscs, and crustaceans, are taken from the wild every year. Extinctions can disrupt vital ecological processes such as pollination and seed dispersal, leading to cascading losses, ecosystem collapse, and a higher extinction rate overall (Sodhi et al., 2009). The risks related to biodiversity and loss of ecosystem services are already impacting agriculture, forestry, and fisheries (European

Commission, 2011a). According to the European Commission (2011a), biodiversity loss is the most critical global environmental threat along with climate change. Pereira et al. (2012) stated that "global biodiversity change is one of the more pressing environmental issues of our time." Also, the Aldesgate Group (2011) pointed out that "the protection of biodiversity and ecosystem services, while complex to value and quantify accurately, is essential for future well-being and economic development." There are also strong ethical arguments for protecting biodiversity (European Commission, 2009b). BF is therefore very important.

However, unlike the presented footprints, the BF is complicated to assess and cannot be calculated easily for the products, businesses, activities, and reporting (Burrows, 2011). A global observation system for monitoring biodiversity changes does not exist yet (Pereira et al., 2012). There is still limited awareness of biodiversity issues among the general public and decision-makers (Secretariat of the Convention on Biological Diversity, 2010). Several attempts have been made to measure the BF, such as using land area appropriated by human activity, the number of threatened species (Burrows, 2011), extinction rate, and the population trend of species (Butchart et al., 2010). Currently, it might be impossible to evaluate the BF at satisfactory levels, but it is possible to identify the hot spots (Burrows, 2011). Standards and frameworks of the BF are currently undergoing development (BIQ Forum, 2014).

LAND FOOTPRINT

Population growth and economic income have been identified as the more important drivers for the LF (Giljum et al., 2013b). More than 75% of the Earth's land (excluding Greenland and Antarctica) is already used by humans (Haberl et al., 2011). Land use ranges from very intensive to very extensive. Approximately 1% of the land is used as infrastructure and urban areas, approximately 10–12% is used as cropland, approximately 26% is used as forestry land, and between 23% (Toderoiu, 2010) and 36% (Haberl et al., 2011) is used as grazing land. Of the remaining land, approximately half is unproductive or covered by rocks, snow, or deserts. The other half includes pristine forests (4.6% of total area), including tropical rainforests and all other forests with almost no signs of human use (Haberl et al., 2011).

The LF usually assesses those land areas that are directly and indirectly required to satisfy the consumption either for specific product(s) or for total consumption (Giljum et al., 2013a). It is a powerful method of illustrating the dependencies of local areas (regions or countries) on foreign land, which is embodied in imports and exports ("virtual land") (Giljum et al., 2013b). Another definition and usage of land footprint is that it is equivalent to EF excluding carbon uptake land, because this is directly related to CO_2 emissions already captured by the CF (Steen-Olsen et al., 2012). Furthermore, it is defined as the amount of biologically productive land required to satisfy the consumption (Weinzettel et al., 2013). Currently, no harmonized definition of the LF exists (Giljum et al., 2013a). It is important to note that LF approaches differ from calculations of the EF, because no weighting of land areas by different bioproductivities is applied.

Because of data restrictions, LF studies have often focused on the agricultural and forestry areas (Giljum et al., 2013a). The summary of LF definition, policies, and reduction options is presented in Table 5.2.

OTHER ENVIRONMENTAL FOOTPRINTS

In addition to footprints that are recognized as key, several other environmental footprints have been introduced to date (June 2014). These footprints are briefly described in this section.

MATERIAL FOOTPRINT

Material footprint is a consumption-based indicator of resource use. It is defined as a global allocation of used raw material extraction to the final demand of an economy (Wiedmann et al., 2013). The material footprint does not record the actual physical movement of materials within and among countries but, instead, enumerates the link between the beginning of a production chain (where raw materials are extracted from the natural environment) and its end (where a product or service is consumed). It opens a new perspective on global material supply chains and on the shared responsibility for the impacts of extraction, processing, and consumption of environmental resources (Wiedmann et al., 2013). It illustrates the global life cycle-wide material extraction and the final consumption of material within a region, occurring either within a country or beyond borders of countries (Giljum et al., 2013a). Material footprint is a newer term for "ecological rucksacks" (Giljum et al., 2013a). Another definition of material footprint is that it is an indicator for measuring and optimizing the resource consumption of products and their ingredients and the production processes along the whole value chain (Lettenmeier et al., 2012).

EMISSION FOOTPRINT

The emission footprint represents the quantities of product or service-created emissions into the air (e.g., SO_2, particles, CO, CO_2), water (e.g., chemical oxygen demand (COD), nitrogen, and phosphorus), and soil (through spillage in the soil). Emission footprints are calculated on a per-area basis. The conversion of emissions is calculated according to the principle that anthropogenic mass flows must not alter the qualities of local compartments. Maximum flows are defined based on the naturally existing qualities of the compartments and their replenishment rate per unit area. For emissions to soil, the replenishment rate is given by the decomposition of biomass to humus (measured by the production of compost by biomass). For groundwater, this is the seepage rate (given by local precipitation). For emissions into the air, the natural exchange of substances between forests and air per unit area is taken as a basis of comparison between natural and anthropogenic flows. Different emissions to air are not weighted, because only the largest dissipation areas are to be considered. Lower area consumption emissions may be dissipated without violating the principle that anthropogenic mass flows must not alter the qualities of local compartments (Sandholzer and Narodoslawsky, 2007).

HUMAN FOOTPRINT

The human footprint measures energy quantities, resources, and products consumed by a human during his/her lifetime and includes, for example, the number of food "pieces," the volumes of fuel and water, and the mass of waste (National Geographic, 2014b). Human footprint evaluates everything humans eat, use, wear, buy, and discard during their lifetime (Kirk, 2011). Human footprint is also a measure of transformation, integrating information regarding human access, settlement, transformation of land use/land cover, and development of energy infrastructure (Trombulak et al., 2010).

WASTE FOOTPRINT

Waste footprint is the amount of waste produced by sourcing ingredients and materials, manufacturing and processing, and transportation (Thinking Ahead, 2011). It also represent the environmental, economic, and social impacts that result from waste that humans create (HEC Global Learning Centre, 2009).

EMERGY FOOTPRINT

Emergy footprint combines the EF calculation and emergy analysis (Zhao et al., 2005). Emergy is calculated by the following equation:

$$\text{Emergy (seJ)} = \text{exergy (J)} \cdot \text{transformity (seJ/J)} \tag{5.15}$$

The concepts of emergy, exergy, and transformity are briefly presented in Table 5.8.

Biocapacity is estimated as a function of the renewable resources available (Brown et al., 2009). Consumption data can be converted into emergy flows and are grouped into the following categories: cropland; forestry; animal husbandry; and energy resources. All the energy flows (J) are transformed into solar emergy (seJ/y) through transformity. Emergy flows are divided by the global emergy density (seJ/gha) to obtain the global average productive hectares—emergy footprint (e.g., gha/capita) (Brown et al., 2009).

Table 5.8 Emergy, Exergy, and Transformity

	Description
Emergy	Ability of energy (exergy) of one kind or another is that it is used in transformations directly and indirectly when making a product or creating a service (Odum, 1996). It measures quality differences between forms of energy. It accounts for different forms of energy and resources, such as natural resources, renewable and non-renewable energy flows, economic goods, labor, and information, and it describes the amount of one kind of energy required to make another form of energy or resource. The unit of emergy is emjoule (emJ)—available energy of one kind consumed in transformations. It is usually expressed in a unit of solar emergy – embodied solar equivalent joules or solar emjoules (seJ) – all direct and indirect solar energy inputs that have been used in creating a service or product. Other units used are coal emjoules, electrical emjoules, and others. The greatest strength of emergy is that all energy and resources are put on a common scale that is objective and scientific (Brown and Angelo, 2010)
Exergy	Exergy is the energy that is available to be used for work. It quantifies maximal theoretical work that can be delivered by bringing an energy source into equilibrium with its environment. Exergy is always destroyed when a process involves a temperature change because energy conversion processes are irreversible. This destruction is proportional to the entropy of the system and its surroundings. The destroyed energy is also called anergy
Transformity	Transformity or energy quality ratio is the emergy (usually solar emergy) required to make a unit of available energy of the product or service, expressed in (seJ)/J. It aims to quantify energy quality and is defined as the ratio of emergy (work put into a product) and useful energy or exergy (value received from the product). Transformities are obtained through the analysis of a production process or by other empirical means (Campbell et al., 2005)

$$\text{Emergy footprint (gha)} = \frac{\text{emergy (seJ)}}{\text{global emergy density (seJ/gha)}} \tag{5.16}$$

The most important feature of emergy footprint is that it enables a comparison of all resources on a fair basis. Emergy footprint calculations also include deserts, glaciers, and open oceans, which are excluded in EF analyses. However, there are several disadvantages: environmental services and negative externalities are excluded from analyses and calculations of transformities are difficult and should be estimated (Brown et al., 2009).

EXERGY FOOTPRINT

The exergy footprint includes the following resource consumption categories: materials; water; energy; food; and, with the additional research underway, human and monetary capital (Caudill et al., 2010). By using exergy, the need to define, normalize, and aggregate various impact categories is avoided. It uses national-level exergy consumption on a per-capita basis as the normalization factor and compares it to a national baseline value (Caudill et al., 2010).

CHEMICAL FOOTPRINT

The chemical footprint is an indication of potential risk posed by a product based on its chemical composition, the human and ecologically hazardous properties of the ingredients, and the exposure potential of the ingredients during its life cycle. Its analysis should include a comprehensive quantification of the chemicals used, consumed, produced, or modified throughout the life cycle of the product of interest and the risks posed (Panko and Hitchcock, 2011).

POLLUTION FOOTPRINT

Pollution footprint is an EF based on pollution absorption. It accounts for pollutants incurred by human activities and is clarified in terms of different classes or types of pollutants (Min et al., 2011). It is based on input–output analysis for measuring the discharge coefficients and amounts of pollutants for the system (Gao and Fan, 2009).

RADIOACTIVE FOOTPRINT

Radioactive footprint is defined as the total geographical area contaminated by radioactive isotopes, including the quantity and radioisotope composition dispersed both on land and water, as a result of a nuclear power plant disaster, atomic bomb blast, or depleted uranium munitions event (Urban Dictionary, 2014).

FOOD FOOTPRINT

Food footprint is also called the material footprint of food production. It includes those areas needed to grow crops, fish, and graze animals, and to absorb carbon emissions from food processing and transportation (Sokoli, 2014); it is expressed in hectares/capita. The global average food footprint is 0.79 ha/capita (Neset and Lohm, 2005). A person's food footprint (foodprint) is also defined as all the emissions that result from the production, transportation, and storage of the food supplied to meet their consumption needs. Focus is especially placed on food supply and is expressed in $t_{CO_2,eq}$/capita (Shrink That Footprint Ltd, 2014). Furthermore, it includes the land required to produce the food and to sequester CO_2 emissions from the food production process (Zeev et al., 2014).

NUTRITIONAL FOOTPRINT

Nutritional footprint is a concept that links environmental, health, and social issues of nutrition and food products together. Nutritional footprint assessment is a quantitatively based approach and evaluates the levels of health and environmental impacts from the life cycle's perspective. It can provide a more environmentally friendly and healthier choice of food products (Lukas et al., 2013). The methodology for evaluating the nutritional footprint consists of two preparation steps and three main steps. The preparation steps include assessment of relevant health (calorie and salt content, content of dietary fiber and saturates, nutrient density, etc.) and environmental indicators (material, CF, WF and LF, biodiversity, erosion and earth movement, etc.), and reference data and assessment levels. Furthermore, the main steps include (i) estimation of health and environmental issues related to food products; (ii) allocation and evaluation of the selected indicators during the life cycle phases; and (iii) identification and quantitative comparison of results. All the indicators have allocated ranges from 1 to 3 to approximately identify hot spots, health and environmental indicators have equal weights, and the overall assessment consists of one number (the lower the better) (Lukas et al., 2013). It is a tool that enables a detailed overview, and it is relatively easy to compare different impacts of food products (Lucas et al., 2013).

CONCLUDING REMARKS

Humanity is in the middle of one of the largest experiments in the history of the Earth (Chapin et al., 2000). In an increasingly resource-constrained world, accurate and effective accounting systems are needed to map supply and demand for ecosystem services (Galli et al., 2014). Regretfully, it seems that humanity is moving further and further away from sustainability (Zhao et al., 2005). Moving toward sustainability requires the redesigning of production, consumption, and waste management, as well as the will to implement it. Reliable definitions and measurements are necessary for achieving these goals. Several tools for measuring sustainability have been developed to evaluate the (un)sustainability of humans, nations, processes, products, or activities. Nevertheless, the definition of a suitable environmental and/or sustainability metric for supporting objective environmental and/or sustainability assessments is still an open issue within the literature.

This chapter briefly presented the metrics for measuring environmental sustainability such as indicators of potential environmental impact, eco-efficiency, eco profit and total profit, eco NPV and total NPV and other combined criteria, sustainability indexes, and environmental footprints. Furthermore, key environmental footprints, such as CF, WF, EF, ENF, NF, PF, BF, and LF have been presented in detail. Finally, other environmental footprints have been assessed that are not widely known and used, and that are not recognized as key environmental footprints.

Key environmental footprints pose significant pressure to the Earth. Several options regarding how to reduce key environmental footprints have been recognized. The importance of evaluating total footprints (burdening and unburdening) instead of only direct (burdening) footprints has been pointed out. However, it should be noted that reducing one footprint can negatively impact other footprints. One such example is biofuel production, which has the potential for improving CF and ENF but could negatively impact the WF, NF, PF, BF, and LF.

The negative consequences of increasing specific footprints have been recognized by governments, businesses, researchers, and the public in general; therefore, the number of footprint evaluations has significantly increased over the past decade. Several policies and targets have been adopted throughout the world to move toward more sustainable development and to mitigate or prevent harmful impacts on the environment, such as climate change, biodiversity loss, eutrophication of water bodies, and many others. It should be noted, however, that measuring environmental footprints and their consequences on the environment is a relatively young discipline; therefore, it is expected that more researchers will join the effort and more frameworks and concepts will be found in the near future with the developed tools.

ACKNOWLEDGMENT

The authors are grateful for financial support from the Slovenian Research Agency (Programs P2-0032 and P2-0377) and from the Hungarian State and the European Union under project TÁMOP-4.2.2/A-11/1/KONV-2012-0072 "Design and optimisation of modernisation and efficient operation of energy supply and utilisation systems using renewable energy sources and ICTs."

REFERENCES

Abbott, J., 2008. What is a carbon footprint? Edinburgh, UK: The Edinburgh Centre for carbon management. <www.timcon.org/CarbonCalculator/Carbon%20Footprint.pdf> (accessed 20.02.2014).

Aldesgate Group, 2011. Pricing the Priceless, The business case for action on biodiversity. <www.aldersgategroup.org.uk/asset/download/472/Business%20and%20Biodiversity.pdf> (accessed 02.05.2014).

Allen, T., 2008. Life cycle tools for sustainable change. Prodesign (96), 52–54.

Attorre, F., 2014. Soqotra Archipelago (Yemen): Toward Systemic and Scientifically Objective Sustainability in Development and Conservation. Edizioni Nuova Cultura, Rome, Italy.

Australian Government, 2013. Water Act 2007, No. 137, 2007 as amended. <www.comlaw.gov.au/Details/C2014C00043> (accessed 02.06.2014).

Australian Government; Department of the Environment, 2014. The Renewable Energy Target (RET) scheme. <www.environment.gov.au/climate-change/renewable-energy-target-scheme> (accessed 05.06.2014).

Azapagic, A., Clift, R., 1999. Life cycle assessment and multiobjective optimisation. J. Clean. Prod. 7, 135–143.

Bertok, B., Barany, M., Friedler, F., 2012. Generating and analyzing mathematical programming models of conceptual process design by P-graph software. Ind. Eng. Chem. Res. 52, 166–171.

Bhaskar, V., Gupta, S.K., Ray, A.K., 2000. Applications of multiobjective optimization in chemical engineering. Rev. Chem. Eng. 16, 1–54.

BIQ Forum, 2014. Biodiversity & Ecosystem Service Footprint Project. <www.biofootprint.org/> (accessed 02.05.2014).

Bojarski, A.D., Laínez, J.M., Espuña, A., Puigjaner, L., 2009. Incorporating environmental impacts and regulations in a holistic supply chains modeling: an LCA approach. Comput. Chem. Eng. 33, 1747–1759.

Borucke, M., Moore, D., Cranston, G., Gracey, K., Iha, K., Larson, J., Lazarus, E., Morales, J.C., Wackernagel, M., Galli, A., 2013. Accounting for demand and supply of the biosphere's regenerative capacity: The National Footprint Accounts' underlying methodology and framework. Ecol. Indic. 24, 518–533.

Brandi, H.S., Daroda, R.J., Souza, T.L., 2011. Standardization: an important tool in transforming biofuels into a commodity. Clean Technol. Environ. Policy 13, 647–649.

British Petroleum, 2011. BP Energy Outlook 2030. <www.bp.com/content/dam/bp/pdf/Energy-economics/Energy-Outlook/BP_Energy_Outlook_Booklet_2011.pdf> (accessed 28.04.2014).

British Petroleum, 2013. BP Statistical Review of World Energy June 2013. <www.bp.com/content/dam/bp/pdf/statistical-review/statistical_review_of_world_energy_2013.pdf> (accessed 28.04.2014).

Brown, M., Sweeney, S., Campbell, D.E., Huang, S.-L., Ortega, E., Rydberg, T., Tilley, D. and Ulgiati, S., 2009. Emergy synthesis: Theory and applications of the emergy methodology. Proceedings from the Fifth Biennial Emergy Conference, Gainesville, FL, USA.

Brown, M.T., Angelo, M.J., 2010. Valuing nature, the challenge of the National Environmental Legacy Act. In: Flournoy, A.C., Driesen, D.M. (Eds.), Beyond Environmental Law: Policy Proposals for a Better Environmental Future. Cambridge University Press, New York, USA.

Burrows, D., 2011. How to measure your firm's biodiversity footprint: the Guardian. <www.theguardian.com/sustainable-business/biodiversity-footprint-new-carbon-measurement-management> (accessed 29.04.2014).

Butchart, S.H.M., Walpole, M., Collen, B., van Strien, A., Scharlemann, J.P.W., Almond, R.E.A., Baillie, J.E.M., Bomhard, B., Brown, C., Bruno, J., Carpenter, K.E., Carr, G.M., Chanson, J., Chenery, A.M., Csirke, J., Davidson, N.C., Dentener, F., Foster, M., Galli, A., Galloway, J.N., Genovesi, P., Gregory, R.D., Hockings, M., Kapos, V., Lamarque, J.-F., Leverington, F., Loh, J., McGeoch, M.A., McRae, L., Minasyan, A., Morcillo, M.H., Oldfield, T.E.E., Pauly, D., Quader, S., Revenga, C., Sauer, J.R., Skolnik, B., Spear, D., Stanwell-Smith, D., Stuart, S.N., Symes, A., Tierney, M., Tyrrell, T.D., Vié, J.-C., Watson, R., 2010. Global biodiversity: indicators of recent declines. Science 328, 1164–1168.

Campbell, D.E., Brandt-Williams, S.L., Meisch, M.E., 2005. Environmental Accounting Using Emergy: Evaluation of the State of West Virginia. US Environmental Protection Agency, Office of Research and Development, National Health and Environmental Effects Research Laboratory, Atlantic Ecology Division, Narragansett, USA.

Carbonfund.org, 2014. How to reduce your carbon footprint. <www.carbonfund.org/reduce> (accessed 15.03.2014).

Carpenter, S.R., Bennett, E.M., 2011. Reconsideration of the planetary boundary for phosphorus. Environ. Res. Lett. 6 (1), 1–12.

Carson, R., 1962. Silent Spring. Houghton Mifflin, New York, USA.

Caudill, R.J., Olapiriyakul, S., Seale, B., 2010. An Exergy Footprint Metric Normalized to US Exergy Consumption per Capita. 2010 IEEE International Symposium on Sustainable Systems and Technology (ISSST), 1–6.

Chait, J., 2009. The 15 best carbon calculators. <www.mnn.com/earth-matters/climate-weather/stories/the-15-best-carbon-calculators> (accessed 19.03.2014).

Chakrabortty, R., Hasin, M., 2013. Solving an aggregate production planning problem by using multi-objective genetic algorithm (MOGA) approach. International Journal of Industrial. Engineering Computations 4, 1–12.

Chapin III, F.S., Zavaleta, E.S., Eviner, V.T., Naylor, R.L., Vitousek, P.M., Reynolds, H. L., Hooper, D.U., Lavorel, S., Sala, O.E., Hobbie, S.E., Mack, M.C., Díaz, S., 2000. Consequences of changing biodiversity. Nature 405, 234–242.

Chen, C.-Z., Lin, Z.-S., 2008. Multiple timescale analysis and factor analysis of energy ecological footprint growth in China 1953–2006. Energy Policy 36, 1666–1678.

Chesapeake Bay Foundation, 2014. Nitrogen Calculator, Your Bay Footprint. <www.cbf.org/news-media/multimedia/nitrogen-calculator> (accessed 28.04.2014).

China.org.cn, 2014. Policies & Announcements. <www.china.org.cn/environment/node_7076102.htm> (accessed 02.05.2014).

Clean Water Act., 2002. Federal Water Pollution Control Act. As Amended Through PL 107-133.

Conserve Energy Future, 2014. Water Pollution Facts. <www.conserve-energy-future.com/various-water-pollution-facts.php> (accessed 10.04.2014).

Cordell, D., 2010. The Story of Phosphorus, Sustainability implications of global phosphorus scarcity for food security (PhD Thesis). Institute for Sustainable Futures, University of Technology, Sydney, Australia, and Department of Water and Environmental Studies, Linköping University, Sweden.

Cordell, D., Drangert, J.-O., White, S., 2009. The story of phosphorus: global food security and food for thought. Global Environ. Change 19, 292–305.

Čuček, L., 2013. Synthesis of sustainable bioprocesses using computer-aided process engineering (PhD Thesis). Maribor, Slovenia University of Maribor; Faculty of Chemistry and Chemical Engineering. <dkum.uni-mb.si/Dokument.php?id = 55146> (accessed 08.05.2013).

Čuček, L., Drobež, R., Pahor, B., Kravanja, Z., 2012a. Sustainable synthesis of biogas processes using a novel concept of eco-profit. Comput. Chem. Eng. 42, 87–100.

Čuček, L., Klemeš, J., Kravanja, Z., 2012b. Carbon and nitrogen trade-offs in biomass energy production. Clean Technol. Environ. Policy 14, 389–397.

Čuček, L., Klemeš, J.J., Kravanja, Z., 2012c. A review of footprint analysis tools for monitoring impacts on sustainability. J. Clean. Prod. 34, 9–20.

Čuček, L., Klemeš, J.J., Varbanov, P.S., Kravanja, Z., 2012d. Reducing the dimensionality of criteria in multi-objective optimisation of biomass energy supply chains. Chem. Eng. Trans. 29, 1231–1236.

Čuček, L., Varbanov, P.S., Klemeš, J.J., Kravanja, Z., 2012e. Total footprints-based multi-criteria optimisation of regional biomass energy supply chains. Energy 44, 135–145.

Čuček, L., Klemeš, J.J., Kravanja, Z., 2013a. A novel concept of eco- and total-net present value applied to the synthesis of biogas processes. Conference ECCE9/ECAB2, The Hague, Netherlands, Number 1250. <www.ecce2013.eu/documenten/wednesday_april_24.pdf> (accessed 04.06.2014).

Čuček, L., Klemeš, J.J., Varbanov, P.S., Kravanja, Z., 2013b. Dealing with high-dimensionality of criteria in multiobjective optimization of biomass energy supply network. Ind. Eng. Chem. Res. 52, 7223–7239.

Čuček, L., Klemeš, J.J., Varbanov, P.S., Kravanja, Z., 2013c. Dimensionality reduction approach for multi-objective optimisation extended to total footprints. Sixth International Conference on Process Systems Engineering (PSE ASIA), Kuala Lumpur, Malaysia. <www.sps.utm.my/download/PSEAsia2013-128.pdf> (accessed 30.04.2014).

Čuček, L., Klemeš, J.J., Kravanja, Z., 2014. Objective dimensionality reduction method within multi-objective optimisation considering total footprints. J. Clean. Prod. 71, 75–86.

Daimler, 2011. Well-to-wheel. <www2.daimler.com/sustainability/optiresource/index.html> (accessed 21.05.2014).

De Benedetto, L., Klemeš, J., 2008. LCA as environmental assessment tool in waste to energy and contribution to occupational health and safety. Chem. Eng. Trans. 13, 343–350.

De Benedetto, L., Klemeš, J., 2009. The environmental performance strategy map: an integrated LCA approach to support the strategic decision-making process. J. Clean. Prod. 17, 900–906.

De Benedetto, L., Klemeš, J., 2010. The environmental bill of material and technology routing: an integrated LCA approach. Clean. Technol. Environ. Policy 12, 191–196.

de Bie, S., van Dessel, B., 2011. Compensation for Biodiversity Loss, Advice to the Netherlands' Taskforce on Biodiversity and Natural Resource. De Gemeynt, Klarenbeek, the Netherlands, <www.gemeynt.nl/en/component/docman/doc_view/8-compensation-for-biodiversity-loss> (accessed 03.05.2014).

de Haes, H.A.U., Heijungs, R., Suh, S., Huppes, G., 2004. Three strategies to overcome the limitations of life-cycle assessment. J. Ind. Ecol. 8, 19–32.

Delft University of Technology, 2012. The Model of the Eco-costs/Value Ratio (EVR). Delft, the Netherlands. <www.ecocostsvalue.com/> (accessed 22.01.2014).

Denholm, P., Margolis, R.M., 2008. Land-use requirements and the per-capita solar footprint for photovoltaic generation in the United States. Energy Policy 36, 3531–3543.

Diaz, R.J., Rosenberg, R., 2008. Spreading dead zones and consequences for marine ecosystems. Science 321, 926–929.

Díaz, S., Fargione, J., Chapin III, F.S., Tilman, D., 2006. Biodiversity loss threatens human well-being. PLoS Biol. 4, 277.

Dilling, L., Doney, S.C., Edmonds, J., Gurney, K.R., Harriss, R., Schimel, D., Stephens, B., Stokes, G., 2003. The role of carbon cycle observations and knowledge in carbon management. Annu. Rev. Environ. Resour. 28, 521–558.

Dong, H., Geng, Y., Sarkis, J., Fujita, T., Okadera, T., Xue, B., 2013. Regional water footprint evaluation in China: a case of Liaoning. Sci. Total Environ. 442, 215–224.

Dourte, D.R., Fraisse, C.W., 2012. What is a Water Footprint?: An Overview and Applications in Agriculture. <edis.ifas.ufl.edu/pdffiles/AE/AE48400.pdf> (accessed 20.04.2014).

Downie, J., Stubbs, W., 2013. Evaluation of Australian companies' scope 3 greenhouse gas emissions assessments. J. Clean. Prod. 56, 156–163.

Energy Independence and Security Act, 2007. <www.gpo.gov/fdsys/pkg/STATUTE-121/pdf/STATUTE-121-Pg1492.pdf> (accessed: 01.06.2014).

Energy Policy Act, 2005. <www.gpo.gov/fdsys/pkg/STATUTE-119/pdf/STATUTE-119-Pg594.pdf> (accessed 01.06.2014).

Ercin, A.E., Hoekstra, A.Y., 2014. Water footprint scenarios for 2050: a global analysis. Environ. Int. 64, 71–82.

Erisman, J.W., Sutton, M.A., Galloway, J., Klimont, Z., Winiwarter, W., 2008. How a century of ammonia synthesis changed the world. Nat. Geosci. 1, 636–639.

European Commission, 2000. Directive 2000/60/EC of the European Parliament and of the Council of 23 October 2000 establishing a framework for community action in the field of water policy. Official Journal of the European Communities; Luxembourg.

European Commission, 2007. Carbon footprint—what it is and how to measure it. <lca.jrc.ec.europa.eu/Carbon_footprint.pdf> (accessed 15.02.2014).

European Commission, 2009a. Directive 2009/28/EC of the European Parliament and of the Council of 23 April 2009 on the promotion of the use of energy from renewable sources and amending and subsequently repealing Directives 2001/77/EC and 2003/30/EC. Official Journal of the European Union, Brussels, Belgium.

European Commission, 2009b. The Message from Athens. Athens, Greece. <ec.europa.eu/environment/nature/biodiversity/conference/pdf/message_final.pdf> (accessed 02.05.2014).

European Commission, 2011a. Communication from the Commission to the European Parliament, the Council, the Economic and Social Committee and the Committee of the Regions, Our life insurance, our natural capital: an EU biodiversity strategy to 2020 Brussels, Belgium. <ec.europa.eu/environment/nature/biodiversity/comm2006/pdf/2020/1_EN_ACT_part1_v7%5B1%5D.pdf> (accessed 02.05.2014).

European Commission, 2011b. The EU biodiversity strategy to 2020. Brussels, Belgium. <ec.europa.eu/environment/nature/info/pubs/docs/brochures/2020%20Biod%20brochure%20final%20lowres.pdf> (accessed 02.06.2014).

European Commission, 2014. Policies. <ec.europa.eu/environment/policies_en.htm> (accessed 02.05.2014).

European Commission DG Environment News Alert Service, 2012. SCU; The University of the West of England; Bristol, Online calculator measures consumers' "nitrogen footprint". <ec.europa.eu/environment/integration/research/newsalert/pdf/278na3.pdf> (accessed 29.04.2014).

Ewing, B., Moore, D., Goldfinger, S., Oursler, A., Reed, A., Wackernagel, M., 2010. Ecological Footprint Atlas 2010. Global Footprint Network, Oakland, New Zeeland, <www.footprintnetwork.org/images/uploads/Ecological_Footprint_Atlas_2010.pdf> (accessed 28.04.2014).

Fang, K., Heijungs, R., de Snoo, G., 2013. The footprint family: comparison and interaction of the Ecological, Energy, Carbon and Water footprints. Metallurgical Res. Technol. 110, 77–86.

Fang, K., Heijungs, R., de Snoo, G.R., 2014. Theoretical exploration for the combination of the ecological, energy, carbon, and water footprints: overview of a footprint family. Ecol. Indic. 36, 508–518.

Ferng, J.-J., 2002. Toward a scenario analysis framework for energy footprints. Ecol. Econ. 40, 53–69.

Finkbeiner, M., 2009. Carbon footprinting—opportunities and threats. Int. J. Life Cycle Assess. 14, 91–94.

Finnveden, G., Hauschild, M.Z., Ekvall, T., Guinée, J., Heijungs, R., Hellweg, S., Koehler, A., Pennington, D., Suh, S., 2009. Recent developments in life cycle assessment. J. Environ. Manage. 91, 1–21.

Fuel-Cell e-Mobility, 2014. Well-to-Wheel—Integral Efficiency Analysis. <www.fuel-cell-e-mobility.com/h2-infrastructure/well-to-wheel-en/> (accessed 21.05.2014).

Galli, A., Wiedmann, T., Ercin, E., Knoblauch, D., Ewing, B., Giljum, S., 2011. Integrating Ecological, Carbon and Water Footprint: Defining the "Footprint Family" and its Application in Tracking Human Pressure on the Planet. Technical Document, Surrey, UK.

Galli, A., Wiedmann, T., Ercin, E., Knoblauch, D., Ewing, B., Giljum, S., 2012. Integrating ecological, carbon and water footprint into a "Footprint Family" of indicators: definition and role in tracking human pressure on the planet. Ecol. Indic. 16, 100–112.

Galli, A., Weinzettel, J., Cranston, G., Ercin, E., 2013. A footprint family extended MRIO model to support Europe's transition to a One Planet Economy. Sci. Total Environ. 461–462, 813–818.

Galli, A., Wackernagel, M., Iha, K., Lazarus, E., 2014. Ecological footprint: implications for biodiversity. Biol. Conserv. 173, 121–132.

Galloway, J.N., Aber, J.D., Erisman, J.W., Seitzinger, S.P., Howarth, R.W., Cowling, E.B., Cosby, B.J., 2003. The nitrogen cascade. BioScience 53, 341–356.

Galloway, J.N., Townsend, A.R., Erisman, J.W., Bekunda, M., Cai, Z., Freney, J.R., et al., 2008. Transformation of the Nitrogen cycle: recent trends, questions, and potential solutions. Science 320, 889–892.

GAMS Development Corporation, 2013. GAMS—A User's Guide. Washington, DC. <www.gams.com/dd/docs/bigdocs/GAMSUsersGuide.pdf> (accessed 05.11.2013).

Gao, J.-x., Fan, X.-s., 2009. Pollution footprint contrastive analysis on industrial import and export in China. China Environ. Sci. 1, 030.

Gerbens-Leenes, W., Hoekstra, A., van der Meer, T., 2008. The Water Footprint of Bio-Energy and Other Primary Energy Carriers, Value of Water Report Series No. 29. UNESCO-IHE, Delft, the Netherlands, <www.waterfootprint.org/Reports/Report29-WaterFootprintBioenergy.pdf> (accessed 02.09.2013).

Giljum, S., Lutter, S., Bruckner, M., Aparcana, S., 2013a. State-of-Play of National Consumption-Based Indicators, A Review and Evaluation of Available Methods and Data to Calculate Footprint-Type (Consumption-Based) Indicators for Materials, Water, Land and Carbon. Sustainable Europe Research Institute (SERI), Vienna, Austria, <ec.europa.eu/environment/enveco/resource_efficiency/pdf/FootRev_Report.pdf> (accessed 03.05.2014).

Giljum, S., Wieland, H., Bruckner, M., de Schutter, L., Giesecke, K., 2013b. Land Footprint Scenarios, A Discussion Paper Including A Literature Review and Scenario Analysis on the Land Use Related to Changes in Europe's Consumption Patterns. Sustainable Europe Research Institute (SERI), Vienna, Austria, <www.foeeurope.org/sites/default/files/publications/seri_land_footprint_scenario_nov2013_1.pdf> (accessed 03.05.2014).

Glavič, P., Lukman, R., 2007. Review of sustainability terms and their definitions. J. Clean. Prod. 15, 1875–1885.

Glavič, P., Lesjak, M., Hirsbak, S., 2012. European training course on eco-efficiency. 15th European Roundtable on Sustainable Consumption and Production, Bregenz, Austria. <vbn.aau.dk/files/66744447/European_Training_Course_on_Eco_Efficiency.pdf> (accessed 03.06.2014).

Gleick, P.H., Ajami, N., Christian-Smith, J., Cooley, H., Donnelly, K., Fulton, J., Ha, M.-L., Heberger, M., Moore, E., Morrison, J., Orr, S., Schulte, P., Srinivasan, V., 2014. The World's Water Volume 8: The Biennial Report on Freshwater Resources. Island Press, Washington, USA.

Global Footprint Network, 2010. Footprint Basics—Overview. <www.footprintnetwork.org/en/index.php/GFN/page/footprint_basics_overview/> (accessed 21.02.2014).

Global Footprint Network, 2011. Frequently asked questions—What is the Ecological Footprint? <www.footprintnetwork.org/en/index.php/gfn/page/frequently_asked_questions/> (accessed 21.04.2014).

Global Footprint Network, 2012. Glossary. <www.footprintnetwork.org/en/index.php/gfn/page/glossary/> (accessed 20.02.2014).

Goodland, R., 1995. The concept of environmental sustainability. Annu. Rev. Ecol. Syst. 26, 1–24.

Government of Canada, 2013. What is eco-efficiency? <www.ic.gc.ca/eic/site/ee-ee.nsf/eng/h_ef00010.html> (accessed 15.04.2014).

GRACE Communications Foundation, 2014. Water footprint calculator. <www.gracelinks.org/1408/water-footprint-calculator> (accessed 07.06.2014).

Grossmann, I.E., Guillén-Gosálbez, G., 2010. Scope for the application of mathematical programming techniques in the synthesis and planning of sustainable processes. Comput. Chem. Eng. 34, 1365–1376.

Gruener, O., 2010. The water footprint: water in the supply chain. Environmentalist 1, 12.

Gu, B., Leach, A.M., Ma, L., Galloway, J.N., Chang, S.X., Ge, Y., Chang, J., 2013. Nitrogen Footprint in China: food, energy, and nonfood goods. Environ. Sci. Technol. 47, 9217–9224.

Guillén-Gosálbez, G., 2011. A novel MILP-based objective reduction method for multi-objective optimization: application to environmental problems. Comput. Chem. Eng. 35, 1469–1477.

Guinée, J.B., Gorrée, M., Heijungs, R., Huppes, G., Kleijn, R., de Koning, A., van Oers, L., Wegener Sleeswijk, A., Suh, S., de Haes, H.A.U., de Bruijn, H., van Duin, R., Huijbregts, M.A.J., Lindeijer, E., Roorda, A.A.H., van der Ven, B.L., Weidema, P.P., 2002. Handbook on Life Cycle Assessment, Operational Guide to the ISO Standards. Kluwer Academic Publishers, Dordrecht, the Netherlands.

Haberl, H., Erb, K.-H., Krausmann, F., Bondeau, A., Lauk, C., Müller, C., Plutzar, C., Steinberger, J.K., 2011. Global bioenergy potentials from agricultural land in 2050: sensitivity to climate change, diets and yields. Biomass Bioenergy 35, 4753–4769.

Haggar, S.E., 2007. Sustainable Industrial Design and Waste Management: Cradle-to-Cradle for Sustainable Development. Elsevier Academic Press, NY, USA.

Haimes, Y.Y., Lasdon, L.S., Wismer, D.A., 1971. On a bicriterion formulation of the problems of integrated system identification and system optimization. IEEE Trans. Syst. Man Cybern. SMC-1, 296–297.

Hanafiah, M.M., Hendriks, A.J., Huijbregts, M.A.J., 2012. Comparing the ecological footprint with the biodiversity footprint of products. J. Clean. Prod. 37, 107–114.

HEC Global Learning Centre, 2009. Waste Footprint. <www.globalfootprints.org/waste> (accessed 22.05.2014).

Hendrickson, C.T., Horvath, A., Joshi, S., Klausner, M., Lave, L.B., McMichael, F.C., 1997. Comparing two life cycle assessment approaches: a process model vs. economic input-output-based assessment. Proceedings of the 1997 IEEE International Symposium on Electronics and the Environment, ISEE-1997, 99. 176–181.

Hermann, B.G., Debeer, L., De Wilde, B., Blok, K., Patel, M.K., 2011. To compost or not to compost: carbon and energy footprints of biodegradable materials' waste treatment. Polym. Degrad. Stabil. 96, 1159–1171.

Hertwich, E.G., Peters, G.P., 2009. Carbon footprint of nations: a global, trade-linked analysis. Environ. Sci. Technol. 43, 6414–6420.

Hoekstra, A., 2009. Human appropriation of natural capital: a comparison of ecological footprint and water footprint analysis. Ecol. Econ. 68, 1963–1974.

Hoekstra, A.Y., 2008. Water Neutral: Reducing and Offsetting the Impacts of Water Footprints, Value of Water Research Report Series No. 28. UNESCO-IHE Institute for Water Education, Delft, the Netherlands, <doc.utwente.nl/77202/> (accessed 06.06.2014).

Hoekstra, A.Y., Chapagain, A.K., 2007. Water footprints of nations: water use by people as a function of their consumption pattern. Water Resour. Manag. 21, 35–48.

Hoekstra, A.Y., Hung, P.Q., 2002. Virtual Water Trade: A Quantification of Virtual Water flows Between Nations in Relation to International Crop Trade. UNESCO-IHE Institute for Water Education, Delft, the Netherlands.

Hoekstra, A.Y., Chapagain, A.K., Aldaya, M.M., Mekonnen, M.M., 2011. The Water Footprint Assessment Manual: Setting the Global Standard: Water Footprint Network. <www.waterfootprint.org/downloads/TheWaterFootprintAssessmentManual.pdf> (accessed 02.09.2013).

Holden, J., 2014. Water Resources: An Integrated Approach. Routhledge, New York, NY.

Hooper, D., 10 Things you can do to help Biodiversity. Western Washington University, Bellingham, WA, USA. <fire.biol.wwu.edu/hooper/10thingsforbiodiversity.pdf> (accessed 18.07.2014).

Huang, H., Lu, J., 2014. Identification of river water pollution characteristics based on projection pursuit and factor analysis. Environ. Earth Sci., 1–9.

Huijbregts, M.A., Hellweg, S., Frischknecht, R., Hungerbühler, K., Hendriks, A.J., 2008. Ecological footprint accounting in the life cycle assessment of products. Ecol. Econ. 64, 798–807.

Hundal, M., 2000. Life Cycle Assessment and Design for the Environment. International Design Conference—Design 2000, Dubrovnik, Croatia. <www.cem.uvm.edu/~mhundal/defrec/lcadfe.pdf> (accessed 06.06.2014).

Hunt, R., Franklin, W., Hunt, R.G., 1996. LCA—How it came about. Int. J. Life Cycle Assess. 1, 4–7.

Huppes, G., Ishikawa, M., 2005. Eco-efficiency and its xsterminology. J. Ind. Ecol. 9, 43–46.

Integrated Sustainability Analysis (ISA), 2014. About BottomLine3. <www.isa.org.usyd.edu.au/consulting/BL3.shtml> (accessed 21.05.2014).

Intergovernmental Panel on Climate Change, 2009. IPCC Expert Meeting on the Science of Alternative Metrics, Meeting report. Oslo, Norway.

International Organisation for Standardisation (ISO), 2006a. ISO 14040, Environmental Management—Life Cycle Assessment—Principles and Framework. Geneva, Switzerland.

International Organisation for Standardisation (ISO), 2006b. ISO 14044, Environmental Management—Life Cycle Assessment—Requirements and Guidelines. Geneva, Switzerland.

Jeswani, H.K., Azapagic, A., 2011. Water footprint: methodologies and a case study for assessing the impacts of water use. J. Clean. Prod. 19, 1288–1299.

Jeswani, H.K., Azapagic, A., Schepelmann, P., Ritthoff, M., 2010. Options for broadening and deepening the LCA approaches. J. Clean. Prod. 18, 120–127.

Jones, C.M., Kammen, D.M., 2011. Quantifying carbon footprint reduction opportunities for U.S. Households and communities. Environ. Sci. Technol. 45, 4088–4095.

Jørgensen, A., Le Bocq, A., Nazarkina, L., Hauschild, M., 2008. Methodologies for social life cycle assessment. Int. J. Life Cycle Assess. 13, 96–103.

Juhász, C., Szőllősi, N., 2008. Environmental management. <www.tankonyvtar.hu/en/tartalom/tamop425/0032_kornyezetiranyitas_es_minosegbiztositas/ch03.html> (accessed 10.04.2014).

Kemira, 2014. Kemira Water Footprint Calculator. <www.waterfootprintkemira.com/> (accessed 07.06.2014).

Kirk, E., 2011. Human Footprint: Everything You Will Eat, Use, Wear, Buy, and Throw Out in Your Lifetime. National Geographic Kids, Washington, DC.

Kitzes, J., Wackernagel, M., 2009. Answers to common questions in ecological footprint accounting. Ecol. Indic. 9, 812–817.

Kitzes, J., Peller, A., Goldfinger, S., Wackernagel, M., 2007. Current methods for calculating national ecological footprint accounts. Sci. Environ. Sustainable Soc. 4, 1–9.

Kitzes, J., Galli, A., Bagliani, M., Barrett, J., Dige, G., Ede, S., Erb, K., Giljum, S., Haberl, H., Hails, C., Jolia-Ferrier, L., Jungwirth, S., Lenzen, M., Lewis, K., Loh, J., Marchettini, N., Messinger, H., Milne, K., Moles, R., Monfreda, C., Moran, D., Nakano, K., Pyhälä, A., Rees, W., Simmons, C., Wackernagel, M., Wada, Y., Walsh, C., Wiedmann, T., 2009. A research agenda for improving national Ecological Footprint accounts. Ecol. Econ. 68, 1991–2007.

Klemeš, J., Cockerill, T., Bulatov, I., Shackely, S., Gough, C., 2006. Engineering Feasibility of Carbon Dioxide Capture and Storage. Carbon Capture and its Storage: An Integrated Assessment. Ashgate Publishing Ltd., Ashgate, UK, 43–82.

Klemeš, J., Friedler, F., Bulatov, I., Varbanov, P., 2010. Sustainability in the Process Industry: Integration and Optimization: Integration and Optimization. McGraw-Hill Companies, New York, NY, USA.

Koskela, M., Vehmas, J., 2012. Defining eco-efficiency: a case study on the finnish forest industry. Bus. Strat. Env. 21, 546–566.

Krajnc, D., Glavič, P., 2003. Indicators of sustainable production. Clean. Technol. Environ. Policy 5, 279–288.

Kravanja, Z., 2010. Challenges in sustainable integrated process synthesis and the capabilities of an MINLP process synthesizer MipSyn., Comput. Chem. Eng. 34, 1831–1848.

Kravanja, Z., 2012. Process systems engineering as an integral part of global systems engineering by virtue of its energy—environmental nexus. Curr. Opin. Chem. Eng. 1, 231–237.

Kravanja, Z., Čuček, L., 2013. Multi-objective optimisation for generating sustainable solutions considering total effects on the environment. Appl. Energy 101, 67–80.

Kravanja, Z., Ropotar, M., Pintarič, Z.N., 2005. Incorporating Sustainability into the Superstructural Synthesis of Chemical Processes. The conference on Industrial pollution and sustainable development—CIPSD, Maribor, Slovenia.

Krotscheck, C., Narodoslawsky, M., 1996. The Sustainable Process Index a new dimension in ecological evaluation. Ecol. Eng. 6, 241–258.

Leach, A.M., Galloway, J.N., Bleeker, A., Erisman, J.W., Kohn, R., Kitzes, J., 2012. A nitrogen footprint model to help consumers understand their role in nitrogen losses to the environment. Environ. Dev. 1, 40–66.

Lettenmeier, M., Göbel, C., Liedtke, C., Rohn, H., Teitscheid, P., 2012. Material footprint of a sustainable nutrition system in 2050—Need for dynamic innovations in production, consumption and politics. Proc. Food Syst. Dyn., 584–598.

Li, J., Wang, X., 2012. Energy and climate policy in China's twelfth five-year plan: a paradigm shift. Energy Policy 41, 519–528.

Lott, J.N.A., Bojarski, M., Kolasa, J., Batten, G.D., Campbell, L.C., 2009. A review of the phosphorus content of dry cereal and legume crops of the world. Int. J. Agric. Resour. Governance Ecol. 8, 351–370.

Lucas, M., Palzkill, A., Rohn, H., Liedtke, C., 2013. The nutritional footprint: an innovative management approach for the food sector. In: Brebbia, C.A., Popov, V. (Eds.), Food and Environment II: The Quest for a Sustainable Future. WIT Press, Southampton, UK.

Lukas, M., Liedtke, C., Rohn, H., 2013. Food Chain Management, The Nutritional Footprint—Assessing Environmental and Health Impacts of Foodstuffs. World Resources Forum, Davos, Switzerland, <www.worldresourcesforum.org/files/WRF2013/Scientific%20Sessions/Melanie.Lukas-Holger.Rohn_.pdf> (accessed 21.05.2014).

Lygoe, R.J., 2010. Complexity Reduction in High-Dimensional Multi-Objective Optimisation. University of Sheffield, Sheffield, UK.

Ma, L., Liu, P., Fu, F., Li, Z., Ni, W., 2011. Integrated energy strategy for the sustainable development of China. Energy 36, 1143–1154.

Maggio, G., Cacciola, G., 2012. When will oil, natural gas, and coal peak? Fuel 98, 111–123.

Magoha, P., 2002. Footprints in the wind?: Environmental impacts of wind power development. Refocus 3, 30–33.

Martineau, R.J., Novello, D.P., 2004. The Clean Air Act Handbook. American Bar Association, Chicago, IL, USA.

Mata, T.M., Martins, A.A., Sikdar, S.K., Costa, C.A.V., 2011. Sustainability considerations of biodiesel based on supply chain analysis. Clean Technol. Environ. Policy 13, 655–671.

McManus, M., 2010. Life Cycle Assessment: An Introduction. University of Bath, Institute for Sustainable Energy and the Environment, Bath, UK, <www.bath.ac.uk/i-see/posters/Life_Cycle_Assessment_An_Introduction_MM_ppp_ISEE_Website.pdf%3e> (accessed 08.02.2013).

Meisterling, K., Samaras, C., Schweizer, V., 2009. Decisions to reduce greenhouse gases from agriculture and product transport: LCA case study of organic and conventional wheat. J. Clean. Prod. 17, 222–230.

Mekonnen, M.M., Hoekstra, A.Y., 2010. The Green, Blue and Grey Water Footprint of Farm Animals and Animal Products. UNESCO-IHE Institute for Water Education, Delft, the Netherlands, <doc.utwente.nl/76912/> (accessed 22.01.2013).

Millennium Ecosystem Assessment, 2005. Ecosystems and Human Well-Being: Synthesis. Island Press, Washington, DC, USA.

Min, Q., Jiao, W., Cheng, S., 2011. Pollution footprint: a type of ecological footprint based on ecosystem services. Resour. Sci. 2, 003.

Minister of Justice, 2014. Species at Risk Act. Canada. <laws-lois.justice.gc.ca/PDF/S-15.3.pdf> (accessed 02.06.2014).

Ministry of Science Technology and The Environment Malaysia, 1998. National Policy on Biological Diversity. Institut Terjemahan Negara Malaysia Berhad, Kuala Lumpur, Malaysia.

Mohr, S.H., Evans, G.M., 2009. Forecasting coal production until 2100. Fuel 88, 2059–2067.

Mohr, S.H., Evans, G.M., 2011. Long term forecasting of natural gas production. Energy Policy 39, 5550–5560.

Motavalli, J., (1999). Dr. Nafis Sadik, The UN's Prescription for Family Planning. <www.emagazine.com/includes/print-article/magazine-archive/8090/> (accessed 28.04.2014).

Munasinghe, M., 2004. Sustainable development: basic concepts and application to energy. In: Cleveland, C.J. (Ed.), Encyclopedia of Energy. Elsevier, New York, NY, USA, pp. 789–808.

Nag, A., 2008. Textbook of Agricultural Biotechnology. PHI Learning Private Limited, New Delhi, India.

National Biodiversity Authority, 2014. The Biological Diversity Act 2002. <www.nbaindia.org/> (accessed 02.06.2014).

National Geographic, 2014a. What is Your Water Footprint? <environment.nationalgeographic.com/environment/freshwater/change-the-course/water-footprint-calculator/> (accessed 07.06.2014).

National Geographic, 2014b. Human Footprint. <www.nationalgeographic.com/xpeditions/lessons/14/g68/HumanFootprint.pdf> (accessed 02.05.2014).

Natural Resource Management Ministerial Council, 2010. Australia's Biodiversity Conservation Strategy 2010–2030. Commonwealth of Australia Canberra, Australia, <www.cbd.int/doc/world/au/au-nbsap-v2-en.pdf> (accessed 02.06.2014).

Neset, T.-S., Lohm, U., 2005. Spatial imprint of food consumption: a historical analysis for Sweden, 1870–2000. Hum. Ecol. 33, 565–580.

Niemetz, N., Kettl, K.H., Eder, M., Narodoslawsky, M., 2012. RegiOpt conceptual planner—Identifying possible energy network solutions for regions. Chem. Eng. Trans. 29, 517–522.

N-Print Team, 2014. Welcome to the N-Print website! <www.n-print.org/> (accessed 12.02.2014).

Odum, H.T., 1996. Environmental Accounting: Emergy and Environmental Decision Making. Wiley, New York, NY, USA.

Oenema, O., Bleeker, A., Braathen, N.A., Budňáková, M., Bull, K., Čermák, P., Geupel, M., Hicks, K., Hoft, R., Kozlova, N., Leip, A., Spranger, T., Valli, L., Velthof, G., Winiwarter, W., 2011. Nitrogen in current European policies,. In: Sutton, M.A., Howard, C.M., Erisman, J.W., Billen, G., Bleeker, A., Greenfelt, P., van Grinsven, H., Grizzetti, B. (Eds.), The European Nitrogen Assessment: Sources, Effects and Policy Perspectives. Cambridge University Press, Cambridge, UK.

Padgett, J.P., Steinemann, A.C., Clarke, J.H., Vandenbergh, M.P., 2008. A comparison of carbon calculators. Environ. Impact Assess. Rev. 28, 106–115.

Palmer, A.R.P., 1998. Evaluating ecological footprints. Electron. Green J. 1 (9), <escholarship.org/uc/item/05k183c9> (accessed 03.03.2014).

Panko, J., Hitchcock, K., 2011. Chemical footprint: ensuring product sustainability. Air Waste Manag. Assoc., 12–15.

Patzek, T.W., Croft, G.D., 2010. A global coal production forecast with multi-Hubbert cycle analysis. Energy 35, 3109–3122.

Pereira, H.M., Navarro, L.M., Martins, I.S., 2012. Global biodiversity change: the bad, the good, and the unknown. Annu. Rev. Environ. Resour. 37, 25–50.

Perry, S., Klemeš, J., Bulatov, I., 2008. Integrating waste and renewable energy to reduce the carbon footprint of locally integrated energy sectors. Energy 33, 1489–1497.

Pezzey, J., 1989. Economic Analysis of Sustainable Growth and Sustainable Development, Environment Department Working Paper No. 15. World Bank, Washington, DC, USA.

Pieragostini, C., Mussati, M.C., Aguirre, P., 2012. On process optimization considering LCA methodology. J. Environ. Manage. 96, 43–54.

Pimm, S.L., Russell, G.J., Gittleman, J.L., Brooks, T.M., 1995. The future of biodiversity. Science 269, 347–350.

Pozo, C., Ruíz-Femenia, R., Caballero, J., Guillén-Gosálbez, G., Jiménez, L., 2012. On the use of principal component analysis for reducing the number of environmental objectives in multi-objective optimization: application to the design of chemical supply chains. Chem. Eng. Sci. 69, 146–158.

Procter & Gamble Science-in-the-Box, 2014. Life Cycle Assessments (LCAS). <scienceinthebox.com/life-cycle-assessment> (accessed 15.03.2014).

Purvis, A., Hector, A., 2000. Getting the measure of biodiversity. Nature 405, 212–219.

Quantis Sustainability counts, 2009. What is Life Cycle Assessment? <www.quantis-intl.com/life_cycle_assessment.php> (accessed 12.05.2014).

Rao, S.S., 2009. Engineering Optimization: Theory and Practice, fourth ed. John Wiley & Sons, Inc., Hoboken, NJ, USA.

Rebitzer, G., Ekvall, T., Frischknecht, R., Hunkeler, D., Norris, G., Rydberg, T., Schmidt, W.P., Suh, S., Weidema, B.P., Pennington, D.W., 2004. Life cycle assessment. Part 1: Framework, goal and scope definition, inventory analysis, and applications. Environ. Int. 30, 701–720.

Rees, W.E., 1992. Ecological footprints and appropriated carrying capacity: what urban economics leaves out. Environ. Urban. 4, 121–130.

Rees, W., 2010. The human nature of unsustainability. In: Heinberg, R., Lerch, D. (Eds.), The Post Carbon Reader: Managing the 21st Century's Sustainability Crises. Watershed Media in collaboration with Post Carbon Institute, Healdsburg, US, pp. 194–206.

Rockström, J., Steffen, W., Noone, K., Persson, Å., Chapin, F.S., Lambin, E.F., Lenton, T. M., Scheffer, M., Folke, C., Schellnhuber, H.J., Nykvist, B., De Wit, C.A., Hughes, T., Van Der Leeuw, S., Rodhe, H., Sörlin, S., Snyder, P.K., Costanza, R., Svedin, U., Falkenmark, M., Karlberg, L., Corell, R.W., Fabry, V.J., Hansen, J., Walker, B., Liverman, D., Richardson, K., Crutzen, P., Foley, J.A., 2009. A safe operating space for humanity. Nature 461, 472–475.

Sabio, N., Kostin, A., Guillén-Gosálbez, G., Jiménez, L., 2012. Holistic minimization of the life cycle environmental impact of hydrogen infrastructures using multi-objective optimization and principal component analysis. Int. J. Hydrogen Energy 37, 5385–5405.

Sandholzer, D., Narodoslawsky, M., 2007. SPIonExcel—Fast and easy calculation of the sustainable process index via computer. Resour. Conserv. Recycl. 50, 130–142.

Saur, K., 1997. Life cycle impact assessment. Int. J. Life Cycle Assess. 2, 66–70.

Schaefer, F., Luksch, U., Steinbach, N., Cabeça, J., Hanauer, J., 2006. Ecological Footprint and Biocapacity, The World's Ability to Regenerate Resources and Absorb Waste in a Limited Time Period. Office for Official Publications of the European Communities, Luxembourg.

Schindler, 2013. Energy and GHG Footprint, A Big Step Forward. <www.schindler.com/com/internet/en/about-schindler/corporate-citizenship/site-ecology/energy-and-ghg-footprint.html> (accessed 13.02.2013).

Schülke, A., 2008. Biodiversity in German Development Cooperation. Kasparek Verlag, Heidelberg, Germany.

Secretariat of the Convention on Biological Diversity, 2010. Global Biodiversity Outlook 3. Montréal, Canada. <www.cbd.int/doc/publications/gbo/gbo3-final-en.pdf> (accessed 02.05.2014).

Shrink That Footprint Ltd., 2014. Shrink Your Food Footprint. <shrinkthatfootprint.com/shrink-your-food-footprint> (accessed 22.05.2014).

Sims, R.E.H., Schock, R.N., Adegbululgbe, A., Fenhann, J., Konstantinaviciute, I., Moomaw, W., Nimir, H.B., Schlamadinger, B., Torres-Martínez, J., Turner, C., Uchiyama, Y., Vuori, S.J.V., Wamukonya, N., Zhang, X., 2007. Energy supply. In: Metz, B., Davidson, O.R., Bosch, P.R., Dave, R., Meyer, L.A. (Eds.), Climate Change 2007: Mitigation, Contribution of Working Group III to the Fourth Assessment Report of the Intergovernmental Panel on Climate Change. Cambridge University Press, Cambridge, UK; New York, NY, USA.

Smil, V., 1997. Global population and the nitrogen cycle. Sci. Am. 277, 76–81.

Smil, V., 2000. Phosphorus in the environment: natural flows and human interferences. Annu. Rev. Energy Environ. 25, 53–88.

Smil, V., 2001. Enriching the Earth: Fritz Haber, Carl Bosch, and the Transformation of World Food Production. The MIT Press, Cambridge, USA.

Sobhani, R., Abahusayn, M., Gabelich, C.J., Rosso, D., 2012. Energy footprint analysis of brackish groundwater desalination with zero liquid discharge in inland areas of the Arabian Peninsula. Desalination 291, 106–116.

Sodhi, N.S., Brook, B.W., Bradshaw, C.J., 2009. Causes and consequences of species extinctions. Princet. Guide Ecol., 514–520.

Sokoli, L., 2014. Sustainable development as the imperative of the twenty-first century; towards alternative approaches on measuring and monitoring. Acad. J. Interdiscipl. Stud. 3, 105–109.

Sorda, G., Banse, M., Kemfert, C., 2010. An overview of biofuel policies across the world. Energy Policy 38, 6977–6988.

Staudt, A., Huddleston, N., Kraucunas, I., 2008. Understanding and Responding to Climate Change: Highlights of National Academies Reports. <www.tribesandclimatechange.org/documents/nccc/nccc20110504_229.pdf> (accessed 10.03.2014).

Steen-Olsen, K., Weinzettel, J., Cranston, G., Ercin, A.E., Hertwich, E.G., 2012. Carbon, land, and water footprint accounts for the European Union: consumption, production, and displacements through international trade. Environ. Sci. Technol. 46, 10883–10891.

Stoeglehner, G., Narodoslawsky, M., 2009. How sustainable are biofuels? Answers and further questions arising from an ecological footprint perspective. Bioresource Technol. 100, 3825–3830.

Stoeglehner, G., Levy, J.K., Neugebauer, G.C., 2005. Improving the ecological footprint of nuclear energy: a risk-based lifecycle assessment approach for critical infrastructure systems. Int. J. Crit. Infrastructures 1, 394–403.

Tallis, B., Azapagic, A., Howard, A., Parfitt, A., Duff, C., Hadfield, C., Pritchard, C., Gillett, J., Hackitt, J., Seaman, M., Darton, R., Rathbone, R., Clift, R., Watson, S.,

Elliot, S., 2002. The Sustainability Metrics, Sustainable Development Progress Metrics Recommended for use in the Process Industries. IChemE, Rugby, UK.

The Central People's Government of the People's Republic of China. 2002. Water Law of the People's Republic of China. <www.gov.cn/english/laws/2005-10/09/content_75313.htm> (accessed 01.06.2014).

The Clean Air Act, 2004. <www.epw.senate.gov/envlaws/cleanair.pdf> (accessed 02.06.2014).

Thinking Ahead, 2011. Waste Footprint and Sustainable Suppliers. <www.usbthinkingahead.com/2011/09/waste-footprint-and-sustainable-suppliers.html> (accessed 02.05.2014).

Toderoiu, F., 2010. Ecological footprint and biocapacity—methodology and regional and national dimensions. Agric. Econ. Rural Dev. 2, 213–238.

Treat, S., Callahan, N.B., 2013. Drivers of Biodiversity Loss: Environmental Literacy Council. <enviroliteracy.org/subcategory.php/352.html> (accessed 29.04.2014).

Trombulak, S., Baldwin, R., Woolmer, G., 2010. The human footprint as a conservation planning tool. In: Trombulak, S.C., Baldwin, R.F. (Eds.), Landscape-scale Conservation Planning. Springer, the Netherlands, pp. 281–301.

U.S. Environmental Protection Agency (EPA), 2014. Basic Information about Nitrate in Drinking Water. <water.epa.gov/drink/contaminants/basicinformation/nitrate.cfm> (accessed 28.04.2014).

UK Parliamentary Office of Science and Technology (POST), 2011. Carbon footprint of electricity generation. London, UK. <www.parliament.uk/documents/post/postpn_383-carbon-footprint-electricity-generation.pdf> (accessed 22.01.2013).

UNEP/SETAC, 2009. Life Cycle Management: How Business Uses it to Decrease Footprint, Create Opportunities and Make Value Chains More Sustainable. <www.unep.org/pdf/DTIE_PDFS/DTIx1208xPA-LifeCycleApproach-Howbusinessusesit.pdf> (accessed 21.01.2013).

Union of Concerned Scientists, 2009. NO SURE FIX, Prospects for Reducing Nitrogen Fertilizer Pollution through Genetic Engineering. <www.ucsusa.org/assets/documents/food_and_agriculture/no-sure-fix.pdf> (accessed 20.03.2014).

United Nations Environment Programme, 2007. Water, Section B, State-and-Trends of the Environment: 1987–2007, Global Environment Outlook 4: Environment for Development, Valetta, Malta, pp. 115–156.

United Nations Framework Convention on Climate Change, 1998. Kyoto Protocol to the United Nations Framework Convention on Climate Change. <unfccc.int/resource/docs/convkp/kpeng.pdf> (accessed 12.04.2014).

Urban Dictionary, 2014. Radioactive Footprint. <www.urbandictionary.com/define.php?term=radioactive%20footprint> (accessed 28.05.2014).

Vaccari, D.A., 2009. Phosphorus: a looming crisis. Sci. Am. 300, 54–59.

van den Bergh, J.C.J.M., Grazi, F., 2014. Ecological footprint policy? Land use as an environmental indicator. J. Ind. Ecol. 18, 10–19.

Vitousek, P.M., Aber, J.D., Howarth, R.W., Likens, G.E., Matson, P.A., Schindler, D.W., Schlesinger, W.H., Tilman, D.G., 1997. Human alteration of the global nitrogen cycle: sources and consequences. Ecol. Appl. 7, 737–750.

Vogtländer, J.G., Bijma, A., Brezet, H.C., 2002. Communicating the eco-efficiency of products and services by means of the eco-costs/value model. J. Clean. Prod. 10, 57–67.

Vogtländer, J.G., Baetens, B., Bijma, A., Brandjes, E., Lindeijer, E., Segers, M, et al., 2010. LCA-Based Assessment of Sustainability: The Eco-costs/Value Ratio (EVR). VSSD, Delft, the Netherlands.

von Blottnitz, H., Curran, M.A., 2007. A review of assessments conducted on bio-ethanol as a transportation fuel from a net energy, greenhouse gas, and environmental life cycle perspective. J. Clean. Prod. 15, 607–619.

Vujanović, A., Čuček, L., Pahor, B., Kravanja, Z., 2014. Multi-objective synthesis of a company's supply-network by accounting for several environmental footprints. Process Saf. Environ. Prot. 92 (5), 456–466.

Wackernagel, M., Rees, W.E., 1996. Our Ecological Footprint: Reducing Human Impact on the Earth. New Society Publishers, Gabriola Island, British Columbia, Canada.

Wackernagel, M., Schulz, N.B., Deumling, D., Linares, A.C., Jenkins, M., Kapos, V., Monfreda, C., Loh, J., Myers, N., Norgaard, R., Randers, J., 2002. Tracking the ecological overshoot of the human economy. Proc. Natl. Acad. Sci. 99, 9266–9271.

Wackernagel, M., Kitzes, J., Moran, D., Goldfinger, S., Thomas, M., 2006. The ecological footprint of cities and regions: comparing resource availability with resource demand. Environ. Urban. 18, 103–112.

Water Footprint Network, 2014. Water Footprint. <www.waterfootprint.org/?page = files/home> (accessed 15.02.2014).

Weinzettel, J., Hertwich, E.G., Peters, G.P., Steen-Olsen, K., Galli, A., 2013. Affluence drives the global displacement of land use. Global Environ. Change 23, 433–438.

Wiedmann, T., Barrett, J., 2011. A greenhouse gas footprint analysis of UK Central Government, 1990–2008. Environ. Sci. Policy 14, 1041–1051.

Wiedmann, T., Lenzen, M., 2007. On the conversion between local and global hectares in ecological footprint analysis. Ecol. Econ. 60, 673–677.

Wiedmann, T., Minx, J., 2008. A definition of "carbon footprint". In: Pertsova, C.C. (Ed.), Ecological Economics Research Trends. Nova Science Publisher, Hauppauge, NY, USA, pp. 1–11.

Wiedmann, T.O., Schandl, H., Lenzen, M., Moran, D., Suh, S., West, J., Kanemoto, K., 2013. The material footprint of nations. Proc. Natl. Acad. Sci. < www.pnas.org/content/early/2013/08/28/1220362110 > (accessed 26.10.2014).

World Commission on Environment and Sustainable Development (WCED), 1987. Our Common Future (The Brundtland Report). Oxford University Press, Oxford Bungay, Suffolk, UK.

World Meteorological Organization, 2009. WMO Greenhouse Gas Bulletin, The State of Greenhouse Gases in the Atmosphere Using Global Observations through 2008. <www.wmo.int/pages/prog/arep/gaw/ghg/documents/GHG-bulletin2009_en.pdf> (accessed 21.03.2014).

World Meteorological Organization, 2014. GAW Greenhouse Gas Research. <www.wmo.int/pages/prog/arep/gaw/ghg/ghgbull06_en.html> (accessed 21.03.2014).

World Wide Fund for Nature, 2002. Living planet report. <www.footprintnetwork.org/images/uploads/lpr2002.pdf> (accessed 20.02.2014).

Worldometers—Real Time World Statistics, 2014. Current World Population. <www.worldometers.info/world-population/wpc.php?utm_expid=4939992-7.scuhn054Q5WXvFD9uRG9Xw.2&utm_referrer=http%3A%2F%2Fwww.google.si%2Furl%3Fsa%3Dt%26rct%3Dj%26q%3D%26esrc%3Ds%26source%3Dweb%26cd%3D2%26ved%3D0CDgQFjAB%26url%3Dhttp%253A%252F%252Fwww.worldometers.info/world-population/

wpc.php?utm_expid=4939992-7.scuhn054Q5WXvFD9uRG9Xw.2&utm_referrer=http%3A%2F%2Fwww.google.si%2Furl%3Fsa%3Dt%26rct%3Dj%26q%3D%26esrc%3Ds%26source%3Dweb%26cd%3D2%26ved%3D0CDgQFjAB%26url%3Dhttp%253A%252F%252Fwww.worldometers.info%252Fworld-population%252F%26ei%3Dp9ViU7mROfLH7Aan0oHwAw%26usg%3DAFQjCNErbPyUCHWnx-PRFhnobEtJRV06Mg%26sig2%3DCnTCiGUgjkEdU5NefXkphA%26bvm%3Dbv.65788261%2Cd.ZGU> (accessed 28.04.2014).

Wright, L.A., Kemp, S., Williams, I., 2011. Carbon footprinting": towards a universally accepted definition. Carbon Manage. 2, 61–72.

Zaccai, E., 2012. Over two decades in pursuit of sustainable development: influence, transformations, limits. Environ. Dev. 1, 79–90.

Zeev, S., Meidad, K., Avinoam, M., 2014. A multi-spatial scale approach to urban sustainability— An illustration of the domestic and global hinterlands of the city of Beer-Sheva. Land Use Policy. 41, 498–505.

Zhao, S., Li, Z., Li, W., 2005. A modified method of ecological footprint calculation and its application. Ecol. Model. 185, 65–75.

CHAPTER

N footprint and the nexus between C and N footprints

Shweta Singh*, **Erin L. Gibbemeyer**[†], **and Bhavik R. Bakshi**[†]

**Department of Agricultural & Biological Engineering and Division of Environmental & Ecological Engineering, Purdue University, West Lafayette, IN, USA*

[†]William G. Lowrie Department of Chemical & Biomolecular Engineering, The Ohio State University, Columbus, OH, USA

MOTIVATION

Nitrogen is the first element of group 15 in the periodic table and is classified as a nonmetal. The stable state of nitrogen is in the form of diatomic gas N_2, which has two atoms combined with a triple covalent bond that gives N_2 an inert form. Even though N_2 is an inert gas (lifeless), the compounds of nitrogen are the key ingredients for supporting life. Nitrogen can exist in multiple oxidation states ranging from −3 in ammonia (NH_3) to +5 in nitrate (NO_3^-) and provides a range of reactive compounds formed with nitrogen. It forms an essential part of protein, approximately 78.09% by volume of air is N_2, it is a major component of fertilizers, and nitrogen exists in a plethora of chemical compounds such as plastics and polymers. The world today would not exist without nitrogen (Schlesinger, 1997). The conversion of nitrogen from an inert state of N_2 to an active state occurs through the biogeochemical cycle of nitrogen that starts with the process of nitrogen fixation. Nitrogen fixation is the process of conversion of N_2 to any other active form either through the natural processes or through anthropogenic processes. Atmospheric lightening and biological fixation are the natural processes that fix nitrogen, whereas the anthropogenic process is the Haber−Bosch process that converts N_2 to NH_3. The invention of the Haber−Bosch process has exponentially increased the nitrogen fixation rate over the natural rate (Galloway et al., 1995). It is partly responsible for bringing the green revolution because it provides the required nutrient for plant growth at an exponentially higher rate, thereby helping meet the growing need of food. Among the industrial uses of nitrogen, it is widely used in the form of inert gas (N_2) as an industrial coolant, in food preparation, and in cryogenics; NH_3 from the Haber−Bosch process is further converted to polymers and plastics. Recently, liquid nitrogen has also been used as an energy storage medium that

Assessing and Measuring Environmental Impact and Sustainability.

has the potential for solving the renewable energy storage problem (Knowlen et al., 1998). Nitrogen is also an essential nutrient for the growth of plants that necessitates use of fertilizers and growing demand of fertilizers. The numerous benefits of nitrogen make it an indispensable resource for the growth and existence of human life on earth. However, nitrogen has also been associated with many negative environmental impacts. The hypoxia zone in the Gulf of Mexico that has degraded the quality of water, destroyed fisheries, and damaged the ocean ecosystem has been caused by excessive nitrogen (Rabalais et al., 2009). This nitrogen is contributed by the runoff from the Upper Mississippi Basin attributable to excessive use of fertilizer and efforts to solve the ecological problem are studied extensively (Alexander et al., 2008) and later (Mitsch et al., 2001). Among other environmental impacts of nitrogen is the contribution of N_2O (nitrous oxide) to the global warming potential. N_2O is a 300 times more potent greenhouse gas than CO_2. It has caught the attention of researchers worldwide, and growing concerns attributable to excessive deposition of nitrogen in environment have been addressed. International Nitrogen Initiative (INI) is a worldwide research coordination network that specifically focuses on the issue of judiciously managing the reactive nitrogen to minimize the negative impact while meeting the demands (INI). The United Nations also has focused its efforts on this crucial nutrient management program through the Global Partnership on Nutrient Management (United Nations Environment Programme, 2014). A comprehensive report was released that outlines the global challenges of nutrient management, suggests nutrient use efficiency as an indicator to reduce impact and improve the benefits, and mandates that the management of N at various scales should also be related to quantification of other benefits such as water, air, climate, soil, and biodiversity (Sutton et al., 2013). The US National Science Foundation has also funded a 5-year project beginning in March 2011 to build a Research Coordination Network called the "Reactive Nitrogen Research Coordination Network" that brings together researchers working on various facets of reactive nitrogen impact (Research Coordination Network on Reactive Nitrogen, National Science Foundation). Galloway et al. independently developed a tool called the Nitrogen Footprint Calculator (N-Print), which is available online to help identify the activities that lead to the highest N losses (Leach et al., 2012). Singh and Bakshi (2013) have developed an inventory that can capture the impact of different economic activities on various components of the nitrogen cycle. This inventory is focused on the 2002 US economy, and the methodology is scalable at different scales provided that the underlying data are collected. The growing efforts for quantification of the cost and benefits of nitrogen will provide a clear path forward to sustainably manage this very critical nutrient for human well-being.

BACKGROUND ON FOOTPRINTS

Footprints have become standard measures for sustainability assessment. The concept of footprint originated in 1992, when William Rees proposed the ecological footprint (Rees, 1992). Ecological footprint is the measure of human impact on the limited land resource. In this approach, human consumption is converted into equivalent land area that is required to provide these resources along with area required to absorb the CO_2 emissions incurred because of consumption activities. This land area is then compared with the available land area to meet the demand that provides an idea of overshoot beyond the available capacity. There are several variations of ecological footprint, and a critical evaluation of different approaches can be found in the work by Zhang et al. (2010b). The original concept of ecological footprint led to the development of carbon footprint (Wiedmann and Minx, 2007), water footprint (Hoekstra et al., 2014); one of the most recent entrants to the footprint family is the nitrogen footprint (Leach et al., 2012).

Carbon footprint is calculated as the total amount of GHG emissions in terms of an equivalent amount of CO_2. The concept of carbon footprint is based on the global warming potential of GHG emissions that can result in an increase in temperature beyond the safe limits for functioning of global systems. Each of the GHG emissions related to GWP is converted to a CO_2 equivalent based on the relative potential for global warming. In a recent study, Gao et al. (2013) compared various carbon footprints and assessment standards. There are several definitions for carbon footprint that are used in assessment methods. Wiedmann and Minx (2007) define it as "total amount of direct and indirect CO_2 emissions for a process, product or system." Thus, this definition limits the carbon footprint to CO_2 emissions only, whereas other definitions include other greenhouse gases like methane (CH_4) and nitrous oxide (N_2O). There is also a large variation in the selection of boundaries for calculation of carbon footprint. The carbon footprint can be defined by personal consumption, process level, product, urban scale (Bellucc et al., 2012), organization, and country (Hertwich and Peters, 2009). The calculation and data collection are dependent on the estimation boundary that varies from process scale to economy scale to global scale. The scale of calculation of footprints is discussed later in this chapter. Matthews et al. (2008) highlight the importance of selection of boundaries for carbon footprint estimation in their discussion of scope 1, scope 2, and scope 3 for carbon footprint calculations. The term "scope" is also referred to as "tier" of analysis. The GHG protocol developed by the World Resources Institute (WRI) and the World Business Council for Sustainable Development (WBCSD) provides a streamlined framework for organizations to report carbon footprint (Greenhouse Gas Protocol). Scope 1 or tier 1 carbon footprint includes the emissions attributable to direct activities by the organization such as CO_2 emissions by a firm's vehicles and factories. Scope 2 or tier 2 carbon footprint includes the CO_2 emissions attributable to energy consumption by the organization such as emissions attributable to electricity consumption by the organization. The last tier, or scope 3, includes all other emissions that occur because of the organization, such as emissions attributable to raw material production, waste disposal, or other indirect activities that produce GHG emissions. Matthews et al. (2008) estimate that for an average sector in the economy, tier 1 and tier 2 only cover 26% of total emissions or carbon footprint, thus emphasizing the necessity of choosing appropriate boundaries while using carbon footprint in decision-making. The method for boundary selection and footprint calculation on different scales is discussed in detail in the Scale of Footprints section (later in this chapter).

Most of the footprint measures originate with a particular concern such as the roots of "carbon footprint" and global warming. Similarly, the water footprint is based on the concern of scarcity of water for human consumption. The water footprint concept captures the total amount of water required to support human consumption (Hoekstra et al., 2014). This metric captures the sustainability of human consumption based on availability of water resource. The water footprint calculation is more involved because it classifies water into three different categories, blue, green, and gray water footprint. The first two categories (blue and green water footprints) are based on the source or origin of water being consumed, whereas the last (gray water footprint) is based on the pollution generated in the process. The blue water footprint is defined as the consumption of water originating in blue water resources such as surface- and groundwater. The consumption for a water source is defined as water withdrawal from a particular catchment area either by being incorporated into a product or by being evaporated in the process and lost to other catchment area. The green water footprint is defined as the consumption of water that originates in sources such as rainwater stored in soil as soil moisture. The green water footprint becomes available for consumption through incorporation in biomass. The last category, gray water footprint, is inspired by the concept of land requirement to sequester emission as part of the ecological footprint of a process. The gray water footprint is the volume of water that will be needed to dilute the pollution of water bodies so that the water bodies meet the water quality standards. It is not an actual consumption category of water; however, it captures the impact on availability of water as a resource for consumption, because the polluted water may not be fit for consumption. The total water footprint is calculated as the sum of blue, green, and gray water footprints. Hoekstra et al. (2014) describe these methodology, calculations, data requirements, and applications in detail as state-of-the-art for water footprint calculations. Other considerations in calculations of water footprints are the scale of study, as in the case of carbon footprint calculations. Water footprint can also be calculated as process, product, consumer, and geographical (such as expanding boundary (Hoekstra, 2011) or impact of international trade (Aldaya and Hoekstra, 2010)), on a national scale (i.e., specifically, trade of national agricultural commodities (Chapagain and Hoekstra, 2008), impact of national consumption patterns on footprint (Hoekstra and Chapagain, 2007), and detailed account of footprints for different nations (Mekonnen and Hoekstra)), and on sectoral or economy and business scales (Ercin et al., 2011). The direct water footprint only includes the flows that occur because of direct consumption at the process or product scale. Indirect water footprint also includes the water footprint of upstream raw materials, downstream pollution, or other consumption impact. The boundaries can be determined similar to any other footprint measures. The section on scale of footprints provides more details regarding the boundary selection and calculation method.

The most recent entrant in the family of footprints is the nitrogen footprint. In contrast to "ecological footprint" and "water footprint," where the scarcity of resources is motivation of calculation of these footprints, in the case of nitrogen footprint it is the negative impact association with reactive nitrogen losses to the environment that is the motivation. Eutrophication is a widely known consequence of excessive loss of nitrogen in the form of organic and inorganic nitrogen to the coastal areas (Rabalais et al., 2009). This negative impact has led to several previous efforts to develop farm scale and regional and national nitrogen balances. Salo et al. (2007) show net nitrogen balances in a region and a national scale focusing on Finland for a time period from 1990 to 2005, whereas Olsthoorn and Fong (1998) studied the nitrogen flows through various systems in the Netherlands and termed it the anthropogenic nitrogen cycle. The work of Salo et al. (2007) was inspired by the work of Slak et al. (1998), who studied the feasibility of national nitrogen balances. In another study, Forkes focuses on an even narrower boundary of the city of Toronto to study the relationship between nitrogen and urban consumption (Forkes, 2007). However, the term "nitrogen footprint" was only recently coined by Leach et al. (2012). It captures the total amount of nitrogen losses attributable to human consumption of various goods and services. This definition of nitrogen footprint only deals with the impact of the output side regarding nitrogen flows by human consumption. However, the interaction of human activities with nitrogen flows also occurs on the input side by nitrogen mobilization, as discussed by Singh and Bakshi (2013). Rockström et al. (2009) identified nitrogen mobilization as one of planetary boundary and suggested reduction of nitrogen mobilization by 25% of the current rate. It is also important to include nitrogen mobilization in the metric for assessing impact on nitrogen. Thus, Singh and Bakshi suggested improvements in the nitrogen footprint (Singh and Bakshi, 2014), as discussed in more detail in the section about nitrogen footprint.

Despite several uncertainties associated with selection of boundary and data quality, footprints provide a simple metric that can be used for decision-making across different systems. Fang et al. (2014) provide a comprehensive review of ecological, energy, carbon, and water footprints with regard to data availability, methodological consistency, and suggestions for improvements. Each of the footprints represents a unique aspect of environmental impact in a complementary sense, and use of a family of footprint should be encouraged in making decisions. This will ensure that decisions based on a single footprint metric do not push the system in a direction that neglects the impact captured by another footprint.

SCALE OF FOOTPRINTS

The scale of footprints is dependent on the scope of the study conducted, as demonstrated by the wide variation in the scale of footprint application discussed previously. The decision can range from comparing two products, processes, economies, or regions (nation). This wide variation of systems undergoing study leads to development of variation in the calculation method for footprint measure. However, all the footprints share this common phenomenon of selection of scale of study to calculate footprint. Therefore, it is important to separately discuss the scale of footprint calculation along with boundary selection, data requirements, and interpretation for decision-making. The common use of footprints is to compare the environmental impacts of products, which is mostly performed on the process scale or economy scale. In several cases, a hybrid study that combines economy scale impact with process scale impact is also preferred for completeness and includes indirect impacts. A detailed description of boundary selection, data requirements, and interpretation of results is provided for each scale of footprint study.

PROCESS SCALE

The most common use of footprints is to compare two processes or two products. In a process scale comparison of footprints, the boundary is well defined by the process undergoing study and the immediate associated flows with the respective footprint measures added together. In Figure 6.1, all the inputs and outputs of a corn farming process can be recorded and specific flows associated with a chosen footprint can be added together for the process scale footprint reporting. For example, in the case of carbon footprint, CO_2 emissions and N_2O emissions in equivalent CO_2 can be added together.

LIFE CYCLE FOOTPRINTS

Life cycle analysis (LCA) is a method that extends the boundary of study beyond the immediate processes associated with the products. The calculation method for the life cycle footprint is similar to the process scale footprint. It only requires selection of an appropriate boundary expansion based on the scope of the LCA. LCA can be cradle-to-gate, gate-to-gate, cradle-to-grave, gate-to-grave, or well-to-wheel. The last physical boundary, well-to-wheel, is only applicable to automotive fuels. Figure 6.1 shows the processes included to study life cycle scale footprints. A cradle-to-gate analysis studies the processes upstream in the production

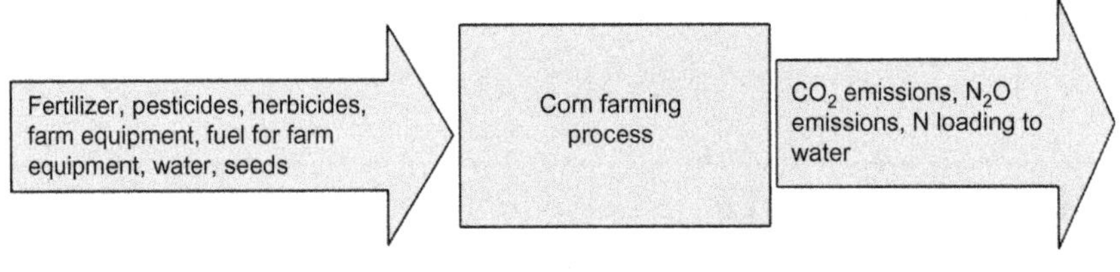

FIGURE 6.1

Process scale footprint calculation.

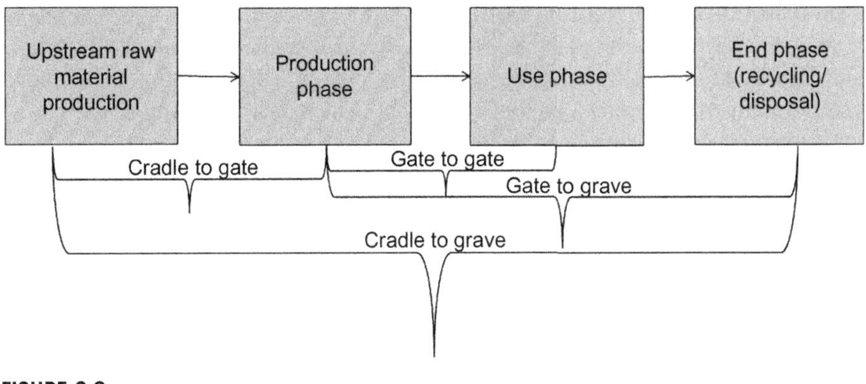

FIGURE 6.2

Boundary selection for life cycle footprint calculations.

chain of a product until the stage at which the product is ready for use. The cradle-to-grave analysis studies the processes from upstream production to the disposal phase at the end of product life cycle. The well-to-wheel boundary is similar to the cradle-to-grave analysis because it studies the impact of fuels from production (including upstream production) to the end use in vehicles. The stages for study of life cycle footprints are identical to the stages used in life cycle impact analysis; the only difference is that the flows for which the data are collected. Using the processes at each stage in the diagram shown for the boundary chosen in life cycle footprint calculation, appropriate flow data can be calculated (Figure 6.2).

ECONOMIC INPUT–OUTPUT FOOTPRINT

The economic input–output (EIO) footprint calculation is based on the EIO model, which is an analytical framework to capture the interconnections among the processes in the economy. In this case, the boundary of study is expanded to include the whole economy. The EIO model was proposed by Leontief (1936) and has been widely used as an environmentally extended input–output (EEIO) model to account for economy scale environmental impacts (Leontief, 1970). The EIO model is given by Eq. (6.1) and is used to calculate total production in the economy (X) that can support the final demand (f) from various sectors of the economy. The interconnections of the economic sectors are captured by the direct requirements matrix (A), which consists of coefficients a_{ij} that are assumed to be constant. These coefficients a_{ij} capture the inputs required from ith sector per unit output from the jth sector and are calculated as $a_{ij} = Z_{ij}/X_j$, where Z_{ij} is the flow from ith sector to jth sector and X_j is the total output from jth sector.

$$X = (I - A)^{-1} f \tag{6.1}$$

In the EEIO model, the EIO model of Eq. (6.1) is augmented by environmental impact information for each sector. The extension of the EIO model to capture total direct and indirect environmental impacts, specifically pollution generation, was originally proposed by Leontief (1970). In his original proposition, the pollution generation was included as an additional sector in the economic model. This helped in the calculation of total pollution generation for final demand from each sector using a similar model as Eq. (6.1). The linearity assumption of constant pollution generation per unit of output from a sector is implicit in these calculations. In the current implementation of the EEIO model, pollution generation is not considered a sector of production in the economy. Equation (6.1) is enhanced with an environmental intervention vector that can capture resource consumption or pollution generation. The EEIO model was used to develop a generalized framework of EIO-LCA that can capture the life cycle impact on an economy scale. The environmental intervention vector (R_i) is a vector of total emissions or total resource requirement per unit output from the ith sector. It is calculated by Eq. (6.2), where X_i is the total dollar output of ith sector.

$$R_i = \text{total emissions or total resouces}/X_i \tag{6.2}$$

To calculate the total resource consumption or emissions, Eq. (6.3) is used, where ΔB is the total change in the specific impact category as captured in the vector R_i.

$$\Delta B = \hat{R}_i\, \Delta X \tag{6.3}$$

ΔX is the total change in the output of various sectors for a given change in the final demand of economic sectors (Δf). Equation (6.4) is used to calculate ΔX.

$$\Delta X = (I - A)^{-1} \Delta f \tag{6.4}$$

The economy scale footprint calculations are derived from this EEIO model to account for total footprint due to both direct and indirect impact. The vector R_i contains the respective footprint variables such as water and carbon, and Eqs. (6.3) and (6.4) are used to calculate the economy scale footprint for the respective processes.

HYBRID SCALE FOOTPRINT

The hybrid scale footprint originates from the boundary selection as in the case of hybrid LCA. Hybrid LCA was developed to combine the advantages of process-based LCA and EIO-LCA. The process-based LCA suffers from the shortcoming of underestimation of environmental impact because of the possibility of truncation of upstream processes, thus missing several indirect impacts of the process or product undergoing consideration. However, EIO-LCA has the disadvantage of using average values for impact attributable to aggregation of sectors. This leads to the difficulty of not being able to compare two products on a fine scale because the products belong to the same sector of the economy; for example, two different categories of electricity belong to the same sector of the power generation sector. Hybrid LCA tends to combine the benefits of process-based LCA and EIO-LCA, thus improving the reliability of results. The types of hybrid LCA methods that can be used for calculation of the hybrid scale footprint are tiered, IO–based hybrid, integrated method, and augmented process-based model (Bullard et al., 1978; Joshi, 2000; Suh et al., 2004; Suh, 2004; Wiedmann et al., 2011; Guggemos, 2003). Tiered hybrid analysis mainly relies on the IO data and utilizes increasing disaggregation of the product system to map the parts to specific IO sectors (Bullard et al., 1978). The IO–based hybrid analysis, however, disaggregates the sector by obtaining detailed information for the sectors and introduces process scale data into the IO model. Based on the data for the product undergoing study, the following approaches can be used as part of the IO–based hybrid analysis: a new sector can be introduced in the economy to study the overall system scale impact; disaggregating the sector to study the interaction; or the introduction of new sectors for product and disaggregation of existing sectors can simultaneously be performed to improve the reliability of studies (Joshi, 2000).

Integrated hybrid analysis (Suh et al., 2004) and also (Suh, 2004) uses a more structured mathematical framework to combine process scale data with the IO scale. The process scale data are used to develop a technology matrix that has an impact on physical units (e.g., emissions or unit for footprint) per operation time for all the processes. The IO and process scale matrix interact through a make matrix and a use matrix. A make matrix captures the output from each process (on a process scale) and sector (on an IO scale), whereas a use matrix captures the input to each process (on a process scale) and sector (on an IO scale). The interaction between the process scale matrix and IO matrix helps to calculate the overall impact or footprint for any output from the process or IO. This method improves the reliability of results because the process scale matrix is built on detailed available data for the life cycle of the product undergoing consideration.

The augmented process-based hybrid method (Guggemos, 2003) utilizes the process scale analysis and augments the lack of data availability with IO data. Each process involved in the life cycle (such as an upstream raw material manufacturing), where process scale modeling is difficult, appropriate IO sector level data are used. For processes that cannot be modeled with IO sectors, process level data are used. A comparison of results obtained from a process-based integrated hybrid analysis and an IO-based hybrid analysis is presented in the work of Wiedmann et al. (2011) The steps involved in various hybrid-based LCA analyses can be combined with the data for specific footprint studies to calculate a hybrid scale footprint.

NATIONAL SCALE

National scale footprint analysis expands the scale of analysis to a national scale. The concept of national footprint analysis arose from the need to compare how various countries are contributing to the environmental impact in various categories. The calculations are straightforward in terms of including all the flows associated with a particular footprint for the nation. For example, a national scale carbon footprint analysis includes all the flows to be included in the carbon footprint, such as carbon dioxide (CO_2) and CO_2 equivalents of other greenhouse gases (N_2O, CH_4). The choice of mass flows to be included depends on the specific category of footprint being evaluated. The Global Footprint Network specifically focuses on the ecological risk for various nations attributable to the consumption pattern that exceeds the biocapacity of the nation. This is derived from the ecological footprint definition. It helps nations assess the risks associated with ecosystem degradation within the nation and outside the nation because the dependence has become global because of open trade. The ecological footprints and biocapacity profiles for 150 nations are available through the Global Footprint Network (2014).

NITROGEN ACCOUNTING AND FOOTPRINT

Unlike other footprints that are motivated by limited resource availability and a possibility of a lack of resources to meet the demands, nitrogen footprint is motivated by the negative impact of reactive nitrogen in the environment. The term "nitrogen footprint" did not come into existence until recently (Leach et al., 2012), and it was preceded by several attempts to include nitrogen balance in systems analysis (Barry et al., 1993; Slak et al., 1998; Van der Hoek, 1998; Salo et al., 2007; OECD and EUROSTAT: Gross Nitrogen Balances, Handbook), along with calculation of virtual nitrogen (Burke et al., 2008). Salo et al. (2007) examined nitrogen balances in agricultural systems as a primary indicator for environmental performance. The approach is a process scale mass balance technique called "gross and net nitrogen balances." The "gross nitrogen balance" estimates the total nitrogen input to soil and total nitrogen output from soil. The gross balances include all emissions to soil, air, and water. In the case of net nitrogen balance, the emissions to air are excluded. This involves exclusion of N volatilization and denitrification from manure and fertilizer to calculate net nitrogen balances. The relationship between gross nitrogen balance and net nitrogen balance is given as: net nitrogen balance = gross nitrogen balance − (emissions from N volatilization and denitrification). Because denitrification emissions are difficult to estimate, the study by Salo et al. (2007) omitted denitrification emissions from the net nitrogen balance calculation. The components that are used for calculation of "gross nitrogen balance" and "net nitrogen balance" are given in Table 6.1. These nitrogen balances can be used as an indicator for potential of environmental impacts associated with nitrogen use. Even though the balance

Table 6.1 Gross and Net Nitrogen Balance Calculations

Components	Inputs/Outputs
Fertilizers (mineral and organic)	Inputs
Livestock manure	Inputs
Biological nitrogen fixation	Inputs
Atmospheric deposition	Inputs
Other inputs (products such as seeds etc)	Inputs
Harvested yield/product	Outputs
GROSS NITROGEN BALANCE	SUM OF COMPONENTS ABOVE
Ammonia volatilization from fertilizers and livestock manure	Output
Denitrification emissions	Output
NET NITROGEN BALANCE	GROSS NITROGEN BALANCE– (volatilization + denitrification)

indicators are defined for agricultural systems, these can also be used for various manufacturing unit processes that have nitrogen-based inputs and outputs. The scale of nitrogen balance can be national, regional, process, or life cycle.

Another method for nitrogen accounting is to examine the total nitrogen budget, which is similar to the material flow analysis in the manufacturing processes. The boundary can be set as desired. Olsthoorn and Fong (1998) used this method to study the nitrogen budget for the Dutch economy with the objective of identifying the anthropogenic impact on the nitrogen cycle and future N_2O emissions. The anthropogenic losses related to various sectors such as industry, agriculture, transportation, and residential are accounted separately to provide an insight into the magnitude of impact of specific human consumption on the nitrogen cycle. This again relates to the mass balance approach used in industrial studies. The categorization into various sectors can be used to study the contribution of different types of manufacturing processes to nitrogen balance. Similarly, a boundary can be drawn across the urban systems and a mass balance approach can be followed to achieve nitrogen accounting (Forkes, 2007).

Even though the mass balance approach provides an idea of nitrogen budget and insights into the system impact on total flows, it does not provide a specific direction for decision-making. The nitrogen budgets have been used without any impact on reducing the negative impact of nitrogen losses. This can be considered as the motivation of the recent introduction of the "nitrogen footprint" as a decision-making indicator.

Nitrogen footprint is calculated as the total reactive nitrogen (Nr) lost to the environment in the form of air, water, and land emissions during a process or manufacturing of a product being consumed. It is formally defined as "...the total amount of Nr released to the environment as a result of an entity's resource consumption, expressed in total units of Nr" (Leach et al., 2012). Based on this definition, Galloway et al. (Online N Footprint Calculator) presented an online calculator

Table 6.2 Input Components for Consumption-Based Nitrogen Footprint Calculation

Input Components for Consumption	Assumption (if any)	Contributing Variables
Food	All food consumed eventually end up as Nr lost	Food production energy N, food production virtual N, and food consumption
Transportation	Emissions from fuel combustion	Plane, public transit, personal car (energy-related nitrogen footprint)
Goods and services	Emissions from fuel combustion for provision of goods & services	
Housing	NA	Electricity and natural gas consumption (energy-related nitrogen footprint)

for individual or community nitrogen footprints that uses food, housing, transportation, and goods and services as the main categories of consumption for calculation of the nitrogen footprint. All these consumption activities contribute to the Nr footprint, either through energy use or through use of nitrogen as a resource that gets lost to the environment during the process of manufacturing products being consumed. Table 6.2 presents the components, assumptions, and scale for inclusion in the Nr footprint, along with the contributing variables for calculation of the Nr footprint (Leach et al., 2012). Table 6.2 should be used as a guideline for input components to be considered in Nr footprint calculation of any process, product manufacturing, individual, or nation. Specifically, in LCA all direct and indirect inputs associated with these components can be considered based on data availability.

ECO-LCA NITROGEN FOOTPRINT

The ecologically based life cycle assessment (Eco-LCA) accounts for the role of various ecological goods and services in supporting economic activities. Recently developed Eco-LCA nitrogen inventory considers the impact of anthropogenic activities on three components: nitrogen mobilization (conversion of inert N to reactive N); nitrogen end products (use of mobilized nitrogen as end product); and nitrogen emissions. Figure 6.3 shows the relation between different categories of nitrogen Eco-LCA inventory and detailed components in each category are mentioned in Appendix 6.1. The Eco-LCA approach to the nitrogen footprint is based on the holistic impact of human activities on the nitrogen cycle instead of only focusing on the nitrogen emissions that cause environmental problems.

This is aligned with the concept of the life cycle view to resource management by identifying the upstream activities that may result in negative impact downstream. A challenge with defining an appropriate nitrogen footprint can be associated with the

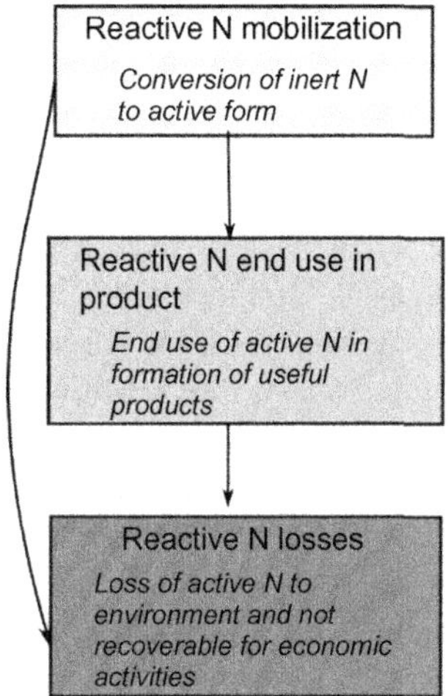

FIGURE 6.3

Eco-LCA nitrogen inventory component.

lack of understanding of limits for resource utilization. Rockström et al. (2009) identified the amount of nitrogen mobilization as the boundary limit for the nitrogen cycle and suggested reducing the mobilization by 25% to meet the safe operating limits of the nitrogen cycle. This also suggests that dependence of human activities on Nr mobilization should be included in the definition of the Nr footprint. Therefore, Singh and Bakshi (2014) utilized the Eco-LCA nitrogen inventory to define nitrogen footprint as both nitrogen mobilization and nitrogen emissions. Using a two-dimensional approach for nitrogen footprint ensures that both aspects related to nitrogen impact—managing nitrogen cycle and reducing negative environmental impacts attributable to reactive nitrogen losses—are captured in decision-making.

ECO-LCA NITROGEN INVENTORY AND CALCULATIONS

The components of nitrogen inventory that are included in the Eco-LCA nitrogen inventory can be collected on all scales (process, life cycle, economy, and national). Thus, the Eco-LCA nitrogen footprint can be calculated on all these defined scales depending on the data availability and objective of footprint calculations. The Eco-LCA nitrogen inventory available in the current Eco-LCA online tool Ecologically Based Life Cycle Assessment (Eco-LCA tool) is on an economy scale and is based on the original Eco-LCA model (Zhang et al., 2010a). The components of inventory form the V_{ph} vector in the Eco-LCA model given by Eq. (6.5).

$$X_{\text{ph}} = (I - G^{\text{T}})^{-1} V_{\text{ph}} \tag{6.5}$$

In Eq. (6.5), $(I - G^{T})^{-1}$ is called the Ghosh inverse and V_{ph} is the physical resource vector. In the case of the nitrogen categories, various components of each category are used to construct this vector. The use of Eq. (6.5) gives total direct and indirect impacts on the flow components of nitrogen inventory of a particular activity. The direct impact intensity is calculated as $\hat{X}^{-1}(I + G^{T})V_{ph}$, where X is the total economic size of the sectors measured in monetary units. Indirect impact intensity is calculated as $\hat{X}^{-1}((G^{T})^2 + (G^{T})^3 + (G^{T})^4 + \cdots)V_{ph}$. Both direct and indirect intensity values are multiplied by the total final demand (f) required to support the system undergoing study, thus resulting in the economy scale direct and indirect impacts associated with the process undergoing study. The Eco-LCA nitrogen inventory contains the intensities for all the components in the US economy and has been used to study the nitrogen profile for various sectors in the 2002 US economy (Singh and Bakshi, 2013).

ECO-LCA NITROGEN FOOTPRINT FOR 2002 US ECONOMY

Figures 6.4 and 6.5 show the Eco-LCA nitrogen footprint for all the economic sectors in the 2002 US economy. Table 6.3 contains the legend for the X axis in Figures 6.4 and 6.5. The X axis is the aggregated economic sector in the 2002 US economy. The aggregation into these coarse 28 sectors is based on similarity of economic activities of 426 economic sectors as identified by NAICS codes and defined by Bureau of Economic analysis (BEA). The aggregation scheme can be found in the supplementary information provided by Singh and Bakshi (2013).

Figure 6.4 shows the "input-side" footprint or the nitrogen mobilization impact on Tg–N of various economic activities. There are clear differences that can be seen when Figures 6.4 and 6.5 are compared. In Figure 6.4, which is the input of nitrogen mobilization supporting various activities, the agriculture, livestock, forestry, and fisheries activities have the highest impact. Among the categories of inputs of nitrogen mobilization to this sector, atmospheric N deposition and industrial N fixation impacts are highest. The atmospheric N deposition is related to the land area as a driving factor for higher deposition; for this reason, the agricultural and forestry sector that dominates land use also has highest contribution from atmospheric N deposition as input. The contribution from industrial N fixation is related to the use of fertilizer. The next activity showing high impact is food, beverage, and tobacco. In this economic activity, most of the impact is indirect. This is because most of the activities in this sector are not directly involved in N mobilization, but they depend on other activities that have direct dependence on N mobilization. Food, beverage, and tobacco activities involve mostly food manufacturing that depends on output of harvested food from farming. This creates indirect dependence on several N mobilization categories. The third highest impact occurs in the resin, rubber, artificial fibers, agricultural chemicals, and pharma sector, which is attributable to fertilizer manufacturing being the largest source of nitrogen fixation or nitrogen mobilization by the Haber–Bosch process.

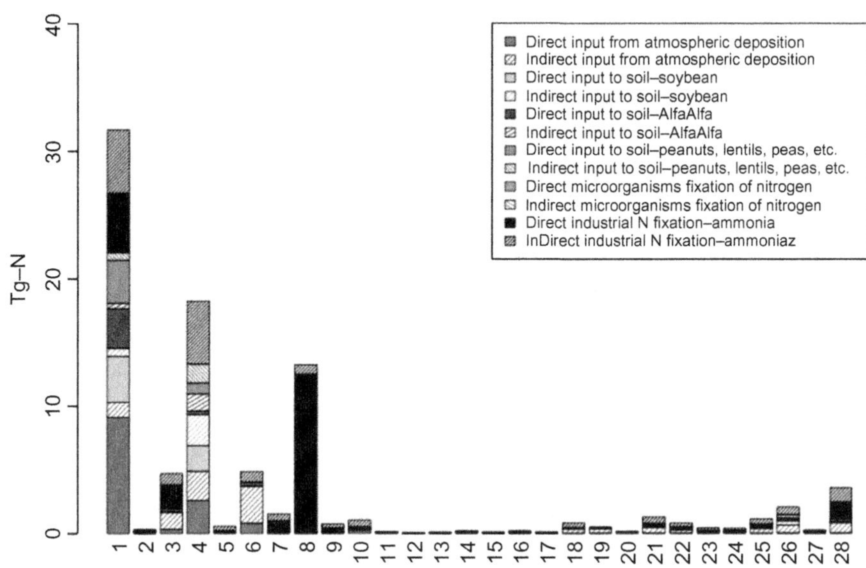

FIGURE 6.4

Input-side Eco-LCA nitrogen footprint (nitrogen mobilization).

Reprinted from Singh and Bakshi (2013).

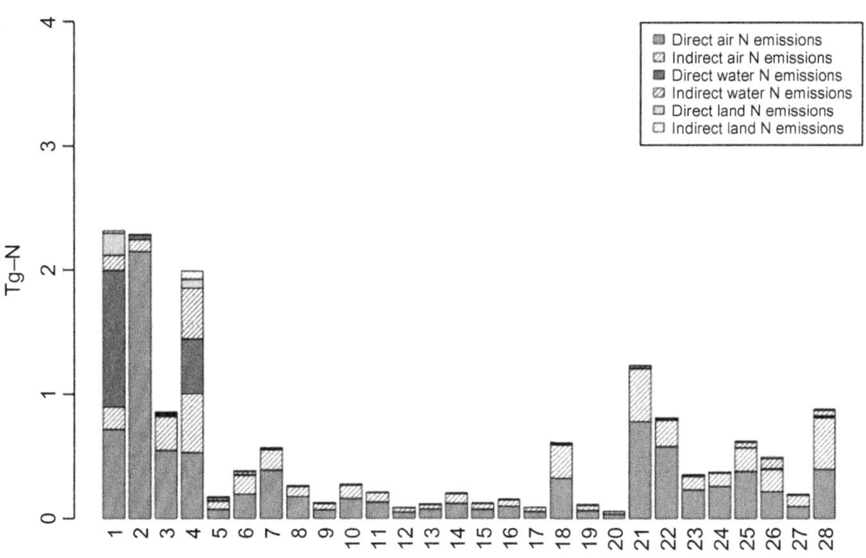

FIGURE 6.5

Output-side Eco-LCA nitrogen footprint (nitrogen emissions).

Reprinted from Singh and Bakshi (2013).

Table 6.3 Legend for *X* Axis in Figures 6.4 and Figure 6.5

Number	Aggregated Sector Name
1	Agriculture, livestock, forestry, and fisheries
2	Mining and utilities
3	Construction
4	Food, beverage, and tobacco
5	Textiles, apparel, and leather
6	Wood, paper, and printing
7	Petroleum and basic chemical
8	Resin, rubber, artificial fibers, agricultural chemicals, and pharma
9	Paint, adhesives, cleaning and other chemicals
10	Plastics, rubber, and nonmetallic mineral products
11	Ferrous and nonferrous metal production
12	Cutlery, handtools, structural and metal containers
13	Other metal hardware and ordnance manufacturing
14	Machinery and engines
15	Computers, audio–video and communications Equipment
16	Semiconductors, electronic equipment, and media reproduction
17	Lighting, electrical components, and batteries
18	Vehicles and other transportation equipment
19	Furniture, medical equipment and supplies
20	Other miscellaneous manufacturing
21	Trade, transportation, and communications media
22	Finance, insurance, real estate, rental and leasing
23	Professional and technical services
24	Management, administrative, and waste services
25	Education and health care services
26	Arts, entertainment, hotels, and food services
27	Other services except public administration
28	Government and special services

However, Figure 6.5 shows a different perspective of the nitrogen footprint for the same economy. The output-side nitrogen footprints or the nitrogen emissions are the usual nitrogen footprint. Because of large nitrogen water loading from farming activities, agriculture, livestock, forestry, and fisheries also show high impact on the output-side nitrogen footprint. The second sector is the mining and utilities sector that originates from impact on nitrogen emissions from fossil fuel combustion. The third highest impact on the output-side nitrogen footprint occurs in the food, beverage, and tobacco sector, with high contributions from the indirect impacts (Figure 6.5).

ECO-LCA NITROGEN FOOTPRINT FOR PRODUCTS

The components of the Eco-LCA nitrogen inventory can also be used to estimate the nitrogen footprint for various products. The footprint for products can be calculated on a process scale, economy scale, or hybrid scale based on the scope of analysis and data availability on the fine scale. Similar to hybrid LCA studies in which fine scale process data are combined with the economy scale data, Eco-LCA nitrogen inventory on the economy scale can be combined with the process scale data for Eco-LCA nitrogen inventory variables. The following steps describe the process of calculating the nitrogen footprint using Eco-LCA inventory:

1. Identify the system boundary for impact calculation of the product under consideration.
2. Collect Eco-LCA nitrogen inventory component data on the process scale for each of the processes in the system boundary.
3. For processes, when the process scale data for nitrogen inventory components are not available, obtain the monetary final demand on the economic sector scale. Use the Eco-LCA economy scale inventory to obtain the corresponding nitrogen impact.
4. Combine the process scale and economy scale data for corresponding inventory components to calculate the "input side" (nitrogen mobilization footprint) and "output side" (nitrogen emissions footprint) for the products undergoing comparison.

These steps can generically be used to study any kind of product and processes for calculation of a nitrogen footprint based on Eco-LCA nitrogen inventory. It is to be recognized that the process described is similar to the augmented process-based hybrid method (Guggemos, 2003). However, purely EIO scale or process scale Eco-LCA nitrogen inventory is also feasible based on data availability for the user and scope of study.

CASE STUDY: COMPARISON OF ECO-LCA NITROGEN FOOTPRINT FOR FUELS

The method for the calculation of a nitrogen footprint based on the Eco-LCA nitrogen inventory discussed here is demonstrated with the help of a case study of three different types of fuels: gasoline, E85 corn, and E85 switchgrass. These are compared on the basis of impact associated with an output of driving 1 km using each of these fuels. The selection of this output is driven by functional unit definition

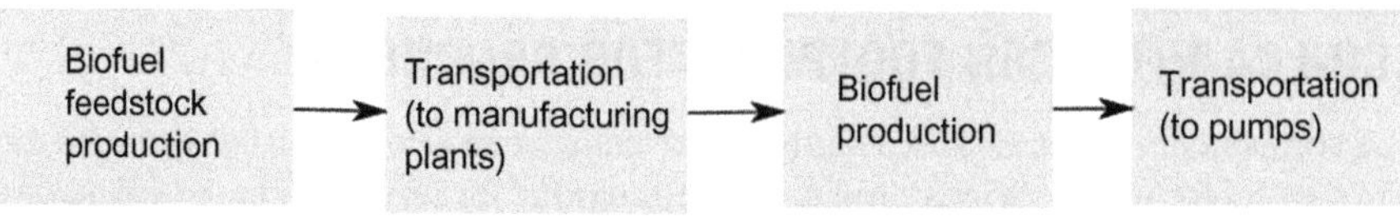

FIGURE 6.6

Cradle-to-gate biofuel footprint analysis.

Table 6.4 Eco-LCA Variables Included at Process Scale in Footprint Calculation of Biofuels

Variables for C and N Footprint Calculation	Source of Data
Carbon Sequestration in Feedstock	GREET
CO_2 Tailpipe	GREET
CO_2 Emitted from Fossil Fuels	GREET
N in N_2O Emissions from Farming : Fertilizer & Residue	Mosier et al., 1998
N in N_2O (Combustion)	GREET
N in NO_X (Combustion)	GREET
N loading to Water	Seitzinger and Kroeze, 1998
N fixation in farming process	Russelle and Birr, 2004

used in LCA studies for comparability of results. Steps demonstrated here can be used as guidelines for the calculation of the Eco-LCA nitrogen footprint for any industrial process, supply chain, or life cycle of a product.

1. System boundary for fuels: The study is a "well-to-pump" or "cradle-to-gate" analysis

   Figure 6.6 shows the production processes included in the study. As compared with Figure 6.2, Figure 6.6 shows the expansion of processes within the "cradle-to-gate" analysis boundary.
2. Process scale data collection for Eco-LCA inventory components.

   The next step in the calculation of the nitrogen footprint is to collect data for the processes in Figure 6.6 for each of the variables included in the Eco-LCA nitrogen inventory. Table 6.4 gives the variables in Eco-LCA inventory that were considered on the process scale because better data were available on the fine scale. Data for these variables come from either the GREET (Greenhouse Gases Regulated Emissions and Energy Use in Transportation) model or from other published studies.
3. Economy Scale Processes

   Table 6.5 give the various economic sectors that are involved in the life cycle of the corresponding fuels and were considered for calculation of the N footprint based on economy scale Eco-LCA N inventory.
4. Input-side and output-side N footprint calculation

Table 6.5 Economic Sectors for Biofuels Footprint Calculation at Economy Scale Using Eco-LCA Inventory

Sector Name	Sector NAICS Code	Gasoline	E85 Corn	E85 Switchgrass
Oil and gas extraction	211000	x	x	x
Coal Mining	212100	x	x	x
Electric power generation, transmission, and distribution	221100	x	x	x
Natural gas distribution	221200	x	x	x
Petroleum refineries	324110	x	x	x
Iron and steel mills and ferroalloy manufacturing	331110	x	x	x
Rail transportation	482000	x	x	x
Pipeline transportation	486000	x	x	x
Fertilizer manufacturing	325310		x	x
Pesticide and other agricultural chemical manufacturing	325320		x	
Cement manufacturing	327310		x	x
Lime and gypsum product manufacturing	3274A0		x	x
Farm machinery and equipment manufacturing	333111		x	x
Water, sewage and other systems	221300		x	x
Other basic organic chemical manufacturing	325190		x	x
Grain farming	1111B0		x	
All other crop farming	1119B0			x
Wet corn milling	311221			x
All other basic inorganic chemical manufacturing	325188			x
Biological product (except diagnostic) manufacturing	325414			x

The corresponding categories of Eco-LCA N inventory components were aggregated into input-side and output-side Eco-LCA N footprints.

Figure 6.7 shows the input-side (N mobilization) and output-side (N emissions) footprints for the fuels undergoing study. From both input and output perspectives, E85 corn has the highest impact, and the next highest impact is from E85 switchgrass. Only looking at the output N footprint, gasoline fuel seems to have a high impact; however, when considering the two-dimensional impact, the focus shifts to the bio-based fuels for reducing impact on the nitrogen cycle.

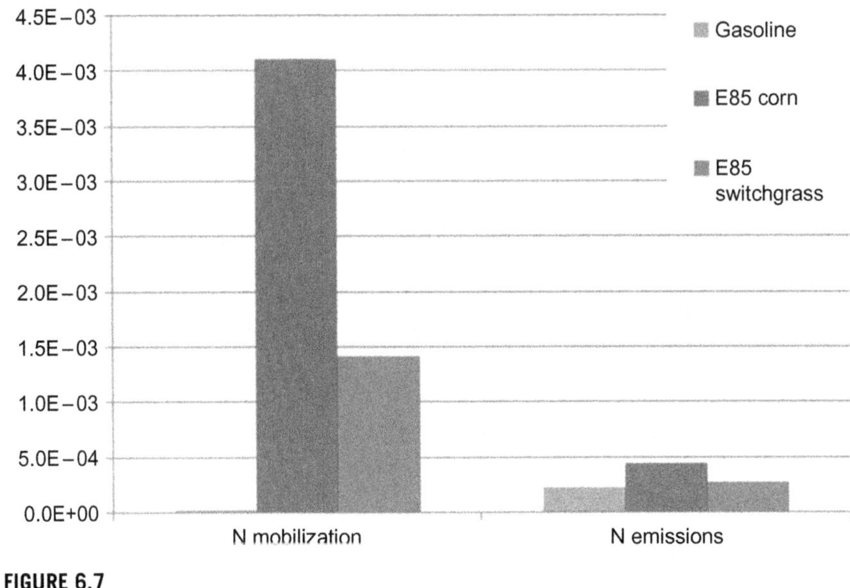

FIGURE 6.7

N footprint for fuels.

NEXUS OF CARBON AND NITROGEN FOOTPRINT

Carbon footprint has been the focus of most analyses for product or process comparisons. However, in recent years, there has been an increased emphasis on the use of multiple footprints for decision-making (Fang et al., 2014). To understand the relationships between carbon footprint and nitrogen footprint, it is important to understand the origin of definition of carbon footprint in a similar fashion as nitrogen footprint, as discussed previously. Similar to the effect of economic activities on the nitrogen cycle through reactive nitrogen mobilization and emissions, the carbon cycle is also hugely disrupted because of economic activities. Carbon forms another basic building block for the existence of life on Earth, along with nitrogen, as discussed previously. Every organism is composed of C, H, and O, along with N, which necessitates the availability of these building blocks for life. Carbon also forms the basis of various products like food, fiber, and fuel, thus supporting economic activities. The natural biogeochemical cycle of carbon mainly relies on the process of photosynthesis to convert the oxidized form of carbon CO_2 into a reduced form such as organic carbon, which gets incorporated into various food products. This process of carbon fixation is the key link for circulation of carbon into useful products. Thus, the flow of "carbon sequestration" that provides input for various resource manufacturing in economic activities can be considered an input-side carbon footprint in a similar way as nitrogen mobilization as an input-side N footprint. The output side of impact on the carbon cycle is the usual carbon footprint definition of examining carbon emissions. Details of the paradigms for footprint development and the foundation for input-side and output-side footprints are discussed by Singh and Bakshi (2014). The nexus of the C—N footprint for the US economy in 2002 is also provided by Singh and Bakshi (2014).

C—N NEXUS FOR PRODUCT SCALE: CASE STUDY OF FUELS

The fuel comparison on the hybrid scale is used to show the methodology for the product scale C—N footprint nexus study. Figure 6.8 shows the C—N input-side nexus for three fuels: gasoline, E85 corn, and E85 switchgrass. The graphs are drawn showing gasoline as the baseline fuel, so the impact of gasoline is shown as 1, whereas the impacts of other fuels are shown relative to gasoline.

It can be seen on the input-side impacts for both carbon and nitrogen that the bio-based fuels dominate in impact. If the C—N impact is considered together, then E85 corn has the highest impact on both carbon and nitrogen cycles, because it has high impact on carbon sequestration and also has the highest impact on N mobilization. Therefore, a switch from gasoline to E85 corn can cause approximately 100 times more stress on carbon sequestration flows and more than 200 times increase in stress on nitrogen mobilization. E85 switchgrass is not much

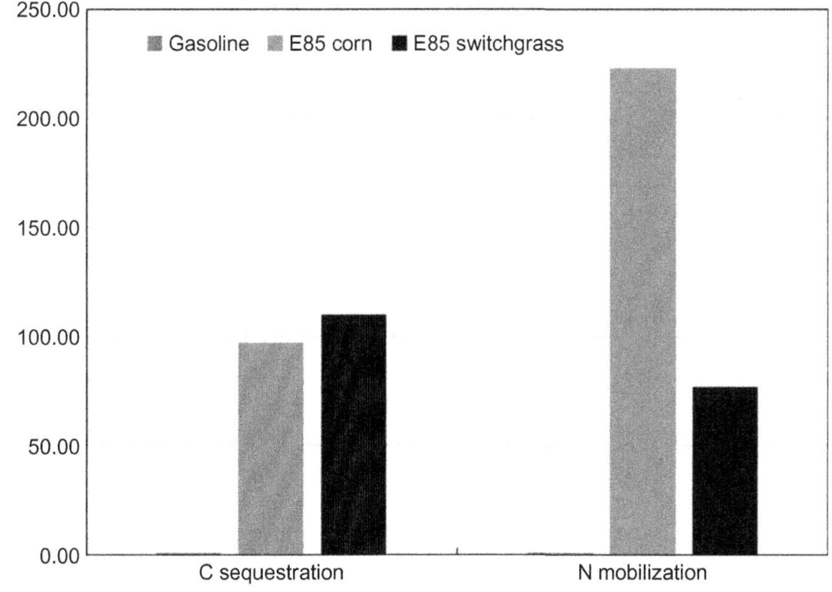

FIGURE 6.8

C sequestration compared with N mobilization (input-side C—N nexus).

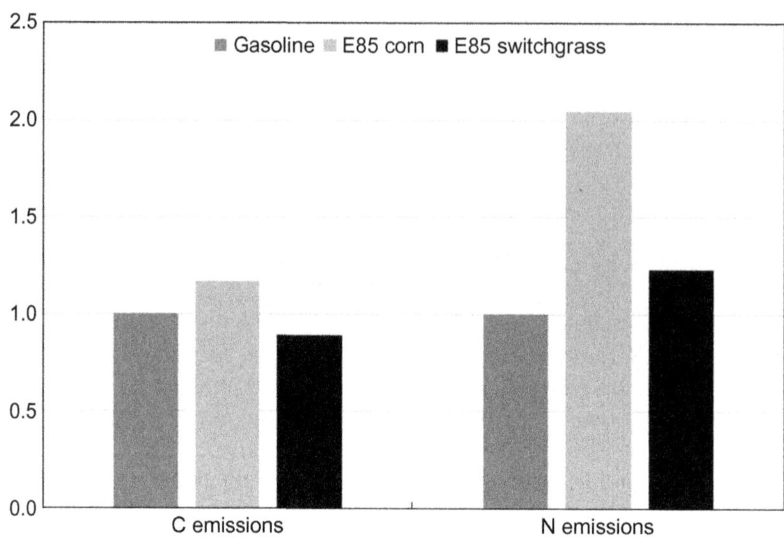

FIGURE 6.9

C emissions compared with N emissions (output-side C–N nexus).

better than E85 corn when compared with gasoline, because the impact on C sequestration is approximately the same as that for E85 corn, but the impact on N mobilization is much lower (approximately 80 times more than gasoline). Nonetheless, focusing on the input-side impacts, both E85 corn and E85 switchgrass have very high impacts on both carbon and nitrogen cycles. The focus on emissions only fails to capture this stress on natural biogeochemical cycles because of the increase in input-side stress.

On the output side of impact (Figure 6.9), the N emissions on the hybrid life cycle scale are insignificant as compared with C emissions. This is because the use phase of the fuels, which forms the majority of N emissions from fuel life cycle, has not been included in the study.

As compared with gasoline on the life cycle scale, E85 corn has slightly higher C emissions, whereas E85 switchgrass has lower C emissions. When comparing N emissions, E85 corn has almost twice more N emissions than gasoline because of N losses related to fertilizer use in the corn farming phase. E85 switchgrass is only slightly higher than gasoline. Therefore, when using the output-side impacts (C and N emissions), E85 switchgrass may be the preferred fuel for reducing C impact but gasoline may still be the preferred fuel for reducing N impacts. These results do not include the use phase emissions.

Finally, when both C and N are considered, E85 corn has highest impact on both input-side and output-side footprints. Considering the nexus at the input and output sides is important to make decisions that can reduce impact on both sides of interaction in carbon and nitrogen cycles.

APPENDIX 6.1

Eco-LCA Nitrogen Inventory Components

Reactive N Mobilization	Reactive N Products	Reactive N Emissions
N fixed by soil microorganisms	N use in farm fertilizer	NO_x and N_2O fuel combustion
Legume N fixation (alfalfa, soybean, other)	N used as nonfarm fertilizer	N_2O emissions–manure management
N by atmospheric deposition	N used as livestock manure	N_2O emissions–agricultural land
Industrial N fixation (Haber–Bosch)	N used as plastics and synthetics manure	N_2O and NO_x emissions–agricultural residue burning
	N in explosives	N_2O–forest fire
	N in animal feed and other chemical manure	N_2O emissions from nitric and adipic acid manure
	N consumed in food (crops and meat)	N_2O emissions from waste water management
		Organic and inorganic N emissions loading to water
		N sewage applied to land

REFERENCES

Aldaya, M.M., Hoekstra, A.Y., 2010. Strategic importance of green water in international crop trade. Ecol. Econ. 69 (4), 887–894.

Alexander, R.B., Smith, R.A., Schwarz, G.E., Boyer, E.W., Nolan, J.V., Brakebill, J.W., 2008. Difference in phosphorus and nitrogen delivery to the Gulf of Mexico from the Mississippi River Basin. Environ. Sci. Technol. 42, 822–830.

Barry, D.A.J., Goorahoo, D., Goss, M.J., 1993. Estimate of nitrate concentrations in groundwater using a whole farm nitrogen budget. J. Environ. Qual. 22, 767–775.

Bellucc, F., Bogner, J.E., Sturchio, N.C., 2012. Greenhouse gas emissions at the urban scale. Elements 8, 445–449.

Bullard, C.W., Penner, P.S., Pilati, D.A., 1978. Net energy analysis, handbook for combining process and input–output analysis. Resour. Energy 1 (3), 267–313.

Burke, M., Oleson, K., McCullough, E., Gaskell, J., 2008. A global model tracking water, nitrogen, and land inputs and virtual transfers from industrialized meat production and trade. Environ. Model. Assess. http://dx.doi.org/10.1007/s10666-008-9149-3.

Chapagain, A.K., Hoekstra, A.Y., 2008. The global component of freshwater demand and supply: as assessment of virtual water flows between nations as a result of trade in agricultural and industrial products. Water Int. 33 (1), 19–32.

Ecologically Based Life Cycle Assessment (Eco-LCA tool). <resilience.eng.ohio-state.edu/eco-lca/index.htm> (accessed 30.05.2014).

Ercin, A.E., Aldaya, M.M., Hoekstra, A.Y., 2011. Corporate water footprint accounting and impact assessment: the case of the water footprint of a sugar-containing carbonated beverage. Water Resour. Manage. 24 (5), 941–958.

Fang, K., Heijungs, R., de Snoo, G.R., 2014. Theoretical exploration for the combination of the ecological, energy, carbon and water footprints: overview of a footprint family. Ecol. Indic. 36, 508–518.

Forkes, J., 2007. Nitrogen balance for the urban food metabolism of Toronto, Canada. Resour. Conserv. Recycl. 52, 74–94.

Galloway, J.N., Schlesinger, W.H., Levy II, H., Michaels, A., Schnoor, J.L., 1995. Nitrogen fixation: anthropogenic enhancement-environmental response. Global Biogeochem. Cycles 9, 235–252.

Gao, T., Liu, Q., Wang, J., 2013. A comparative study of carbon footprint and assessment standards. Int. J. Low-Carbon Technol. 0, 1–7.

Global Footprint Network, <www.footprintnetwork.org/> (accessed 22.04.2014).

Global Partnership on Nutrient Management, United Nations Environment Programme, <www.gpa.unep.org/index.php/global-partnership-on-nutrient-management> (accessed 31.05.2014).

Greenhouse Gas Protocol, <www.ghgprotocol.org/> (accessed 31.05.2014).

Guggemos, A., 2003. Environmental Impacts of On-Site Construction: Focus on Structural Frames (PhD Thesis). Department of Civil and Environmental Engineering, University of California, Berkeley, CA, USA.

Hertwich, E.G., Peters, G.O., 2009. Carbon footprint of nations: a global, trade-linked analysis. Environ. Sci. Technol. 43, 6414–6420.

Hoekstra, A.Y., 2011. The global dimension of water governance: why the river basin approach is no longer sufficient and why cooperative action at global level is needed. Water 3 (1), 21–46.

Hoekstra, A.Y., Chapagain, A.K., 2007. Water footprint of nations: water use by people as a function of their consumption pattern. Water Resour. Manage. 21 (1), 35–48.

Hoekstra, A.Y., Chapagain, A.K., Aldaya M.M., Mekonnen, M.M. Water Footprint Manual, State of the Art 2009. <www.waterfootprint.org/downloads/WaterFootprintManual2009.pdf> (last accessed 24.07.2014).

International Nitrogen Initiative, <initrogen.net/about_ini> (accessed 31.05.2014).

Joshi, S., 2000. Product environmental life-cycle assessment using input–output techniques. J. Ind. Ecol. 3 (2&3), 95–120.

Knowlen, C., Mattick, A.T., Bruckner, A.P., Hertzberg, A., 1998. High Efficiency Energy Conversion Systems for Liquid Nitrogen Automobilers. Society of Automotive Engineers Inc., <large.stanford.edu/publications/coal/references/docs/sae98.pdf> (accessed 14.06.2014).

Leach, A.M., Galloway, J.N., Bleeker, A., Erisman, J.W., Kohn, R., Kitzes, J., 2012. A nitrogen footprint model to help consumers understand their role in nitrogen losses to the environment. Environ. Dev. 1, 40–66.

Leontief, W., 1936. Input–Output Economics. Oxford University Press, New York, NY, USA.

Leontief, W., 1970. Environmental repercussions and the economic structure: an input-output approach. Rev. Econ. Stat. 52, 262–271.

Matthews, H.S., Hendrickson, C.T., Weber, C.L., 2008. The importance of carbon footprint estimation boundaries. Environ. Sci. Technol. 42, 5839–5842.

Mekonnen, M.M., Hoekstra, A.Y. National Water Footprint Accounts: The Green, Blue and Grey Water Footprint of Production and Consumption, Volume 1, Main Report. Value of Water Research Report Series No. 50, UNESCE-IHE, Delft, The Netherlands.

Mitsch, W.J., Day, J.W., Gillian, J.W., Groffman, P.M., Hey, D.L., Randall, G.W., Wang, N., 2001. Reducing nitrogen loading to the Gulf of Mexico from the Mississippi River Basin: strategies to counter a persistent ecological problem. BioScience 51 (5).

Mosier, A., Kroeze, C., Nevison, C., Oenema, O., Seitzinger, S., van Cleemput, O., 1998. Closing the global N2O budget: nitrous oxide emissions through the agricultural nitrogen cycle. Nutr. Cycling. Agroecosyst. 225–248.

OECD and EUROSTAT:Gross Nitrogen Balances, Handbook: <www.oecd.org/greengrowth/sustainable-agriculture/40820234.pdf> (accessed 30.05.2014).

Olsthoorn, C.S.M., Fong, N.F.K., 1998. The anthropogenic nitrogen cycle in the Netherlands. Nutr. Cycling Agroecosyst. 52, 269–276.

Online N Footprint Calculator: <n-print.org/system/files/footprint_tz/index.html#/home> (accessed 30.05.2014).

Rabalais, N.N., Turner, R.E., Diz, R.J., Justic, D., 2009. Global Change and Eutrophication of Coastal Waters. International Council for the Exploration of the Sea, Oxford Journals.

Rees, W.E., 1992. Ecological footprints and appropriated carrying capacity: what urban economics leaves out. Environ. Urban. 4, 121, http://dx.doi.org/10.1177/095624789200400212.

Research Coordination Network on Reactive Nitrogen, National Science Foundation. <nitrogennorthamerica.org/projects.html> (accessed 30.03.2014).

Rockström, J., Steffen, W., Noone, K., Persson, Å., Chapin III, F.S., Lambin, E.F., Lenton, T.M., Scheffer, M., Folke, C., Schellnhuber, H.J., Nykvist, B., de Wit, C.A., Hughes, T., van der Leeuw, S., Rodhe, H., Sörlin, S., Snyder, P.K., Costanza, R., Svedin, U., Falkenmark, M., Karlberg, L., Corell, R.W., Fabry, V.J., Hansen, J., Walker, B., Liverman, D., Richardson, K., Crutzen, P., Foley, J.A., 2009. A safe operating space for humanity. Nature 461, 472–475.

Russelle, M.P., Birr, A.S., 2004. Biological nitrogen fixation: large-scale assessment of symbiotic dinitrogen fixation by crops. Agronomy J. 96, 1754–1760.

Salo, T., Lemola, R., Esala, M., 2007. National and regional net nitrogen balances in Finland in 1990–2005. Agric. Food Sci. 16 (4).

Schlesinger, W.H., 1997. The Global Cycles of Nitrogen and Phosphorus in—Biogeochemistry: An Analysis of Global Change. Academic Press, Elsevier, NY, USA.

Seitzinger, S.P., Kroeze, C., 1998. Global distribution of nitrous oxide production and N inputs in freshwater and coastal marine ecosystems. Global Biogeochem. Cycles 12 (1), 93–113.

Slak, M.F., Commagnac, L., Lucas, S., 1998. Feasibility of national nitrogen balances. Environ. Pollut. 102, 235–240.

Singh, S., Bakshi, B.R., 2013. Accounting for the biogeochemical cycle of nitrogen in input–output life cycle assessment. Environ. Sci. Technol. 47 (16), 9388–9396.

Singh, S., Bakshi, B.R., 2014. Footprints of Carbon and Nitrogen: Revisiting the Paradigm and Exploring their Nexus for Decision Making, In Review, Ecological Indicators.

Suh, S., 2004. Functions, commodities and environmental impacts in an ecological–economic model. Ecol. Econ. 48, 451–467.

Suh, S., Lenzen, M., Treloar, G.J., Hondo, H., Horvath, A., Huppes, G., Jolliet, O., Klann, U., Krewitt, W., Moriguchi, Y., Munksgaard, J., Norris, G., 2004. System boundary selection in life-cycle inventories using hybrid approaches. Environ. Sci. Technol. 38 (3), 657–664.

Sutton, M.A., Bleeker, A., Howard, C.M., Bekunda, M., Grizzetti, B., de Vries, W., van Grinsven, H.J.M., Abrol, Y.P., Adhya, T.K., Billen, G., Davidson, E.A., Datta, A., Diaz, R., Erisman, J.W., Liu, X.J., Oenema, O., Palm, C., Raghuram, N., Reis, S., Scholz, R.W., Sims, T. and Zhang, F.S., 2013. with contributions from Ayyappan, S., Bouwman, A.F., Bustamante, M., Fowler, D., Galloway, J.N., Gavito, M.E., Garnier, J., Greenwood, S., Hellums, D.T., Holland, M., Hoysall, C., Jaramillo, V.J., Klimont, Z., Ometto, J.P., Pathak, H., Plocq Fichelet, V., Powlson, D., Ramakrishna, K., Roy, A., Sanders, K., Sharma, C., Singh, B., Singh, U., Yan, X.Y. and Zhang, Y. Our Nutrient World: the challenge to produce more food and energy with less pollution. Global Overview of Nutrient Management. Centre for Ecology and Hydrology, Edinburg on behalf of the Global Partnership on Nutrient Management and the International Nitrogen Initiative. <www.unep.org> (last accessed 27.07.2014).

Van der Hoek, K.W., 1998. Nitrogen efficiency in global animal production. Environ. Pollut. 102 (1), 127–132.

Wiedmann, T., Minx J., 2007. A definition of "carbon footprint." ISA Res. Rep, 7, 1–7.

Wiedmann, T.O., et al., 2011. Application of hybrid life cycle approaches to emerging energy technologies—the case of wind power in the UK. Environ. Sci. Technol. 45, 5900–5907.

Zhang, Y., Baral, A., Bakshi, B.R., 2010a. Accounting for ecosystem services in life cycle assessment, Part II: Toward an ecologically based LCA. Environ. Sci. Technol. 44 (7), 2624–2631.

Zhang, Y., Singh, S., Bakshi, B.R., 2010b. Accounting for ecosystem services in life cycle assessment, Part 1: a critical review. Environ. Sci. Technol. 44, 2232–2242.

CHAPTER

The water footprint of industry

7

Arjen Y. Hoekstra
University of Twente, Twente Water Centre Enschede, The Netherlands

INTRODUCTION

The World Economic Forum has listed water scarcity as one of the three global systemic risks of highest concern in an assessment based on a broad global survey of risk perception among representatives from business, academia, civil society, governments, and international organizations (WEF, 2014). Freshwater scarcity manifests itself in the form of declining groundwater tables, reduced river flows, shrinking lakes, and heavily polluted waters, and also in increasing costs of supply and treatment, intermittent supplies, and conflicts over water (Hoekstra, 2014a). Future water scarcity will grow as a result of various drivers such as population and economic growth, increased demands for animal products and biofuels, and climate change (Ercin and Hoekstra, 2014). The private sector is becoming aware of the problem of freshwater scarcity but is facing the challenge of formulating effective responses. Even companies operating in water-abundant regions can be vulnerable to water scarcity, because the supply chains of most companies stretch across the globe. An estimated 22% of global water consumption and pollution relates to the production of export commodities (Hoekstra and Mekonnen, 2012). Countries such as the United States, Brazil, Argentina, Australia, India, and the People's Republic of China are significant virtual water exporters, meaning that they intensively use domestic water resources for producing export commodities. In contrast, countries in Europe, North Africa, and the Middle East as well as Mexico and Japan are dominated by virtual water import, meaning that they rely on import goods produced with water resources elsewhere. The water use behind those imported goods is often not sustainable, because many of the export regions overexploit their resources.

Increasingly, companies start exploring their water footprint (WF) by looking at both their operations and supply chain. Key questions that industry leaders pose themselves are: where is my WF located?; what risks does water scarcity impose to my business?; how sustainable is the WF in the catchments where my operations and supply chain processes are located?; where and how can water use efficiency be increased?; and what is good water stewardship? The demand for new types of data emerges, types of data that were usually not collected. The focus shifts from relatively simple questions regarding whether the company has sufficient water abstraction permits and whether wastewater disposal standards are met to the more pressing question regarding how the company actually contributes to the overexploitation and pollution of water resources, not only through its own facilities but also through its supply chain. Sustainability is not implied by having permits and meeting standards. Most experience with collecting the new types of data required and with addressing questions about good water stewardship is within the food and beverage sector, which most clearly depends on water. In other industries, the connection with water is not always clear, because the connection is indirect and mostly through the supply chain. The aim of this chapter is to review experiences with WF accounting in different sectors of the economy and to reflect on the question of what good water stewardship is.

First, I discuss what new perspective the WF concept brings to the table compared with the traditional way of looking at water use. Second, I discuss and compare three methods to trace resource use and pollution over supply chains: environmental footprint assessment (EFA), life cycle assessment (LCA), and environmentally extended input–output analysis (EE-IOA). Third, I review some of the recent literature on direct and indirect WFs of different sectors of the economy. Finally, I discuss the emerging concept of water stewardship for business and the challenge of creating greater product and business transparency.

THE WF CONCEPT

The WF is a measure of freshwater appropriation underlying a certain product or consumption pattern. Three components are distinguished: the blue, green, and gray WF (Hoekstra et al., 2011). The blue WF measures the volume of water abstracted from the ground or surface water system minus the volume of water returned to the system. It thus refers to the sum of the water flow that evaporates during the process of production, the water incorporated into a product, and the water released in another catchment. The blue WF differs from the conventional way of measuring freshwater use by looking at net rather than gross water withdrawal. This is done because it makes more sense to look at net water withdrawal if one is interested in the effect of water use on water scarcity within a catchment. Return flows can be reused within the catchment, unlike the water flow that evaporates or is captured within a product. The green WF refers to the volume of rainwater consumed in a production process. This is particularly relevant in agriculture and forestry, where it refers to the total rainwater evapotranspiration (from fields and plantations) plus the water incorporated into the harvested crop or wood. The gray WF is an indicator of freshwater pollution and defined as the volume of freshwater that is required to assimilate a load of pollutants based on natural background concentrations and existing ambient water quality standards. The advantage of expressing water pollution in terms of the water volume required for assimilating the pollutants, rather than in terms of concentrations of contaminants, is that this brings water pollution into the same unit as consumptive use. In this way, the use of water as a drain and the use of water as a resource, two competing uses, become comparable. The WF thus refers to both consumptive water use [of rainwater (the green WF) and of surface and groundwater (the blue WF)] and degenerative or degradative water use (the gray WF).

As a measure of freshwater use, the WF differs from the classical measure of "water withdrawal" in several ways. The term "water withdrawal," also called "water abstraction" or often simply "water use", refers to the extraction of water from the groundwater or a surface water body like a river, lake, or artificial storage reservoir. It thus refers to what we call *blue* water use. The WF is not restricted to measuring blue water use; it also measures the use of green water resources (the green WF) and the volume of pollution (the gray WF). Another difference between the WF and the classical way of measuring water use was mentioned previously: the classical measure of "water use" always refers to gross blue water abstraction, whereas the blue WF refers to net blue water abstraction. Another difference between the classical way of measuring water use and the WF is that the latter concept can be used to measure water use over supply chains. When we talk about the WF of a product, we refer to the water consumption and pollution in all stages of the supply chain of the product. When we speak about the WF of a producer or a consumer, we refer to the full WF of all the products produced or consumed.

Thus, the WF offers a wider perspective on how a product, producer, or consumer relates to the use of freshwater systems. It is a volumetric measure of water consumption and pollution. WF accounts give spatiotemporally explicit information on how water is appropriated for various human purposes. The local environmental impact of a certain amount of water consumption and pollution depends on the vulnerability of the local water system and the number of water consumers and polluters who make use of the same system. The WF within a catchment needs to be compared with the maximum sustainable WF in the catchment to understand the sustainability of water use. The WF of a specific process or product needs to be compared with a WF benchmark based on the best available technology and practice to understand the efficiency of water use. The WF per capita for a community can be compared with the WF of other communities to understand the degree of equitable sharing of limited water resources. WF accounts can thus feed the discussion about the sustainability, efficiency, and equitability of water use and allocation (Hoekstra, 2013, 2014b).

The definition of the green and blue WF can best be understood by considering the water balance of a river basin (Figure 7.1). The total annual water availability in a catchment area is given by the annual volume of precipitation, which will leave the basin partly through evapotranspiration and partly through runoff to the sea. Both the evaporative flow and the runoff can be appropriated by humans. The green WF refers to the human use of the evaporative flow from the land

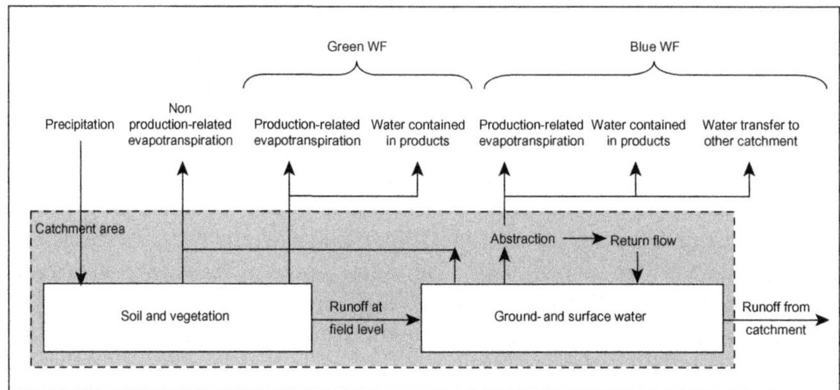

FIGURE 7.1

Definition of the green and blue WF in relation to the water balance of a catchment area.

From Hoekstra et al. (2011) with permission from the publishers.

surface, mostly for growing crops or production forest. The blue WF refers to the consumptive use of the runoff flow, i.e., the net abstraction of runoff from the catchment. The term "water consumption" can be confusing, because many people, particularly those not aware of the big difference between gross and net water abstraction, use the term for gross water abstraction. Specialists, though, define water consumption as net blue water abstraction (gross abstraction minus return flow). Evaporation is generally considered as a loss to the catchment. Even though evaporated water will always return in the form of precipitation on a global scale, this will not alleviate the water scarcity in the catchment during the period that the river is emptied because of net water abstractions. Moisture recycling on smaller spatial scales is generally only modest.

The definition of the gray WF is clarified in Figure 7.2. The basis for the calculation is the anthropogenic load of a substance into a freshwater body (groundwater, river, lake), i.e., the additional load caused by a human activity (e.g., a production process). We should acknowledge that the effluent from an industry might contain certain amounts of chemicals that were already in the water abstracted. Therefore, we should look at the *additional* load to a freshwater body as a result of a certain activity. Furthermore, we should look at the load of a substance that really enters the river, lake, or groundwater, which means that if an effluent is treated before disposal, then we have to consider the load of chemicals in the effluent that remains *after* treatment. The critical load in a freshwater body is defined as the difference between the maximum acceptable and natural concentration of a chemical for the receiving water body multiplied by the renewal rate of the freshwater body. Note that for the maximum allowable concentration, we have to use the ambient water quality standard for the receiving freshwater body, not the effluent standard (Franke et al., 2013). In a river, the renewal rate is equal

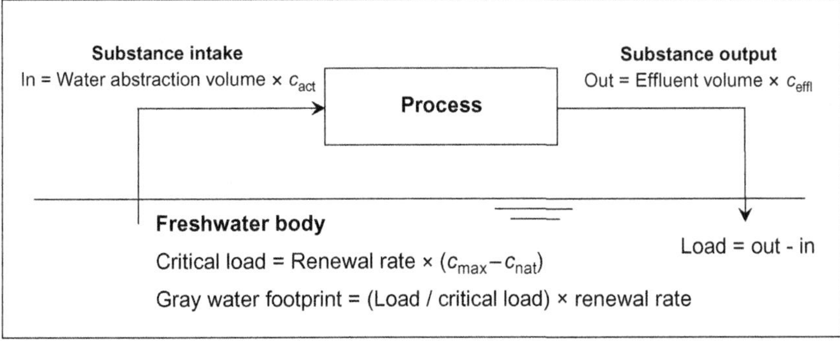

FIGURE 7.2

Definition of the gray WF based on the load of a chemical into a freshwater body. The symbols c_{act}, c_{nat}, and c_{max} refer to the actual, natural, and maximum allowable concentration of the chemical in the freshwater body; c_{effl} refers to the concentration of the chemical in the effluent.

to runoff; in a groundwater reservoir, the renewal rate is equal to groundwater recharge, which (over the longer-term) is the same as groundwater runoff. In a lake, the renewal rate equals the flow through the lake. The gray WF is calculated as the pollutant load to a freshwater body divided by the critical load multiplied by the renewal rate of that freshwater body. Defined in this way, it means that when the gray WF onto a freshwater body becomes as big as the renewal rate of this freshwater body, the assimilation capacity has been fully used. When the size of the gray WF in a catchment exceeds the size of runoff from this catchment, pollution is bigger than the assimilation capacity, resulting in a violation of the maximum acceptable concentration. When an effluent contains different types of pollutants, as is usually the case, the gray WF is determined by the pollutant that is most critical, i.e., the one that gives the largest pollutant-specific gray WF. Thermal pollution can be dealt with in a way similar to that of pollutants, whereby the load consists of heat and the assimilation capacity depends on the accepted temperature increase of the receiving water body (Hoekstra et al., 2011).

METHODS TO TRACE NATURAL RESOURCES USE AND POLLUTION OVER SUPPLY CHAINS

Different methods have been developed to analyze direct and indirect natural resources use and emissions in relation to products or economic sectors. They have all been applied specifically to trace direct and indirect water use and pollution over supply chains as well. I discuss the methods of EFA, LCA, and EE-IOA. Each of the three methods has its specific goal, approach, and focus, but there are commonalities across the methods as well. They all focus on understanding natural resource use and emissions along supply or value chains. EFA focuses on macro-questions about resource use sustainability, efficiency, equitability, and security. LCA concentrates on the comparative analysis of environmental impacts of products. EE-IOA focuses on understanding how natural resource use and environmental impacts can be traced throughout the economy.

The field of EFA comprises methods to quantify and map land, water, material, carbon, and other environmental footprints and to assess the sustainability of these footprints as well as the efficiency, equitability, and security of resource use (Hoekstra and Wiedmann, 2014). WF assessment (WFA) can be regarded as a specific branch of this field and refers to the full range of activities to quantify and locate the WF of a process, product, producer, or consumer or to quantify in space and time the WF in a specified geographic area; assess the environmental sustainability, economic efficiency, and social equitability of WFs; and formulate a response strategy (Hoekstra et al., 2011). Broadly speaking, the goal of assessing WFs is to analyze how human activities or specific products relate to issues of water scarcity and pollution, and to see how consumption, production, trade, and specific products can become more sustainable from a water perspective.

LCA is a method for estimating and assessing the environmental impacts attributable to the life cycle of a product, such as climate change, stratospheric ozone depletion, tropospheric ozone (smog) creation, eutrophication, acidification, toxicological stress on human health and ecosystems, the depletion of resources, water use, land use, and noise (and others) (Rebitzer et al., 2004). The assessment includes all stages of the life cycle of a product from cradle-to-grave (from material extraction to returning of wastes to nature). An LCA study includes four phases: setting a goal and scope; inventory accounting; impact assessment; and interpretation. Water use and pollution can be considered specific impact categories within LCA (Kounina et al., 2013). LCA focuses on *comparing* the environmental impacts of alternative processes, materials, products, or designs.

EE-IOA is a method for studying the relation between different sectors of the economy and indirect natural resource use and environmental impacts. It combines the classical monetary input–output formalism with satellite accounts containing data on resource use and emissions into the environment. Over the past decade, we have seen quite a number of applications of EE-IOA to analyze "embodied" water flows through the economy (Daniels et al., 2011). Applications have been performed, e.g., for Australia (Lenzen and Foran, 2001), Spain (Duarte et al., 2002; Cazcarro et al., 2013), the United Kingdom (Yu et al., 2010; Feng et al., 2011b), the People's Republic of China (Zhao et al., 2009; Zhang and Anadon, 2014), and the city of Beijing (Zhang et al., 2011). Input–output models basically show monetary flows between sectors within the economy; environmentally extended input–output models usually express water use in terms of liters per dollar (or other currency). Most environmentally extended input–output models also have some form of accounting of product flows in physical units, but because of the aggregation of specific economic activities into sectors it remains difficult to reach the same high level of detail as achieved in a process-based WFA or LCA. Both WFA and LCA enable an analysis of water use in all processes of the value chain and attribution of the water use along value chains to specific products. A promising path in this respect is the method of so-called *hybrid* environmentally extended input–output modeling, in which physical flows are integrated into the model (Ewing et al., 2012; Steen-Olsen et al., 2012).

Process-based WFA and LCA are generally constrained by the fact that parts of the value chain have to be omitted from the analysis for practical reasons. This problem does not occur in input–output modeling. Therefore, there is a development to enhance process-based WFA and LCA with the advantage of input–output modeling. In the case of LCA, this results in the so-called hybrid LCA approach (Finnveden et al., 2009). In hybrid LCA, the environmental impacts of flows that were not included in the process-based LCA are estimated with an environmentally extended input–output model. In the case of WFA, a similar development can be expected (Feng et al., 2011a).

The difference between EFA and LCA is the focus on sustainability of production and consumption at a macro-level of the former and the focus on comparing potential environmental impact at process and product level of the latter (Box 7.1). Typical questions in EFA studies relate to how different processes and products contribute to the overall footprints at larger scales, how different consumption patterns influence the overall footprint, whether footprints at the larger scales remain within their maximum sustainable levels, how footprints can be reduced by better technology, whether different people have equitable shares in the total footprint of humanity, and what externalization of footprints may imply for resource security (Hoekstra and Wiedmann, 2014). LCA is designed to compare the potential environmental impact of one product over its full value chain with the potential impact of another product, or to compare the differences in potential impact between different product designs or alternative production processes.

At the level of basic data, EFA and LCA require similar data. The data collection and analysis required in the accounting stage of a product-focused WFA (as opposed to a geographic-focused or consumption-focused WFA) is very similar to what is needed in the inventory stage of a water-focused LCA (Boulay et al., 2013).

EFA, LCA, and EE-IOA are not static analytical methods, but rather are young fields undergoing development. We can observe a development in the past few years in which a fruitful exchange between the three fields leads to the adoption of approaches from one field to the other. In EFA studies, we have seen the adoption of life cycle accounting procedures from LCA and the exploration of using input–output models to calculate national and sector footprints, in addition to the already existing bottom-up and top-down trade balance approaches. In LCA we recently observed, fed by experiences in EFA, an interest to develop methods to perform an LCA for a whole organization instead of for a product, and to perform LCAs for consumer lifestyles or for national consumption as a whole (Hellweg and Milà i Canals, 2014). Additionally, based on experiences in EE-IOA, the LCA community is exploring hybrid LCA methods, as already

BOX 7.1 THE SUSTAINABILITY OF CUTTING TREES—THE FUNDAMENTAL DIFFERENCE BETWEEN LCA AND EFA

Is it sustainable to cut a tree? Although a relevant question, it is impossible to answer this question in isolated form. It is difficult to argue that cutting just one tree is not sustainable. After a tree has been cut, a new one will grow, so it is sustainable. However, if one takes this insight on the sustainability of cutting one tree to conclude that one can cut all forests, one cannot maintain that this is sustainable. The reason why answering a simple question like this tree-cutting question causes a fundamental problem is that sustainability is a concept that cannot be applied at the level of single activities, but only at the level of a system as a whole. Still, there is a strong wish among people to measure the sustainability of single activities, because individuals undertake single activities and consume goods and services that relate to series of single activities to produce them. The methods of LCA and EFA deal with this problem in fundamentally different ways. In LCA, the approach is to leave the larger question of sustainability and look at *comparative* contributions of different activities to natural resource appropriation, emissions, and potential impacts on the larger scale. In other words, LCA addresses the question of how cutting one tree compares with cutting two trees, a question that is not difficult to answer. In EFA, the approach is to estimate humanity's total natural resource appropriation and emissions and to compare that with the Earth's carrying or assimilation capacity. Both methods struggle in a similar way with how to compare apples and pears, e.g., how to compare cutting trees with polluting water. The approach in LCA is to *weigh* different types of primary resource use or emissions according to their potential final impact on human health and ecosystem health. The approach in EFA is to compare the different types of resource use and pollution with their respective maximum sustainable levels. The great similarity between LCA and EFA is that resource use and emissions are analyzed per process (activity) and per product (by analyzing the processes along supply chains). The difference comes when LCA starts weighing different types of resource use and emissions based on their potential impact and comparing alternative processes or products according to their overall potential environmental impact. In contrast, EFA adds the resource use and emissions of different activities to get a complete picture, analyze the sustainability of the whole, and study the relative contribution of different processes, products, and consumers to the total. In many applications, though, the difference between LCA and EFA is not so clear. By comparing the footprints of two different processes or products, EFA also allows for comparative analysis. However, the comparative analysis is partial in this case, because different footprints are not weighted and added to get a measure of "overall potential environmental impact." One can also extend an LCA from comparing products to comparing consumption patterns, which is on the larger scale typically for EFA. The fundamental difference between LCA and EFA in the way they treat the tree-cutting question, however, remains.

mentioned. The EE-IOA practice improves in the direction of hybrid methods that include physical accounting and have greater granularity in the analysis, fed by the practices in the EFA and LCA fields. This mutual enrichment and, to some extent, convergence of approaches does not imply that the three methods will grow into one. They may develop into a more consistent framework of coherent methods, but the fact that different sorts of questions will remain implies that different approaches will continue to be necessary.

All three methods—EFA, LCA, and EE-IOA—have a focus on environmental issues, leaving out social issues (like labor conditions, human rights). Principally, there is nothing that necessarily restricts the methods to environmental issues.

Broadly speaking, one can trace all sorts of process characteristics along supply chains. The oldest forms of accounting along supply chains are the accounting of monetary added value and the accounting of material flows and energy use along supply chains. Material flow analysis (MFA) or substance flow analysis (SFA) aims at the quantification of stocks and flows of materials or substances in a well-defined system, drawing mass balances for each subsystem and the system as a whole. Energy flow analysis aims at quantifying the energy content of flows within an economy. The innovation of EFA, LCA, and EE-IOA lies in the attribution of resource use, emissions, or impacts along supply chains to products and final consumption. In this context, one speaks about the embodied, embedded, indirect or virtual land, water, and energy in a product or consumption pattern, or the indirect emissions. When doing so, the method of EE-IOA is linked to traditional economic accounting, which is a strong point of this method. The methods of EFA and LCA are linked to physical accounting, which is their strength. In all three fields, we observe efforts to enhance the methods and broaden the scope with an increasing number of hybrid approaches.

DIRECT AND INDIRECT WFs OF DIFFERENT SECTORS OF THE ECONOMY

THE IMPORTANCE OF WATER USE IN THE PRIMARY SECTOR

Usually, economic activities are categorized into three different sectors. The primary sector of the economy, the sector that extracts or harvests products from the Earth, has the largest WF on Earth. This sector includes activities like agriculture, forestry, fishing, aquaculture, mining, and quarrying. The green WF of humanity is nearly entirely concentrated within the primary sector. It has been estimated that approximately 92% of the blue WF of humanity is just in agriculture alone (Table 7.1).

The secondary sector covers the manufacturing of goods in the economy, including the processing of materials produced by the primary sector. It also includes construction and the public utility industries of electricity, gas, and water. Sometimes, the public utility industries are also mentioned under the tertiary (service) sector, because they not only produce something (electricity, gas, purified water) but also supply it to customers (as a service). Water utilities could even partly fall under the primary sector, because part of the activity is the abstraction of water from the environment (rivers, lakes, and groundwater). The work of water utilities comprises water collection, purification, distribution and supply, wastewater collection (sewerage), wastewater treatment, materials recovery, and wastewater disposal. It is rather common to categorize the whole water utility sector under the secondary sector. The tertiary sector is the service industry and covers services to both businesses and final consumers. This sector includes activities like retail and wholesale sales, transportation and distribution,

Table 7.1 Global WF within Different Water-Using Categories during 1996–2005

Economic Sector	Water Use Category	Global WF (10^9 m^3/year)					Remark
		Green	Blue	Gray	Total	%	
Primary sector	Crop farming	5,771	899	733	7,404	81.5	
	Pasture	913	–	–	913	10.0	
	Animal farming	–	46	–	46	0.5	Water for drinking and cleaning
	Agriculture total	6,684	945	733	8,363	92.0	
	Aquaculture	?	?	?	?	?	No global data
	Forestry	?	?	?	?	?	No global data
	Mining, quarrying	?	?	?	?	?	No global data
Secondary sector	Industry (self-supply)	–	38	363	400	4.4	Water use in manufacturing, electricity supply, and construction
	Municipal water supply	–	42	282	324	3.6	Water supply to consumers and (small) users in primary, secondary, and tertiary sectors
Tertiary sector	Self-supply	?	?	?	?	?	No global data
Consumers	Self-supply	?	?	?	?	?	No global data
	Total	6,684	1,025	1,378	9,087	100	

Note that the blue WF figure for crop farming relates to evapotranspiration of irrigation water at field level; it excludes losses from storage reservoirs and irrigation canals. The blue WF figure for "industry" presented here includes water use in mining, which is part of the primary sector. The figure excludes water lost from reservoirs for hydroelectric generation. All gray WF figures are conservative estimates. Forestry is not included as a water use sector because of a lack of data.
From Mekonnen and Hoekstra (2011) for crop farming; Mekonnen and Hoekstra (2012) for pasture and animal farming; Hoekstra and Mekonnen (2012) for industry and municipal water supply.

entertainment, restaurants, clerical services, media, tourism, insurance, banking, health care, defense, and law. Even though sometimes categorized into another quaternary sector, one can also list activities related to government, culture, libraries, scientific research, education, and information technology. The secondary and tertiary sectors have much smaller WFs than the primary sector.

It is difficult to get water use statistics organized along the same structure of economic sector classifications. Many countries and regions have their own classification of economic activities, distinguishing main sectors and subsectors. One of the international standard classifications is the Industrial Classification of All Economic Activities of the United Nations (UN, 2008). Conventional water use statistics mostly show gross blue water withdrawals and distinguish three main categories: agricultural, industrial, and municipal water use (FAO, 2014). WF statistics also distinguish between the agricultural, industrial, and municipal sector. These three sectors cannot be mapped one-to-one onto the primary, secondary, and tertiary sector. "Agricultural water use" obviously is about water use in the primary sector, whereas "industrial water use" is about water use in the secondary sector. However, water use in mining—part of the primary sector—will generally be categorized under "industrial water use" as well. Industrial water use refers to self-supplied industries not connected to the public distribution network. It includes water for the cooling of thermoelectric plants, but it does not include hydropower (which is often left out of the water use accounts altogether). Municipal water use—often alternatively called domestic water use or public water supply—refers to the water use by water utilities and distributed through the public water distribution network. Water utilities provide water directly to consumers, but also to water users in the primary, secondary, and tertiary sector.

The mismatch between the three main categories in water use statistics and the different sectors as usually distinguished in the economy can be quite confusing. The "water supply sector" as distinguished in economic classifications refers to water utilities delivering municipal water to households and others connected to the public water supply system. Unfortunately, the category of municipal water use lumps water use for a great variety of water users: final consumers (households) and users in all economic sectors. Specifications by type of user are not always available. Additionally confusing is that even though the "water supply sector" serves all types of users, the sector refers to only a minor fraction of total water use. Most of the water use in agriculture, the largest water user, is not part of the "water supply sector." Furthermore, water self-supply by industries does not fall within this sector, and neither does self-supply in the tertiary sector or self-supply by final consumers. Given that only an estimated 3.6% of the total WF of humanity relates to what we call the "water supply sector" (Hoekstra and Mekonnen, 2012), the sector receives disproportionate attention in public debates about water use and scarcity, diverting the necessary attention to water use in agriculture and industry.

An additional problem is that the contribution of agriculture to water scarcity is underestimated by conventional water use statistics, which show gross blue water abstractions. In agriculture, most of the gross water use will evaporate from storage reservoirs, irrigations canals, or from the field. The water abstracted for irrigation in agriculture is thus largely unavailable for reuse within the basin. In industrial water use, the ratio of net to gross abstraction is estimated at less than 5%. In municipal water use, this ratio varies from 5% to 15% in urban areas and from 10% to 50% in rural areas (FAO, 2014). Water that returns to the catchment after use can be reused. Presenting gross or net water abstractions thus makes a huge difference for industries and households and less of a difference in agriculture.

Even though the primary sector is the largest water user, governmental programs to create public awareness of water scarcity often focus on public campaigns calling for water-saving at home. This is not very effective at large given the fact that the major share of water use in most places relates to agriculture and, secondarily, to industry. Water scarcity is thus generally caused mostly by excessive water use in agriculture. Installing water-saving showerheads and dual-flush toilets in households will have barely any impact on mitigating water scarcity, but still this is what most water-saving campaigns advocate. It would be more useful to make people aware of the water use and pollution underlying the food items and other products they buy and to advocate product labels that show the sustainability of the WF of a product.

AGRICULTURE, FISHING, AND FORESTRY

The WF in the agricultural sector has been studied in great detail by a variety of authors. Most studies focus on the WF of crops. The first global study of green–blue WFs of crops per country was performed by Hoekstra and Hung (2002), followed by a study by Hoekstra and Chapagain (2007). The first global grid-based study was performed by Rost et al. (2008), who applied the LPJmL model. Later grid-based studies include those by Liu and Yang (2010), who used the GEPIC model; Siebert and Döll (2010), who used the GCWM model; Hanasaki et al. (2010), who applied the H08 model; and Fader et al. (2011), who used the LPJmL model again. The different studies were performed with different models, but for partially different periods and using different underlying land, soil, climate, and crop data as well, so it is difficult to compare the results. The study by Mekonnen and Hoekstra (2011) is the only global grid-based study that includes an assessment of the gray WF of crops. Fewer studies address the WF of animal products. The first global study was by Chapagain and Hoekstra (2003), who distinguished eight animal groups, three farming systems (grazing, mixed, industrial), and different feed composition per animal group, farming system, and country. Oki and Kanae (2004) published data for Japan, and Pimentel et al. (2004) published data for the United States. The best global dataset currently available is that provided by Mekonnen and Hoekstra (2012); it has the same details as the study by Chapagain and Hoekstra (2003) but also includes gray WFs and follows a number of methodological improvements. General findings are that WFs of both crops and livestock products show a great variation depending on production circumstances and that, in general, the WF per kilogram or kilocalorie is smaller for crops than for animal products (Table 7.2). In the case of animal products, the feed conversion efficiency, feed composition, and feed origin are the most important determinants (Hoekstra, 2012).

Table 7.2 The Global Average WF of Some Selected Food Products

Origin	Food Item	WF per kg (L/kg)				Nutritional Content			WF per Unit of Nutritional Value		
		Green	Blue	Gray	Total	Calorie (kcal/kg)	Protein (g/kg)	Fat (g/kg)	Calorie (L/kcal)	Protein (L/g protein)	Fat (L/g fat)
Vegetable origin	Sugar crops	130	52	15	197	285	0.0	0.0	0.69	0.0	0.0
	Vegetables	194	43	85	322	240	12	2.1	1.34	26	154
	Starchy roots	327	16	43	387	827	13	1.7	0.47	31	226
	Fruits	726	147	89	962	460	5.3	2.8	2.09	180	348
	Cereals	1,232	228	184	1,644	3,208	80	15	0.51	21	112
	Oil crops	2,023	220	121	2,364	2,908	146	209	0.81	16	11
	Pulses	3,180	141	734	4,055	3,412	215	23	1.19	19	180
	Nuts	7,016	1,367	680	9,063	2,500	65	193	3.63	139	47
Animal origin	Milk	863	86	72	1,020	560	33	31	1.82	31	33
	Eggs	2,592	244	429	3,265	1,425	111	100	2.29	29	33
	Chicken	3,545	313	467	4,325	1,440	127	100	3.00	34	43
	Butter	4,695	465	393	5,553	7,692	0.0	872	0.72	0.0	6.4
	Pork	4,907	459	622	5,988	2,786	105	259	2.15	57	23
	Sheep/goat meat	8,253	457	53	8,763	2,059	139	163	4.25	63	54
	Beef	14,414	550	451	15,415	1,513	138	101	10.2	112	153

Source: From Mekonnen and Hoekstra (2012).

The WF of fish primarily depends on four factors: the type of water in which it grows (saltwater, brackish water, or freshwater systems); whether it lives in natural waters or is cultivated in aquaculture; its feed composition and origin; and its feed conversion efficiency. A saltwater fish naturally feeding itself, not cultivated but caught in open water, does not have any freshwater footprint. This is not to say that this fish may not be accompanied by other environmental concerns (like overfishing, problems related to bycatch, and damage caused by fishing techniques applied), but it means that this fish puts no claims on the limited global freshwater resources. The WF of this fish available at the retailer will refer only to the WF of materials and energy involved in fishing, transport, and packaging. This WF is small when compared with the WF that fish can have when fed with land-based and, thus, freshwater-based feed. According to Naylor et al. (2009), the range of plant feedstuffs in aquafeeds currently includes barley, rapeseed, maize, cottonseed, peas/lupines, soybean, and wheat. The ratio of plant-based protein in aquafeeds is increasing, so the question about the WF of fish becomes increasingly relevant. Fish grown in open ponds also have a WF related to the evaporation losses from those ponds.

With an average feed conversion efficiency of approximately 2 (i.e., 2 kg of feed per kg of fish), fish is more efficient than chicken, so the feed-related WF of fish will generally be lower than that of chicken, even with very high fractions of plant-based material in the aquafeeds. Fish grown in open ponds, however, will additionally have a blue and gray WF related to evaporation and water pollution from those ponds. According to Verdegem et al. (2006), a fish pond with evaporation plus seepage losses of 3,500 mm/y and an annual production of 1,000 kg/ha/y loses 35 m^3 of water through evaporation and seepage per kilogram of fish produced. If the pond is drained and filled once per year, then total water consumption equals 45 m^3/kg of fish produced. The blue WF will be smaller, however, because only the evaporation counts as consumptive water use. This will be of the order of 1,000–2,000 mm/y, depending on climatic conditions, and thus implies a blue WF of 10–20 m^3/kg of fish. An important factor is the fish production per hectare. The previously mentioned 1,000 kg/ha/y refers to extensive systems; in intensively mixed systems, the productivity can be 100 times higher, and the blue WF per kilogram of fish related to open-pond evaporation thus can be 100 times smaller (100–200 L/kg). Water from fish ponds is generally highly polluted, thus causing a gray WF. No estimates of that are available yet.

Regarding the WF of wood, Van Oel and Hoekstra (2012) found WFs of harvested wood varying between 200 and 1,100 m^3 of water per m^3 of wood. The global average found is approximately 500 m^3/m^3. Important determinants are the evapotranspiration rate of a production forest (in mm/year) and the average wood yield (in m^3) per hectare per year. The former depends on climate and tree species, and the latter depends on tree species and forestry practice. Eucalyptus is a fast-growing, thus high-yielding, species, but grows in warmer climates with high evapotranspiration rates.

MINING AND QUARRYING

Mining is the process of extracting buried material below the earth surface. Quarrying refers to extracting materials directly from the surface. In mining and quarrying, water is used and gets polluted in a range of activities, including mineral processing, dust suppression, and slurry transport. In addition, water is subtracted from the environment in the process of dewatering, the process of pumping away the water that naturally flows into the pit or tunnels of the mine. When disposed, this water may also carry pollutants. The mining and quarrying sector includes mining of fossil fuels (coal and lignite mining, oil and gas extraction), mining of metal ores, quarrying of stone, sand, and clay, and mining of phosphate and other minerals. A rich data source of water use in the mining of conventional and unconventional oil and gas, coal, and uranium is provided in the work of Williams and Simmons (2013).

Mudd (2008) provides a useful review of gross blue water use in different types of mining (Table 7.3). In general, he found that the higher the ore throughput, the more likely that, through economies of scale, the unit water use per kilogram of ore is lower. Furthermore, he found that as metallic ore grades decline, there is a strong probability of an increase in water use per unit of metal. Gold has the highest water use per kilogram of metal, with platinum closely behind; this is presumably attributable to the very low grade of gold and platinum ores (i.e., parts per million compared with percent for base metals). It is noted here that net blue water use, the blue WF, will be substantially lower than the figures presented in Table 7.3, because most of the water will remain within the catchment.

Table 7.3 Gross Blue Water Use in Mining

	Gross Blue Water Use Per Unit of Ore Throughput		Gross Blue Water Use Per Unit of Ore Grade	
Mineral/Metal	**Average**	**SD**	**Average**	**SD**
Bauxite (L/kg bauxite)	1.09	0.44	–	–
Black coal (L/kg coal)	0.30	0.26	–	–
Copper (L/kg ore; L/kg Cu)	1.27	1.03	172	154
Copper–gold (L/kg ore; L/kg Cu)	1.22	0.49	116	114
Diamonds (L/kg ore; L/carat)	1.32	0.32	477	170
Gold (L/kg ore; L/kg Au)	1.96	5.03	716,000	1,417,000
Zinc ± lead ± silver ± copper ± gold (L/kg ore; L/kg Zn ± Pb ± Cu)	2.67	2.81	29.2	28.1
Nickel (sulfide) (L/kg ore; L/kg Ni)	1.01	0.26	107	87
Platinum group (L/kg ore; L/kg PGM)	0.94	0.66	260,000	162,000
Uranium (L/kg ore; L/kg U3O8)	1.36	2.47	505	387

The figures refer to the sum of water abstractions and recycling volumes. SD = standard deviation.
Source: From Mudd (2008).

Peña and Huijbregts (2014) made a detailed estimate of the operational and supply chain blue WF for the extraction, production, and transport to the nearest seaport of high-grade copper refined from two types of copper ore—copper sulfide ore and copper oxide ore—in the Atacama Desert of northern Chile, one of the driest places on earth. The total blue WF (direct and upstream consumption) for the sulfide ore refining process was 96 L/kg of copper cathode. The first step in the process, the extraction from the open pit mine, accounts for 5% of the total blue WF; the second step, comminution (crushing, grinding), accounts for 3%; the third step, the concentrator plant, accounts for 59%; the fourth step, the smelting plant, contributes 10%; and the last two steps, electrorefinery and the sulfuric acid plant, contribute 3% and 1%. The supply chain contributes 19%: approximately 9% related to materials and 10% related to electricity. In the case of the copper oxide ore-refining process, the blue WF was 40 L/kg of copper cathode. The first step, extraction, accounts for 2%; the second step, comminution and agglomeration, contributes 18%; the third step, the heap leaching process, accounts for 44%; the fourth step, solvent extraction, contributes nothing; and the last step, electrowinning, accounts for 10%. The supply chain contributes 26%: approximately 6% related to materials and 20% related to electricity.

Generally, mining has a significant gray WF, but it is difficult to obtain quantitative data for this. The first source of pollution can come from the "overburden," the waste soil and rock that has to be removed before the ore deposit can be reached and that has to be stored somewhere after removal. The "strip ratio", the ratio of the quantity of overburden to the quantity of mineral ore extracted, can be much higher than one. The overburden material, sometimes containing significant levels of toxic substances, is usually deposited on-site in piles on the surface or as backfill in open pits, or within underground mines (ELAW, 2010). Through erosion, runoff, and seepage, these toxic substances may reach groundwater or surface water bodies. The second source of pollution comes from the pit itself, where similar processes may spread toxic chemicals into the wider environment. In addition, mine dewatering can bring polluted water from the mine to the streams into which the water is released. The third source of pollution comes from the waste material that remains after concentration of the valuable mineral from the extracted ore and that often contains various toxic substances (like cadmium, lead, and arsenic). This waste, the so-called tailings, is generally stored in tailings ponds, which may leak. Also, there are numerous incidents of tailings reservoir dam breaks, after which the content of the reservoir released itself into the environment. A fourth source of pollution can come from the process of heap leaching. With leaching, finely ground ore is deposited in a large pile (called a "leach pile") on top of an impermeable pad, and a solution containing cyanide is sprayed on top of the pile. The cyanide solution dissolves the desired metals and the "pregnant" solution containing the metal is collected from the bottom of the pile using a system of pipes, a procedure that brings significant environmental risk (ELAW, 2010). Finally, a form of mining that typically results in significant water pollution is the so-called placer mining, in which bulldozers, dredges, or hydraulic jets of water are used to extract the ore from a stream bed or flood plain (ELAW, 2010). Placer mining is a common method to obtain gold from river sediments.

MANUFACTURING

The manufacturing sector is the most diverse of all economic sectors. I reflect on the WF of just a few specific subsectors: food and beverage; textile and apparel; paper; computers; and motor vehicles.

Food and beverage products

The food and beverage sector is the manufacturing sector with the largest WF (maybe not the largest operational WF, but definitely the largest supply chain WF). The reason is that the food and beverage sector is the largest client of the agricultural sector, which is responsible for the largest share in global water consumption (Table 7.1). WF studies performed in the beverage sector include the studies performed by SABMiller (SABMiller and WWF-UK, 2009; SABMiller, GTZ and WWF, 2010), Coca-Cola (TCCC and TNC, 2010; Coca-Cola Europe, 2011) and the Beverage Industry Environmental Roundtable (BIER, 2011). Some good examples of WF studies in the food sector include those regarding Unilever (Jefferies et al., 2012), Dole (Sikirica, 2011), Mars (Ridoutt et al., 2009), and Barilla (Ruini et al., 2013).

Traditionally, the beverage industry focuses on the so-called water use ratio (WUR), which is defined as the total water use divided by the total production at a bottling facility, expressed in terms of liter of water used per liter of beverage produced. Water use here represents gross blue water abstraction, not net blue water abstraction (blue WF). In a global benchmarking study for the period 2009–2011, BIER (2012) reported a WUR of 1.2–2.2 L/L (with an average of 1.5) for bottled water, a WUR of 1.5–4.0 (average 2.1) for carbonated soft drinks, a WUR of 3.2–6.6 (average 4.3) for beer breweries, a WUR of 8–126 (average 36) for distilleries, and a WUR of 2.0–18.5 (average 4.4) for wineries. The WUR is of limited value because the operational WF of bottling factories is very small when compared with the full WF of a beverage, as shown by Ercin et al. (2011) regarding carbonated soft drinks. They showed that the WF of a half-liter bottle of a soft drink resembling cola can range between 150 and 300 L, of which 99.7–99.8% refers to the supply chain.

Textile and apparel

According to Wang et al. (2013), the blue operational WF of the People's Republic of China's textile industry was, on average, 0.8×10^9 m^3/y during the period 2001–2010 (increasing over time from 0.5 to 1.0×10^9 m^3/y). Based on the loads of COD (chemical oxygen demand) to freshwater, accounting for the treatment of wastewater before disposal, they compute a gray WF of approximately

10×10^9 m^3/year on average during the same period (again increasing over time). Without current levels of treatment, the gray WF would have been five times larger. The gray WF calculation was based on a maximum acceptable biochemical oxygen demand (BOD) of 100 mg/L in textile effluents and the assumption of zero background concentrations in the receiving water bodies. Using the effluent standard as a reference leads to an underestimation of the gray WF, because effluent standards are generally less strict than ambient water quality standards. Gray WF guidelines of WFN, for example, recommend an ambient water quality standard for COD of 30 mg/L (Franke et al., 2013). The assumption of zero background concentrations leads to an overestimation of gray WF, because natural concentrations of COD are not zero. The figures reported by Wang et al. (2013) are totals for three subsectors: manufacture of textiles; manufacture of textile apparel, footwear, and hats; and manufacture of chemical fibers. The manufacture of textiles contributed the largest part to the gray WF of the textile sector as a whole. The gray WF per unit of output value of the textile manufacturing industry decreased from 70 to 20 L/USD.

Chico et al. (2013) estimated the WF of a pair of jeans made in Spain (assuming a weight of 780 g per pair of trousers) by considering two different fibers (cotton and Lyocell fiber) and five different production methods for spinning, dying, and weaving. Including water use in the full supply chain (cotton growing for cotton lint and wood growth for Lyocell fibers), they reported a total WF of 2,800–4,900 L for one pair of cotton trousers (on average 8% green, 86% blue, and 6% gray WF) and a total WF of 1,200–1,900 L for one pair of Lyocell trousers (on average 95% green, 2% blue, and 2% gray WF). In the case of cotton trousers, cotton-growing contributed the largest share to the total, whereas for Lyocell trousers it was the growing of the wood that contributed the largest share. Cotton-growing often heavily relies on irrigation and, therefore, blue water (Chapagain et al., 2006), and wood relies mainly on green water. The WF of wood mainly varies depending on the origin of the wood and forest type (Van Oel and Hoekstra, 2012). According to Chico et al. (2013), ginning of cotton had a blue WF of 30–60 L/kg and a zero gray WF, whereas spinning and weaving had a blue WF of 54–134 L/kg and gray WF of 0–0.06 L/kg. Lyocell fiber production from pulp would have a blue WF of 1 L/kg and a gray WF of 4–272 L/kg, whereas spinning and weaving would have a blue WF of 105 L/kg and a gray WF comparable with that for the case of cotton.

There can be large differences in the supply chain WF of the textile and apparel sector, depending on the type of fibers used and the source region of the fibers. The WF of cotton fibers is substantially larger than most other plant fibers. To honestly compare, we can compare cotton lint, which is the cotton fiber separated from the cottonseed, with other plant fibers (Hoekstra, 2013). According to Mekonnen and Hoekstra (2011), the global average WF of seed cotton is 4,030 L/kg (the sum of green, blue, and gray). The seed cotton is split into cottonseed (63% of the weight, 21% of the economic value) and cotton lint (35% of the weight, 79% of the economic value). The WF of the cotton lint thus can be calculated as $(0.79/0.35) \times 4{,}030 = 9{,}100$ L/kg. In the process from cotton lint to final

Table 7.4 Global Average WF of Different Plant Fibers during 1996–2005

	Global Average WF (L/kg)			
Product	**Green**	**Blue**	**Gray**	**Total**
Abaca fiber	21,529	273	851	22,654
Cotton lint	5,163	2,955	996	9,113
Sisal fiber	6,791	787	246	7,824
Agave fiber	6,434	9	106	6,549
Ramie fiber	3,712	201	595	4,507
Flax fiber	2,866	481	436	3,783
Hemp fiber	2,026	0	693	2,719
Jute fiber	2,356	33	217	2,605

Source: From Mekonnen and Hoekstra (2011) and from Hoekstra (2013).

cotton fabric, there are again some weight losses and by-products, so that the WF of cotton fabric is again a bit larger. In this way, we arrive at 10,000 L/kg. For the purpose of a fair appraisal, we can compare WFs in L/kg either at the level of the fibers or at the level of the final textile. For the outcome it will make little difference, because the big differences in water use are in the growth of the plants, not in the water use for processing of fibers into final textile. Here, we compare the WF of cotton lint with the WF of the fibers of other plants. An overview is given in Table 7.4. From this overview, it is clear that, on average, the WF of cotton fibers is a bit larger than the WF of sisal and agave fibers, much larger than that of ramie and flax fibers, and very much larger than the WFs of hemp and jute fibers. We should be careful to immediately conclude that we should replace cotton fibers by, for example, hemp fibers, because fibers are different and textiles made from different fibers have different characteristics. However, it shows that it is worth investigating how cotton compares with hemp and other fibers in other respects and to what extent and in which applications cotton can be substituted by other plant fibers. It would also make sense to compare the performance of plant fibers with animal fibers (like different types of wool) and synthetic fibers (often made from petroleum), whereby, again, the claim on water resources of a fiber can be just one of a more extended set of criteria.

Paper

The WF of any wood product is the sum of the WFs in the forestry and the industrial stage. We focus here on the pulp and paper industry. A pulp mill converts wood chips or other plant fiber sources into thick fiberboard that can be shipped to a paper mill for further processing into final paper products. The blue WF in the industrial stage can be estimated by summing the evaporation flows from the pulp and paper mills, the amount of water incorporated in the products delivered by the mills, and the volume of water contained in solid residuals (Hoekstra, 2013).

The gray WF depends on the loads of different chemicals contained in the mill effluents discharged into the environment. Paper industries are known for their large water demand and for producing polluted effluents, which, if not properly treated, can cause significant ecological damage in the streams into which the effluents are disposed. The pulp and paper industry in the United States withdraws approximately $5,500 \times 10^9$ L of water annually from surface and groundwater sources (NCASI, 2009). A major part of the water used, however, returns to the catchments from where the water has been taken, so that consumptive water use is much less than total abstraction: an estimated volume of 507×10^9 L of water annually evaporates from pulp and paper mills in the United States, and 10×10^9 L of water leaves the mills (and the catchments) incorporated in products. Probably more important than the consumptive use of water in pulp and paper mills is the pollution that comes from those mills. Although mechanical pulping is applied as well, chemical pulping is the most commonly used pulping process. Chemical pulps are made by cooking the raw materials and adding a mixture of chemicals. After pulping, the pulp is generally bleached to make it whiter. Different sorts of chemicals are used in this process, including, for example, chlorine, sodium hypochlorite, and chlorine dioxide. The use of elemental chlorine or chlorine compounds particularly results in high concentrations of undesired compounds in effluents. Water pollution from pulp and paper mills mostly stems from the organic matter contained in the effluents, which generally include several chlorinated organic compounds like dioxins and other adsorbable organic halides (usually abbreviated as AOX). The organic matter content in effluents from pulp and paper mills is measured by the BOD in the effluent; a large BOD in effluents can lead to oxygen depletion and fish kills in rivers. High concentrations of AOX can also lead to toxicity and fish kills. According to Hoekstra (2013), the WF for one final A4 sheet of copy paper (80 g/m^2) ranges from 2 to 20 L of water, which covers the water use in both the forestry and industry stages. The two major variables that influence the size of the WF of paper and that can be relatively easily influenced are the paper recycling rate and the amount of chemicals in effluents discharged into the environment (Hoekstra, 2013).

As an example, we discuss a study by the UPM-Kymmene Corporation, a Finnish pulp, paper, and timber manufacturer. They assessed the operational and supply chain WF of their Nordland paper mill in Germany (Rep, 2011). The majority of chemical pulp used at this paper mill comes from three pulp mills: the Kaukas and Pietarsaari pulp mills in Finland and the Fray Bentos pulp mill in Uruguay. In the Finnish pulp mills, three different types of tree are used: broadleaves, pine, and spruce. In the pulp mill in Uruguay, eucalyptus trees are used as the raw resource. The Nordland paper mill in Germany produces two paper grades: wood-free coated paper (150 g/m^2) and wood-free uncoated paper (80 g/m^2). Wood-free paper is paper made from chemical pulp instead of mechanical pulp. Chemical pulp is made from pulpwood and is considered wood-free because most of the lignin is removed and separated from the cellulose fibers during processing, in contrast to mechanical pulp, which retains most of its wood components and therefore can still be described as wood-containing. It was found that the total WF of one A4 sheet of paper leaving the Nordland paper mill is 13 L for wood-free uncoated paper and 20 L for wood-free coated paper. The color composition of that total WF is 60% green, 39% gray, and 1% blue. Approximately 99% of the total WF originates from the raw material supply chain (forestry stage and pulp mills in Finland and Uruguay) and the remaining 1% originates from the production processes within the Nordland paper mill in Germany. The gray WF assessment showed that AOX was the most critical indicator from an environmental impact perspective, requiring the biggest volume of water to dilute to acceptable concentrations.

Computers

The semiconductor manufacturing process requires high-purity water, which is generally produced on-site from municipal water. For the fabrication of a silicon wafer, Williams et al. (2002) reported water use figures between 5 and 58 L/cm^2 of silicon. With a typical value of 20 L/cm^2 and a surface of 1.6 cm^2, this means that producing a single 2-g 32-MB DRAM (dynamic random access memory) chip requires 32 L per chip in the fabrication stage.

In a life cycle study of personal computers for Hewlett-Packard, Alafifi (2010) estimated the blue water use of a desktop computer at 10,000 L, 59% of which would relate to electricity in the use stage of the computer (assuming a life span of 5 years). Approximately 34% was related to manufacturing of the components (22% for producing the LCD monitor, 8% for manufacturing printed circuit boards, and 3% for fabrication of semiconductors), 7% was related to the extraction of raw materials, and 1% was related to assembly. It is noted that different electricity generation mixes have significant influence on the total water use due to the different water use intensities of various energy sources. A major drawback of the water use figures presented by Alafifi (2010) is that it remains unclear how to interpret the numbers, which is typical for present-day studies of water use in industry. Sources of data are often unclear about whether water use figures refer to gross water use or consumptive water use. In practice, many studies use the terms "water use" and "water consumption" interchangeably, but both generally refer to gross water use. Even though Alafifi (2010) presented 10,000 L for a desktop computer as the "blue WF" of the personal computer, most of the underlying data used probably refer to gross water use, not consumptive water use, which makes a great difference.

Regarding computer monitors, Socolof et al. (2001) studied the life cycle impacts of desktop computer displays and found water use of 13,100 L per CRT (cathode ray tube) and 2,820 L per LCD (liquid crystal display).

Motor vehicles

A number of car companies have performed WF studies, but little has been made publicly available. An interesting public report is one regarding the WF of a few facilities of TATA Motors in India for the year 2012 (Unger et al., 2013). Tata Motors has approximately 1,000 suppliers, accounting for the majority of its overall WF. The highest inside-the-fence water consumption in the facilities studied was from the paint shop and forging. Among the facilities, the largest direct blue WF was found for a facility in Lucknow in the state of Uttar Pradesh that produces heavy and medium commercial vehicles, mainly buses, with 5.5 m^3 per equivalent vehicle. The base model used for equivalent vehicle calculations was the 1,612 (load-bearing capacity of 16 tons, 120 hp). In Pune, in the state of Maharashtra, the direct blue WF was 4.75 m^3 per equivalent vehicle, again with the 1,612 as the base model, and in Jamshedpur, in the state of Jharkhand, it was 3.3 m^3 per equivalent vehicle with the same base model. In Pantnagar, in the state of Uttarakhand, the direct blue WF was 4.9 m^3 per equivalent vehicle, this time with the ACE Goods Carrier as the base model, which is a 1-ton mini-truck. The smallest direct blue WF was found in another facility in Pune that produced passenger cars, with 1.7 m^3 per equivalent vehicle and with the Tata Indica diesel as the base model for the equivalent calculations. Direct gray WFs were smaller than the blue WFs in the six facilities studied. For the heavy vehicles, the direct gray WF varied from 2.9 m^3 per equivalent vehicle in Jamshedpur and 2.4 m^3 in Pune to 0.4 m^3 in Lucknow. For the mini-trucks from Pantnagar, the gray WF was 2.0 m^3 per equivalent vehicle. For the passenger cars from Pune, the direct gray WF was 0.7 m^3 per equivalent vehicle. The number of parameters included in the gray WF calculations varied, with up to nine parameters. In all cases, effluents are treated before disposal.

Another interesting study is one by Berger et al. (2012) regarding the blue WF of three car models of Volkswagen over their full life cycle. They estimated that the water consumption along the life cycles of the three cars studied amounts to 52 m^3 (Polo 1.2 TDI), 62 m^3 (Golf 1.6 TDI), and 83 m^3 (Passat 2.0 TDI). In all three cases, 95% of the total water consumption lies in the production stage of the car (as opposed to the use and end-of-life stages). In the case of the Golf 1.6 TDI, the largest contributions to the total life cycle water consumption come from steel and iron (approximately 34%), polymers (another 34%), and precious metals like gold, silver, and platinum (20%). The latter figure is high given the fact that it takes less than 1 kg of precious metals per car. The reported figure for precious metals is probably an overestimate because the study assumed 100% primary material, and Volkswagen has been operating a catalyst recycling program for years, which helps to recover and recycle PGM in a closed-loop system. Water consumption for manufacturing, the final assembly, was 0.36 m^3 per car. Apart from the water consumption for final assembly, there was also water consumption at the car production sites for other activities, like injection moulding of polymer components and hot stamping of steel components. Altogether, approximately 10% of the total life cycle water consumption occurs directly at the car production sites in Pamplona, Wolfsburg, and Emden, mainly resulting from painting and evaporation of cooling water.

WATER SUPPLY

One would expect that the WF of the "water supply" sector is most significant of all sectors, but this is not the case. On a global level, the WF of the municipal water supply has been estimated to be 3.6% of the total WF of humanity (Table 7.1). Wuppertal Institute (2011) reports a material intensity factor for drinking water of 1.3 L/L, referring to gross blue water use. As noted, the ratio of net to gross abstraction has been estimated to be 5–15% in urban areas and 10–50% in rural areas (FAO, 2014). This means that the blue WF of drinking water from the tap can be as little as 0.065 L/L, or 0.65 L/L in the worst case, assuming that the rest of the water returns to the water system from which it was abstracted. The gray WF related to municipal water supply depends on the extent of treatment of the wastewater. The gray WF of municipal water supply will generally be larger than the blue WF, even in the case of treatment before disposal, because the concentrations of nitrogen, phosphorous, and other substances in the wastewater after treatment will still be beyond the concentrations in the intake water (Figure 7.2).

CONSTRUCTION

The direct WF of the construction industry is small compared with the indirect WF related to the mining and manufacturing of materials used in construction. McCormack et al. (2007) illustrate this with a number of Australian nonresidential case studies using an input–output-based hybrid embodied-water analysis. Regarding the water embodied in construction, they found values between 5 and 20 m^3 of water per m^2 of gross floor area. The lower values represented refurbishment projects rather than complete new construction projects. According to this study, steel had the largest contribution in this total of embodied water, followed by concrete. Carpet had the third largest contribution, which is important because it is often replaced every 10 years or even more frequently in prestigious commercial buildings. The direct water use in the construction process was small compared with the total, maximally 1 m^3 of water per m^2 gross floor area. It has to be noted that the figures cited here refer to gross blue water use, not net consumptive water use (blue WF).

TRANSPORT

Transport is always considered an important sector in carbon footprint assessment, because transport can significantly contribute to the overall carbon footprint of a final product, measured over its full supply chain. In the case of the WF of a final product, the contribution of transport will generally be relatively small, because not much freshwater is being consumed or polluted during transport. It is worth considering the indirect WF of transport related to materials (trucks, trains, boats, airplanes) and energy used, but materials will generally contribute very little because the WF of a transport vehicle can be distributed over all goods transported over the lifetime of the vehicle. The WF of energy may be more relevant, but even that can be small compared with the other components of the WF of goods, particularly in the case of agricultural goods. The key determinant in the WF of transport is probably the energy source (King and Webber, 2008; Gerbens-Leenes et al., 2009a). The WF of bioenergy in terms of cubic meter per GJ is generally two to three orders of magnitude larger than that for energy from fossil fuels or wind or solar power. However, in all energy categories, WFs per unit of energy can widely vary, depending on the precise source and production technology. The technique of hydraulic fracturing (fracking) to mine natural gas or petroleum reserves, for example, has a larger blue and gray WF than that for mining reserves that are more easily accessible using more conventional techniques. In the case of bioenergy, it matters greatly whether one speaks about biodiesel from oil crops, bioethanol from sugar or starch crops (Gerbens-Leenes et al., 2009b; Dominguez-Faus et al., 2009), biofuel from cellulosic fractions of crops or waste materials (Chiu and Wu, 2012), or bioelectricity. In the latter case, it makes a large difference what is burned, for example biomass grown for the purpose or organic waste. As an illustration of the large differences between different bioenergy forms, Table 7.5 gives the WF of different modes of passenger and freight transport when based on first-generation biofuel produced in the European Union. Governmental policies to replace substantial percentages of fossil fuels by biofuels will lead to a rapid growth of the WF of the transport sector (Gerbens-Leenes et al., 2012).

WHOLESALE, RETAIL TRADE, AND SERVICES

There has been little investigation regarding the WF of the wholesale, retail trade, and services sectors. The reason is that the direct WF of these sectors will be generally small compared with their indirect WF, i.e., the WF of the goods bought for use or sale. Particularly in the wholesale and retail trade sectors, all that matters is the WF of the goods purchased to sell. Wholesale and retail companies can play an important role in WF reduction, not because of the significance of their operational WF but rather because they form a point where many products from a great number of producers come together to be distributed over a large number of consumers. Wholesale companies and retailers can influence the WF of the products on their store shelf by using sustainability criteria in their purchasing choices.

In the service sector, the major determinant in the total WF will generally be the WF related to consumables, like paper, computers, printers, machineries, vehicles, other materials, and energy. The WF of the construction materials of office buildings may play a minor role. One component will often dominate: the food served in the company restaurant, even though this is obviously not part of the primary business of a company. As an example, we discuss here a study by Factor-X, an environmental consultancy firm in Belgium, which was one of the first companies in the service sector to estimate its operational and supply chain

Table 7.5 The WF of Different Modes of Passenger and Freight Transport When Based on First-Generation Biofuel Produced in the European Union

Transport Mode	Energy Source	Green + Blue WF of Passenger Transport (L/passenger km)	Green + Blue WF of Freight Transport (L/1,000 kg of freight per km)
Airplane	Biodiesel from rapeseed	142–403	576–1,023
	Bioethanol from sugar beet	42–89	169–471
Car (large)	Biodiesel from rapeseed	214–291	–
	Bioethanol from sugar beet	138–289	–
Car (small)	Biodiesel from rapeseed	65–89	–
	Bioethanol from sugar beet	24–50	–
Bus/lorry	Biodiesel from rapeseed	67–126	142–330
	Bioethanol from sugar beet	20–58	–
Train	Biodiesel from rapeseed	15–40	15–40
Ship (inland)	Biodiesel from rapeseed	–	36–68
Ship (sea, bulk)	Biodiesel from rapeseed	–	8–11
Electric train	Bioelectricity from maize	3–12	2–7
Electric car	Bioelectricity from maize	4–7	–
Walking	Sugar from sugar beet	3–6	–
Bike	Sugar from sugar beet	1–2	–

The total WF of transport based on first-generation biofuel mainly relates to the water volumes consumed in growing the crop.
Source: From Gerbens-Leenes and Hoekstra (2011).

WF. The scope of the study included both direct and indirect water use during the approximate 225 work days during the year 2011 (Factor-X, 2011). The study included food consumption by employees and the use of electricity for computers and internet, telephones, paper for printing, and office equipment. The study did not include domestic and international travel, clothing, and use of mobile phones, or the construction of the office. The direct blue, green, and gray WFs were estimated at 115, 4, and 320 L/employee per work day, respectively. The gray WF referred to the pollution from organic matter in the wastewater. The indirect blue–green WF related to food was estimated at 3,420 L/employee per work day, the indirect blue WF related to professional activities (heating of the building, paper, electricity use, etc.) was estimated to be 140 L, and the indirect blue WF related to the manufacture of office equipment (computers, printers, desks, chairs, cupboards, lockers, plastic) was estimated to be 5 L/employee per work day. Factor-X concluded that the WF related to food consumption of their workers is dominant over their direct operational WF or their indirect WF related to energy use or office equipment, and that promoting vegetarian food among their workers is probably the most effective measure. However, the company recognizes other measures, like using dry toilets, moving to a paperless office, and reducing energy consumption. Also, it was noted that improved wastewater treatment in the country would help.

WATER STEWARDSHIP AND TRANSPARENCY

There is an increasing call for good water stewardship and transparency in the private sector that is driven by increased public awareness, demands from investors, and perceived water risks by the sector itself. Water stewardship is a comprehensive concept that includes the evaluation of the sustainability of water use across the entire value chain, the formulation of water consumption and pollution reduction targets for both the company's operations and supply chain, the implementation of a plan to achieve these targets, and proper reporting of targets and achievements (Hoekstra, 2014a). In priority catchments, the pursuit of collective action and community engagement is required (Sarni, 2011). High-priority river basins are, for example, the Colorado and San Antonio basins in North America; the Lake Chad, Limpopo, and Orange basins in Africa; the basins of the Jordan, Tigris, Euphrates, Indus, Ganges, Krishna, Cauvery, Tarim, Yellow River and Yongding River in Asia; and the Murray–Darling basin in Australia (Hoekstra et al., 2012). For most companies, moving toward a sustainable supply chain is a much bigger challenge than greening their own operations, because the WF of the supply chain is often up to 100 times bigger than the company's operational footprint and can be influenced only indirectly. Common reduction targets in the beverage industry, such as going from 2 to 1.5 L of water use in the bottling plant per liter of beverage, have little effect on the larger-scale given that the supply chain WF of most beverages is approximately 100 L of water per liter of beverage or even more (Hoekstra, 2013).

The increasing interest in how companies relate to unsustainable water use calls for greater transparency on water consumption and pollution. Openness is required at different levels: the company, product, and facility level. Driven by environmental organizations and the investment community, businesses are increasingly urged to disclose relevant data at a company level regarding how they relate to water risks (Deloitte, 2013). Simultaneously, there is an increasing demand for product transparency through labeling or certification. Despite the plethora of existing product labels related to environmental sustainability, none of these includes criteria on sustainable water use. Finally, there is a movement to develop principles and certification schemes for sustainable site or facility management, such as the initiatives of the European Water Partnership and the Alliance for Water Stewardship. Despite progress in awareness, barely any companies in the world report on water consumption and pollution in their supply chain or reveal information about the sustainability of the WF of their products.

Much confusion exists regarding what needs to be measured and reported. Traditionally, companies have focused on monitoring gross water abstractions and compliance with legal standards. However, net water abstractions are more relevant than gross abstractions, and meeting wastewater quality standards is not enough to discard the contribution to water pollution made by a company. Regarding terminology and calculation standards, the Water Footprint Network—a global network of universities, nongovernmental organizations, companies, investors, and international organizations—developed the global WF standard (Hoekstra et al., 2011). The International Organization for Standardization developed a reporting standard based on LCA (ISO, 2014). Both standards emphasize the need to incorporate the temporal and spatial variability in WFs and the need to consider the WF in the context of local water scarcity and water productivity. In practice, companies face a huge challenge in tracing their supply chain. Apparel companies, for example, have generally little idea of where their cotton is grown or processed, yet the growing and processing of cotton are notorious water consumers and polluters. It is difficult to see quick progress in the field of supply chain reporting if governments do not force companies to do it.

The indirect blue and gray WF of many industries is often many times greater than their direct, operational WF. Nevertheless, most industries restrict their efforts to reducing their operational WF, leaving the supply chain WF out of scope. Studies performed by companies like Coca-Cola, PepsiCo, SABMiller, and Heineken have shown that the supply chain WF for beverage companies can easily be more than 99% of their total WF. Nevertheless, all these companies apply a "key performance indicator" for water that refers to the water use in their own operations only. Investments are geared to perform better in this respect, which means that, under the goal of sustainability, investments are made that aim to reduce that 1% of their total WF. It is difficult to imagine that these investments will be most cost-effective if sustainability is the actual goal. Incorporating sustainability principles into a company's business model would include the adoption of mechanisms to secure sustainable water use in the supply chain.

WFs per unit of product strongly vary across different production locations and production systems. Therefore, we need to establish WF benchmarks for water-intensive products such as food and beverages, cotton, cut flowers, and biofuels. The benchmark for a product will depend on the maximum reasonable water consumption in each step of the product's supply chain based on the best-available technology and practice. Benchmarks for the various water-using processes along the supply chain of a product can be taken together to formulate a WF benchmark for the final product. An end-product point of view is particularly relevant for the companies, retailers, and consumers who are not directly involved in the water-using processes in the early steps of the supply chains of the products they are manufacturing, selling, or consuming but are still interested in the water performance of the product over the chain as a whole. WF benchmarks will offer a reference for companies to work toward and a reference for governments in allocating WF permits to users. Manufacturers, retailers, and final consumers on the lower end of the supply chain get an instrument to compare the actual WF of a product with a certain reference level. Business associations within the different sectors of economy can develop their own regional or global WF benchmarks, although governments can take initiatives in this area as well, including the development of regulations or legislation. The latter will be most relevant to completely ban worst practices.

Companies should strive toward zero WF in industrial operations, which can be achieved through nullifying evaporation losses, full water recycling, and recapturing chemicals and heat from used water flows. The problem is not the fact that water is being used, but that it is not fully returned to the environment or not returned clean. The WF measures exactly that (the consumptive water use and the volume of water polluted). As the last steps toward zero WF may require more energy, it may be necessary to find a balance between reducing the water and the carbon footprint. Furthermore, companies should set reduction targets regarding the WF of their supply chain, particularly in areas of great water scarcity and in cases of low water productivity. In agriculture and mining, achieving a zero WF will generally be impossible, but in many cases the water consumption and pollution per unit of production can be reduced easily and substantially (Brauman et al., 2013).

When formulating WF reduction targets for processes in their operations or supply chain, companies should look not only at the numbers but also at the geographic locations where their WF is sited. Priority is to be given to WF reduction in catchments in which the overall footprint exceeds the carrying capacity or assimilation capacity of the catchment. It has been argued that reduction in water-abundant catchments does not make sense (Pfister and Hellweg, 2009), but this is based on a misunderstanding. Because the WF (m^3/product unit) is simply a reverse of water productivity (product units per m^3), it is difficult to see why one would not set targets regarding the reduction of the WF of a product, which is the same as setting targets regarding the increase of water productivity. The relevance of increased water productivities worldwide, also in water-abundant places, can be illustrated with the following example (Hoekstra, 2013). Suppose the hypothetical case of two river basins with the same surface (Table 7.6). Basin A is relatively dry and has, on an annual basis, 50 water units available. This is the maximum sustainable WF, which is, however, exceeded by a factor of two. Farmers in the basin consume 100 water units per year to produce 100 crop units.

Table 7.6 Example of How Overexploitation in a Water-Stressed River Basin (A) Can Be Solved by Increasing Water Productivity in a Water-Abundant Basin (B)

		Current Situation		Possible Solution	
Parameter	**Unit**	**Basin A**	**Basin B**	**Basin A**	**Basin B**
Max. sustainable WF	Water units/unit of time	50	250	50	250
WF	Water units/unit of time	100	200	50	200
Production	Product units/unit of time	100	100	50	200
WF per product unit	Water units/ product unit	1	2	1	1
Water productivity	Product units/ water unit	1	0.5	1	1

Source: From Hoekstra (2013).

Basin B has more water, 250 water units per year, available. Water is more abundant than in the first basin, but water is used less efficiently. Farmers in the basin consume 200 water units per year to produce 100 crop units, the same amount as in the first basin but using two times more water per crop unit. A geographic analysis shows that in basin B, the WF (200) remains below the maximum level (250), so this is sustainable. In basin A, however, the WF (100) by far exceeds the maximum sustainable level (50), so this is clearly unsustainable. The question is, should we categorize the crops originating from basin A as unsustainable and the crops from basin B as sustainable? From a geographic perspective, the answer is affirmative. In basin A, the WF of crop production needs to be reduced, and that seems to be the crux. However, from a product perspective, we observed that the WF per crop unit in basin B is two times larger than in basin A. If the farmers in basin B would use their water more productively and reach the same water productivity as in basin A, then they would produce twice as many crops without increasing the total WF in the basin. It may be that farmers in basin A cannot easily further increase their water productivity, so—if the aim is to keep global production at the same level—the only solution is to reduce the WF in basin A to a sustainable level by cutting production by half while enlarging production in basin B by increasing the water productivity. If basin B manages to achieve the same water productivity level as in basin A, then the two basins together could increase global production while halving the total WF in basin A and keeping it at the same level in basin B.

A final concern regarding good water stewardship is the extent to which a company pays for the full cost of its water use. Water use is subsidized in many countries, either through direct governmental investments in water supply infrastructure or indirectly by agricultural subsidies, promotion of crops for bioenergy, or fossil energy subsidies to pump water. Water scarcity and pollution remain unpriced (Hoekstra, 2013). To give the right price signal, users should pay for their pollution and consumptive water use, with a differentiated price in time and space based on water vulnerability and scarcity.

CONCLUSION

Spatial patterns of water depletion and contamination are closely tied to the structure of the global economy. As currently organized, the economic system lacks incentives that promote producers and consumers to move toward wise use of our limited freshwater resources. To achieve sustainable, efficient, and equitable water use worldwide, we need greater product transparency, international cooperation, WF ceilings per river basin, WF benchmarks for water-intensive commodities, water pricing schemes that reflect local water scarcity, and some agreement about equitable sharing of the limited available global water resources among different communities and nations.

ACKNOWLEDGMENT

In writing this chapter, I have made use of pieces of text from previous publications. In particular, for the introductory section and the section on water stewardship and transparency, I have heavily drawn on a commentary in *Nature Climate Change* (Hoekstra, 2014a).

REFERENCES

Alafifi, A.H., 2010. Water Footprint of HP Personal Computers (MSc thesis). University of Surrey, Guildford, UK.

Berger, M., Warsen, J., Krinke, S., Bach, V., Finkbeiner, M., 2012. Water footprint of European cars: potential impacts of water consumption along automobile life cycles. Environ. Sci. Technol. 46 (7), 4091–4099.

BIER, 2011. A Practical Perspective on Water Accounting in the Beverage Sector. Beverage Industry Environmental Roundtable. <www.waterfootprint.org/Reports/BIER-2011-WaterAccountingSectorPerspective.pdf> (last accessed 04.08.2014).

BIER, 2012. Water Use Benchmarking in the Beverage Industry: Trends and Observations 2012. Beverage Industry Environmental Roundtable, Anteagroup, St. Paul, MN, USA.

Boulay, A.M., Hoekstra, A.Y., Vionnet, S., 2013. Complementarities of water-focused life cycle assessment and water footprint assessment. Environ. Sci. Technol. 47 (21), 11926–11927.

Brauman, K.A., Siebert, S., Foley, J.A., 2013. Improvements in crop water productivity increase water sustainability and food security: a global analysis. Environ. Res. Lett. 8 (2), 024030.

Cazcarro, I., Duarte, R., Chóliz, J.S., 2013. Multiregional input–output model for the evaluation of Spanish water flows. Environ. Sci. Technol. 47 (21), 12275–12283.

Chapagain, A.K., Hoekstra, A.Y., 2003. Virtual water flows between nations in relation to trade in livestock and livestock products, Value of Water Research Report Series No. 13, UNESCO-IHE, Delft, The Netherlands.

Chapagain, A.K., Hoekstra, A.Y., Savenije, H.H.G., Gautam, R., 2006. The water footprint of cotton consumption: an assessment of the impact of worldwide consumption of cotton products on the water resources in the cotton producing countries. Ecol. Econ. 60 (1), 186−203.

Chico, D., Aldaya, M.M., Garrido, A., 2013. A water footprint assessment of a pair of jeans: the influence of agricultural policies on the sustainability of consumer products. J. Cleaner Prod. 57, 238−248.

Chiu, Y.W., Wu, M., 2012. Assessing county-level water footprints of different cellulosic-biofuel feedstock pathways. Environ. Sci. Technol. 46, 9155−9162.

Coca-Cola Europe, 2011. Water Footprint Sustainability Assessment: Towards Sustainable Sugar Sourcing in Europe. Brussels, Belgium.

Daniels, P.L., Lenzen, M., Kenway, S.J., 2011. The ins and outs of water use—a review of multi-region input−output analysis and water footprints for regional sustainability analysis and policy. Econ. Syst. Res. 23 (4), 353−370.

Deloitte, 2013. Moving Beyond Business as Usual: A Need for a Step Change in Water Risk Management, CDP Global Water Report 2013. CDP, London, UK.

Dominguez-Faus, R., Powers, S.E., Burken, J.G., Alvarez, P.J., 2009. The water footprint of biofuels: a drink or drive issue? Environ. Sci. Technol. 43 (9), 3005−3010.

Duarte, R., Sánchez-Chóliz, J., Bielsa, J., 2002. Water use in the Spanish economy: an input−output approach. Ecol. Econ. 43 (1), 71−85.

ELAW, 2010. Guidebook for Evaluating Mining Project EIAs. Environmental Law Alliance Worldwide, Eugene, OR.

Ercin, A.E., Hoekstra, A.Y., 2014. Water footprint scenarios for 2050: a global analysis. Environ. Int. 64, 71−82.

Ercin, A.E., Aldaya, M.M., Hoekstra, A.Y., 2011. Corporate water footprint accounting and impact assessment: the case of the water footprint of a sugar-containing carbonated beverage. Water Resour. Manage 25 (2), 721−741.

Ewing, B.R., Hawkins, T.R., Wiedmann, T.O., Galli, A., Ercin, A.E., Weinzettel, J., Steen-Olsen, K., 2012. Integrating ecological and water footprint accounting in a multi-regional input-output framework. Ecol. Indic. 23, 1−8.

Factor-X, 2011. Water footprint, Factor-X Climate Consulting Group SPRL. Braine-l'Alleud, Belgium.

Fader, M., Gerten, D., Thammer, M., Heinke, J., Lotze-Campen, H., Lucht, W., Cramer, W., 2011. Internal and external green-blue agricultural water footprints of nations, and related water and land savings through trade. Hydrol. Earth Syst. Sci. 15 (5), 1641−1660.

FAO, 2014. Aquastat Database. Food and Agriculture Organization of the United Nations, Rome, Italy. <www.fao.org/nr/aquastat> (last accessed 04.08.2014).

Feng, K., Chapagain, A., Suh, S., Pfister, S., Hubacek, K., 2011a. Comparison of bottom-up and top-down approaches to calculating the water footprints of nations. Econ. Syst. Res. 23 (4), 371−385.

Feng, K., Hubacek, K., Minx, J., Siu, Y.L., Chapagain, A., Yu, Y., Guan, D., Barrett, J., 2011b. Spatially explicit analysis of water footprints in the UK. Water 3 (1), 47−63.

Finnveden, G., Hauschild, M.Z., Ekvall, T., Guinée, J., Heijungs, R., Hellweg, S., Koehler, A., Pennington, D., Suh, S., 2009. Recent developments in life cycle assessment. J. Environ. Manage. 91 (1), 1−21.

Franke, N.A., Boyacioglu, H., Hoekstra, A.Y., 2013. Grey Water Footprint Accounting: Tier 1 Supporting Guidelines, Value of Water Research Report Series No. 65, UNESCO-IHE, Delft, Netherlands.

Gerbens-Leenes, P.W., Hoekstra, A.Y., Van der Meer, T.H., 2009a. The water footprint of energy from biomass: a quantitative assessment and consequences of an increasing share of bio-energy in energy supply. Ecol. Econ. 68 (4), 1052–1060.

Gerbens-Leenes, P.W., Hoekstra, A.Y., Van der Meer, T.H., 2009b. The water footprint of bioenergy. Proc. Natl. Acad. Sci. 106 (25), 10219–10223.

Gerbens-Leenes, W., Hoekstra, A.Y., 2011. The water footprint of biofuel-based transport. Energy Environ. Sci. 4 (8), 2658–2668.

Gerbens-Leenes, P.W., Van Lienden, A.R., Hoekstra, A.Y., Van der Meer, Th.H., 2012. Biofuel scenarios in a water perspective: the global blue and green water footprint of road transport in 2030. Global Environ. Change 22, 764–775.

Hanasaki, N., Inuzuka, T., Kanae, S., Oki, T., 2010. An estimation of global virtual water flow and sources of water withdrawal for major crops and livestock products using a global hydrological model. J. Hydrol. 384, 232–244.

Hellweg, S., Milà i Canals, L., 2014. Emerging approaches, challenges and opportunities in life cycle assessment. Science 344 (6188), 1109–1113.

Hoekstra, A.Y., 2012. The hidden water resource use behind meat and dairy. Anim. Front. 2 (2), 3–8.

Hoekstra, A.Y., 2013. The Water Footprint of Modern Consumer Society. Routledge, London, UK.

Hoekstra, A.Y., 2014a. Water scarcity challenges to business. Nat. Clim. Change 4 (5), 318–320.

Hoekstra, A.Y., 2014b. Sustainable, efficient and equitable water use: the three pillars under wise freshwater allocation. WIREs Water 1 (1), 31–40.

Hoekstra, A.Y., Chapagain, A.K., 2007. Water footprints of nations: water use by people as a function of their consumption pattern. Water Resour. Manage 21 (1), 35–48.

Hoekstra, A.Y., Hung, P.Q., 2002. Virtual Water Trade: A Quantification of Virtual Water Flows between Nations in Relation to International Crop Trade, Value of Water Research Report Series No. 11. UNESCO-IHE, Delft, The Netherlands.

Hoekstra, A.Y., Chapagain, A.K., Aldaya, M.M., Mekonnen, M.M., 2011. The Water Footprint Assessment Manual: Setting the Global Standard. Earthscan, London, UK.

Hoekstra, A.Y., Mekonnen, M.M., 2012. The water footprint of humanity. Proc. Natl. Acad. Sci. 109 (9), 3232–3237.

Hoekstra, A.Y., Mekonnen, M.M., Chapagain, A.K., Mathews, R.E., Richter, B.D., 2012. Global monthly water scarcity: blue water footprints versus blue water availability. PLoS ONE 7 (2), e32688.

Hoekstra, A.Y., Wiedmann, T.O., 2014. Humanity's unsustainable environmental footprint. Science 344 (6188), 1114–1117.

ISO, 2014. ISO 14046: Environmental Management—Water Footprint—Principles, Requirements and Guidelines. International Organization for Standardization, Geneva, Switzerland.

Jefferies, D., Muñoz, I., Hodges, J., King, V.J., Aldaya, M., Ercin, A.E., Milà i Canals, L., Hoekstra, A.Y., 2012. Water footprint and life cycle assessment as approaches to assess potential impacts of products on water consumption: key learning points from pilot studies on tea and margarine. J. Cleaner Prod. 33, 155–166.

King, C.W., Webber, M., 2008. Water intensity of transportation. Environ. Sci. Technol. 42 (21), 7866–7872.

Kounina, A., Margni, M., Bayart, J.-B., Boulay, A.-M., Berger, M., Bulle, C., Frischknecht, R., Koehler, A., Milà I Canals, L., Motoshita, M., Núñez, M., Peters, G., Pfister, S., Ridoutt, B., Van Zelm, R., Verones, F., Humbert, S., 2013. Review of methods addressing freshwater use in life cycle inventory and impact assessment. Int. J. Life Cycle Assess. 18 (3), 707–721.

Lenzen, M., Foran, B., 2001. An input–output analysis of Australian water usage. Water Policy 3 (4), 321–340.

Liu, J., Yang, H., 2010. Spatially explicit assessment of global consumptive water uses in cropland: green and blue water. J. Hydrol. 384, 187–197.

McCormack, M., Treloar, G.J., Palmowski, L., Crawford, R., 2007. Modelling direct and indirect water requirements of construction. Build. Res. Inf. 35 (2), 156–162.

Mekonnen, M.M., Hoekstra, A.Y., 2011. The green, blue and grey water footprint of crops and derived crop products. Hydrol. Earth Syst. Sci. 15 (5), 1577–1600.

Mekonnen, M.M., Hoekstra, A.Y., 2012. A global assessment of the water footprint of farm animal products. Ecosystems 15 (3), 401–415.

Mudd, G.M., 2008. Sustainability reporting and water resources: a preliminary assessment of embodied water and sustainable mining. Mine Water Environ. 27, 136–144.

Naylor, R.L., Hardy, R.W., Bureau, D.P., Chiu, A., Elliott, M., Farrell, A.P., Forster, I., Gatlin, D.M., Goldburg, R.J., Hua, K., Nichols, P.D., 2009. Feeding aquaculture in an era of finite resources. Proc. Natl. Acad. Sci. 106 (36), 15103–15110.

NCASI, 2009. Water Profile of the United States Forest Products Industry. National Council for Air and Stream Improvement, National Triangle Park, NC, USA.

Oki, T., Kanae, S., 2004. Virtual water trade and world water resources. Water Sci. Technol. 49 (7), 203–209.

Peña, C.A., Huijbregts, M.A.J., 2014. The blue water footprint of primary copper production in Northern Chile. J. Ind. Ecol. 18 (1), 49–58.

Pfister, S., Hellweg, S., 2009. The water "shoesize" vs. footprint of bioenergy. Proc. Natl. Acad. Sci. 106 (35), E93–E94.

Pimentel, D., Berger, B., Filiberto, D., Newton, M., Wolfe, B., Karabinakis, E., Clark, S., Poon, E., Abbett, E., Nandagopal, S., 2004. Water resources: agricultural and environmental issues. BioScience 54 (10), 909–918.

Rebitzer, G., Ekvall, T., Frischknecht, R., Hunkeler, D., Norris, G., Rydberg, T., Schmidt, W.P., Suh, S., Weidema, B.P., Pennington, D.W., 2004. Life cycle assessment Part 1: framework, goal and scope definition, inventory analysis, and applications. Environ. Int. 30, 701–720.

Rep, J., 2011. From Forest to Paper, the Story of Our Water Footprint. UPM-Kymmene, Helsinki, Finland.

Ridoutt, B.G., Eady, S.J., Sellahewa, J., Simons, L., Bektash, R., 2009. Water footprinting at the product brand level: case study and future challenges. J. Cleaner Prod. 17, 1228–1235.

Rost, S., Gerten, D., Bondeau, A., Lucht, W., Rohwer, J., Schaphoff, S., 2008. Agricultural green and blue water consumption and its influence on the global water system. Water Resour. Res. 44 (9), W09405.

Ruini, L., Marino, M., Pignatelli, S., Laio, F., Ridolfi, L., 2013. Water footprint of a large-sized food company: the case of Barilla pasta production. Water Resour. Ind. 1–2, 7–24.

SABMiller, WWF-UK, 2009. Water Footprinting: Identifying & Addressing Water Risks in the Value Chain, SABMiller, Woking, UK/WWF-UK. Goldalming, UK.

SABMiller, GTZ, WWF, 2010. Water Futures: Working Together for a Secure Water Future, SABMiller, Woking, UK/WWF-UK, Goldalming, UK.

Sarni, W., 2011. Corporate Water Strategies. Earthscan, London, UK.

Siebert, S., Döll, P., 2010. Quantifying blue and green virtual water contents in global crop production as well as potential production losses without irrigation. J. Hydrol. 384 (3–4), 198–207.

Sikirica, N., 2011. Water Footprint Assessment Bananas and Pineapples. Dole Food Company, Soil & More International, Driebergen, The Netherlands.

Socolof, M.L., Overly, J.G., Kincaid, L.E., Geibig, J.R., 2001. Desktop Computer Displays a Life-Cycle Assessment, Report EPA-744-R-01-004a. United States Environmental Protection Agency, Washington, DC. <www.epa.gov/dfe/pubs/comp-dic/lca> (last accessed 04.08.2014).

Steen-Olsen, K., Weinzettel, J., Cranston, G., Ercin, A.E., Hertwich, E.G., 2012. Carbon, land, and water footprint accounts for the European Union: consumption, production, and displacements through international trade. Environ. Sci. Technol. 46 (20), 10883–10891.

TCCC, TNC, 2010. Product Water Footprint Assessments: Practical Application in Corporate Water Stewardship. The Coca-Cola Company, The Nature Conservancy, Atlanta, GA; Arlington, TX.

UN, 2008. International Standard Industrial Classification of All Economic Activities, Revision 4, Statistics Division. Department of Economic and Social Affairs, United Nations, New York, NY, USA.

Unger, K., Zhang, G., Mathews, R., 2013. Water Footprint Assessment Results and Learning: Tata Chemicals, Tata Motors, Tata Power, Tata Steel. International Finance Corporation, New Delhi, India.

Van Oel, P.R., Hoekstra, A.Y., 2012. Towards quantification of the water footprint of paper: a first estimate of its consumptive component. Water Resour. Manage 26 (3), 733–749.

Verdegem, M.C.J., Bosma, R.H., Verreth, J.A.V., 2006. Reducing water use for animal production through aquaculture. Water Resour. Dev. 22 (1), 101–113.

Wang, L., Ding, X., Wu, X., 2013. Blue and grey water footprint of textile industry in China. Water Sci. Technol. 68 (11), 2485–2491.

WEF, 2014. Global Risks 2014. World Economic Forum, Geneva, Switzerland.

Williams, E.D., Ayres, R.U., Heller, M., 2002. The 1.7 kilogram microchip: energy and material use in the production of semiconductor devices. Environ. Sci. Technol. 36 (24), 5504–5510.

Williams, E.D., Simmons, J.E., 2013. Water in the Energy Industry: An Introduction. BP International, London, UK.

Wuppertal Institute, 2011. Material Intensity of Materials, Fuels, Transport Services, Food, Updated 3 February 2014. Wuppertal Institute, Wuppertal, Germany.

Yu, Y., Hubacek, K., Feng, K., Guan, D., 2010. Assessing regional and global water footprints for the UK. Ecol. Econ. 69 (5), 1140–1147.

Zhang, C., Anadon, L.D., 2014. A multi-regional input–output analysis of domestic virtual water trade and provincial water footprint in China. Ecol. Econ. 100, 159–172.

Zhang, Z., Shi, M., Yang, H., Chapagain, A., 2011. An input–output analysis of trends in virtual water trade and the impact on water resources and uses in China. Econ. Syst. Res. 23 (4), 431–446.

Zhao, X., Chen, B., Yang, Z.F., 2009. National water footprint in an input–output framework—a case study of China 2002. Ecol. Modell. 220 (2), 245–253.

CHAPTER

Life cycle sustainability aspects of microalgal biofuels

George G. Zaimes and Vikas Khanna

Department of Civil and Environmental Engineering, University of Pittsburgh, Pittsburgh, PA, USA

INTRODUCTION

The production of liquid transportation fuels derived from microalgae has gained significant attention over the past several decades. Microalgae's promising characteristics such as high photosynthetic yields, high lipid content, and the ability to utilize waste resources such as wastewater (Zhou et al., 2014) or waste carbon dioxide (CO_2) (Bhatnagar et al., 2011) from industrial activities make them a favorable alternative to traditional terrestrial biofeedstocks for conversion to liquid fuel. Additionally, while the use of some biofeedstocks such as corn or soybean may generate competition between food and fuel uses for crops, algae-derived biofuels do not directly displace or put market pressure on food crops, because high-rate algal ponds can be sited on non arable and marginal lands (Nonhebel, 2012). Furthermore, microalgae can be grown using different growth media, including brackish, saline, and wastewater, which may prove critical for reducing the high-synthetic fertilizer and water footprint associated with microalgae cultivation. Moreover, concerns over fossil fuel price volatility, petroleum supply constraints, and issues of global climate change are key drivers that motivate ongoing research in microalgal biofuel systems (Ferrell and Sarisky-Reed, 2010).

BACKGROUND

From 1978 to 1998, the US Department of Energy's (DOE) aquatic species program extensively studied the process dynamics of algae-to-fuel systems including algal biochemistry, strain selection, and techno economic evaluation of pilot-scale microalgae production (Oswald, 1996). Current issues of peak fossil fuel supply and mounting concerns over global climate change induced via anthropogenic-derived carbon emissions have remotivated research in large-scale biofuel production from microalgae. Common industrial practice for commercial microalgal

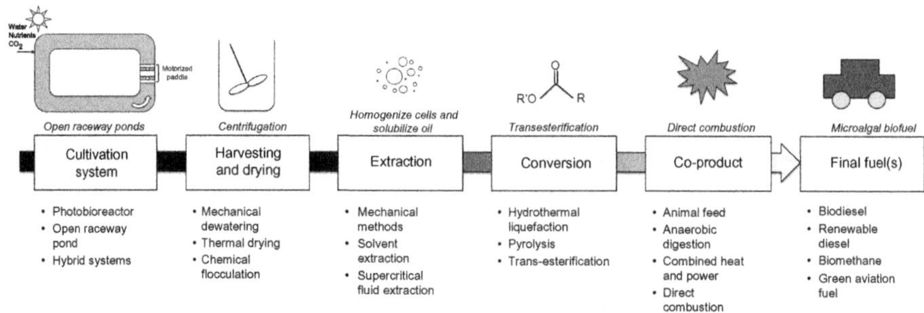

FIGURE 8.1

Generalized microalgal biofuel production chain. Common unit operations/technologies are considered for each stage in the process chain.

biofuel production includes cultivation followed by harvesting, dewatering/drying, extraction, conversion, and co-product utilization. Extensive research and development (R&D) has been invested in technologies and unit operations utilized at each stage in the biomass-to-fuel process chain (Figure 8.1).

As evident from Figure 8.1, there are a host of different harvesting, extraction, and conversion options available to produce various biofuels from microalgal feedstock. Photobioreactors (PBRs) and open raceway ponds (ORP) have emerged as the two predominant systems for mass cultivation of microalgae. PBRs are closed systems and have a high degree of control over reactor operating and environmental conditions. As such, PBRs typically have higher volumetric growth rates, lower risk of contamination and monoculture species degradation, and can cultivate algal strains with higher lipid content as compared with open systems. However, PBRs have exceedingly high operating and capital costs as compared with ORPs (Oswald, 1996) and thus may be better-suited for production of high-valued products such as biopharmaceuticals. Furthermore, studies have suggested that PBRs have higher environmental burdens as compared with ORPs (Stephenson et al., 2010). Accordingly, ORPs appear to be the economically and environmentally favored option for mass cultivation of microalgae for biofuel production. Past research has explored different strategies for microalgal biomass cultivation such as periodically starving the microalgae culture of a nitrogen source; this nitrogen starvation results in increased lipid concentration in the biomass (Cho et al., 2011). Multiple synergistic opportunities including utilizing wastewater or saline/brackish water as the growth media for algae cultivation and the use of industrial flue gas as a source of carbon for algae growth are being investigated. Furthermore, numerous dewatering technologies including centrifugation, filtration, and chemical flocculation have also been explored for microalgae cultivation/harvesting (Weschler et al., 2014). Research has focused on the production of biodiesel from microalgal biomass via transesterification of algal lipids. However, algal lipids can be converted to infrastructure-compatible biofuels (i.e., biofuels that act as drop-in replacements for conventional petroleum fuels) via established hydroprocessing technologies. Furthermore, recent research has investigated the production of green aviation fuel from microalgal feedstock (Fortier et al., 2014). It is important to note that the process options evaluated for microalgal biofuel production vary regarding technological maturity and commercial scalability. Furthermore, ongoing research is investigating which combinations of reactor configuration and processing routes provide the greatest environmental and economic benefits (Handler et al., 2014; Soh et al., 2014).

POLITICAL MOTIVATION

In an effort to combat global climate change, increase domestic energy security, and mitigate the high reliance on foreign oil, the United States congress passed the Energy Independence and Security Act (EISA) in 2007 (Sissine, 2007). This legislation established yearly volume requirements for renewable fuels starting in 2008 and increasing until the year 2022; it is graphically represented in Figure 8.2.

This federal legislation stipulates that a fraction of the total volumetric renewable fuel requirement should be derived from conventional, cellulosic, biomass-based diesel, and advanced biofuels, and it mandates that these biofuels have specific reduction in life cycle greenhouse gas (GHG) emissions relative to petroleum fuels. EISA defines cellulosic biofuel as any renewable fuel derived from cellulose, hemicellulose, or lignin, which reduces GHG emissions by 60% as compared with baseline petroleum fuels. Biomass-based diesel is defined as a renewable transportation fuel, transportation fuel additive, heating oil, or jet fuel that meets the definition of either biodiesel or non ester renewable diesel that

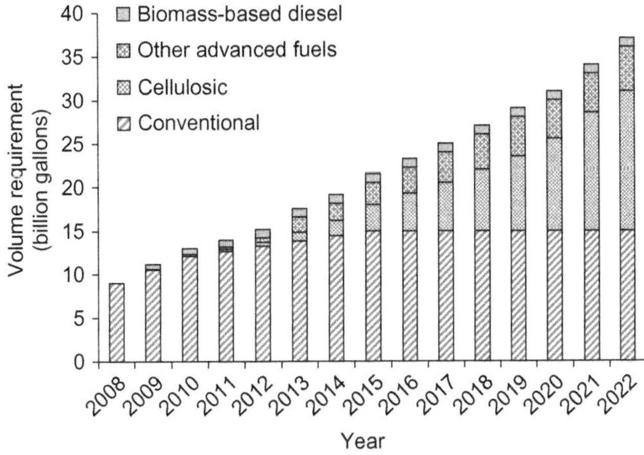

FIGURE 8.2

Renewable fuel standard (RFS2) volume requirement.

reduces GHG emissions by 50% or more. Advanced biofuels are defined as any renewable fuel other than ethanol derived from corn that reduces GHG emissions by a minimum of 50%. Microalgal biofuels represent a promising option for meeting US regulatory renewable fuel targets. However, comprehensive life cycle assessment (LCA) is integral for determining microalgal biofuel pathways that meet the GHG reduction thresholds as set by the Renewable Fuel Standard (RFS2), and for identifying the potential environmental ramifications of commercial microalgal biofuel production before its widespread adoption.

LCA FRAMEWORK

LCA is a tool/technique used to quantify the direct and indirect environmental impacts of a product or service throughout all stages of its life cycle, from raw materials extraction to production and final use (Figure 8.3).

LCA allows for a comprehensive understanding of the environmental impacts that occur at each stage of the production chain. This systems-level analysis allows for the LCA practitioner to identify processes/inputs responsible for highest environmental burden, and thus to target these areas for process improvement and identify process or component alternatives that may provide environmental benefits. LCA can be used to quantify the anticipated impacts of a product or service before its widespread adoption, thus identifying and avoiding potential environmental pollutants, wastes, and environmental damages before they become

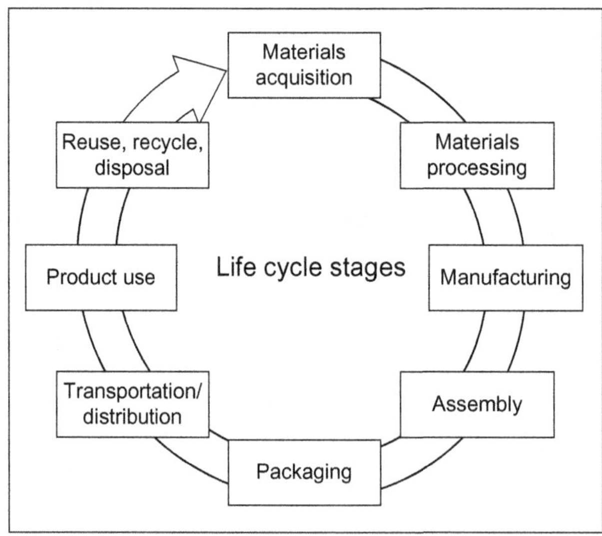

FIGURE 8.3

Primary stages in a product life cycle.

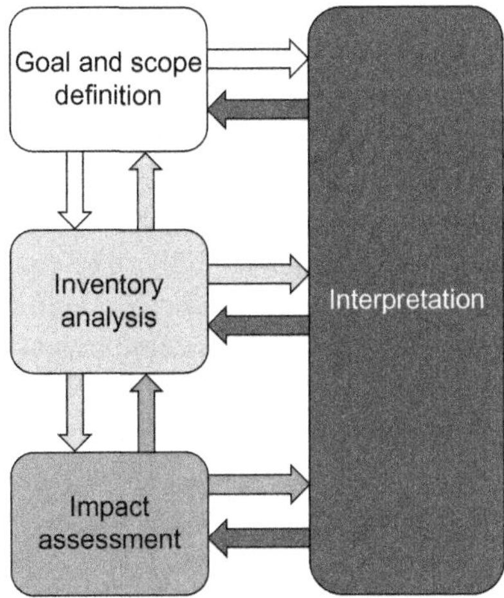

FIGURE 8.4

Life cycle framework.

embedded within the supply chain. Furthermore, LCA is often used to compare the environmental performance between two products with the same functionality, and thus it can be used to inform environmentally conscious decision-making. There are two main types of LCA: attributional LCA and consequential LCA. Attributional LCA (ALCA) quantifies and reports the environmental impacts and burdens associated with a product, service, or process. However, ALCA does not provide information regarding how the physical, material, and energy flows to and from the technosphere/economy may change in response to a change in the product system. Consequential LCA quantifies the environmental impacts and burdens of a decision or a change to a product/system. The choice of attributional or consequential LCA should reflect the underlying goals and objectives of the study. LCA is comprised of four steps: goal and scope definition; life cycle inventory; life cycle impact assessment (LCIA); and interpretation. These steps are discussed in detail in the following sections. An overview of the LCA methodology is provided in Figure 8.4.

GOAL AND SCOPE DEFINITION

The first steps of performing an LCA include defining the purpose of the study, its scope, and the functional unit, as well as the temporal, spatial, and production chain boundaries of the system, process, or product for which the LCA is being conducted. The system boundary definition includes defining all elements, processes, and life cycle stages to be included in the analysis. Several LCAs of algae-to-energy systems have restricted the scope of the analysis to consider only the cultivation, harvesting, and drying/dewatering of microalgae feedstock

(Clarens et al., 2010). However, by restricting the scope of the analysis, these previous studies were able to provide detailed evaluation of the environmental tradeoffs between different processing options and cultivation/harvesting unit operations. As such, the LCA practitioner may choose to restrict the scope of the study via considering only submodules of a larger system to increase the detail and resolution of the analysis. Another important consideration in performing an LCA is defining the functional unit of the study. The functional unit provides a quantitative measure of the function of a product, system, or service and, as such, provides a common basis for which the system/product may be compared or benchmarked. For biofuel systems, the functional unit is typically defined in terms of volume of fuel product per MJ biofuel, VKT, or other relevant quantity.

LIFE CYCLE INVENTORY ANALYSIS

The life cycle inventory analysis step consists of itemizing all inputs (materials and energy resources) and outputs (emissions and wastes to the environment) to and from the product system or process undergoing study. Input and output data are collected and documented for each process contained in the system boundary, including flows of raw materials, energy, products, co-products, wastes, and emissions to air, soil, and water. Data collection can be particularly time-intensive and resource-intensive because it must include all upstream processes (resources extraction, production, and transport) as well as downstream processes (product use and disposal); however, some process data may be available in public or commercial databases, such as Ecoinvent, the Greenhouse Gases Regulated Emissions and Energy Use in Transportation (GREET) model, and US LCI. Once the data are compiled, aggregate resource use and pollutant emissions can be calculated to determine environmental loads and material/energy flows per functional unit.

LIFE CYCLE IMPACT ASSESSMENT

The LCIA stage translates the energy, resource, and emissions flows identified in the LCI into their potential consequences for human health and the environment. It consists of a two-step process of impact classification and quantitative characterization. The classification step links each LCI flow with its related impacts on resource use, human health, and the environment. The characterization step calculates the magnitude of the associated impacts in terms of a reference unit for each category via multiplying the related resource, material, or energy flows with their respective impact factors. Translating the environmental impacts to a reference unit provides a common basis or measure for the generated impact so different emissions and resources can be compared and aggregated using a common unit. For example, fugitive CH_4 emissions from anaerobic digestion of algal biomass may be evaluated for its impact on global warming. The global warming potential (GWP) for CH_4 is measured in respect to its equivalent CO_2 emissions specified over a given time horizon. It is calculated by multiplying the mass of CH_4

released by its corresponding impact factor. Classifying, characterizing, and aggregating related results into a series of indicators provide a basis for quantifying and comparing the impacts of a product or service across multiple domains. For example, the US Environmental Protection Agency (EPA) developed the tool for the reduction and assessment of chemical and other environmental impacts (TRACI) that translates the environmental loads identified by the life cycle inventory into 11 specific impact categories:

- Ozone depletion
- Global warming
- Acidification
- Eutrophication
- Tropospheric ozone (smog) formation
- Ecotoxicity
- Human health criteria-related effects
- Human health cancer effects
- Human health non cancer effects
- Fossil fuel depletion
- Land-use effects.

The choice of LCIA methodology should reflect the underlying goal and scope of the study. Additionally, for processes that produce more than one product, a fraction of the total environmental burdens/impacts may be assigned to the co-product streams via allocation procedures or utilizing system boundary expansion (SBE).

ALLOCATION IN LCA

For a system or process that produces multiple products, there is no generally accepted method regarding how to apportion the environmental impacts between the products (Luo et al., 2009). ISO standards for LCA state that, when possible, the SBE method should be used to handle the joint production of products/co-products (International Organization for Standardization, 2006). In SBE, co-products generated in a process are assumed to displace existing products within the market. The life cycle impacts corresponding to the displaced products are then subtracted from total life cycle impacts. In this regard, the boundaries of the product system are expanded such that it is credited for the environmental impacts that are avoided. However, SBE cannot be performed for products that have no alternatives or substitutes. Furthermore, uncertainty in the substitutability and displacement rate of the co-products can influence the results of SBE. When SBE cannot be used, the environmental impacts should be partitioned between the system's "different products or functions in a way which reflects the underlying physical relationships between them" (International Organization for Standardization, 2006). Several common allocation methods exist, such as weighting the environmental burdens for each product based on its fraction of total mass, energy, or market value. However, the choice of allocation methodology is subjective and can often have a drastic influence on total life cycle impacts. Additionally, the effect of different allocation methodologies may be heightened when there exists a disproportionality between the mass, market value, energy content, or other underlying physical relationship of said co-products (Zaimes et al., 2013). Thus, a hybrid approach using both allocation and SBE may be preferred for systems in which multiple allocations procedures are necessary. It is important to note that SBE is the favored method of the US EPA for determining fuels eligible for the renewable fuel standard (RFS2) (EPA, 2010). Thus, the LCA practitioner's choice of allocation scheme may depend on the regulatory fuel program or legislation as defined by the governing political body.

Microalgal biofuel production provides a unique case study of allocation in LCA attributable to the myriad of co-product options and the inherent differences in the physical properties, energy content, and market value of the co-products. Consequently, the use of some allocation methods may produce distorted results, and not all allocation schemes are relevant to the product system. For example, energy-based allocation may give misleading results for scenarios that produce biofertilizers as co-products, because these products are not valued for their energy content. As such, their heating value is small or negligible as compared with that of other co-products and final fuels. Market-based allocation overcomes some of these shortcomings; however, market imperfections can distort the results of the analysis and current price data are not always available for all products. SBE may be used to avoid allocation; however, SBE cannot be performed for products with no alternatives or substitutes. Furthermore, uncertainty in the substitutability and displacement rate of the co-products can influence the results of SBE. When allocation is unavoidable, the allocation scheme chosen by the LCA practitioner should reflect the intrinsic value or functionally of the product (i.e., for a process producing multiple energy products, energy-based allocation may be preferred). However, if multiple attributes can be ascribed to the product/co-product, then a sensitivity analysis should be conducted to determine the effect of different allocation schemes on the LCA results.

INTERPRETATION

During the interpretation phase of the LCA process, the process is evaluated and validated, conclusions are drawn from the preceding LCI and LCIA, and recommendations are made based on inventory and impact assessment data. In addition, the interpretation phase should identify the data that provide the greatest percent contribution to the impact indicator results, quantify the variability/uncertainty in the LCIA data and the related uncertainty in the impact results, and determine how changes in the LCI or LCIA methods impact the conclusions of the LCA. Validation of the results is essential and may include analysis of variability and uncertainty in the model using a variety of qualitative sensitivity analysis techniques as well as quantitative statistical procedures. The results from the LCA may identify processes that are responsible for the greatest environmental impacts, and

Table 8.1 Summary of Key Parameters from Select Prior Microalgal Biofuel LCAs

Study	Functional Unit	Life Cycle Data Sources	Co-products	Impacts
Stephenson et al. (2010)	1 t biodiesel	US LCI	Electricity	GWP, energy use, water use
Campbell et al. (2011)	1 km diesel truck	Australian LCI	Electricity	GWP, energy use, land use
Jorquera et al. (2010)	1 t dry solids	Literature review	None	Energy use
Clarens et al. (2010)	317 GJ biomass	EcoInvent	None	GWP, land use, eutrophication, water use, energy use
Lardon et al. (2009)	1 MJ fuel	EcoInvent	Glycerol	Abiotic depletion, acidification, eutrophication, GWP, ODP, human toxicity, marine toxicity, land use, ionizing radiation, and photochemical oxidation
Zaimes and Khanna (2013b)	1 MJ fuel	EcoInvent & US LCI	Biofertilizer, electricity, heat	GWP, energy use

From Clarens and Colosi (2013).

target these areas for process improvement to increase the environmental performance of the product system. Additionally, comparing the LCA results regarding existing published literature is pivotal for validating the results and for appropriately benchmarking the environmental performance of the product system. However, biofuel LCAs often utilize diverse functional units and different life cycle data sources, consider different process and co-product scenarios, and may evaluate different environmental impact categories (Table 8.1). Thus, the LCA results may not be directly comparable between studies.

SUSTAINABILITY METRICS

ENERGY METRICS

A host of energy metrics has been utilized in the fuel/energy literature to quantify the viability of fuel production. These metrics quantify the amount of *primary energy* consumed throughout the fuel supply chain. Primary energy refers to any naturally occurring renewable or non renewable energy form, such as crude oil, coal, natural gas, uranium, solar, wind, tidal, biomass, and geothermal energy, that has not been subjected to any conversion or transformation process. Several principle energy metrics are defined and provided in Table 8.2.

One key metric often considered in biofuel analysis is net energy balance (NEB). NEB is defined as the difference between the energy value of the output fuel and the total primary energy consumed in producing the fuel. As such, a positive NEB is one important criterion for an environmental sustainable transportation fuel, because it indicates that more energy is produced than is consumed via the system. Energy return on investment (EROI) and net energy ratio (NER) are two other widely reported metrics and represent the ratio of the energy of the final fuel to the direct and indirect primary energy required for its production. Thus, if EROI and NER values are less than unity, then the system has a negative NEB. A variation of EROI known as fossil energy ratio (FER) or $EROI_{fossil}$ considers only the consumption of primary fossil energy throughout the fuel supply chain, and thus measures how much fuel product is generated per unit investment of primary fossil resources. As such, FER provides a surrogate measure for the *renewability* of the biofuel. Accordingly, FER values more than 1 are desirable, because more energy is produced via the biofuel than the fossil energy consumed throughout the supply chain. It is important to note that the denominator in EROI, FER, and NER does not include the feedstock energy (i.e., for algal biofuel production, the feedstock energy corresponds to direct sunlight utilized in biomass growth, whereas for petroleum fuels the feedstock energy corresponds to the energy of crude oil). Research has suggested that the large-scale adoption of alternative fuels with lower EROI relative to traditional transportation fuels could have long-standing societal implications (Hall et al., 2014; Lambert et al., 2014), because a

Table 8.2 Key Energy Metrics Utilized in Prior Microalgal Biofuel LCAs

Metric	Abbreviation	Equation	LCA Study
Net Energy Balance	NEB	$\text{Energy}_{output} - \text{Primary Energy}_{input}$	Batan et al. (2010), Razon and Tan (2011)
Energy Return on Investment[a]	EROI	$\frac{\text{Energy}_{output}}{\text{Primary Energy}_{input}}$	Clarens et al. (2011), Zaimes and Khanna (2013a,b)
Net Energy Ratio	NER	$\frac{\text{Energy}_{output}}{\text{Primary Energy}_{input}}$	Batan et al. (2010)
Fossil Energy Ratio[a]	FER	$\frac{\text{Energy}_{output}}{\text{Primary Fossil Energy}_{input}}$	Xu et al. (2011), Zaimes and Khanna (2013a)

[a]*EROI that considers only the primary fossil energy consumed in fuel product (typically denoted as $EROI_{fossil}$) is equivalent to FER.*

larger portion of useful work must be diverted from the economy for biofuel production and thus cannot be used to sustain other economic actives. Furthermore, previous studies have proposed that a liquid technical fuel must have a minimum EROI of 3 to support the US transportation system (Hall et al., 2009).

CARBON FOOTPRINT

Carbon footprint for a fuel is the sum of direct and indirect GHG emissions that occur throughout all stages of the fuel life cycle. GHG emissions are measured in terms of their CO_2 equivalents, and are calculated over a specific time horizon (typically 100 y). It is important to note that the combustion of biofuels releases biogenic carbon (i.e., atmospheric carbon that is captured via biomass in the carbon cycle). As such, the "use phase" or combustion of biofuels is considered to be carbon-neutral because it does not contribute to the net release of CO_2, that is, it does not increase the overall CO_2 concentration in the atmosphere. Contrarily, the combustion of petroleum fuels emits carbon that has previously been stored in geologic formations, thus increasing the total CO_2 concentration in the atmosphere. A biofuels carbon footprint is often discussed regarding its GHG reduction potential, defined as the percent reduction in life cycle GHG emissions relative to baseline petroleum fuels or another reference fuel source or product.

WATER FOOTPRINT (THE DETAILED DESCRIPTION AND ASSESSMENT HAVE BEEN PROVIDED IN CHAPTER 7)

Water resources are critical to the proper functioning of ecological services and are vital to human welfare and global economic well-being. In recent years, increased population growth, human activity, and economic development have resulted in heightened water consumption and degradation of freshwater resources (Vörösmarty et al., 2000). As such, issues of water scarcity, water consumption, and water quality have raised significant environmental and sustainability concerns. However, multiple definitions of water footprint exist, and understanding the differences between these terms is crucial for a broader and deeper understanding of the current methodological approach in quantifying water footprint.

Water footprint is a measure of the direct and indirect freshwater used in producing products and/or services. The water footprint is expressed in terms of the total volume of freshwater that is *consumed* and/or *polluted* throughout the entire supply chain. *Water consumption* is defined as the volume of freshwater that is evaporated or incorporated into a product and includes any abstracted surface or groundwater that is not returned to the same water resource system from which it was withdrawn. Researchers express the water footprint in terms of three separate components: the blue water footprint, green water footprint, or gray water footprint.

- Blue water footprint: The volume of surface and groundwater *consumed* throughout the production of a good or service.
- Green water footprint: The volume of rainwater *consumed* throughout the production of a good or service. For agricultural and bio-based products, the green water footprint refers to the total rainwater evapotranspiration and additional rainwater that may be incorporated into the biomass.
- Gray water footprint: The volume of freshwater that is required to dilute the load of pollutants, such that the resulting water meets or exceeds existing ambient water quality standards.

The total water footprint is equal to the sum of the blue, green, and gray water footprints. Furthermore, the water footprint is spatially explicit and is dependent on the system boundary of the analysis. Studies have indicated that the production of biomass for both food and fiber accounts for more than 85% of worldwide freshwater use (Worldwatch Institute, 2007). Increasing demands for food/crops as well as a global shift toward incorporating a greater share of biofuels into the current transportation fuel mix could place additional strain on global watersheds and water resources. Figure 8.5 plots the water footprint and algal growth rates for producing algal biomass via an ORP for different cultivation locations. Analysis shows that the areal biomass growth rates as well as the corresponding water footprint are highly spatially dependent. It is important to note that regions with high areal growth rates (such as Phoenix, AZ) may also have a correspondingly high water footprint and may be prone to issues of water scarcity/availability. Thus, relationships between algal growth potential, water footprint, and regional water scarcity should be considered when siting and developing large-scale microalgal biorefineries.

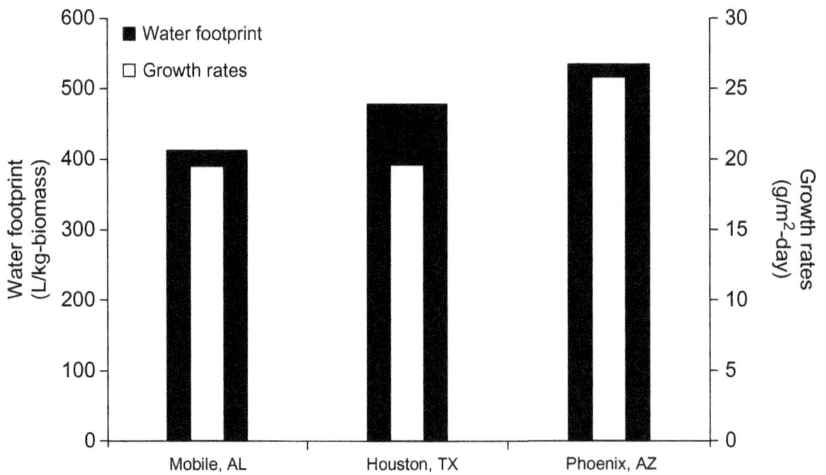

FIGURE 8.5

Location-specific biomass growth rates and water footprint. Based on Zaimes and Khanna (2013a).

CASE STUDY

A comparative well-to-wheel LCA was conducted to investigate the effects of different combinations of co-product scenarios and allocation schemes on the environmental sustainability of microalgal-derived biodiesel and renewable diesel produced via an integrated ORP biorefinery. The functional unit was assumed to be 1 MJ of microalgal fuel. Infrastructure-related environmental impacts were not considered in this analysis as prior studies have shown that their contribution is negligible as compared with operational impacts. When possible life cycle data were obtained from US Life Cycle Inventory (LCI), other life cycle data were obtained from the Ecoinvent database. Additionally, Intergovernmental Panel on Climate Change characterization factors assuming a 100-y time horizon were used to determine the life cycle GHG emissions of material and energy flows developed in the life cycle inventory, whereas the cumulative energy demand (CED) methodology was used to quantify primary energy consumption, and the TRACI was used to quantify the impacts of biofuel production on the following eight impact categories: ozone depletion; smog formation; acidification; eutrophication; carcinogenics; non carcinogenics; respiratory effects; and ecotoxicity.

The major components of the modeled algae-to-fuel production chain include cultivation, harvesting, lipid extraction, and fuel upgrading/co-product scenarios. The cultivation of freshwater microalgae *Chlorella vulgaris* is modeled in polyvinyl chloride (PVC)-lined ORPs. Synthetic fertilizers and industrial flue gas delivered via a co-located natural gas power plant provide were considered as sources of nutrients and carbon for algal growth, respectively. The molecular composition of the biomass was obtained from prior peer-reviewed literature (Lardon et al., 2009); the mean fractionated biomass composition was assumed to be 25% lipids, 28% carbohydrates, and 47% proteins. Algal growth rates were constructed based on values of solar insolation and meteorological data averaged over a 30-y period obtained from the National Solar Radiation Database and National Oceanic and Atmospheric Association (NOAA), with an average growth rate of 23.51 g/m^2 per day over an 8-month production cycle. The harvesting submodel consists of flocculation via the addition of aluminum sulfate followed by dewatering via chamber filter presses. It was assumed that the resultant biomass has a solids concentration of 25% weight per weight (w/w). After harvesting, the lipid portion of the algal biomass is separated from the remaining protein and carbohydrate fractions via counter-current circulation of *n*-hexane. It is important to note that there is high technological uncertainty regarding wet extraction of algal biomass because this process option has yet to be demonstrated on a commercial scale. As such, this fuel pathway represents optimistic or projected future algal conversion technologies. Algal lipids are then either hydroprocessed to produce algal-derived renewable diesel II and co-product propane or transesterified to produce algal biodiesel as well as glycerin. Several co-product options are evaluated for the residual deoiled algal biomass (RDB) including: use as an animal feed; anaerobic digestion of RDB to produce biogas and biofertilizer; and combustion of RDB via combined heat and power (CHP) to provide heat and electricity. Additionally, it was assumed that the liquid portion of the anaerobic digestate would be recycled back to the ORP to reduce the amount of synthetic fertilizers required for biofuel production. The modeled production chain is shown in Figure 8.6.

Market allocation, energy-based allocation, and SBE were considered for allocating and apportioning the environmental impacts between primary fuel products and co-product streams. For market-based allocation, price data for each of the products were normalized over a 12-month period to mitigate the effects of price/market fluctuations. When current price data were not available, a price inflator was used to adjust existing price data to the 2013 market value. Additionally, because no commercial algal biorefinery is currently operational in the United States, the market price of algal fuel and co-products were estimated based on the economic value of existing products that provide the same functionality. For example, the economic value of algal-derived animal feed was constructed based on the market value of soybean meal and the relative protein content of the two feedstocks. For energy-based allocation, the energy content of the fuels and co-products were obtained from technical literature as well as direct calculations via heat and energy balances. For SBE, it was assumed

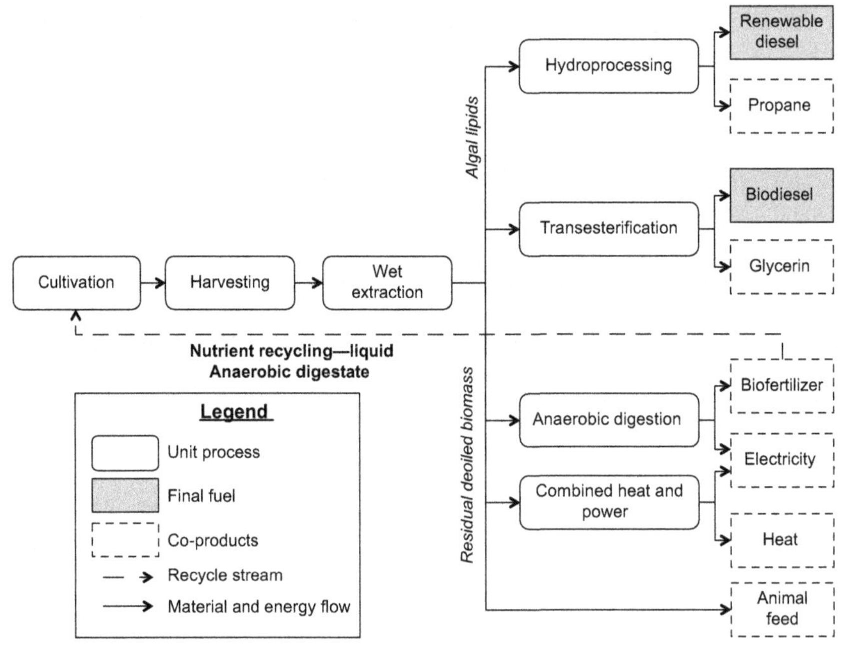

FIGURE 8.6

Modeled algae-to-fuel system.

that algal-derived animal feed would displace soybean meal, co-product glycerin would displace traditional crude petroleum glycerin, and co-product propane would displace liquid petroleum gas (LPG). Additionally, co-product electricity was assumed to displace the US average electricity generation mix. Heat generated from CHP was assumed to displace heat that would have otherwise been generated via natural gas. Additionally, the solids portion of the anaerobic digestate is used as a nitrogen (N) and phosphorous (P) fertilizer supplement. It was assumed that the N and P biofertilizers would displace urea and superphosphate, respectively.

RESULTS

The environmental impacts of producing microalgal-derived renewable diesel and biodiesel for several production scenarios consisting of a combination of fuel conversion scenarios, co-product options, and allocation schemes were evaluated. The environmental impacts were evaluated for the following 10 categories: ozone depletion; global warming; smog; acidification; eutrophication; carcinogens; non carcinogens; respiratory effects; ecotoxicity; and energy use. Additionally, $EROI_{fossil}$ and life cycle GHG emissions were quantified for each production pathway. To increase the transparency of the results, the impacts for microalgal fuels for the 10 impact categories were normalized relative to petroleum diesel, therefore highlighting the relative energy intensity and environmental burden associated with microalgal fuel production. Thus, for the specified environmental impact categories, values greater than unity indicate that microalgal fuels have a larger impact in that category as compared with petroleum diesel. The results of the analysis are provided in Table 8.3.

The results from Table 8.3 indicate that microalgal-derived BD and RD have higher impacts in 7 out of the 10 examined impact categories relative to petroleum diesel across all examined production pathways. The results reveal that microalgal biofuels have significantly higher impacts in ozone depletion and eutrophication as compared with petroleum diesel; this is primarily attributable to the high upstream impacts of N fertilizer production (analysis not shown). The results indicate that microalgal biofuels may provide minor benefits in ecotoxicity and global warming relative to petroleum fuels; however, these results are highly sensitive to production pathway and allocation scheme. Additionally, the results show that for a fixed fuel and co-product option, the choice of allocation scheme can have a significant impact on the LCA results. The results from Table 8.3 indicates that only 2 out of the 18 examined pathways have the requisite GHG reductions as set by the RFS2, and that life cycle GHG emission range from 35 to 141 g CO_2 eq/MJ fuel over the span of scenarios analyzed. Additionally, only four out of the 18 examined possible scenarios have an $EROI_{fossil}$ more than 1; however, these scenarios are only marginally net energy-positive. Additionally, the $EROI_{fossil}$ for biofuel production ranges from 0.50 to 2.00 over the pathways considered in this analysis. Given the large variability in the environmental and energetic performance of algae, careful consideration of the

Table 8.3 Environmental Impacts and Sustainability Performance Metrics for Microalgae Biofuel Production

Parameters	Fuel	Co-product	Allocation	Ozone Depletion[a]	Global Warming[a]	Smog[a]	Acidification[a]	Eutrophication[a]	Carcinogenics[a]	Non carcinogenics[a]	Respiratory Effects[a]	Ecotoxicity[a]	Energy Use[a,b]	EROI $_{fossil}$	Life Cycle GHG Emissions[c]
Scenario 1	RD	AF	Disp.	2,662.1	1.4	2.5	4.8	−87.3	14.5	25.2	19.3	1.6	8.6	0.51	117
	BD	AF	Disp.	3,164.4	1.2	2.0	4.4	−85.4	14.6	25.2	18.9	1.6	8.7	0.50	108
	RD	AF	Market	1,428.7	0.7	1.6	2.5	33.3	5.6	4.3	8.1	0.5	3.7	1.16	62
	BD	AF	Market	1,515.7	0.7	1.6	2.5	33.7	5.6	4.3	8.2	0.5	3.8	1.13	63
	RD	AF	Energy	1,651.7	0.8	1.8	2.9	38.5	6.4	5.0	9.4	0.6	4.3	1.00	72
	BD	AF	Energy	1,733.9	0.8	1.8	2.9	38.6	6.4	4.9	9.4	0.6	4.3	0.98	72
Scenario 2	RD	AD	Disp.	1,694.4	1.1	2.1	3.8	49.2	11.9	11.1	13.9	1.1	5.0	0.83	95
	BD	AD	Disp.	2,197.4	1.0	1.5	3.4	51.0	12.0	11.1	13.4	1.1	5.2	0.81	85
	RD	AD	Market	2,525.2	1.6	3.3	5.6	71.8	12.1	9.5	17.6	1.1	7.9	0.54	141
	BD	AD	Market	2,694.2	1.6	3.3	5.6	72.0	12.0	9.4	17.6	1.1	8.0	0.53	141
	RD	AD	Energy	2,183.9	1.4	2.8	4.9	62.1	10.5	8.2	15.3	1.0	6.9	0.63	122
	BD	AD	Energy	2,302.4	1.4	2.8	4.8	61.5	10.3	8.0	15.0	1.0	6.9	0.63	121
Scenario 3	RD	CHP	Disp.	296.3	0.5	2.0	3.7	54.7	12.3	11.3	14.9	1.1	2.7	2.00	44
	BD	CHP	Disp.	800.2	0.4	1.4	3.3	56.5	12.4	11.3	14.5	1.1	2.9	1.87	35
	RD	CHP	Market	3,091.0	1.5	3.2	5.3	71.9	12.0	9.3	17.6	1.1	7.9	0.54	133
	BD	CHP	Market	3,244.9	1.5	3.3	5.3	72.1	11.9	9.2	17.5	1.1	8.0	0.53	133
	RD	CHP	Energy	1,845.7	0.9	1.9	3.2	43.0	7.2	5.6	10.5	0.7	4.7	0.90	79
	BD	CHP	Energy	1,933.3	0.9	1.9	3.2	42.9	7.1	5.5	10.4	0.6	4.8	0.89	79

[a]*Impacts are normalized to petroleum diesel. Thus, values more than 1 indicate that microalgal biofuels have higher impacts in the specific catergory relative to petroleum diesel.*

[b]*Considers both primary renewable and nonrenewable energy consumption.*

[c]*Life cycle GHG emissions are reported in units of gCO_2 eq/MJ biofuel.*

algae-to-fuel supply chain and appropriate LCA allocation procedures are required to accurately benchmark and guide the sustainable development of emerging large-scale microalgal biofuel systems.

WATER FOOTPRINT OF BIOFUELS

Prior studies have shown that the production of biofuels/bioenergy is water resource-intensive. Therefore, large-scale cultivation and commercial production of biofuels and bioproducts may deplete freshwater resources that would otherwise be used for food and crop production. As such, understanding and quantifying the water footprint of emerging biofuels/bioproducts are critical for guiding the environmentally sustainable development of the nascent biofuels/bioenergy industry. Furthermore, because regions with high water scarcity are likely to be more affected by stress on water resources, quantifying the impacts of biofuel production on water resources at multiple spatial resolutions is critical for holistic understanding of the broader implications of biofuel/bioproducts on water and ecological resources. An overview of the water footprint for select biofeedstocks, biofuels, and petroleum fuels is provided in Table 8.4.

The results from Table 8.4 show that there is a large range in the reported water footprint of different bioenergy crops as well as biofuels. The results suggest that the water footprint for microalgae biofuels is comparable with other

Table 8.4 Water Footprint for Select Biofeedstocks and Fuels

Feedstock	Water Footprint (L/MJ)	Reference
Fossil-petroleum		
Extraction	2.77–11	Harto et al. (2010)
Refining	22.2–41.6	Harto et al. (2010)
Bioethanol		
Sugarcane	38–156	Mulder et al. (2010)
Corn	73–346	Mulder et al. (2010)
Wheat	40–351	Mulder et al. (2010)
Sugar beet	71–188	Harto et al. (2010)
Lignocellulosic ethanol	11–171	Harto et al. (2010)
Biodiesel		
Rapeseed	100–175	Mulder et al. (2010)
Microalgae biomass[a]	20.8–38.8	Zaimes and Khanna (2013a)
Soybean biodiesel	1.5–91.74	Harto et al. (2010)
Algal biodiesel	23.7–62.3	Subhadra and Edwards (2011)

[a]*Considers only the direct component of the water footprint (i.e., the embodied water consumption for material and energy inputs is not included in the analysis).*

leading biofuels such as soybean biodiesel and lingocellulosic ethanol. However, wastewater reclamation/water reuse strategies have the potential to dramatically reduce the water footprint of algal biofuels.

COMPARISON OF PRIOR MICROALGAL BIOFUEL LCAs

Prior LCA studies of algae-to-fuel systems have shown mixed results regarding the environmental sustainability and overall performance of microalgae-derived fuels—often a consequence of variation in system boundaries, model assumptions, processing technologies, design parameters, and allocation schemes. To benchmark the results of the case study, the maximum and minimum EROI and life cycle GHG emissions as reported by prior microalgal biofuel LCA were evaluated and are graphically represented in Figures 8.7 and 8.8. The results from Figure 8.7 indicate that there is high variability in the range of EROI values reported from prior microalgal biofuel LCA studies. The results suggest that the EROI for microalgal fuels may be as high as 5.88. However, the overwhelming majority of prior studies have shown that in the best-case scenario, the EROI for microalgal fuels does not exceed 3, which is the critical value required to sustain the US transportation system (Hall et al., 2009). The results shown in Figure 8.8 indicate that there is also high

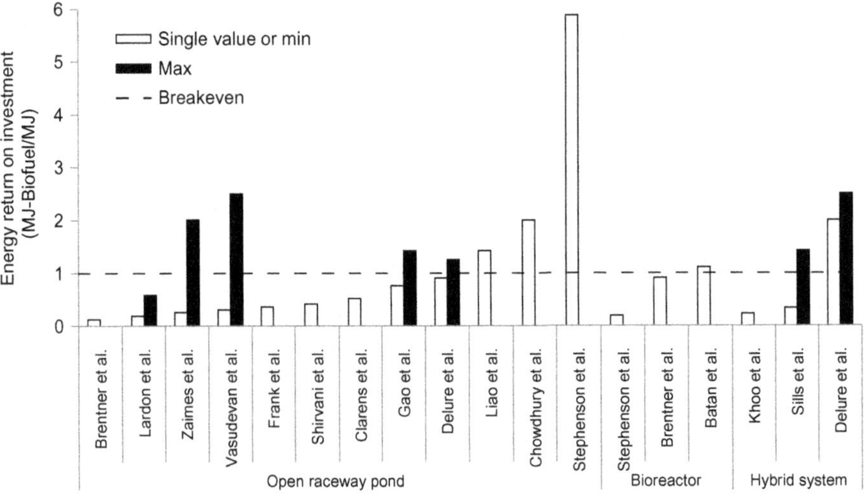

FIGURE 8.7

Reported EROI from select microalgae biofuel LCAs.
Values for EROI are adapted from Gao et al. (2013).
Prior LCA studies include Lardon et al. (2009), Stephenson et al. (2010), Chowdhury et al. (2012), Zaimes and Khanna (2013b), Gao et al. (2013), Vasudevan et al. (2012), Delrue et al. (2012), Shirvani et al. (2011), Yanfen et al. (2012), Clarens et al. (2011), Brentner et al. (2011), Batan et al. (2010), Sills et al. (2012), and Khoo et al. (2011).

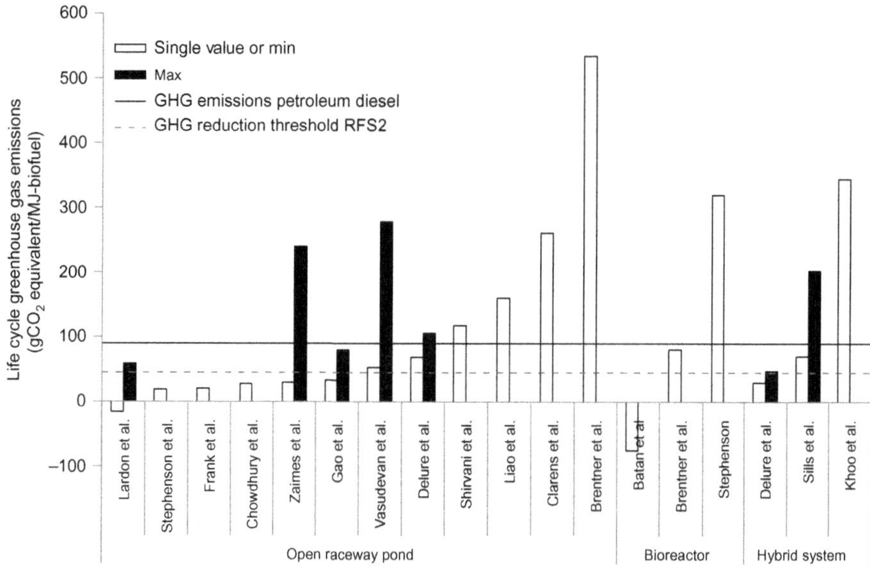

FIGURE 8.8

Reported life cycle GHG emissions from select microalgae biofuel LCAs.
Values for life cycle GHG emissions are adapted from Gao et al. (2013).
Prior LCA studies include Lardon et al. (2009), Stephenson et al. (2010), Chowdhury et al. (2012), Zaimes and Khanna (2013b), Gao et al. (2013), Vasudevan et al. (2012), Delrue et al. (2012), Shirvani et al. (2011), Yanfen et al. (2012), Clarens et al. (2011), Brentner et al. (2011), Batan et al. (2010), Sills et al. (2012), and Khoo et al. (2011).

variability in the reported life cycle GHG emissions as estimated by prior microalgae biofuel LCAs. The results reveal that only two of the prior LCA studies have shown that microalgae biofuels are carbon-negative. Additionally, the results suggest that in the best-case scenario, microalgal biofuels have the potential to meet the GHG reduction requirements as set by the RFS2. However, it is important to note that the best-case scenarios typically consider futuristic or optimal process technologies and, as such, have high technological uncertainty. These comparative results correlate well with the results of the case study. Furthermore, as identified in the case study, the high variability in the EROI and life cycle GHG emissions of prior biofuel studies may be attributed, in part, to different co-product options and allocation schemes considered in the analysis.

CONCLUSIONS

This work reviewed the application of LCA for environmental evaluation of emerging microalgal biofuel systems. Analysis of the biomass-to-fuel supply chain from a life cycle perspective is pivotal for identifying the potential environmental impacts of emerging fuel platforms. Furthermore, by comparing the impacts of emerging microalgal systems regarding key LCA-based energy, water, and climate change performance indicators, the environmental sustainability of microalgal systems can be benchmarked and compared with other leading commercial biofuels and petroleum fuels. The life cycle methodology has been applied to US federal legislation, such as the renewable fuels standard (RFS2), for setting specific sustainability thresholds for emerging renewable fuels. As such, LCA has emerged as a core tenant of environmental sustainability discourse.

REFERENCES

Batan, L., Quinn, J., Willson, B., Bradley, T., 2010. Net energy and greenhouse gas emission evaluation of biodiesel derived from microalgae. Environ. Sci. Technol. 44, 7975–7980.

Bhatnagar, A., Chinnasamy, S., Singh, M., Das, K.C., 2011. Renewable biomass production by mixotrophic algae in the presence of various carbon sources and wastewaters. Appl. Energy 88, 3425–3431.

Brentner, L.B., Eckelman, M.J., Zimmerman, J.B., 2011. Combinatorial life cycle assessment to inform process design of industrial production of algal biodiesel. Environ. Sci. Technol. 45, 7060–7067.

Campbell, P.K., Beer, T., Batten, D., 2011. Life cycle assessment of biodiesel production from microalgae in ponds. Bioresour. Technol. 102 (1), 50–56.

Cho, S., Lee, D., Luong, T.T., Park, S., Oh, Y.K., Lee, T., 2011. Effects of carbon and nitrogen sources on fatty acid contents and composition in the green microalga, *Chlorella* sp 227. J. Microbiol. Biotechnol. 21, 1073–1080.

Chowdhury, R., Viamajala, S., Gerlach, R., 2012. Reduction of environmental and energy footprint of microalgal biodiesel production through material and energy integration. Bioresour. Technol. 108, 102–111.

Clarens, A., Colosi, L., 2013. Life cycle assessment of algae-to-energy systems. Advanced Biofuels and Bioproducts. Springer Science + Business Media, New York, NY, USA.

Clarens, A.F., Resurreccion, E.P., White, M.A., Colosi, L.M., 2010. Environmental life cycle comparison of algae to other bioenergy feedstocks. Environ. Sci. Technol. 44, 1813–1819.

Clarens, A.F., Nassau, H., Resurreccion, E.P., White, M.A., Colosi, L.M., 2011. Environmental impacts of algae-derived biodiesel and bioelectricity for transportation. Environ. Sci. Technol. 45, 7554–7560.

Delrue, F., Setier, P.A., Sahut, C., Cournac, L., Roubaud, A., Peltier, G., Froment, A.K., 2012. An economic, sustainability, and energetic model of biodiesel production from microalgae. Bioresour. Technol. 111, 191–200.

EPA, 2010. Renewable Fuel Standard Program (RFS2) Regulatory Impact Analysis. US Environmental Protection Agency, Washington, DC, USA.

Ferrell, J. Sarisky-Reed, V., 2010. National algal biofuels technology roadmap, A technology roadmap resulting from the National Algal Biofuels Workshop. 140.

Fortier, M.-O.P., Roberts, G.W., Stagg-Williams, S.M., Sturm, B.S.M., 2014. Life cycle assessment of bio-jet fuel from hydrothermal liquefaction of microalgae. Appl. Energy 122, 73–82.

Gao, X., Yu, Y., Wu, H., 2013. Life cycle energy and carbon footprints of microalgal biodiesel production in Western Australia: a comparison of byproducts utilization strategies. ACS Sustain. Chem. Eng. 1 (11), 1371–1380.

Hall, C.A., Balogh, S., Murphy, D.J., 2009. What is the minimum EROI that a sustainable society must have? Energies 2, 25–47.

Hall, C.A.S., Lambert, J.G., Balogh, S.B., 2014. EROI of different fuels and the implications for society. Energy Policy 64, 141–152, http://dx.doi.org/10.1016/j.enpol.2013.05.049.

Handler, R.M., Shonnard, D.R., Kalnes, T.N., Lupton, F.S., 2014. Life cycle assessment of algal biofuels: influence of feedstock cultivation systems and conversion platforms. Algal Res. 4, 105–115.

Harto, C., Meyers, R., Williams, E., 2010. Life cycle water use of low-carbon transport fuels. Energy Policy 38 (9), 4933–4944.

International Organization for Standardization, 2006. ISO 14040 Environmental management—Life cycle assessment—Principles and framework. Bern, Switzerland.

Jorquera, O., Kiperstok, A., Sales, E.A., Embiruçu, M., Ghirardi, M.L., 2010. Comparative energy life-cycle analyses of microalgal biomass production in open ponds and photobioreactors. Bioresour. Technol. 101 (4), 1406–1413.

Khoo, H.H., Sharratt, P.N., Das, P., Balasubramanian, R.K., Naraharisetti, P.K., Shaik, S., 2011. Life cycle energy and CO(2) analysis of microalgae-to-biodiesel: preliminary results and comparisons. Bioresour. Technol. 102, 5800–5807.

Lambert, J.G., Hall, C.A.S., Balogh, S., Gupta, A., Arnold, M., 2014. Energy, EROI and quality of life. Energy Policy 64, 153–167, http://dx.doi.org/10.1016/j.enpol.2013.07.001.

Lardon, L., Hélias, A., Sialve, B., Steyer, J.-P., Bernard, O., 2009. Life-cycle assessment of biodiesel production from microalgae. Environ. Sci. Technol. 43, 6475–6481.

Luo, L., van der Voet, E., Huppes, G., Udo de Haes, H., 2009. Allocation issues in LCA methodology: a case study of corn stover-based fuel ethanol. Int. J. Life Cycle Assess. 14, 529–539.

Mulder, K., Hagens, N., Fisher, B., 2010. Burning water: a comparative analysis of the energy return on water invested. Ambio 39 (1), 30–39.

Nonhebel, S., 2012. Global food supply and the impacts of increased use of biofuels. Energy 37, 115–121.

Oswald, B.J., 1996. Systems and Economic Analysis of Microalgae Ponds for Conversion of CO_2 to Biomass—Final Report. Pittsburgh, PA, USA.

Razon, L.F., Tan, R.R., 2011. Net energy analysis of the production of biodiesel and biogas from the microalgae: Haematococcus pluvialis and Nannochloropsis. Appl. Energy 88 (10), 3507–3514.

Shirvani, T., Yan, X., Inderwildi, O.R., Edwards, P.P., King, D.A., 2011. Life cycle energy and greenhouse gas analysis for algae-derived biodiesel. Energy Environ. Sci. 4, 3773–3778.

Sills, D.L., Paramita, V., Franke, M.J., Johnson, M.C., Akabas, T.M., Greene, C.H., Tester, J.W., 2012. Quantitative uncertainty analysis of life cycle assessment for algal biofuel production. Environ. Sci. Technol. 47, 687–694.

Sissine, F., 2007. Energy Independence and Security Act of 2007: A Summary of Major Provisions. Washington, DC, USA: Congressional Research Service Report for Congress, Order Code RL34294.

Soh, L., Montazeri, M., Haznedaroglu, B.Z., Kelly, C., Peccia, J., Eckelman, M.J., Zimmerman, J.B., 2014. Evaluating microalgal integrated biorefinery schemes: empirical controlled growth studies and life cycle assessment. Bioresour. Technol. 151, 19–27.

Stephenson, A.L., Kazamia, E., Dennis, J.S., Howe, C.J., Scott, S.A., Smith, A.G., 2010. Life-cycle assessment of potential algal biodiesel production in the United Kingdom: a comparison of raceways and air-lift tubular bioreactors. Energy Fuels 24, 4062–4077.

Subhadra, B.G., Edwards, M., 2011. Coproduct market analysis and water footprint of simulated commercial algal biorefineries. Appl. Energy 88 (10), 3515–3523.

Vasudevan, V., Stratton, R.W., Pearlson, M.N., Jersey, G.R., Beyene, A.G., Weissman, J.C., Rubino, M., Hileman, J.I., 2012. Environmental performance of algal biofuel technology options. Environ. Sci. Technol. 46, 2451–2459.

Vörösmarty, C.J., Green, P., Salisbury, J., Lammers, R.B., 2000. Global water resources: vulnerability from climate change and population growth. Science 289, 284.

Weschler, M.K., Barr, W.J., Harper, W.F., Landis, A.E., 2014. Process energy comparison for the production and harvesting of algal biomass as a biofuel feedstock. Bioresour. Technol. 153, 108–115.

Worldwatch Institute, 2007. Biofuels for Transport: Global Potential and Implications for Sustainable Energy and Agriculture. Earthscan, Worldwatch Inst., Washington, DC, USA.

Xu, L., Brilman, D.W.F., Withag, J.A.M., Brem, G., Kersten, S., 2011. Assessment of a dry and a wet route for the production of biofuels from microalgae: energy balance analysis. Bioresour. Technol. 102 (8), 5113–5122.

Yanfen, L., Zehao, H., Xiaoqian, M., 2012. Energy analysis and environmental impacts of microalgal biodiesel in China. Energy Policy 45, 142–151.

Zaimes, G.G., Khanna, V., 2013a. Microalgal biomass production pathways: evaluation of life cycle environmental impacts. Biotechnol. Biofuels 6, 88.

Zaimes, G.G., Khanna, V., 2013b. Environmental sustainability of emerging algal biofuels: a comparative life cycle evaluation of algal biodiesel and renewable diesel. Environ. Prog. Sustain. Energy 32, 926–936.

Zaimes, G., Borkowski, M., Khanna, V., 2013. Life-cycle environmental impacts of biofuels and co-products. In: Gupta, V.K., Tuohy, M.G. (Eds.), Biofuel Technologies. Springer, Berlin/Heidelberg, Germany.

Zhou, W., Chen, P., Min, M., Ma, X., Wang, J., Griffith, R., Hussain, F., Peng, P., Xie, Q., Li, Y., Shi, J., Meng, J., Ruan, R., 2014. Environment-enhancing algal biofuel production using wastewaters. Renew. Sustain. Energy Rev. 36, 256–269.

CHAPTER

Methods and tools for sustainable chemical process design

Carina L. Gargalo*, Siwanat Chairakwongsa†, Alberto Quaglia*, Gürkan Sin*, and Rafiqul Gani*

**CAPEC, Department of Chemical and Biochemical Engineering, Technical University of Denmark, Lyngby, Denmark †The Petroleum and Petrochemical College, Chulalongkorn University, Bangkok, Thailand*

NOMENCLATURE

Abbreviations

AF	Accumulation factor (dimensionless)
AP	Acidification potential (H+ equivalent)
ATP	Aquatic toxic potential (1/LC 50)
CR	Bioethanol production form cassava rhizome (dimensionless)
CR-A, B	Bioethanol production from cassava rhizome, options A and B (dimensionless)
DC	Demand cost ($/y)
EAF	Energy accumulation factor
ET	Eco-toxicological potential (kg 2,4-dichlorophenoxyacetic acid equivalent)
EWC	Energy waste cost ($/y)
GOI	Gross operating income ($/y)
GREV	Gross revenues ($/y)
GWP	Global warming potential (CO_2 equivalent)
HTC	Human toxicity carcinogens (kg of benzene equivalent)
HTNC	Human toxicity noncarcinogenic impacts (kg toluene equivalent)
HTPE	Human toxicity potential from dermal exposure (1/TWA)
HTPI	Human toxicity potential from ingestion and inhalation (1/LD 50)
ISA	Indicator sensitivity analysis (dimensionless)
LC50	Lethal concentration 50 refers to the concentration of a chemical in water that kills 50% of the test animals during the observation period (often expressed in terms of ppm)
LD50	Amount of a material, given all at once, which causes the death of 50% (one half) of a group of test animals (often expressed in terms of mg-min/m^3)

MIX	Bioethanol production from a combined lignocellulosic feedstock
MIX-A, B, C, D	Bioethanol production from a combined lignocellulosic feedstock, options A, B, C, and D
MVA	Material value added ($/y)
NPV	Net present value (10^6/y)
ODP	Ozone depletion potential (CFC-11 equivalent)
OPEX	Operational investment ($/y)
PCOP	Photochemical oxidation or smog formation potential (C_2H_2 equivalent)
RQ	Reaction quality
TDC	Total demand cost ($/y)
TTP	Terrestrial toxic potential (1/LD50)
TVA	Total value added ($/y)
TWA	Total weight average over a specified period of time

Variables

x_1	Raw material usage (kg raw material/kg of product)
x_2	Energy usage (GJ/kg of product)
x_3	Fresh water usage (kg of water/kg of product)
x_4	Operating cost ($/kg of product)
x_5	Carbon Footprint (kg of CO_2 equivalent/kg of product)
i	$\{1,2,\ldots,N\}$ different criteria used (dimensionless)
w_i	Weights given to the metrics/criteria (dimensionless)
K	Total different environmental metrics (dimensionless)
L	Total different economic metrics (dimensionless)
M	Total different resources usage metrics (dimensionless)
N	Total different types of criteria considered (dimensionless)
$F_{\mathbf{m,obj}}$	Multicriteria objective function ($/kg of product)
$\mathbf{Obj}_i$	Metrics times the respective weight (dimensionless)
$\mathbf{Env}_k$	Environmental metrics (dimensionless)
$\mathbf{Res}_\mathbf{m}$	Resources usage metrics (dimensionless)
$\mathbf{Eco}_\mathbf{l}$	Economic metrics K—total different environmental metrics (dimensionless)
$F^f_{i,k,kk}$	Component i flow f from process intervals k to process intervals kk (mass unit/unit of time)
$F^M_{i,kk}$	Component flow after mixing (mass unit/unit of time)
$F^{\mathrm{out1}}_{i,kk}$	Component flow leaving process intervals kk through primary outlet (mass unit/unit of time)
$F^{\mathrm{out2}}_{i,kk}$	Component flow leaving process intervals kk through secondary outlet (mass unit/unit of time)
$F^R_{i,kk}$	Component flow after reaction (mass unit/unit of time)
$F^{out}_{i,kk}$	Component flow after waste separation (mass unit/unit of time)
H^{in}_k	Enthalpy flow in interval k (energy unit/unit of time)
H^{out}_k	Enthalpy flow out interval k (energy unit/unit of time)
H^{ut}_k	Enthalpy flow associated to utility flow to interval k (energy unit/unit of time)
Q_k	Net enthalpy flow to interval k (energy unit/unit of time)

$\overline{Q}_{j,k}$	Partition variable for flow in piecewise linear approximation of capital cost constraints (dimensionless)
$R_{i,kk}$	Utility flow (mass unit/unit of time or energy unit/unit of time)
y_k	Selection of process intervals k (binary) (dimensionless)
$w_{j,k}$	Binary variable for piecewise linear approximation of capital cost constraints (dimensionless)

Parameters

$\delta_{i,kk}$	Waste fraction (dimensionless)
$\xi^{P}_{k,kk}$	Equal to 1 if a primary connection may exist between interval k and interval kk, zero otherwise (dimensionless)
$\xi^{S}_{k,kk}$	Equal to 1 if a secondary connection may exist between interval k and interval kk, zero otherwise (dimensionless)
$\emptyset_{i,kk}$	Raw material component flow (dimensionless)
$\alpha_{i,kk}$	Fraction of utility flow mixed with process stream (dimensionless)
$\gamma_{i,kk,rr}$	Reaction stoichiometry (molar unit)
$\sigma_{i,kk}$	Separation split factor (dimensionless)
$\theta_{\mathbf{react},kk,rr}$	Conversion of key reactant (dimensionless)
$\mu_{i,i,kk}$	Specific utility consumption (dimensionless)
$v_{st,k}$	Allocation of interval k to step st (dimensionless)
h^{f}_{k}	Mass enthalpy associated with flow f in interval k (energy unit/unit of mass)
h^{ut}_{i}	Mass enthalpy associated with utility i (for mixing utilities) (energy unit/unit of mass)
$\Delta h^{\mathrm{ut}}_{i}$	Change of mass enthalpy associated with utility i (for nonmixing utilities) (energy unit/unit of mass)
$P1_{i,kk}$	Raw material price (monetary unit/unit of mass)
$P2_{i}$	Utility price (monetary unit/unit of mass or monetary unit/unit of energy)
$P3_{i,kk}$	Product price (monetary unit/unit of mass)
p_{kk}	Coefficient for capital cost estimation (dimensionless)
q_{kk}	Coefficient for capital cost estimation (dimensionless)
$\overline{Q}^{0}_{i,kk}$	Grid for piecewise linear approximation of capital cost (dimensionless)

INTRODUCTION

Because of increasing environmental and economic concerns and governmental policies, chemical and biochemical processing industries need to develop and adopt more environmental friendly as well as technologically and economically competitive solutions for their processes. To find optimal and more sustainable solutions, the environmental constraints must be assessed in a comprehensive way, side by side with economic and technical criteria. In this chapter, the main concepts, methodologies, and computer-aided tools applied according to an established systematic work flow to achieve a more sustainable chemical process design are presented.

One of the well-known definitions of sustainable development—"meets the needs of the present without compromising the ability of future generations to meet their own needs" (Bruntland, 1987)—has been incorporated within the "triple bottom line" concept in terms of economic profitability, environmental conscience, and social development (Azapagic et al., 2002). Although different perspectives and leading ideas have been proposed over the years to identify and describe the concept of sustainability, it is only recently that sustainability has been widely accepted as a guiding principle, and one of the main challenges is to determine and measure the sustainability performance of products and processes (Zamagni et al., 2012). Although the methods within the environmental and economic domain are considered mature and well defined, the social indicators remain a topic that needs further development and scientific progress (García-Serna et al., 2007).

Sustainable process design has been implemented over the years through engineering expertise, and the main focus has been mostly on waste reduction strategies. A more complete approach to minimize the impact of a process on the global system is therefore needed. The life cycle concept was proposed by Novick (1960) in a report focusing on life cycle analysis of cost. The concept progressed from life cycle cost analysis to the first waste and energy analysis, and from there to the environmental life cycle assessment (LCA) as we understand it today (Curran, 2012). However, from the start it was clear that it should cover the supply chain, the use stage, waste processing, and end-life of the product in focus (Curran, 2012). The conceptual development toward environmental LCA was made in the 1980s and formalized in the 1990s with the work of SETAC and the 14040 Series standardization (Curran, 2012).

The Society of Environmental Toxicology and Chemistry (SETAC) defined LCA as a method to evaluate the environmental burdens related to a process or product, identify and quantify their impact and, therefore, point to possible opportunities for overall improvement (Azapagic, 1999). The International Organization of Standardization (ISO) initiated similar studies establishing principles and guidelines for LCA methodology (Azapagic, 1999). In principle, LCA follows a "cradle-to-grave" approach with two main objectives when applied to product/process design: to quantify and evaluate the environmental performance of a process/system, helping the decision-makers to choose among options, and to provide a tool for assessing the possible improvements by modification or redesign of a system to decrease the environmental impacts from the raw materials extraction until its final disposal (García-Serna et al., 2007). Various efforts have been made over the years to make the LCA a suitable method for a complete sustainability analysis, although a model that fairly reflects the sustainability question in all its complexity has not yet been developed (Zamagni et al., 2012).

Andersen et al. (1998) studied the feasibility of combining the sustainability principles and the methodology of LCA. Upham (2000) proposed a sustainability theory internationally for environmental management, including implicit reasoning and value judgment as well as science. Hunkeler and Rebbitzer (2005) introduced an LCA framework and procedure to model a product's life cycle, providing an overall picture of the available methods and tools for life cycle inventory (LCI). Klöpfer (2006) presented the advantages and the intent to develop an integrated framework for life cycle sustainability analysis.

The life cycle perspective has been seen as the best approach to provide holistic, reliable, and robust results, and it has received far-reaching recognition as a method that enables the quantification of the environmental burden over the complete life cycle of a product, process, or activity. Although LCA has been mostly applied to products, it can greatly support the identification of more sustainable options in process selection, design, and/or optimization (Azapagic, 1999). A common concept frequently used to measure only the sustainable performance of a process is the eco-efficiency ratio, defined as the ratio between economic (creation) and environmental impact (destruction). It has been described as a management theory that promotes competitive prices, satisfies human needs, and reduces the environmental burden and resources consumption throughout the life cycle of a product or system (Saling et al., 2002).

However, after analyzing the process performance and possibly identifying the process critical points (bottlenecks) with respect to a sustainability evaluation, it follows that the designer is faced with the need to make changes to improve the actual process (retrofit design) or the early-stage design of a new process (Carvalho et al., 2008). Retrofitting reflects the addition of new features and/or technologies to existing processes/systems.

LCA: METHODOLOGIES

LCA is commonly divided into four distinct parts: goal and scope definition; LCI; life cycle impact assessment (LCIA); and interpretation.

LCI quantifies the inputs and outputs of a system, material, and energy flows. When producing more than one product, an allocation problem rises. What share of the environmental burdens should be allocated to the product in question, i.e., in the LCI? Different solutions have been proposed, with the most well known being ISO 14041 (Vanegas, 2003). With respect to LCIA, it converts the inventory into simple indicators. One of two methods is followed: problem-oriented methods (midpoint) and damage-oriented methods (endpoint). The problem-oriented methods aim to simplify a very complex set of data, dividing the flows into a few environmental burden classes (Azapagic, 1999). The damage-oriented methods also separate the flows into environmental burden classes but take it further by modeling and cataloging their effects according to the effect on human health, ecosystem, and resources (Azapagic, 1999).

Some of the well-known and accepted methodologies are, among others, ISO 14040, UNEP/SETAC, and International Life Cycle Data System (ILCD) (Azapagic, 1999).

COMPUTER-AIDED TOOLS FOR SUSTAINABLE CHEMICAL PROCESS DESIGN

Even though general guidelines for sustainable process design have been proposed, such as the framework for sustainable design proposed by Hacking and

Guthrie (2008), covering a wide and qualitatively broad spectrum of processes, they are not sufficient for a complete sustainable chemical process design. They need to be complemented with methods and tools that, by following specific indicators, efficiently lead to the design of sustainable chemical processes. To design a sustainable chemical process, various tools, methodologies, and frameworks, including computer-aided tools, have been proposed, focusing separately on modeling a particular sustainability domain, such as economics, environmental assessment, and social impact, at a given point in time. For instance, ones based on:

1. The concept of waste reduction (Singh and Falkenburg, 1993).
2. Multiobjective optimization techniques, simulation, and uncertainty analysis for chemical and material selection, management, and planning (Diwekar, 2003).
3. Hierarchical approach to evaluate new or actual processes flow sheets, incorporating the stakeholders at each level (Lapkin et al., 2004).
4. Methodologies that integrate the environmental constraints in the early stage of design, along with more traditional economic and technical criteria.

Azapagic (1999) presented a review of the available techniques and pointed out the advantages of coupling and multiobjective optimization with LCA. Buxton et al. (1996) introduced the methodology of environmental impact minimization (MEIM) to reaction path synthesis. Mass and energy usage minimization were introduced by retrofitting techniques (Carvalho et al., 2008, 2013). Defined as an integrated problem-solver based on mathematical formulations, these model-based computer-aided tools give the designer a fast and systematic approach to the design problem. A partially complete set of computer-aided tools applicable to process design that can be found in the open literature is given in Table 9.1.

The central problem in the sustainability analysis is in regard to how to make the link between the different dimensions and their different levels of complexity.

To address the sustainability question, the sustainable design method of Carvalho et al. (2008) has been extended and adopted into a computer-aided framework. In this section, the framework and its implemented methods and tools are presented as a potential means for identifying innovative and more sustainable solutions over the whole process/product life cycle. The design work flow implemented in the framework has four main steps: the design problem is defined; the base case is identified through superstructure generation and screening of alternatives; the base case is subjected to a sustainability analysis to determine the critical points in the process; and alternatives that address the identified critical points are generated and screened and the most sustainable process design is determined. Note that the principle concept here is that alternatives that address the identified process critical points of a base case are, by design, more sustainable than the base case. The case study selected to highlight the application of the framework together with its implemented methods and tools involves the production of bioethanol from various renewable resources.

Table 9.1 Computer-Aided Tools to Support Sustainable Process Design

Topic Being Covered	Name of the Software	Reference
Human and environmental hazard/impact	LCSoft	Kalakul et al. (2013)
	Quantis SUITE 2.0	Humbert et al. (2012)
	SimaPro8	Goedkoop et al. (2008)
Environmental and economics	Umberto	Walter Klöpfer and Renner (1995)
	BEES	Boyles and Lippiat (2001)
	TEAM	Julius and Scheraga, 2000
Value chain, flow, and cost modeling	BLCC	Kabassi and Cho (2011)
	ECON	Saengwirun (2011)
	HOMER	Lambert et al. (2004)
	SustainPro	Carvalho et al. (2013)
Impacts on cost, environmental, and social (three sustainability pillars)	Gabi4	Schuller et al. (2013)
	SEEbalance®	Saling et al. (2005)

SUSTAINABLE PROCESS SYNTHESIS AND DESIGN FRAMEWORK

Process design is performed when developing new technologies, when creating new facilities, or when retrofitting existing production processes. From data gathering and process flow sheet synthesis to detailed design and optimization steps, the traditional approach in process design considers only the economic and technical aspects as the main criteria. With increasing demands for sustainable growth, more performance measures have been proposed and incorporated into process design to achieve the sustainability goal. In this section, the framework for sustainable process synthesis and design is presented together with its implemented computer-aided methods and tools.

The framework allows the designer to assess the sustainability potential of a specified production process and provides computer-aided methods and tools to generate and compare more sustainable processing alternatives that match the process product specifications. A number of metrics (techno-economic, LCA, safety, etc.) have been used for assessing the level of sustainability and for identifying more sustainable options. The design methodology comprises a sequence of steps with various computer-aided tools that are used to perform calculations of the metrics (also known as sustainability indicators) systematically and efficiently. The framework enables the designer to link between sustainability potential, process de-bottlenecking, improvements, and possible trade-offs. Raw material, water, and energy usage, as well as LCA and economic impact, are some of the metrics that are used for multicriteria evaluation of generated design alternatives.

The framework includes methods tools for the identification of critical points (bottlenecks) within the process. Another set of computer-aided methods tools are used for the generation of retrofit (or new) design options through a heuristic strategy. The next stage comprises the testing and screening of the generated options in terms of feasibility (matching of process specification) and sustainability (matching of improvement targets). That is, the best options to overcome the identified critical points matching the design targets are evaluated in detail in terms of environmental feasibility, sustainability, and economic profitability.

Therefore, this framework should be able to serve as a decision-making tool that allows the assessment of the production plant performance with respect to the objectives embedded in the idea of sustainable development and as a means for screening among feasible process options for the production of the desired product. The framework overview is illustrated in Figure 9.1.

The framework is divided into four parts:

- Part 1: Problem definition
- Part 2: Superstructure generation, base case(s) identification
- Part 3: Process simulation, sustainability analysis, and bottleneck identification
- Part 4: Generation and screening of new design alternatives

PART 1—PROBLEM DEFINITION

Part 1 of the framework is presented in Figure 9.1 and consists of the following steps.

Step 1.1—Problem and objective function (F_{obj}) definition

In this step the user states the goals, objectives, and scope of the problem. The objective function for the optimization problem is defined based on the main goals of the synthesis and design problem.

Therefore, several terms in the objective function can be selected, such as the gross operating income (GOI), earning before interests and tax (EBIT), and net present value (NPV) (Perry and Green, 2008). Along with the objective function, all the indicators representing the engineering, economic, and environmental performance (the LCA and sustainability indicators) are selected. The framework currently covers the indicators/metrics listed in Table 9.2. Note also that the problem could be formulated as a retrofit design problem or as the design of a new processing network. Table 9.3 lists the data in, the tools used, and the data out for the present step.

Step 1.2—Literature survey for raw materials and process technologies

In this step, information on raw materials and process technologies to produce the main product are obtained through a focused literature survey. Table 9.3 lists the data in, the tools that need to be used, and the data out generated by the step.

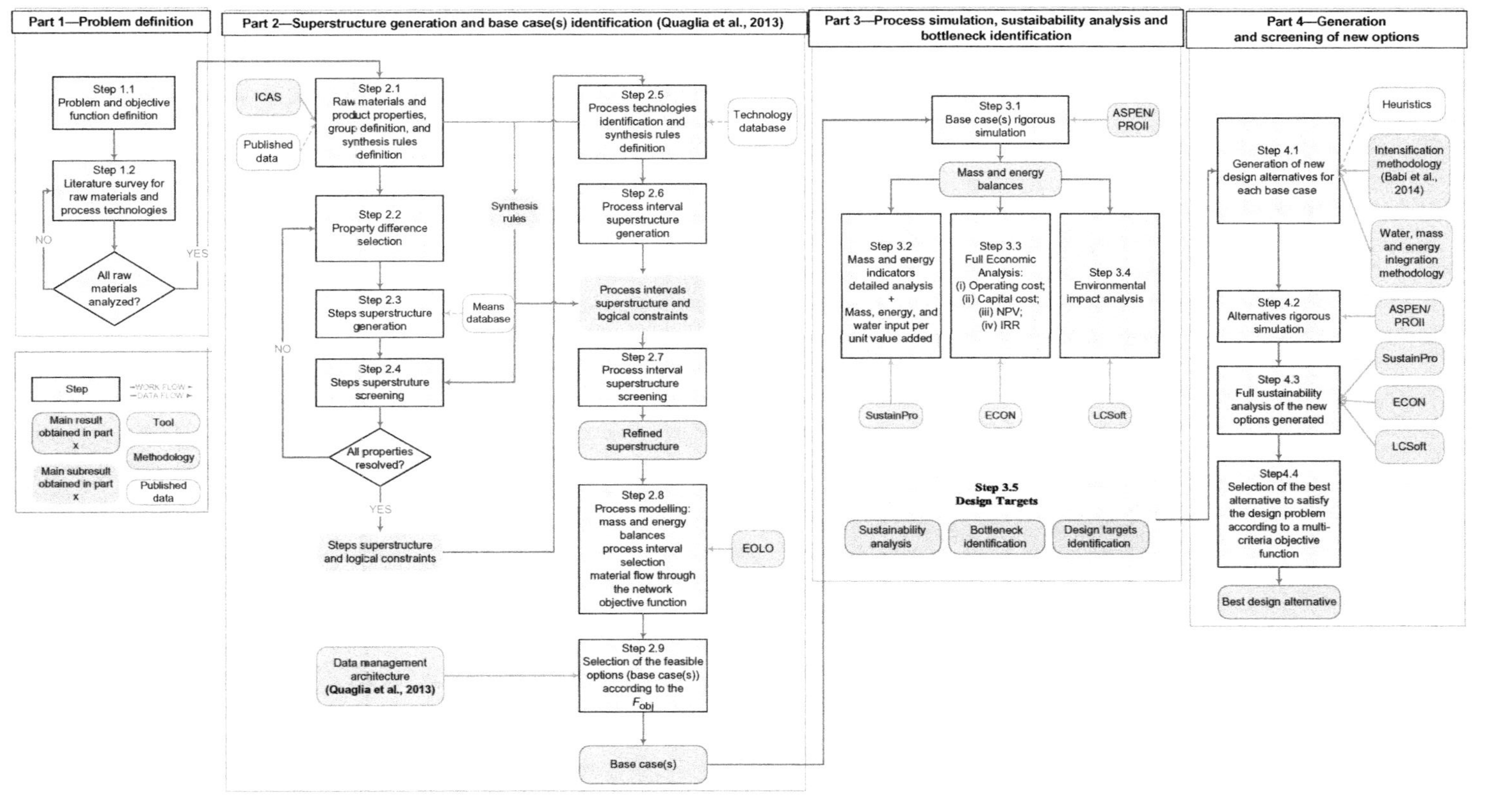

FIGURE 9.1

Overview of a computer-aided framework.

Table 9.2 Sustainability Metrics Estimated by the Presented Framework

Engineering	Economic Impact	Environmental Impact
Yield (from raw material to main product)	Value added	Carbon footprint
Degree of conversion	Net present value	Resource usage (raw materials, water, fossil fuels)
(Bio)catalyst used/needed	Internal rate of return (IRR)	Global warming (GWP)
Specific energy consumption	Payback time	Ozone depletion (ODP)
	Capital cost	Acidification (AP)
	Operating cost	Eutrophication (EP)
	Operating profit	Photochemical oxidation potential (PCOP)
		Human toxicity
		Ecotoxicity
		Solid waste
		Human toxicity potential by ingestion (HTPI)
		Human toxicity potential by dermal exposure (HTPE)
		Human toxicity carcinogens (HTC)
		Human toxicity noncarcinogens (HTNC)
		Terrestrial toxicity potential (TTP)

Table 9.3 Description of the Data In, Tools Used, and Data Out for the Steps of Part 1 of the Framework

Step	Data In	Tools Used	Data Out
1.1	–	–	• Main synthesis problem identification • Objective function that describes the synthesis problem
1.2	–	Literature survey	• Product portfolio • Feedstock portfolio • Process Technologies portfolio

PART 2—SUPERSTRUCTURE GENERATION

Here, the definition of the design space and its representation by means of a superstructure (Quaglia et al., 2013) are presented.

To support the formulation and solution of the design problem, three methods for superstructure synthesis could be used: (i) alternative collection; (ii) combinatorial synthesis; and (iii) insight-based synthesis. The insight-based synthesis method for superstructure generation as developed by (Quaglia, 2013) is implemented in this part of the framework (Figure 9.1). A step-by-step explanation of the work flow used to determine the design space by means of a superstructure is given here.

Step 2.1—Raw material and product properties, group definition, and synthesis rules definition

Properties of the products and raw materials are collected in published data or through ProPred, a software tool for property prediction (a more detailed description is provided later). To establish the conditions that the superstructure needs to fulfill, all significant commercial and engineering (technical) information is translated into synthesis rules. It allows the user to clarify the problem by collecting and systematizing all the information. More detailed information about how to establish the synthesis is presented elsewhere (Quaglia et al., 2013). Table 9.4 lists the data in, the tools used, and the data out for this step.

Step 2.2—Property difference definition and selection

The framework uses differences in properties of materials and products as a guide to synthesize the process superstructure. To this end, the hierarchical approach proposed by Siirola (1996) is used, whereby the identified property differences related to the species are placed first, followed by differences in amount, composition, temperature, pressure, size, and geometry. All the process steps that are able to solve/eliminate the differences between the raw materials and products are systematically added to the superstructure. Each one of the registered differences is interactively resolved by applying the next steps of the method (see the flow diagram of part 2 in Figure 9.1). Table 9.4 lists the input data, the tools used, and the output data obtained from this step, and examples of property differences are given in the first (left) column of Table 9.5.

Step 2.3—Superstructure: process-step generation

In this step, the knowledge base is searched and all means/process steps to address the property differences are identified. After finding every process step corresponding to every property difference, they are organized and added to the superstructure, representing all possible configurations. The list of data in, tools used, and data out for this step are given in Table 9.4, whereas in Table 9.5, the property differences and the corresponding steps are listed.

Step 2.4—Superstructure screening

The objective here is to screen among the possible feasible alternatives included in the superstructure and to eliminate the nonfeasible options by using a set of synthesis rules defined in step 2.1. The elimination of a nonfeasible option is performed by eliminating a process step, a connection from the superstructure, or by defining logical constraints. Note that steps 2.3–2.5 should be repeated until every property difference has been addressed, the final superstructure has been

Table 9.4 Description of the Data In, Tools Used, and Data Out for the Steps of Part 2 of the Framework

Step	Data In	Tools Used	Data Out
2.1	Description of the process technologies	Published data ProPred	• Raw materials properties • Products properties • Synthesis rules
2.2	Raw materials properties	Hierarchical approach by Siirola (1996)	• List of property differences that need to be addressed to eliminate/solve the differences between the raw materials and products
2.3	List of property differences from step 2.2	Engineering knowledge	• Process step for each of the property differences identified • The process steps are linked in all possible configurations
2.4	Configurations from step 2.3	Synthesis rules (defined in step 2.1)	• Reduce the number of configurations (nonfeasible options are eliminated) • Logical constraints
2.5	Configurations from step 2.4	Published data Technology database	• Knowledge base for each process interval specifying the type of step and their main characteristics (see Table 9.4) • Identification of secondary input/output streams originating new process intervals • Synthesis rules defined at the process-interval level
2.6	List of process intervals Configurations from step 2.5	Synthesis rules defined at the process-interval level	• Nonrefined superstructure
2.7	Nonrefined superstructure from step 2.6	Synthesis rules	• Refined superstructure
2.8	Refined superstructure (from step 2.7); mass and energy balances models; process-interval selection model; objective function (from step 1.1) model	Data management architecture EOLO	• Mixed integer nonlinear programming (MINLP) formulation of the synthesis and design problem ready to be solved in GAMS
2.9	Objective function values for every feasible option identified in the superstructure	–	• Base case(s)

Table 9.5 Knowledge Base for a Means-Ends Analysis

Property Difference	Process Step
Species identity	Reaction
Amount	Dosage, flow splitting
Concentration	Separation
Phase	Vaporization, condensation
Temperature	Heating, cooling
Pressure	Compression, expansion
Size and geometry	Agglomeration, milling
location	Transportation

Source: Adapted from Quaglia (2013).

built, and all the corresponding logical constraints have been defined. Table 9.4 lists the data in, the tools used, and the data out for this step.

Step 2.5—Process technologies identification

Based on the built knowledge base for each process interval specifying the type of step and the main characteristics (according to the type of step being specified), it becomes possible to identify the alternatives to perform each of the steps. Examples of the structure of the knowledge base are given by Quaglia (2013).

Each of the process technologies in each process interval is evaluated here with the identification of the related secondary input–output streams. The rules created here act as a supplement to the list created in step 2.1 at process-step level. Table 9.4 lists the data in, the tools used, and the data out for this step.

Step 2.6—Process-interval superstructure generation

Each of the process steps of the superstructure is occupied by one or more process intervals linked in all possible configurations, generating a nonrefined superstructure. A nonrefined superstructure therefore represents a superstructure that includes all paths (feasible as well as infeasible), some of which will be eliminated by the synthesis rules. Table 9.4 lists the data in, the tools used, and the data out for this step.

Step 2.7—Process-interval superstructure screening and generation of the "refined" superstructure

The alternatives generated in the previous step are screened by applying the synthesis rules. Configurations that are not feasible according to the synthesis rules and/or logical constraints are eliminated from the search space by taking out one or more process intervals or connections from the superstructure. This approach allows the user to keep a clear and exhaustive documentation of all the data and conditions used to generate the superstructure while considering all the feasible options impartially. In this way, innovative solutions can be identified. Table 9.4 lists the data in, the tools used, and the data out for this step.

Step 2.8—Process modeling

A process interval is modeled as *sequence of elementary process tasks* by formulating functional descriptors representing the transformation occurring in a stream in a generic manner such as mixing, utility dosage, reaction, stream division, separation, and waste separation (Quaglia et al., 2014). Each process task is represented by a simple mathematical model, a simplified mass balance (see Eqs. (9.1)–(9.5) in Table 9.6). The models are combined to create the total mass balance model to describe each of the process intervals. The net energy balance for each process task is also modeled (Eq. (9.6)). Therefore, the complete mathematical formulation is obtained after combining mass and energy balance models together with a superstructure model that describes the logic of the process-interval selection (Eqs. (9.12) and (9.13)), the materials flow through the network (Eqs. (9.10), (9.11), and (9.14)), and the model of the objective function already selected in part 1, step 1.1, by Eqs. (9.15)–(9.18). Table 9.4 lists the data in, the tools used, and the data out for this step.

The formulation of the synthesis and design problem demands collecting and specifying a large number of multisource data. From a user standpoint, this usually results in a frustrating and time-consuming task, and the probability of committing error is substantial. Therefore, to avoid compromising the results, the framework includes a *data management architecture*, which efficiently provides a structure/system for the collection and management of all the information related to the problem. More detailed information can be found in the PhD Thesis of Quaglia (2013), and a more concise version of the models is presented by Quaglia et al. (2014). This multilayer data management structure was implemented in C# developing the problem formulation software called EOLO. EOLO incorporates automatic data specification and coherence checks. Moreover, EOLO is coupled with an Excel-based database in which data and problem formulation can be stored and quickly accessed.

Step 2.9—Base case(s) identification

The problem formulated in step 2.8 represents an MILP (Mixed Integer Linear Programming) / MINLP (Mixed Integer Non-Linear Programming) problem that can be solved through solvers like GAMS (Quaglia et al., 2014). For different objective functions selected (in step 1.1), different optimal network solutions may be obtained, resulting in a number of base case definitions for the processing network. Table 9.4 lists the data in, the tools used, and the data out for this step.

Table 9.6 MINLP Formulation of the Network Design Problem Using Generic Process-Interval Model (Quaglia et al., 2013)

Raw material assignment	$F^{out}_{i,kk,ss} = \phi_{i,kk} \forall kk \in \text{raw}$	(9.1)
Utility consumption	$F^{M}_{i,kk} = \sum_{k}(F_{i,k,kk}) + \alpha_{i,kk} \cdot \mu_{i,kk} \cdot \sum_{i,k}(F_{i,k,kk})$	(9.2)
Reaction	$F^{R}_{i,kk} = F^{M}_{i,kk} + \sum_{rr,\text{react}}(\gamma_{i,kk,rr} \cdot \theta_{\text{react},kk,rr} \cdot F^{M}_{\text{react},kk})$	(9.3)
Wastes separation	$F^{out}_{i,kk} = F^{R}_{i,kk} \cdot (1 - \delta_{i,kk})$	(9.4)
Product–product separation	$F^{out1}_{i,kk} = F^{out}_{i,kk} \cdot \sigma_{i,kk}; \quad F^{out2}_{i,kk} = F^{out}_{i,kk} \cdot (1 - \sigma_{i,kk})$	(9.5)
Energy balance	$H^{in}_{k} + H^{ut}_{k} - H^{out}_{k} + Q_k = 0$	(9.6)
Enthalpy calculation	$H^{in}_{kk} - \sum_{f,i,k}(F^{1}_{i,k,kk} \cdot h^{1}_{i,k}) + \sum_{f,i,k}(F^{2}_{i,k,kk} \cdot h^{2}_{i,k})$	(9.7)
	$H^{ut}_{kk} = \left(\mu_{i,kk} \cdot \sum_{i,k}(F_{i,k,kk})\right) \cdot (\alpha_{i,kk} \cdot h^{ut}_{i} + (1 - \alpha_{i,kk}) \cdot \Delta h^{ut}_{i})$	(9.8)
	$H^{out}_{k} = \sum_{i}(F^{out1}_{i,k} \cdot h^{1}_{i,k}) - \sum_{i}(F^{out2}_{i,k} \cdot h^{2}_{i,k})$	(9.9)
Superstructure flow model	$F^{1}_{i,k,kk} \le F^{out1}_{i,kk} \cdot \xi^{P}_{k,kk}; \quad F^{2}_{i,k,kk} \le F^{out2}_{i,kk} \cdot \xi^{S}_{k,kk}$	(9.10)
Superstructure logic model	$\sum_{k} F^{1}_{i,k,kk} = F^{out1}_{i,kk}; \quad \sum_{k} F^{2}_{i,k,kk} = F^{out2}_{i,kk}$	(9.11)
	$F^{R}_{i,kk} \le M \cdot y_{kk}$	(9.12)
	$\sum_{k}(y_k \cdot v_{st,k}) \le 1$	(9.13)

(Continued)

Table 9.6 MINLP Formulation of the Network Design Problem Using Generic Process-Interval Model (Quaglia et al., 2013) Continued

Throughput limitations	$\sum_{i} F^{R}_{i,kk} \leq F^{\max}_{kk}$ (9.14)
Objective function	$\max \text{EBIT} = \sum_{i,kk,ss}(P3_{i,kk} \cdot F^{\text{out}}_{i,kk}) - \sum_{i,kk,ss}(P2_{kk} \cdot R_{i,kk,ss}) - \sum_{i,kk,ss}(P1_{i,kk} \cdot F^{\text{out}}_{i,kk}) - W^{\text{Price}} \cdot \sum_{i,kk,ss}(F^{R}_{i,kk} \cdot \delta_{i,kk}) - \frac{\text{CAPEX}}{t}$ (9.15)
	$\max \text{GOI} = \text{GREV} - \text{OPEX} = \sum_{k \in \text{prod}(k)} (F^{i,k}_{\text{out}} \cdot \pi^{i,k}_{P}) - R_{\text{cost}} - U_{\text{cost}}$ (9.16)
Capital cost model	$\text{CAPEX} = \sum_{kk}\left[\sum_{j}(\alpha_{j,kk} \cdot w_{j,kk} + \beta_{j,kk} \cdot Q_{j,kk})\right]$ (9.17)
	$F^{\text{Thr}}_{kk} = \sum_{j} Q_{j,kk}; \sum_{j} w_{j,kk} = 1; \; Q^{o}_{j,kk} \cdot w_{j,kk} \leq Q_{j,kk} \leq Q^{o}_{j+1,kk} \cdot w_{j,kk}$ (9.18)

PART 3—PROCESS SIMULATION, SUSTAINABILITY ANALYSIS, AND BOTTLENECK IDENTIFICATION

For the base cases defined in part 2, a detailed sustainability analysis is performed here. The overview of the framework in part 3 is highlighted in Figure 9.1.

Step 3.1—Base case(s) rigorous simulation

The objective here is to perform a rigorous simulation of the production process using an appropriate process simulator, e.g., ASPEN or PRO II. Even though available information was collected in part 1, some details are usually missing; therefore, more data may be necessary. For example, the number of trays of the distillation columns or/and the reflux rate in the typical purification step may be missing. Optimizing and achieving the best conditions when the data are not available represent another significant challenge for the early-stage process synthesis problem. After convergence of the simulation problem, all the fundamental data needed to perform the next steps of the methodology through the framework are available. Table 9.7 lists the data in, the tools used, and the data out for this step.

Step 3.2—Mass and energy indicators

The objective here is to calculate the mass and energy indicators for the base case using SustainPro. For this calculation, more information is needed, such as equipment duty and pure component temperature-dependent properties like thermal capacity, density, and enthalpy of vaporization. Table 9.7 lists the data in, tools used, and the data out for this step.

Step 3.3—Full economic analysis

To perform a full economic evaluation, two of the most important parameters that need to be estimated are the operating and capital costs. To perform the installation cost calculation, a module factor is needed. Fixed capital costs are calculated for all the plant equipment, and this is performed by the summation of bare module cost, contingency cost, building cost, land, and services.

The operating costs, which are incurred annually in the production of the chemical, should be considered when the alternative process routes are being evaluated because they can have significant influence on the final process choice. They are divided into two groups, fixed costs and variable costs. The former includes laboratory costs, operating labor, and capital repayment; these costs do not depend on the production rate, and they must be paid even if the chemical is not being produced. Variable costs such as raw materials, utilities, and services in general depend only on the amount of chemical that is being produced or planned to be produced (Ray and Sneesby, 1989). Both capital and operating costs are calculated using ECON software (Saengwirun, 2011). Table 9.7 lists the data in, the tools used, and the data out for this step.

Table 9.7 Description of the Data In, Tools Used, and Data Out for Steps of Part 3 of the Framework

Step	Data In	Tools Used	Data Out
3.1	Production rate; raw materials rate; freshwater flow rate; reaction parameters; reactor conditions; thermodynamic constraints; distillation columns data; catalyst requirement; solvent requirement; product recoveries; stream composition; operational conditions	PROII ASPEN Commercial process simulator	• Flow sheet • Mass and energy balances • Equipment connectivity • Equipment duty • Component chemical and physical properties versus pressure and temperature
3.2	Detailed mass and energy balance data; number of streams; number and type of unit operations; component chemical and physical properties	SustainPro	• Mass and energy closed and opened paths • Ordered list of indicators that allows the user to have a quick view of the process critical points regarding raw material, water, and energy usage • Mass, energy, and water per unit value added • Indicators sensitivity analysis • Operational variables sensitivity analysis
3.3	Main equipment sizing Energy balance Main equipment duty Type of material used Utilities rate	ECON	• Capital cost • Operating cost • Net present value • IRR • Payback time
3.4	Mass balance; equipment duty and origin; utilities type and origin	LCSoft	• Carbon footprint • 11 Potential environmental impacts (PEIs)

Step 3.4—Environmental impact analysis

Reduction of greenhouse gas emissions is one of the key requirements for sustainable production and consumption, but while some industries have been very successful in reducing emissions to water and air, and while non-CO_2 greenhouse gas emissions have been minimized, reduction of CO_2 emissions has been less successful. The industry itself forecasts that further reduction of CO_2 emissions will be minimal. However, concerns about global warming are increasing, but at the same time the chemical industry increases its commitment toward sustainability and its three domains. Determining the carbon footprint of a chemical plant and of its products will help to identify more possibilities to reduce the emissions, and it is a necessary step for further reduction of the chemical industry's environmental impact (Stein and Khare, 2009).

The PEI can be predicted based on several qualitative and quantitative metrics that give a general perspective on the overall system. Two different methodologies could be applied to estimate the PEI metrics: waste reduction algorithm (WAR) (Cabezas et al., 1999) and LCA. Both methodologies give the desired environmental metrics; however, the latter presents the results per functional unit of product (1 kg of product), whereas WAR gives the results in a qualitative way by a score system approach.

The LCA is performed to translate the collected emissions and consumptions into environmental and/or health effects and is commonly expressed by representative impact category indicators (Morais and Delerue-Matos, 2010).

In the framework, the LCSoft software (Kalakul et al., 2013) is used. Table 9.7 lists the data in, the tools used, and the data out for this step.

Step 3.5—Design targets

Results from step 3.2 are cross-checked with results from steps 3.3 and 3.4 in this step. If the critical points identified by SustainPro match with the results of the critical points identified by ECON and LCSoft, then a global process bottleneck has been found and this can be selected as a potential design target. That is, if a new process alternative can be found that matches this design target, then a non-trade-off solution that is more sustainable would be obtained (see part 4). If a global bottleneck is not clearly and easily identified, then the next best target is selected for process alternative generation. Table 9.7 lists the data in, the tools used, and the data out for this step.

PART 4—GENERATION AND SCREENING OF NEW OPTIONS

The work flow of part 4 of the framework is highlighted in Figure 9.1.

Step 4.1—Generation of new design alternatives for the base case(s)

Guidelines mostly composed by retrofitting techniques and heuristics, process integration (mass, water, and energy integration), and process intensification are proposed here to overcome the identified bottlenecks and therefore to obtain a more sustainable design. Methods included the Pinch design method that provides a consistent prediction of energy integration and energy saving (Linnhoff and Flower, 1978); water integration (Wan Alwi and Manan, 2013); mass integration (El-Halwagi, 1997); and process intensification (Lutze et al., 2013; Babi et al., 2014).

Table 9.8 lists the data in, the tools used, and the data out for this step.

Step 4.2—Alternatives rigorous simulation

It is important to note that after every change made in the base case design, a new alternative is generated. Therefore, it must be simulated through a rigorous

Table 9.8 Description of the Data In, Tools Used, and Data Out for the Steps of Part 4 of the Framework

Step	Data In	Tools Used	Data Out
4.1	Ordered list of Identified bottlenecks; detailed mass and energy balances; equipment duties; cooling media properties; component chemical and physical properties	Pinch design method Water Pinch analysis Process intensification	• New and sustainable process design options
4.2	New options from step 4.1	ASPEN/PROII	• Processes flow sheet • Mass and energy balances • Equipment duties • Component chemical and physical properties versus pressure and temperature
4.3	Processes flow sheet Mass and energy balances Equipment duties Component chemical and physical properties versus pressure and temperature	SustainPro ECON LCSoft	• Resources consumption • Ordered list of mass and energy indicators that allow an overall picture of the process • Carbon footprint • 11 PEIs • Capital and operating cost • NPV, IRR, and payback time
4.4	Sustainability metrics for the base case(s) and the respective new alternatives generated	Multicriteria objective function	• Best design alternative to address the design problem stated in step 1.1

process simulator that provides the data needed to perform the sustainability analysis, thereby allowing the comparison and the screening of alternatives. Table 9.8 lists the input data, the tools used, and the output data obtained from this step.

Step 4.3—Full sustainability analysis of the new options generated in step 4.1

A complete evaluation of the alternatives generated in step 4.1, using data from step 4.2, is performed using the corresponding software tools (SustainPro, LCSoft, and ECON). As a result, a multicriteria checklist is compiled, including resources consumption, critical points and economic and environmental impact criteria. All the information needed to compare the base case design against the new design options is now available. Table 9.8 lists the data in, the tools used, and the data out for this step.

Step 4.4—Screening and selection of the best alternative through a multicriteria multiobjective decision

The objective here is to perform the final decision based on a deterministic multicriteria evaluation. Table 9.8 lists the input data, the tools used, and the output data obtained from this step.

A multiobjective function is suggested as a decision tool (Eqs. (9.19)–(9.21)); it consists of three fundamental criteria/objects (Obj_1, Obj_2, and Obj_3): (i) environmental (total process carbon footprint); (ii) economics (operating and capital cost); and (iii) resources (energy, water, and raw materials) usage.

The different weights to different criteria and the different metrics within the criteria can be selected according to the established priorities and purposes; however, the final goal is to minimize the objective function. Therefore, several weighting scenarios could be tested and if a single value for all the alternatives is obtained (score system), then a ranking of the options would be possible.

$$F_{\text{m.obj}} = \sum_{i=1}^{i=N} \text{Obj}_i \times w_i \quad i = 1, 2, 3, \ldots, N \tag{9.19}$$

$$F_{\text{m.obj}} = \text{Obj}_1 \times w_1 + \text{Obj}_2 \times w_2 + \text{Obj}_3 \times w_3 \tag{9.20}$$

$$F_{\text{m.obj}} = \left(\sum_{k=0}^{K} \text{Env}_k \times w_k\right) \times w_1 + \left(\sum_{l=0}^{L} \text{Eco}_l \times w_l\right) \times w_2 + \left(\sum_{m=0}^{M} \text{Res}_m \times w_m\right) \times w_3 \tag{9.21}$$

$F_{\text{m.obj}}$ represents the multicriteria objective function (\$/kg of EtOH), N represents total different types of criteria that are being considered, K total represents different environmental metrics, L total represents different economic metrics, and M represents the total of different resources usage metrics. The multiobjective function is presented as a weighted sum (w_1, w_2, and w_3) and is calculated by adding the product of each normalized criteria and, within this, the respective metrics are also weighted (w_k, w_l, and w_m, freely given by the user).

Because not all the metrics are displayed in the same unit, a normalizing constant for different metrics has to be applied (Flores-Alsina et al., 2008) and all the metrics are normalized to \$/kg of EtOH by using different constants (Table 9.9).

The results for each option (A_j) undergoing evaluation can be formulated as a vector of performance scores ($x_{j,i}$), [$A_j = (x_{j,i}, \ldots, x_{j,t})$], where, j represents the number of options the user is considering for evaluation and $x_{j,i}$ represents each criteria value within a certain option.

SOFTWARE TOOLS AVAILABLE/INVOLVED THROUGH THE FRAMEWORK

Superstructure information

The framework uses a superstructure-based method to generate and identify processing networks. A brief overview of the superstructure notation used in this chapter is given for a clearer understanding of the methods used in part 2 of the framework.

Table 9.9 Metric Units Normalization Procedure

	Current Unit	Constant
Raw material usage	kg of RM/kg of EtOH	\$/kg of RM (price of RM)
Energy usage	GJ/kg of EtOH	\$/GJ (average price of energy)
Fresh water usage	kg of water/kg of EtOH	\$/kg of water (water price)
Carbon footprint	kg of CO_2 eq/kg of EtOH	a. Calculate kg of CO_2 produced per kg of EtOH burned b. Average CO_2 capture price (0.075 \$/kg of CO_2)

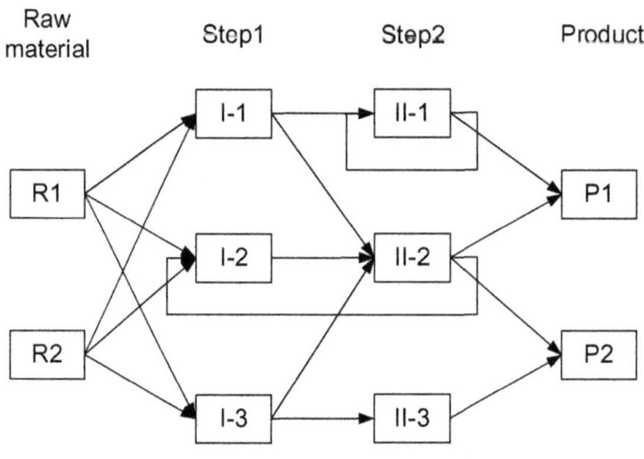

FIGURE 9.2

Example of a process superstructure.

Adapted from Quaglia et al., 2013.

Superstructure notation

Some guidelines/nomenclature are currently used and generally accepted for superstructure representation, raw materials and products occupy the first and last column, respectively, and the processing of materials flows from the left to the right. *Process steps* represent the conversion of the raw materials into products, and they are expressed as columns in the superstructure. Each process step contains one or more *process intervals* (rectangular boxes in Figure 9.2), and it consists of a technical alternative to perform the action that defines a certain process step. The connections between process intervals represent possible material flows, resulting in a network of process intervals.

Incremental superstructure synthesis

The screening procedure is simple to perform. However, when a large number of raw materials and products are to be considered, the task becomes highly complex because of the combinatorial explosion of the number of alternatives that need to be generated, screened with synthesis rules, and incorporated in the superstructure. Therefore, to deal with this difficulty, it is possible to perform the process step of the superstructure synthesis by executing it in an incremental manner. More detailed information on this can be found in the work by Quaglia (2013).

PROPRED

ProPred is part of the ICAS suite—in-house software developed at CAPEC—allowing property estimation of molecules based on various group contribution methods. For example, the Marrero and Gani method (Marrero and Gani, 2001) and the Constantinou and Gani method (Constantinou and Gani, 1994). It provides property values of molecules by using molecular structure as input information. For polymers, *ProPred* also uses a group contribution approach for prediction of various properties of polymer repeat units. A large database containing available experimental data of a wide range of molecules is included in *ProPred*, which also contains a data-fitting tool for regression of functional property model parameters.

SUSTAINPRO

SustainPro (Carvalho et al., 2013) is a software tool that applies a retrofit methodology that proposes new sustainable design alternative for a base case design and is based on a set of mass and energy indicators. Figure 9.3 presents the main SustainPro user interface.

The software has input data mass and energy balances obtained from steady-state process simulation, connectivity, compound chemical properties, compounds, and utility costs.

First, the user has to make the input of the mass and energy balance, along with the connectivity data among the different equipment. The software deconstructs the given mass and energy balance into open and closed paths. It is important to notice that a path represents a compound that follows a certain route; in terms of mass, a closed path means that a certain compound is being recycled in a given path, and an open path means that a compound is entering and leaving the system and that it is not being recycled.

SustainPro gives an ordered list of indicators placing those that have the highest potential at the top and those that have the least potential at the bottom. Each indicator represents an open or closed path and points to bottlenecks with respect to mass and/or energy that are wasted or that go around in a recycle loop (i.e., trapped and cannot get out). Therefore, the indicators *indicate* the potential

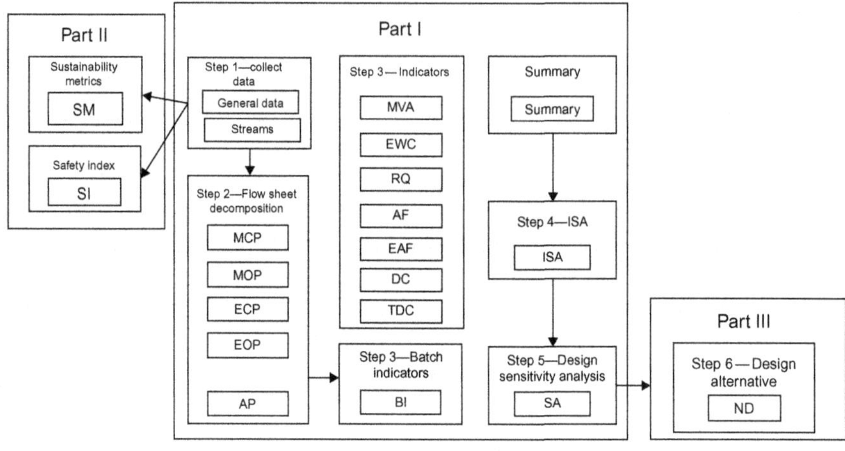

FIGURE 9.3

Work-flow in SustainPro.

for generating more sustainable solutions. A brief summary of the meaning of each indicator is listed in Table 9.10.

The indicators with the highest potential are then designated as design targets for generation of new design alternatives. Based on the knowledge of the definition of the mass and energy indicators and on the general behavior of process systems, it is possible to qualitatively judge if it is feasible to make changes on the operational variables related to the targeted indicators.

ECON

ECON was developed in Visual Basic and contains nine sections, namely, equipment cost calculation, capital cost calculation, operating cost calculation, PIE chart analysis, sensitivity analysis, and alternative comparison. Two of the most important parameters obtained through this software are the operational and the investment costs, along with NPV, rate of return, and payback time. ECON requires as input data the equipment sizing information, including details of the utilities and type of material used. Figure 9.4 presents the main ECON user interface.

LCSOFT

The present framework includes LCSoft software that systematically calculates the important LCA factors (Kalakul et al., 2013). LCSoft is used to estimate the carbon footprint and the PEIs. Figure 9.5 shows the main user's interface of LCSoft.

Table 9.10 Mass and Energy Indicators Summary

Indicator	Meaning
Material value added (MVA)	Value generated between the entrance and exit of a certain compound in a given path (if negative, it must be increased). A negative value of MVA means that the compound in that specific path is losing monetary value. Possible way of improving the process sustainability: recycle (implementation).
Energy waste cost (EWC)	Represents the maximum theoretical amount of energy that can be saved in each path, open or closed (if positive it must be decreased). A positive value of EWC means that, in a certain path, there is a large amount of energy being wasted, energy that is being added to the system, or energy produced *in situ*. Either way, a highly positive value represents possible sustainability improvement with heat integration.
Total value added (TVA)	Describes the economic influence that a compound has in a certain path. It is given by TVA = MVA − EWC (if this value is highly negative it must be increased). It could be the result of a negative MVA, a highly positive EWC, or a combination of both.
Reaction quality (RQ)	Measures the influence of a compound in a given path. The user must provide this information after performing an exhaustive evaluation on the compounds effect (in a certain path) on the process overall productivity. If negative, then the compound represents a negative impact on the process productivity; if zero, then it means that the compound has no impact on the process productivity; if positive, then it means that the compound is contributing positively to the system productivity. Possible way of improving the process sustainability and then decreasing the negative impact on the process productivity: implement purge to decrease the amount of compound on the system or improve the chemical synthesis.
Demand cost and total demand cost (DC and TDC)	Applied only to open paths once it tracks the energy flows across the system. DC represents the energy flow of a compound from the entrance to the exit. TDC represents the total energy flow for every compound in a certain path (both give a quantitative evaluation in monetary units per kg). For each path the user can evaluate the weight that a compound has on the global TDC value for a certain path.
Mass accumulation factor and energy accumulation factor (AF and EAF)	Applied only to closed paths once it describes/determines the accumulative behavior of a compound in that path (AF) or the accumulative behavior of the energy in that same path (EAF). Both are due to the amount that is being recycled relative to the input to the system. High values represent high possibility for improvements.

Source: From Carvalho et al. (2013).

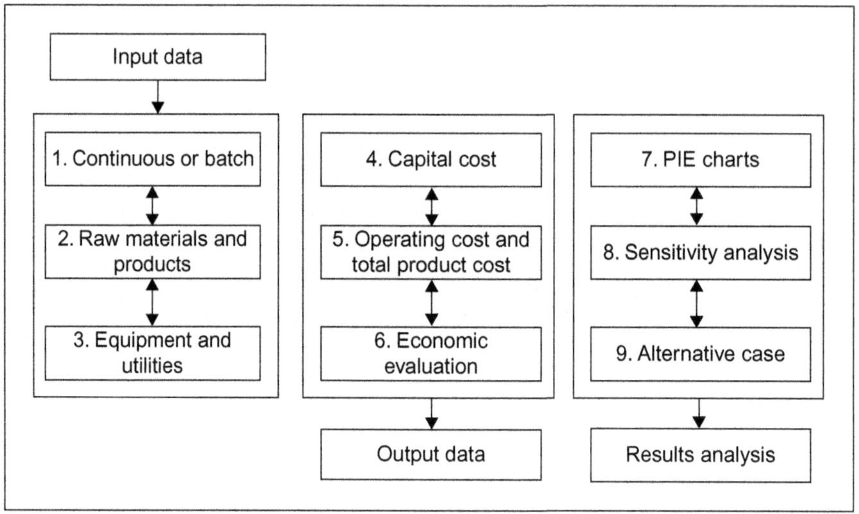

FIGURE 9.4

Work-flow in ECON.

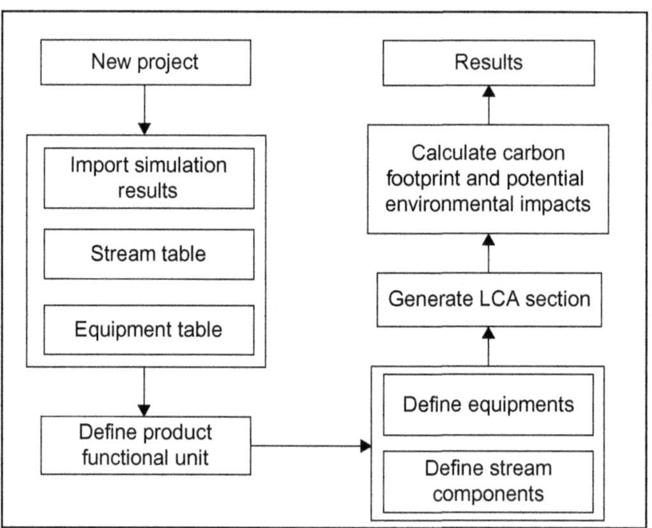

FIGURE 9.5

Work-flow in LCSoft.

As input data, the software needs a detailed mass balance of the production process and the duties of all equipment in the process. Additionally, a careful analysis of the system boundaries must be performed because one needs to specify the origin of each compound present in the system. In other words, one needs to know the origin of every compound that is shown in the mass balance, whether it is a product, by-product, chemical, or raw material. Table 9.11 lists the 11 PEIs calculated by LCSoft, together with a brief description of each one.

ICAS

ICAS is an integrated computer-aided system consisting of several toolboxes that help to solve a wide range of problems in an efficient manner (Gani et al., 1998). Several toolboxes are available, such as the CAPEC Database, ProPred, ProCAMD (computer-aided molecular design), and MoT (modeling toolbox).

PROII

PROII (Invensys Systems, 2007) performs rigorous mass and energy balances for chemical processes. It offers a wide variety of thermodynamic models to optimize plant performance by improving process design.

ASPEN-PLUS

Aspen-plus (Aspen-Tech, 2013) performs rigorous mass and energy balances for chemical processes. It offers a wide variety of thermodynamic models to optimize plant performance by improving process design.

CASE STUDY

A biorefinery case study involving bioethanol production from different renewable resources highlights the features/advantages of the framework. To satisfy the demanding targets set by worldwide governments, bioethanol is being considered as a possible fuel replacement.

Bioethanol can be produced from various renewable raw materials. The most common among them are the lignocellulosic feedstock (first generation) and sugary biomass (second generation) and, because of climatic and geographic variances, accessibility, composition, and price, these feedstocks are different (Kumar et al., 2006; Gargalo and da, 2013). Note that in this study, when selecting the optimal biorefinery process configuration, geographical constraints regarding price and accessibility of raw materials have been considered. The bioethanol production problem is solved and analyzed through the framework described in section 2.

Table 9.11 Brief Description of the PEIs

Environmental Metrics	
HTPI (1/LD 50)	Human toxicity from ingestion and inhalation is used as a measure to estimate the toxicity potential because they take into account the primary routes of exposition to a chemical. The compound is analyzed at 0°C and normal pressure. The concentration that caused death in 50% of a test population of rats by oral ingestion was used as HTPI estimate. Molecular methods were used in certain cases when LD_{50} data were not available.
HTPE (1/TWA)	Human toxicity from dermal exposure, similar to HTPI in terms of significance. To estimate HTPE, time-weighted averages of the threshold limit values (TLV) were used, obtained from OSHA, ACGIH, NIOSH, and represent occupational safety exposure limits. Those values are currently being used as measurement values in LCSoft database.
GWP (CO_2 eq.)	Global warming potential is determined by comparing the extent to which a compound absorbs infraradiation over its atmospheric lifetime and the extent that CO_2 absorbs infrared radiation over its respective lifetimes. The compounds half-life time was also considered for this measure calculation; 100 years was the base timeframe chosen by LCSoft.
ODP (CFC-11eq.)	Ozone depletion potential is estimated by comparison the rate at which a unit mass of chemical reacts with ozone to form molecular oxygen. For a compound is considered to have impact on this field, it must contain chlorine or bromine and remain in the atmosphere long enough to reach the stratosphere.
PCOP (C_2H_2 eq.)	Photochemical oxidation or smog formation potential is estimated by comparison of the rate at which a unit of mass reacts with a hydroxyl radical with the rate at which a unit mass of ethylene reacts with the same radical.
AP (H+ eq.)	Acidification potential or acid rain potential is calculated by comparing the rate of release of H^+ to the atmosphere as promoted by a chemical to the rate of release of H^+ into the atmosphere as promoted by SO_2.
ATP (1/LC 50)	Aquatic toxic potential is estimated based on the concentration of a compound in freshwater that causes death in 50% of the animal population being tested.
TTP (1/LD 50)	Terrestrial toxic potential is estimated based on the concentration of a compound in land sites that causes death of 50% of the animal population being tested.
HTC (kg of benzene eq.)	Human toxicity carcinogens include chemical emissions to urban air, rural air, agricultural soil, and natural soil.
HTNC (kg toluene eq.)	Human toxicity noncarcinogenic impacts include chemical emissions to urban air, rural air, agricultural soil, and natural soil.
ET (kg 2,4-D eq)	Eco-toxicological toxicity includes impacts for emissions to urban air, rural air, freshwater, and agricultural soil.

Source: Adapted from Cabezas et al. (1999).

PART 1—PROBLEM DEFINITION

Four geographic locations (China, the United States, Thailand, and Brazil) and seven feedstocks have been considered in this case study. Four lignocellulosic raw materials (hardwood chips, corn stover, cassava rhizome, and switch grass), together with a combination of three different lignocellulosic feedstocks (45.4% cassava rhizome, 5.04% corn stover, 49.55% sugarcane bagasse) and two sugary biomass (sugarcane and corn), are considered. Raw material availability and price data have been collected according to the respective geographies. The respective feedstock's composition (%w/w) is given in Appendix A.

Lignocellulosic ethanol production process is mainly composed of five process steps: biomass management and pretreatment; hydrolysis (saccharification); fermentation; concentration; and dehydration. The bioethanol production process from lignocellulosic feedstocks is extensively described by Quintero et al. (2011). Alvarado-Morales et al. (2009) presented a more focused work on the bioethanol production from hardwood chips. Pretreatment methods are usually classified into three groups: chemical; physical; and biological. In this study the dilute acid and ammonia belonging to the type of chemical pretreatment are considered; steam explosion and liquid hot water are also used as complementary physical pretreatments.

The main goals of the first step are to break the lignin shell protecting the cellulose and hemicellulose content, to decrease the crystallinity of the cellulose, and to increase the porosity of the biomass. Multiple molecular actions take place during the pretreatment and hydrolysis steps, such as partial degradation of hemicellulose into acetic acid, formation of phenolic compounds from the lignin content, and weak acids and furans products of pentose and hexose sugar degradation (Chakraborty et al., 2013).

Depending on the pretreatment and hydrolysis intensity, various side products may be produced, often inhibiting the consecutive microbiological hydrolysis and/or fermenting steps. The use of dilute sulfuric acid as hydrolysis catalyst is generally found acceptable; it improves the recovery of hemicellulose sugars and the enzymatic hydrolysis on the solid residue, and it cuts the production of the inhibitory compounds. Uses of concentrated acids are not ideal because they are corrosive and a strong alkali solution is needed to neutralize the resulting hydrolysate. The enzymatic complex used to hydrolyze the cellulose fraction is produced *in situ* or acquired from commercial enzyme manufacturers.

The fermentation stage for the sugary biomass is simple and it does not face serious by-product contamination and consequent inhibition of the microorganism. It is processed in only one stage and represents a mature technology. However, the fermentation of the lignocellulosic hydrolysates is more demanding; it fights against challenges like inhibition and low conversion rate of C_5 and C_6 sugars. As by-products, mainly CO_2 and furfural are produced.

Differing in the alignment of the hydrolysis C_5 fermentation and C_6 fermentation, several fermentation configurations are possible, and the most commonly used are: the simultaneous hydrolysis and fermentation (SSF) in which C_6 hydrolysis and C_6 fermentation are performed in only one step and the C_5 hydrolysis and fermentation are also performed in one step, but separately from the C_6 processing, and simultaneous saccharification and co-current fermentation (SSCF) where, directly after the pretreatment, the SSCF of C_5 and C_6 sugars are performed simultaneously. The latter is the production process that is accepted as the best-developed ethanol production process and, because of the significant technology simplification, it reduces the investment costs. The next step is the downstream purification, two distillation columns; the first (beer column) is where the ethanol content is raised and the second is the rectification column, where the ethanol concentration reaches the azeotropic composition (~92.5%m/m). The last step is the dehydration by vapor phase molecular sieves adsorption until the final composition of 99.5% (%m/m) ethanol is obtained.

Regarding the corn feedstock, there are two types of processes that can be used for bioethanol production: the wet milling process or the dry milling process. The dry milling process is the most applied processing route for bioethanol synthesis because it requires lower investment, lower production and operational cost, and only has DDGS (dried distiller grains with solubles) as by-product. Detailed information about the biotechnological routes for bioethanol production from corn and starchy feedstocks can be found elsewhere (Bothast and Schlicher, 2005). Quintero et al. (2008) presented a thorough economic and environmental assessment for the bioethanol production from corn as feedstock.

The process of converting sugarcane to bioethanol is considered to be in its mature state, being used for many years at approximately constant development pace. In an autonomous distillery, first the sugarcane juice is extracted and the bagasse is obtained. The main side product is the *cachaza*. *Cachaza* is marketed as a component for animal feed, for composting, or for energy production. The downstream purification scheme for bioethanol production from corn or sugarcane is the same as described before for the previous feedstocks.

After the data for all the configurations and production schemes have been collected, outlined, and classified, the next step is to define the objective function. Here, the maximization of the GOI has been selected (Eq. (9.16)) as the objective function.

PART 2—SUPERSTRUCTURE GENERATION AND BASE CASE(S) IDENTIFICATION

Systematically following steps 2.1–2.7 of the framework, one obtains the refined superstructure presented in Figure 9.6, where the alternatives that remain from the synthesis rules screening are listed. Step 2.8 corresponds to the process modeling and by using the tool EOLO, the mass and energy balances were estimated along with the respective objective function (Eq. (9.16)); the objective function values calculated for each alternative are presented in the last column of Table 9.12.

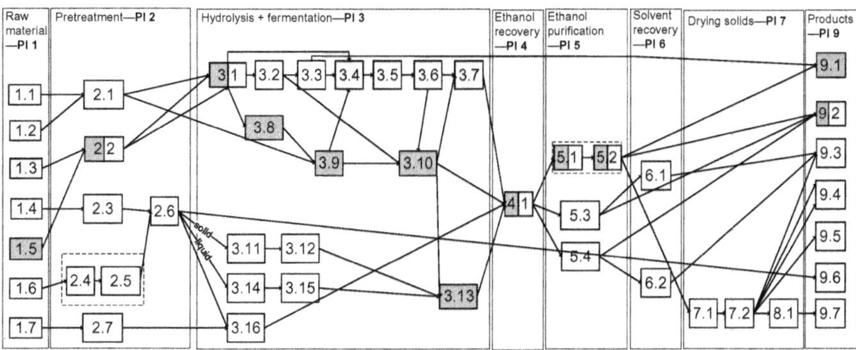

FIGURE 9.6

Refined superstructure obtained from step 2.7 for the bioethanol production. The optimal process configuration (option 14) is shown in dark gray, and the second best is shown in light gray (option 8) (Quaglia et al., 2014).

From these results, it was possible to perform step 2.9. Two feasible alternatives were selected because of the proximity of the objective function values, options 14 and 8, highlighted in dark gray and light gray in Figure 9.6. Therefore, these two options were selected to be the base cases to proceed to part 3 of the framework; they correspond to the bioethanol production in Thailand from a combined lignocellulosic feedstock (14) and from cassava rhizome (8).

Note that option 8 is referred to as CR-BC and option 14 is referred to as MIX-BC in the remainder of this chapter.

PART 3—PROCESS SIMULATION, SUSTAINABILITY ANALYSIS, AND BOTTLENECK IDENTIFICATION

Step 3.1

The rigorous simulations of options MIX-BC and CR-BC, identified as the base cases in part 2, were performed using PROII. Figures 9.7 and 9.8 show the overall mass–energy balances for base cases CR-BC and MIX-BC. Figures 9.9 and 9.10 show the flow sheet of the base cases CR-BC and MIX-BC.

Step 3.2

SustainPro is used to evaluate base cases CR-BC and MIX-BC, which provide as output an ordered list of indicators, ranked from top to bottom in terms of potential for improving the process sustainability if changes were made in those paths. The related critical points identified through this list and stated to be addressed reflect the process bottlenecks with respect to resources consumption. A summary of the list of the main indicators is given in Appendix A. The results of resources usage, of water, raw materials, and energy for base cases MIX-BC and CR-BC

Table 9.12 Results for the 14 Flow Sheet Alternatives Considered for the Bioethanol Production Case Study

#	Feedstock	Country	Products and By-products	Total Net Primary Energy Usage Rate per kg of Product	Total Net Primary Energy Usage Rate per unit value added	Total Raw Materials Used per kg of Product	Total Raw Materials per Unit Value Added	Net Water Consumed per Unit Mass of Product	Net Water Consumed per Unit Value Added	RM Cost per kg of Product	Utilities Cost per kg of Product	Operating Cost per kg of Product	Gross Revenue per kg of Product	Gross Operating Margin per kg of Product
				MJ/kg of EtOH	GJ/$	kg/kg of EtOH	kg/$	kg/kg of product	kg/$	$/kg of EtOH	$/kg of EtOH	$/kg of EtOH	$/kg EtOH	$/kg EtOH
1	HC	China	EtOH	434	0.62	9.16	13.09	8.82	12.60	0.27	1.930	2.21	0.70	**−2.26**
2	HC	China	EtOH, Furfural	434	0.62	9.16	13.09	8.82	12.60	0.27	1.930	2.21	0.81	**−1.77**
3	HC	China	EtOH, Fertilizer	434	0.62	9.16	13.09	8.82	12.60	0.27	1.930	2.21	0.91	**−1.66**
4	HC	China	EtOH, Furfural, Fertilizer	434	0.62	9.16	13.09	8.82	12.60	0.27	1.930	2.21	1.02	**−1.55**
5	CS	United States	EtOH	36	0.07	5.00	9.22	0.39	0.71	0.27	0.224	0.49	0.54	**−0.40**
7	CR	Thailand	EtOH	28	0.05	3.10	5.17	3.52	5.87	0.05	0.111	0.16	0.60	**0.39**
8	CR	Thailand	EtOH, Furfural	28	0.05	3.10	5.17	3.52	5.87	0.05	0.111	0.16	0.67	**0.46**
9	SG	China	EtOH	6176	8.82	4.11	5.87	25.55	36.50	0.12	10.01	10.13	0.70	**−15.08**
10	Corn	United States	EtOH	13	0.02	2.88	4.03	0.00	0.00	0.49	0.444	0.93	0.71	**−0.22**
11	Corn	United States	EtOH, DDGS	13	0.02	2.88	4.03	0.00	0.00	0.49	0.444	0.93	0.71	**−0.22**
12	Sugarcane	Brazil	EtOH	97	0.11	16.8	18.71	4.31	4.79	0.17	0.369	0.54	0.90	**0.31**
13	Sugarcane	Brazil	EtOH, Bagasse	97	0.11	16.8	18.71	4.31	4.79	0.17	0.369	0.54	0.91	**0.31**
14	Mix	Thailand	EtOH	1	0.00	3.67	6.11	0.90	1.50	0.05	0.004	0.06	0.60	**0.51**

Source: Adapted from (Quaglia et al., 2014).

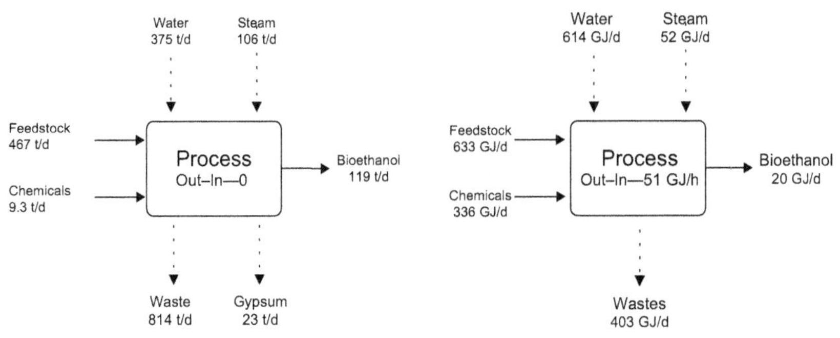

FIGURE 9.7

Mass and energy balance for bioethanol production from cassava rhizome, CR-BC.

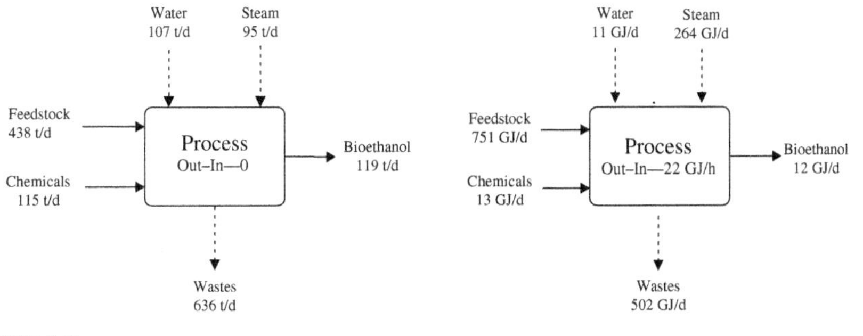

FIGURE 9.8

Mass and energy balances for bioethanol production from a combined lignocellulosic feedstock, MIX-BC.

are reported in Table 9.12. Regarding the energy consumption, the bioethanol production from the mix lignocellulosic feedstock in Thailand has lower energy and water consumption. This is justified by the fact that the pretreatment stage has only one equipment (ammonia conditioning), which makes it less utility demanding than CR-BC. Because the raw material consumption is strongly related to the feedstock composition and pretreatment efficiency, CR-BC stands out by consuming fewer raw materials than the MIX-BC per unit value added.

Step 3.3

ECON is used to evaluate and estimate the economic criteria for base cases CR-BC and MIX-BC. The project lifetime is considered to be 20 years, with an operating year having 7,920 h. A selection of the economic analysis results is given in Table 9.13. Although, according to the analysis from step 3.2, the utilities consumption in CR-BC is lower than the one from MIX-BC, MIX-BC has a higher

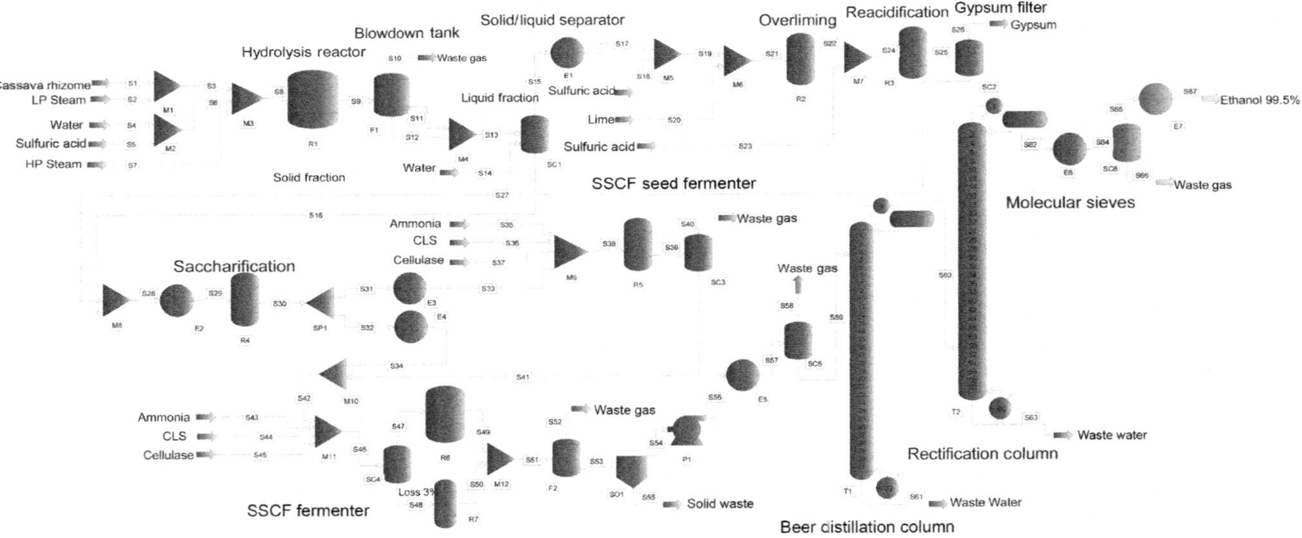

FIGURE 9.9

Bioethanol production from cassava rhizome, CR-BC (option 8).

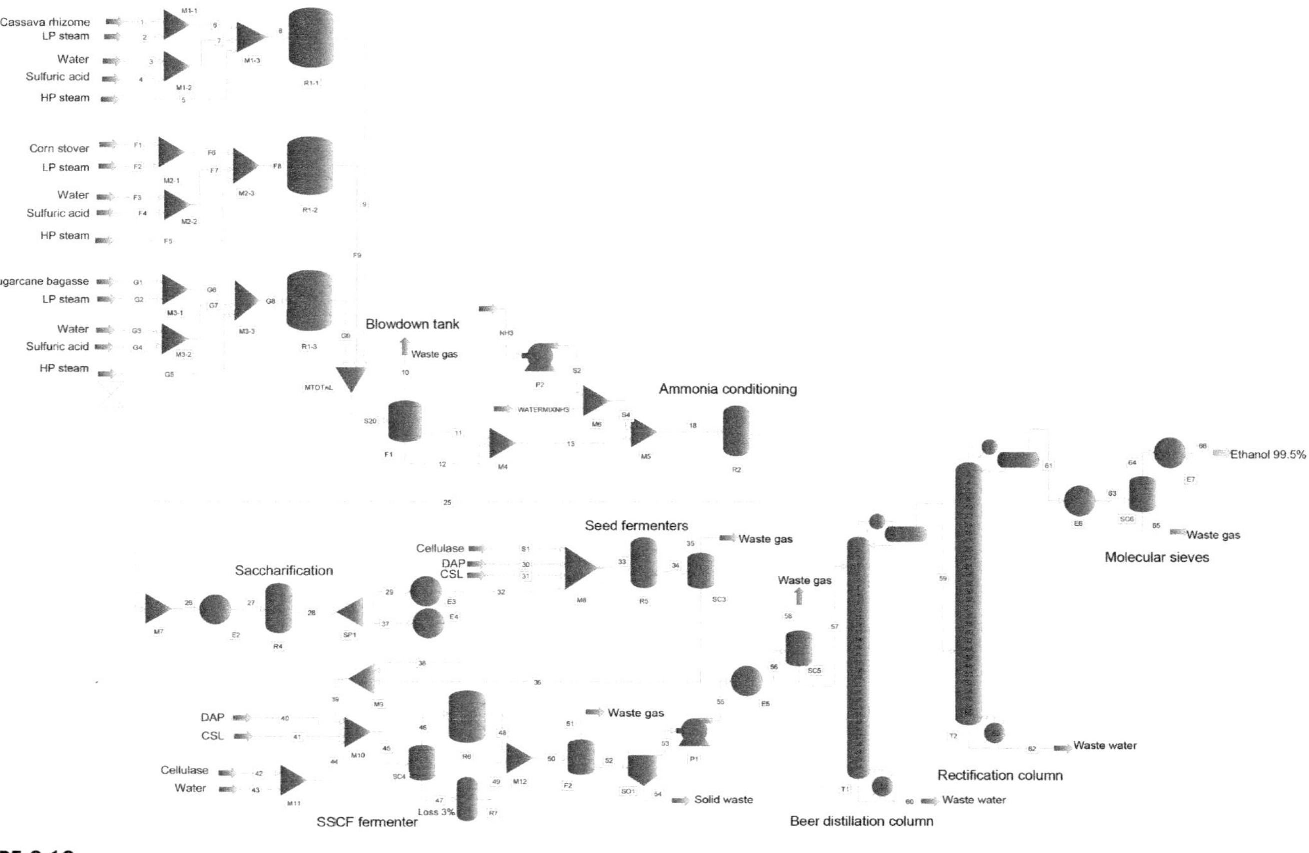

FIGURE 9.10

Bioethanol production from a combined lignocellulosic feedstock, MIX-BC (option 14).

Table 9.13 Summary of the Economic Metrics for Both Base Cases

	CR-BC	MIX-BC
Operating cost (\$/kg EtOH)	0.53	0.78
Capital cost (\$/kg EtOH)	0.07	0.13
Minimum EtOH selling price (\$/kg EtOH)	0.60	0.92
Net present value (10^6\$/y)	102	31
Payback time (y)	2	5
IRR (%)	50	17

Table 9.14 Summary of the Critical Points Identified and the Design Target to Overcome Them

	CR-BC	MIX-BC
Critical points	i. *High net fresh water inlet*, later released as waste ii. *High energy usage in the heat exchange equipment* iii. *High gypsum waste production*	i. *High net fresh water inlet*, later released as waste ii. *High energy usage in the heat exchange equipment*

operating cost because of the demand on ammonia consumption (and the respective waste treatment). Therefore, the minimum selling price of bioethanol in the MIX-BC is found to be approximately 50% higher than the one obtained for CR-BC.

Step 3.4

LCSoft is used to evaluate and estimate the environmental metrics for base cases CR-BC and MIX-BC. The overall carbon footprint obtained is 5.9 and 0.8 kg CO_2 eq/kg of EtOH. There is a reduction of 86% between the carbon footprint of processes CR-BC and MIX-BC. This is explained by the fact that seven equipment pieces within the pretreatment stage that were representative of solid/liquid separation, overliming and reacidification on CR-BC were replaced by only the ammonia-conditioning equipment, thereby leading to less waste of chemicals in pretreatment (such as gypsum) and lower utilities consumption.

Step 3.5—Design targets

The results collected on the critical points given by step 3.2 are compared with the results from steps 3.3 and 3.4; if (and where) there is a match, then a global process bottleneck has been found. Table 9.14 gives a detailed list of the main critical points identified for further process improvements.

Table 9.15 Alternatives Generated for Each Base Case Tested

	CR	MIX
Alternatives generated	CR-A Water recycle + energy cogeneration → to overcome the first two identified critical points	MIX-A Heat integration
	CR-B Overliming and the reacidification step are replaced by only one step, NH_3 conditioning → to overcome the third critical point that was identified	MIX-B Heat integration + water recycle MIX-C Heat integration + water recycle + solid combustion

PART 4—GENERATION AND SCREENING OF NEW OPTIONS

Step 4.1—Generation of new alternatives

Based on the critical points identified in step 3.5, five different alternatives were generated: two new options for bioethanol production from cassava rhizome and three new options regarding bioethanol production from a combined feedstock.

Table 9.15 gives the list of changes that were selected to overcome the identified bottlenecks.

Step 4.2—Rigorous simulation of the alternatives generated

PROII was used to simulate the alternatives generated in the previous step.

Step 4.3—Full sustainability analysis of the generated alternatives

A summary of the main results obtained is given in Table 9.16.

According to NREL studies of bioethanol production from corn stover (Aden et al., 2002; Dutta et al., 2011), the IRR is approximately 10% after taxes in both studies and the minimum selling price is 0.72 and 0.36 $/kg of EtOH. Based on these indications, the results obtained (listed in Table 9.16) are considered to be validated, at least from a qualitative standpoint.

Step 4.4—End step: screening and selection of the best option to satisfy the design problem

After the sustainability evaluation of the alternatives, the user is guided through the last step of the framework, where the alternatives are ranked according to a multicriteria objective function (Eq. (9.21)). Because the objective function is selected to be minimized, the alternative that presents the lowest value is the best solution for bioethanol production in the search space spectrum.

The three different functions considered are weighted equally (one-third each) here, and four different sets (cases 1–4) of weighting factors for the metrics (four

Table 9.16 Summary of the Main Results Obtained for CR-BC and MIX-BC and the Respective Alternatives Generated in Step 4.1

Sustainability Analysis		CR			MIX			
		CR-BC	CR-A	CR-B	MIX-BC	MIX-A	MIX-B	MIX-C
Resources usage	Raw material (kg/kg EtOH)	3.94	3.10	3.70	3.70	3.69	3.69	3.69
	Energy (GJ/kg EtOH)	0.02	0	0.02	0.02	0.01	0.01	0.01
	Fresh water (kg/kg EtOH)	3.17	4.29	1.83	1.83	1.84	0.34	1.14
Economic metrics	Operating cost ($/kg EtOH)	0.53	0.26	0.78	0.78	0.73	0.78	0.70
	Capital cost ($/kg EtOH)	0.07	0.18	0.13	0.13	0.13	0. 14	0.13
	Minimum EtOH selling price ($/kg EtOH)	0.60	0.29	0.92	0.92	0.88	0.92	0.92
	NPV ($10^6$$/year)	102	104	21.31	21.31	31.39	20.24	39.21
	IRR (%)	50	25	17	17	21	16.5	27
	Payback time (years)	2	2	4.48	4.48	3.80	4.65	2.54
Environmental metrics	Carbon footprint (kg CO_2 eq/kg of EtOH)	5.92	4.19	0.18	0.78	0.48	0.49	0.48

different scenarios) were tested to obtain a more clear view of the system (Table 9.17).

As mentioned in section 2.4 (step 4.4), because the metrics are not all in the same units of measure, normalization has been used to make a meaningful comparison.

It can be seen from the graphical representation of the results (Figure 9.11) that the alternative that minimizes the multiobjective function is option CR-B,

Table 9.17 Four Different Sets of Weighting Factors

		Case 1	Case 2	Case 3	Case 4
Resources usage	Raw material usage, x_1	33	20	60	20
	Energy usage, x_2	33	20	20	60
	Freshwater, x_3	33	60	20	20
Economic metrics	Operating cost, x_4	80	80	80	80
	Capital cost, x_5	20	20	20	20
Environmental metrics	Carbon footprint, x_6	100	100	100	100

Results presented as percentages.

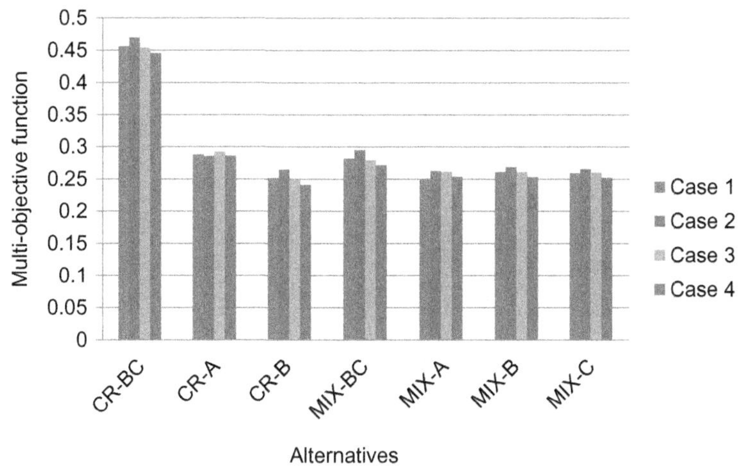

FIGURE 9.11

Graphical representation of the different sets of weighting parameters and the F_{obj} value for each one of the sets.

with respect to all sets of weighting factors. Therefore, one can say that by changing the weights that different metrics have over the decision criteria, alternative CR-B is the one that still stands out.

CONCLUDING REMARKS

A computer-aided multicriteria/level framework for sustainable chemical process design has been presented together with its implemented methods and tools. The framework aims to provide a robust decision-making tool through a step-by-step methodology for process design, allowing optimization of different scenarios based on specific selections of technical, economic, and environmental metrics. The application of the framework has been highlighted through a case study involving the production of bioethanol from different renewable resources. Tools for economic analysis, sustainability analysis, and LCA have been integrated with process simulation, process synthesis, and design algorithms. The framework together with its implemented tools is able to identify sustainable options for current and future chemical processes. Current work involves the extension of the capabilities of the framework as well as development of new case studies. Future work will need to involve the same.

APPENDIX A

Table A.9.1 Compositions of the Simple Different Feedstocks Used for the Bioethanol Case Study

Lignocellulosic Biomass	Cellulose %	Hemicellulose %	Lignin %
Hardwood chips	43	19	28
Corn stover	37	21	18
Witch grass	45	32	23
Cassava rhizome	31	45	24
Sugarcane bagasse	55	25	20

Sugary Biomass	*Glucose %*	*Starch %*	*Cellulose %*	*Carbon %*
Corn	10	72	4	9
Sugarcane	47	0	0	25

Source: Adapted from Gargalo et al. (2014).

Table A.9.2 Summary of the Most Important Indicators (Critical Points) Given by SustainPro Regarding Alternative CR-BC

Path	MVA	Probability	Path	EWC	Probability	Path	TVA	Probability
OP 322	−553	High	OP 67	795	Medium	OP 322	−553	High
OP 206	−506	High	OP 43	271	Low	OP 206	−506	High
OP 15	−440	High	OP 50	145	Low	OP 15	−473	High
OP 318	−376	High	OP 35	103	Low	OP 318	−376	High
OP 326	−300	High	OP 62	75	Medium	OP 326	−300	High
OP 222	−134	High	OP 44	42	Low	OP 43	−271	Low
OP 139	−127	High	OP 15	33	High	OP 50	−145	Low
OP 173	−81	High	OP 36	32	Low	OP 222	−134	High
OP 211	−68	High	OP 209	18	High	OP 139	−127	High
OP 199	−56	High	OP 51	17	Low	OP 35	−103	Low

Table A.9.2 Summary of the Most Important Indicators (Critical Points) Given by SustainPro Regarding Alternative CR-BC *Continued*

Path	MVA	Probability	Path	EWC	Probability	Path	TVA	Probability
OP 14	−49	High	OP 21	7	Low	OP 173	−81	High
OP 10	−40	High	OP 14	6	High	OP 211	−68	High
OP 155	−34	High	OP 39	5	Low	OP 199	−56	High
OP 325	−30	High	OP 6	5	Low	OP 14	−55	High
OP 189	−21	High	OP 225	5	High	OP 44	−42	Low
OP 106	−20	High	OP 142	5	High	OP 10	−41	High
OP 227	−18	High	OP 42	4	Low	OP 155	−34	High
OP 144	−17	High	OP 34	3	Low	OP 36	−32	Low
OP 215	−15	High	OP 22	3	Low	OP 325	−30	High
OP 132	−14	High	OP 176	3	High	OP 189	−21	High
OP 178	−11	High	OP 46	3	Low	OP 106	−20	High
OP 130	−10	High	OP 45	2	Low	OP 209	−20	High
OP 166	−9	High	OP 202	2	High	OP 227	−18	High

Table A.9.3 Summary of the Most Important (Critical Points) Given by SustainPro Regarding Alternative MIX-BC

Path	MVA	EWC	TVA	Probability
OP 177	−27.39036	973.28873	−1000.6791	High
OP 356	−454.75406	0.01482	−454.7688	High
OP 351	−397.48919	16.77849	−414.26768	High
OP 33	−255.55569	70.72156	−326.27726	High
OP 176	−1.12861	319.05145	−320.18006	High
OP 263	−7.897462	293.56217	−301.45963	High
OP 29	−234.27523	64.26179	−298.53702	High
OP 135	−7.57396	281.40983	−288.98379	High
OP 281	−5.53624	205.79196	−211.32820	High
OP 153	−5.47540	203.43814	−208.91354	High
OP 368	−181.33841	0.67500	−182.01341	Low
OP 357	−170.56374	0.01103	−170.57477	High
OP 17	0	153.03881	−153.03881	Low
OP 21	0	132.69867	−132.69867	Low
OP 170	−3.04185	109.39207	−112.43392	High
OP 359	−39.91504	59.97138	−99.88642	Low
OP 27	0	98.754098	−98.75409	Low
OP 262	−0.32541	92.52496	−92.85037	High
OP 128	−0.84113	31.61241	−32.45354	High
OP 162	−0.90335	30.50541	−31.40877	High

REFERENCES

Aden, A., Ruth, M., Ibsen, K., Jechura, J., Neeves, K., Sheehan, J., Slayton, A., 2002. Lignocellulosic Biomass to Ethanol Process Design and Economics Utilizing Co-Current Dilute Acid Prehydrolysis and Enzymatic Hydrolysis for Corn Stover Lignocellulosic Biomass to Ethanol Process Design and Economics Utilizing Co-Current Dilute Acid Prehyd. Golden, CO. <http://dx.doi.org/NREL/TP-510-32438>.

Alvarado-Morales, M., Terra, J., Gernaey, K.V., Woodley, J.M., Gani, R., 2009. Biorefining: computer aided tools for sustainable design and analysis of bioethanol production. Chem. Eng. Res. Des. 87 (9), 1171–1183.

Andersen, K., Eide, M.H., Lundqvist, U., Mattsson, B., 1998. The feasibility of including sustainability in LCA for product development. J. Clean. Prod. 6 (3–4), 289–298.

Aspen-Tech, 2013. Aspen-plus. <www.aspentech.com/> (accessed 30.07.2014).

Azapagic, A., 1999. Life cycle assessment and its application to process selection, design and optimisation. Chem. Eng. J. 73 (1), 1–21.

Azapagic, A., Howard, A., Parfitt, A., Tallis, B., Duff, C., Hadfield, C., 2002. Sustainable Development Progress Metrics. IChemE, <nbis.org/nbisresources/metrics/triple_bottom_line_indicators_process_industries.pdf> (accessed 10.08.2014).

Babi, D.K., Lutze, P., Woodley, J.M., Gani, R., 2014. A process synthesis-intensification framework for the development of sustainable membrane-based operations. Chem. Eng. Process.: Process Intensification <http://dx.doi.org/10.1016/j.cep.2014.07.001>.

Bothast, R.J., Schlicher, M.A., 2005. Biotechnological processes for conversion of corn into ethanol. Appl. Microbiol. Biotechnol. 67 (1), 19–25.

Boyles, A.S., Lippiat B.C., 2001. Building for Environmental and Economic Sustainability (BEES): Software for Selecting Cost-Effective Green Building Products. CIB World Building Congress, April 2001, Wellington, New Zealand, (April), 1–8.

Brundtland, G., 1987. Our Common Future: Report of the 1987 World Commission on Environment and Development. Oxford University Press, Oxford, UK.

Buxton, A., Pistikopoulos, E.N., Livingston, A.G., Stefanis, S.K., 1996. A methodology for environmental impact minimization: solvent design and reaction path synthesis issues. Comput. Chem. Eng 20 (2), S1419–S1424.

Cabezas, H., Bare, J.C., Mallick, S.K., 1999. Pollution prevention with chemical process simulators: the generalized waste reduction (WAR) algorithm—full version. Comput. Chem. Eng. 23 (4–5), 623–634.

Carvalho, A., Gani, R., Matos, H., 2008. Design of sustainable chemical processes: systematic retrofit analysis generation and evaluation of alternatives. Process Saf. Environ. Prot. 86 (5), 328–346.

Carvalho, A., Matos, H.A., Gani, R., 2013. SustainPro—a tool for systematic process analysis, generation and evaluation of sustainable design alternatives. Comput. Chem. Eng. 50, 8–27.

Chakraborty, S., Mondal, R.D., Mukherjee, D., Bhattacharjee, C., 2013. Production of bio-based fuels, bioethanol and biodiesel. In: Plemonte, V., De Falco, M., Basille, A. (Eds.), Sustainable Development in Chemical Engineering, Innovative Technologies, first ed. John Wiley & Sons, West Bengal, India; Rende, Italy, pp. 153–180.

Constantinou, L., Gani, R., 1994. New group contribution method for estimating properties of pure compounds. AIChE J. 40 (10), 1697–1710.

Curran, M.A., 2012. Life Cycle Assessment Handbook: A Guide for Environmentally Sustainable Products. Scrivener Publishing.

Diwekar, U.M., 2003. Greener by design. Environ. Sci. Technol. 37 (23), 5432–5444.

Dutta, A., Talmadge, M., Nrel, J.H., Worley, M., Harris, D.D., Barton, D., Idaho, J.R.H., 2011. Process Design and Economics for Conversion of Lignocellulosic Biomass to Ethanol Process Design and Economics for Conversion of Lignocellulosic Biomass to Ethanol Thermochemical Pathway by Indirect. <www.nrel.gov/biomass/pdfs/51400.pdf> (accessed 06.07.2013).

El-Halwagi, M.M., 1997. Pollution Prevention Through Process Integration: Systematic Design Tools. Academic Press, San Diego, CA.

Flores-Alsina, X., Rodríguez-Roda, I., Sin, G., Gernaey, K.V., 2008. Multi-criteria evaluation of wastewater treatment plant control strategies under uncertainty. Water Res. 42 (17), 4485–4497.

Gani, R., Jensen, A.K., Russel, B.M., Hostrup, M., Harper, P., 1998. An Integrated Computer Aided System (ICAS) for Educational Purposes. An Integrated Computer Aided System (ICAS) for Educational Purposes, 1998. AIChE.

García-Serna, J., Pérez-Barrigón, L., Cocero, M.J., 2007. New trends for design towards sustainability in chemical engineering: green engineering. Chem. Eng. J. 133 (1–3), 7–30.

Gargalo, C.L., da, C.L., 2013. Bio-Ethanol Production Process: Techno-Economic, Sustainability & Environmental Impact Approach. Instituto Superior Técnico, Lisbon, Portugal; Denmark Technical University, Copenhagen, Denmark.

Gargalo, C.L., Carvalho, A., Matos, H.A., Gani, R., 2014. Techno-Economic, Sustainability & Environmental Impact Diagnosis (TESED) Framework. Budapest, Hungry: ESCAPE14.

Goedkoop, M., Oele, M., de Schryver, A., Vieira, M., 2008. SimaPro Database Manual—Methods library. <www.pre-sustainability.com/download/manuals/DatabaseManualMethods.pdf> (accessed 18.07.2014).

Hacking, T., Guthrie, P., 2008. A framework for clarifying the meaning of triple bottom-line, integrated, and sustainability assessment. Environ. Impact Assess. Rev. 28 (2–3), 73–89.

Humbert, S., de Schryver, A., Bengoa, X., Margni, M., Jolliet, O., 2012. IMPACT 2002 + User Guide: Quantis, Sustainability Counts, vol. 21.

Hunkeler, D., Rebbitzer, G., 2005. The future of life cycle assessment. Int. J. Life Cycle Assess. 10 (5), 305–308.

Invensys Systems, 2013. SimSci PRO/II. <software.invensys.com/products/simsci/design/pro-ii/> (accessed 12.08.2014).

Julius, S.H., Scheraga, J.D., 2000. The TEAM model for evaluating alternative adaptation strategies. In: Haim, Y.Y. (Ed.), Research and Practice in Multiple Criteria Decision Making. Springer-Verlag, Heidelberg; Berlin, Germany, p. 319.

Kabassi., K., Cho, Y.K., 2011. BLCC analysis derived from BIM and energy data of zero net energy test home. In: ICSDC 2011: Integrating Sustainability Practices in the Construction Industry, pp. 292–298. <http://dx.doi.org/10.1061/41204(426)37>.

Kalakul, S., Malakul, P., Siemanond, K., Gani, R., 2013. Software Integration of Life Cycle Assessment and Economic Analysis for Process Evaluation. <www.sps.utm.my/download/PSEAsia2013-149.pdf> (Ed.), pp. 25–27. Kuala Lumpur, Malaysia.

Klöpfer, W., 2006. The role of SETAC in the development of LCA. Int. J. Life Cycle Assess. 11 (1), 116–122.

Klöpfer, W., Renner, I., 1995. UBA Texte 23/95: methodology of impact assessment in the context of product life cycle assessments, taking into account not only difficult to quantify or environmental categories (Methodik der Wirkungsbilanz im Rahmen von Produkt-Ökobilanzen unter Berücksichtigung nicht oder nur schwer quantifizierbarer Umwelt-Kategorien). Berlin, Germany, (in German) <http://dx.doi.org/183076222>.

Kumar, L., Dhavala, P., Mathei, S., 2006. Liquid biofuels in South Asia: resources and technologies. Asian Biotechnol. Dev. 8, 31–49.

Lambert, P.D., Lambert, T.W., Gilman, P., 2004. Computer modeling of renewable power systems. Encyclopedia of Energy, 1(NREL Report No. CH-710-36771), 633–647.

Lapkin, A., Joyce, L., Crittenden, B., 2004. Framework for evaluating the greenness of chemical processes: case studies for a Novel VOC Recovery Technology. Environ. Sci. Technol. 38 (21), 5815–5823.

Linnhoff, B., Flower, J.R., 1978. Synthesis of heat exchanger networks. AIChE J. 24 (4), 632–642.

Lutze, P., Babi, D.K., Woodley, J.M., Gani, R., 2013. A phenomena based methodology for process synthesis incorporating process intensification. Ind. Eng. Chem. Res. 52, 7127–7144.

Marrero, J., Gani, R., 2001. Group-contribution based estimation of pure component properties. Fluid Phase Equilib. 183–184, 183–208.

Morais, S.A., Delerue-Matos, C., 2010. A perspective on LCA application in site remediation services: critical review of challenges. J. Hazard. Mater. 175 (1–3), 12–22.

Novick, D., 1960. The federal budget as an indicator of government intentions and the implications of intentions. J. Am. Stat. Assoc. 55, 290.

Perry, R.H., Green, D.W., 2008. Perry's Chemical Engineers' Handbook. McGraw/Hill, New York, NY, USA.

Quaglia, A., 2013. An Integrated Business and Engineering Framework for Synthesis and Design of Processing Networks. Denmark Technical University, Copenhagen, Denmark.

Quaglia, A., Sarup, B., Sin, G., Gani, R., 2013. A systematic framework for enterprise-wide optimization: synthesis and design of processing networks under uncertainty. Comput. Chem. Eng. 59, 47–62.

Quaglia, A., Gargalo, C., Sin, G., Gani, R., 2014. Systematic network synthesis and design: problem formulation, superstructure generation, data management and solution. Comput. Chem. Eng. (special issue: Ignacio Grossman). http://dx.doi.org/10.1016/j.compchemeng.2014.03.007.

Quintero, A., Rinco, L.E., Cardona, C.A., 2011. Production of bioethanol from agroindustrial residues as feedstocks. In: Elsevier (Ed.), Biofuels: Alternative Feedstocks and Conversion Processes. Universidad Nacional de Colombia Sede Manizales, Manizales, Colombia, pp. 251–285.

Quintero, J.A., Montoya, M.I., Sánchez, O.J., Giraldo, O.H., Cardona, C.A., 2008. Fuel ethanol production from sugarcane and corn: comparative analysis for a Colombian case. Energy 33 (3), 385–399.

Ray, M.S., Sneesby, M.G., 1989. Chemical Engineering Design Project: A Case Study Approach, second ed. Curtin University of Technology/Godorn and Breach Science Publishers, Western Australia.

Saengwirun, P., 2011. Cost calculations and economic analysis (MS Dissertation), Chulalongkorn University, Bangkok, Thailand.

Saling, P., Kicherer, A., Dittrich-kriimer, B., Wittlinger, R., Zombik, W., Schmidt, I., Schmidt, S., 2002. Eco-efficiency analysis by BASF: the method. Int. J. Life Cycle Assess. 7 (4), 203–218.

Saling, P., Maisch, R., Silvani, M., König, N., 2005. Life cycle management assessing the environmental-hazard potential for life cycle assessment, Eco-Efficiency and SEEbalance®. Int. J. Life Cycle Assess. 10 (5), 364–371.

Schuller, O., Hassel, F., Kokborg, M., Thylmann, D., Stoffregen, A., Schöll, S., Rudolf, M., 2013. GaBi Database & Modelling Principles. PE INTERNATIONAL, Echterdingen, Germany.

Siirola, J.J., 1996. Strategic process synthesis: advances in the hierarchical approach. Comput. Chem. Eng. 20, 1637–1643.

Singh, N., Falkenburg, D.R., 1993. A Green Engineering Framework for Concurrent Design of Products and Processes. Wayne State University, Detroit, MI, USA.

Stein, M., Khare, A., 2009. Calculating the carbon footprint of a chemical plant: a case study of Akzonobel. J. Environ. Assess. Policy Manage. 11, 291.

Upham, P., 2000. An assessment of the natural step theory of sustainability. J. Clean. Prod. 8 (6), 445–454.

Vanegas, J.A., 2003. Road map and principles for built environment sustainability. Environ. Sci. Technol. 37 (23), 5363–5372, <www.ncbi.nlm.nih.gov/pubmed/14700321> (accessed 21.03.2014).

Wan Alwi, S.R., Manan, Z.A., 2013. Water pinch analysis for water management and minimisation: an introduction. In: Klemeš, J.J. (Ed.), Handbook of Process Integration (PI). © Woodhead Publishing Limited/Elsevier, Cambridge, UK, pp. 353–380.

Zamagni, A., Guinee, J., Heijungs, R., Masoni, P., 2012. Life cycle sustainability analysis. Life Cycle Assessment Handbook: A Guide for Environmentally Sustainable Products. John Wiley & Sons, NJ, USA, pp. 453–474.

CHAPTER

Life cycle assessment as a comparative analysis tool for sustainable brownfield redevelopment projects: cumulative energy demand and greenhouse gas emissions

10

Thomas Brecheisen and Thomas Theis

Institute for Environmental Science and Policy, University of Illinois at Chicago, Chicago, IL, USA

INTRODUCTION AND PURPOSE

The United States Environmental Protection Agency (USEPA) has defined a brownfield as "real property, the expansion, redevelopment, or reuse of which may be complicated by the presence or potential presence of a hazardous substance, pollutant, or contaminant" (USEPA, 2002). Many brownfield sites have the potential to become economically viable and host new businesses that create new jobs. However, some level of public assistance has often been required to achieve this potential, especially for sites that did not attract private redevelopers because the anticipated economic return on the investment did not justify the capital investment (Bartsch, 1999).

Sustainable development of brownfields reflects a fundamental, yet logical, shift in thinking and policy-making regarding pollution prevention. Brownfield redevelopment is inherently more sustainable than conventional development. Given the potentially synergistic effects of the clean-up and reuse of land combined with the creation of new economic opportunities and improved social welfare, there is an increasing level of interest in designing brownfield projects that have sustainable characteristics. The conformance of various brownfield redevelopment practices to the sustainability paradigm is complicated by the fact that there is no universally accepted definition of sustainability. Thus, the evaluation of sustainable designs for complex brownfield redevelopment projects must approach the problem in an adaptive and relativistic manner through the comparison of redevelopment practices among multiple sites using a common set of indices to discern the *comparative directionality* (i.e., more or less sustainable) for alternative practices and outcomes.

Life cycle assessment (LCA) is a tool that can be used to assist in determining the conformity of brownfield development projects to environmentally sustainable practices. According to the Society of Environmental Toxicology and Chemistry (SETAC), the LCA is an objective process to evaluate the environmental burdens associated with a product, process, or activity (Consoli et al., 1993). The LCA process is completed by identifying and quantifying energy and material usage, along with the associated environmental releases, to assess the impact of those energy and material uses and releases on the environment. The final stage of the LCA process is to evaluate and implement opportunities to effect environmental improvements. The LCA includes the complete life of the product, process, or activity: extraction and processing of raw materials; manufacturing, transportation, and distribution; use, reuse, and maintenance; recycling; and final disposal (Bishop, 2000). LCA can entail an iterative procedure that commences with initial scoping requirements that can be adapted later as more data become available (Goedkoop et al., 2008).

The purpose of the LCAs conducted for these brownfield redevelopment projects was to compare the cumulative energy and associated greenhouse gas (GHG) emissions required to redevelop two brownfields, including all site preparation activities, environmental assessment and remediation activities, the rehabilitation and/or construction of the site buildings, and a decade's worth of operational energy. Cumulative energy includes the sum of a building's operational energy and its embodied energy. Embodied energy is the sum of all energy required to produce a product (i.e., building product), including raw materials acquisition, processing and manufacturing, transportation, and installation. The LCAs conducted for these brownfield redevelopment projects were based on actual as-built and operational data.

RELEVANCE

The world today is faced with serious environmental concerns over climate change, ozone depletion, waste accumulation, and natural resource depletion. Of the many environmental impacts of development, climate change has the highest profile. The increased emission of GHGs is the result of the burning of fossil fuels, deforestation, and land use changes. The largest contributor to GHG emissions is the built environment, which accounts for up to 50% of global carbon dioxide emissions and consumes 40% of the materials entering the global economy. Sustainable development requires methods and tools to measure and compare the environmental impacts of human activities for the production of various goods and services (Sharma et al., 2011), including the construction, operation, demolition, and disposal of buildings. Hou and Al-Tabbaa (2014) stated that a number of tools for sustainability evaluation have been developed, but these tools are generally in their infancy and are still in need of improvements as well as further research. Lange et al. (2014) further stated that because brownfields have many stakeholders, there is a clear need for support tools that consider the intentions of the decision-makers. LCA is a powerful tool for the evaluation of the environmental impacts of buildings and it has the potential to make a strong contribution to the goal of sustainable development (Khasreen et al., 2009).

Wedding and Crawford-Brown (2007) asserted that "indicators for the sustainability of the built environment, especially for brownfields, are lacking and assessment methods for brownfields outcomes are needed." Even though metrics for green buildings have not been widely considered when assessing the overall success of brownfield redevelopments, Cherokee Investment Partners, a leading brownfield redevelopment organization, acknowledged that the vertical construction portion of these brownfield redevelopment projects represents more than 75% of overall development cost and value at build out and warrants postdevelopment consideration. Therefore, the lack of a comprehensive suite of postdevelopment brownfield metrics can result in missed opportunities for altering the built environment whose design will have significant environmental, economic, and social impacts for decades (Wedding and Crawford-Brown, 2007).

RELATED LCA STUDIES

Many LCAs have been conducted for residential and commercial buildings for a variety of purposes. Bastos et al. (2013) performed LCAs for three residential building types in Portugal and found that the use life cycle stage was dominant, accounting for 69–83% of the total energy requirement and GHG Emissions over a 75-year life cycle. Adalberth et al. (2001) compared four multi-family buildings over a life cycle of 50 years to determine which life cycle stage had the highest contributed the most significant environmental impact. The study found the occupation phase of the buildings' life cycles accounted for 70–90% of the total environmental impact. Angela and Hutzler (2005) used LCA to analyze the use of water in various multi-occupant residential and commercial buildings over a 25-year operational life cycle. The results of this study found that the use of natural gas to heat the water would have resulted in an $80,000 life cycle savings compared with using electricity as a heat source. Scheuer et al. (2003) performed an LCA on a six-story commercial building and found that heating, ventilation, and air conditioning (HVAC) and electricity accounted for 94.4% of the primary energy consumption. Other studies have asserted that for conventional buildings in northern and central Europe, the life cycle energy is distributed as 10–20% embodied energy for building products, whereas 80–90% corresponds to energy consumption during the operational phase and $<1\%$ is associated with end-of-life treatments and disposal (Kotaji et al., 2003).

Sartori and Hestnes (2007) completed an analysis of 60 case studies of life cycle energy use of buildings in 9 countries. The study further evaluated the performance of "low-energy" (i.e., energy efficient) buildings based on the definition of having an annual heating requirement less than 252 $MJ/m^2/y$. In all of the cases, the operating energy was the dominant life cycle stage and a linear relationship between the operational energy and the total life cycle energy existed. They pointed out that buildings in similar climates might also have very different characteristics in terms of primary energy because of the various energy carriers available for thermal purposes (e.g., natural gas vs. electricity) or because of the various ways to produce electricity. For example, Norway uses 98% hydropower; Sweden relies on 49% nuclear power and 44% hydropower; and the United States uses approximately 40% coal, 30% natural gas, 19% nuclear power, and 12% renewable energy (USEIA, 2013).

Two studies that compared different versions of the same building (e.g., conventional vs. low energy) showed that the amount of embodied energy used to construct a low-energy building was higher than the embodied energy required to construct a conventional building (Sartori and Hestnes, 2007). One study (Winther and Hestnes, 1999) analyzed six versions of a residential unit in Germany, whereas the other five analyzed versions of a residential unit in Norway (Feist, 1996). However, only some of the buildings were actually built and several of the cases were hypothetical versions of the same buildings. Over a life cycle of 80 years, it was estimated that with an incremental increase in initial embodied energy equivalent to approximately 1 year of operational energy, a low-energy building could be constructed that would result in a threefold decrease in the total life cycle energy. Therefore, it was concluded that the reduced operating energy demand was the most important aspect of designing buildings that are more energy-efficient over their life cycle (Sartori and Hestnes, 2007).

By developing more energy-efficient buildings, the percentage of life cycle energy associated with building products is expected to increase (Kotaji et al., 2003). The products required for buildings use great quantities of raw materials and also require large quantities of energy for processing, but these building products also help determine the long-term energy consumption. Bribian et al. (2011) evaluated the impacts of construction materials most commonly used in the building sector in comparison with different "green" building materials based on the LCA. The study highlighted some of the most energy-intensive building products, such as steel, aluminum, copper, reinforced concrete, PVC, and glass because of their high energy consumption and raw materials in the numerous production processes that make up their life cycle, especially aluminum, which has a higher electricity energy demand that increases its impact on the global warming potential. The primary embodied energy of wood products was mainly from biomass potential energy, which represented 69–83% of the total primary energy demand because the processing energy for wood products is relatively low. The study concluded that it was important to harmonize existing inventory databases of construction materials to the characteristics of the construction industries in each country.

Jackson (2005) found that seven primary building components comprise a building's embodied energy: wood; paint; asphalt; glass; stone and clay; iron and steel; and non-ferrous metals. Asif et al. (2007) found that concrete accounted for 65% of the embodied energy for a residential building in Scotland. As improvements in the operational energy efficiency of buildings are made, the relative significance of embodied energy forms a higher proportion of the total energy over the life cycle of the building (Yohanis and Norton, 2002).

Peuportier (2001) developed a life cycle simulation tool to compare three different single-family homes. Theoretical homes were compared on the basis of LCAs in terms of their overall energy consumption. The building materials data were based on published standards, and the energy consumption data were predicted using a thermal simulation tool. The study indicated that many uncertainties and limitations were associated with data and indicators. The study suggested that the application of LCA to buildings was difficult and encouraged improvement of the assessment methodology.

Few LCA studies have evaluated building rehabilitation projects (Cabeza et al., 2013). Brecheisen and Theis (2013) used LCA to compare the life cycle stages for a brownfield redevelopment project that involved the rehabilitation of an existing building. The study concluded that the preservation and rehabilitation of the existing building, the installation of renewable energy systems (geothermal and photovoltaic) on-site, and the use of more sustainable building products resulted in 72 TJ of avoided energy impacts, which was equivalent to 14 years of operational energy for the site.

RATIONALE FOR LCA RESEARCH

Based on the cited literature, data availability has been an impediment in performing LCAs on buildings. Of all the challenges in applying LCA to buildings, the main problems were the buildings themselves. The production process is complicated and the life cycle is long with future phases based on numerous assumptions. Because there is little standardization within the building sector, there is a clear lack of data inventory. There is a need for the completion of LCAs based on actual as-built data such as construction blueprints and actual operational energy expenditures.

The United States Green Building Council (2009), in their postoccupancy study of LEED (Leadership in Energy and Environmental Design)-certified projects in Illinois, concluded "A building's best benchmark is its own performance." In that study, measured energy performances of buildings were compared with theoretical modeling results that were predicted according to ASHRAE 90.1 Standards. The study concluded "design models were not a reliable indicator of performance" (USGBC, 2009). LCAs performed using actual building operational data would, at a minimum, narrow the level of uncertainty associated with LCAs based on theoretical energy modeling.

Among the cited literature, no studies could be compared directly because of differences in goal and scope, methodology, and data used. More studies have calculated the embodied impacts associated with building materials than the complete process of building construction and use; thus, there is a need to conduct LCA studies to establish the effect of alternative materials on the energy performance of buildings. There is limited research published regarding a complete LCA of buildings and there are no comprehensive LCA studies that included

Table 10.1 Brownfield Site Characteristics

Attribute	CCGT	The Sigma Group
Location	Chicago, IL	Milwaukee, WI
Completion date	2002	2003
Site size	1.4 ha	1.1 ha
Building floor area	2,600 m^2	2,600 m^2
Property use	Office	Office
Stormwater management	Yes (retention swale)	Yes (grass swale)
Construction type	Rehabilitation	New construction
LEED status	LEED platinum	Non-LEED
Renewable energy	Yes (solar and geothermal)	No
Project development costs	$14.4 M	$3.2 M

environmental site assessments (ESAs), remediation, construction, and operational phases of a brownfield redevelopment project.

The purpose of this study was to compare two brownfield redevelopment projects based on performance data and identical scope and boundary. The sites chosen were the Chicago Center for Green Technology (CCGT), a rehabilitated brownfield located in Chicago, IL, USA, and The Sigma Group site (Sigma site), a newly constructed building located in the Menomonee Valley brownfield in Milwaukee, WI, USA. The two sites are similar in terms of size and function but differ with respect to LEED certification and use of renewable energy (Table 10.1).

BROWNFIELD SITE HISTORIES

CHICAGO CENTER FOR GREEN TECHNOLOGY

The CCGT is an approximate 1.4-ha former brownfield site located at 445 North Sacramento Boulevard in Chicago, IL. The site was characterized, remediated, and redeveloped into Chicago's first LEED platinum-certified site (Brecheisen and Theis, 2013); it is now renowned as an integrated model of energy efficiency and sustainable design (De Sousa and D'Souza, 2012). A $100 M settlement to the city of Chicago from the Commonwealth Edison Company's violation of their franchise agreement was used to redevelop the site. The total project cost was $14.4 M; $9 M was used to clean the site while the building construction and renovation costs required an additional $5.4 M (Building Green Inc., 2010). Most of the clean-up costs were associated with the removal of 382,500 m^3 of illegally stockpiled construction and demolition (C&D) debris. The CCGT site is shown in Figure 10.1A and B.

THE SIGMA SITE

The Sigma site is an approximately 1.1-ha former brownfield site located at 1300 West Canal Street in Milwaukee, WI, USA. This site was redeveloped into

FIGURE 10.1

(A and B) The CCGT site before (left) and after (right) development.

Courtesy of City of Chicago.

FIGURE 10.2

(A and B) The Sigma site before (left) and after (right) development.

Courtesy of The Sigma Group.

a commercial office building in 2003, and it now houses multiple civil and environmental engineering firms. The redevelopment of the Sigma site presented a variety of soil, groundwater, methane, and geotechnical challenges during the planning, design, orientation, and construction of its building. The site and its building incorporated numerous green design features including natural day lighting, a high-efficiency HVAC system, and a natural grass swale for stormwater management (De Sousa, 2012). In recognition of design excellence, the site was awarded the 2003 Mayor's Design Award for having added value to its neighborhood by restoring the site in a way that respected the urban fabric and character of its surroundings (Sigma, 2004a). The site was awarded a $155,000 brownfields grant from the Wisconsin Department of Commerce that was used for soil and groundwater remediation (WDNR, 2009). The total project cost was $3.2 M (Sigma, 2004b). Approximately $400,000 was obtained through various grants and the remainder of the project costs were financed conventionally. The Sigma site is depicted in Figure 10.2A and B.

ENVIRONMENTAL ASSESSMENT AND REMEDIATION

Both of these brownfield sites were eligible for and participated in their respective state's voluntary clean-up program. The CCGT was enrolled in the Illinois Environmental Protection Agency (IEPA) voluntary Site Remediation Program (SRP) for the express purpose of obtaining a comprehensive no further remediation letter (NFR letter). The Sigma site participated in the Wisconsin Department of Natural Resources (WDNR) Voluntary Party Liability Exemption (VPLE) 292 Program with the goal of obtaining a certificate of closure. The environmental assessment and remediation activities for the brownfield sites were performed systematically as described in the following text.

CHICAGO CENTER FOR GREEN TECHNOLOGY

In April 1999, a Phase I ESA conducted for the CCGT site revealed the site was formerly utilized as a foundry, dating back until at least 1896, in concert with a former gas plant adjacent to the site. Visual evidence of an underground storage tank (UST) was observed at the site; therefore, the Phase I ESA concluded that a Phase II ESA was warranted (Patrick Engineering, Inc., 1999a).

In June 1999, a 38,000 L heating oil UST was removed from the site. Soil samples collected from the excavation and analyzed for BETX (benzene, ethylbenzene, toluene, and xylenes), PNAs (polynuclear aromatic hydrocarbons), and heavy metals confirmed a release from the UST (Patrick Engineering, Inc., 1999b). In July 1999, a Phase II ESA was performed to characterize the site. The results of the Phase II ESA revealed that certain PNAs and heavy metals were detected in the site's soil at levels exceeding the allowable levels for industrial/commercial land use. A supplemental investigation was completed in August 2000 to delineate the full nature and extent of the impacted soil (ESE, 2000a,b).

After the full nature and extent of soil impacts were delineated, the remedial action plan was developed to outline the remedial action needed to ensure there was not an unacceptable risk to human health or the environment. Because the soil impacts were limited to relatively low levels of PNAs and heavy metals, the remedial action plan proposed only the limited excavation and disposal of the most severely impacted area of the site, whereas the majority of residual soil impacts were managed in-place by constructing an *in situ* cap known as an engineered barrier. In addition, institutional controls were proposed that would restrict the site to industrial/commercial land use. The remedial action plan was submitted to the IEPA for approval prior to its implementation (ESE, 2000b).

Upon IEPA approval of the remedial action plan, approximately 2,450 m^3 of contaminated soil was excavated and replaced with clean fill (Harding ESE, Inc., 2002). In addition, an impervious engineered barrier including the building foundation and the paved parking lots was constructed to mitigate human exposure to the underlying contaminants. The results of the remedial activities were submitted to the IEPA in a *Remedial Action Completion Report*. The IEPA subsequently approved the remedial activities and issued the NFR letter for the site (Harding ESE, Inc., 2002).

THE SIGMA SITE

In July 1991, when the Sigma site was still owned by the city of Milwaukee, a soil contamination assessment and foundation investigation were conducted. The file review and site inspection indicated that the site had been last owned by Milwaukee Dressed Beef Co., was used as a slaughterhouse from 1970 to 1980, and had been vacant until 1987, after which time the site was used by the Milwaukee Electric Co. as a parking lot. Eight to 10 feet of fill material was observed as an environmental concern. Laboratory analysis of soil samples confirmed the presence of PNAs and heavy metals, which were limited to the uppermost 4 feet of the site's soils. Additionally, the foundation investigation, based on the results of two soil borings that were drilled to depths of 20 m, concluded with a recommendation for a deep foundation consisting of displacement pipe piles filled with concrete (Singh and Associates, 1991).

In November 1993, Sigma performed a Phase II ESA at the site and confirmed the existence of soil and groundwater contamination beneath the site (Sigma, 1994). In January 1999, Sigma performed a Phase I ESA and a Phase II ESA. The Phase I ESA revealed that the site was previously used by the Milwaukee Fuel and Dock Co., and it was also used as a coal and lumber yard. The Phase II ESA included the advancement of 14 soil borings and the installation of 6 monitoring wells; soil and groundwater sampling again confirmed the existence of PNA-impacted soil and groundwater beneath the site (Sigma, 1999a). Later in 1999, Sigma performed a methane study and a supplemental site investigation to delineate the extent of soil and groundwater impacts. The methane study verified the existence of methane-producing soils beneath the site (Sigma, 2000). A former boat slip, 15 m wide by 30 m long, had been filled with a thick layer of organic rich soil and was suspected to be the source of methane-producing soils beneath the building footprint (Sigma, 1999b).

The proposed remedial action addressed soil contamination, groundwater contamination, and methane-producing soils. To address the impacted soil, the remedial action plan required the limited excavation and disposal of the most severely impacted soil and the replacement with clean fill, coupled with the use of the site building and associated parking lot as an *in situ* cap. For the vegetated areas of the site, a 6-inch layer of clean topsoil underlain by a geotextile warning layer also served as an *in situ* cap. To address the impacted groundwater, natural attenuation and quarterly groundwater monitoring were proposed. To address the methane-producing soils, a passive methane venting and abatement system was proposed for construction beneath the building foundation (Sigma, 1999b).

Before the construction of the site building, Sigma was required to obtain an exemption from the WDNR to construct on an abandoned landfill (Sigma, 2002). The remedial activities were completed during site development activities in 2003. Upon completion of the remedial action activities, Sigma submitted a *VPLE case summary and site closure request* (VPLE case summary) to the WDNR in January 2006. The VPLE case summary documented the completion of the appropriate remedial activities (Sigma, 2006). The WDNR subsequently approved the remedial activities and issued the final case closure on April 14, 2007, which documented that no further investigation or remediation was required at the site.

SITE DEVELOPMENT AND BUILDING DESIGN FEATURES

CHICAGO CENTER FOR GREEN TECHNOLOGY

Site development at the CCGT site involved the renovation of an existing building on-site. The building renovation activities included improvements to the building envelope and interior, including new walls, new ceilings, floors, doors, and paint. A geothermal heating and cooling system, consisting of 28 vertical wells drilled to a depth of 60 m and 6 high-efficiency heat pumps (160 kW cooling capacity), was constructed to provide 100% of the cooling and 90% of the heating requirements (IBC Engineering Services, Inc., 2010). A 970 m^2 photovoltaic array was constructed, which was designed to provide approximately 490,000 MJ of electricity annually (Building Green Inc., 2010). A 230-m^2 extensive green roof of low-growing sedum was constructed to reduce the urban heat island effect (Zvenyach and Littman, 2006). Four rainwater cisterns were installed to provide 45,425 L of rainwater storage that was used to irrigate a 0.40 ha retention swale and reduce runoff from the site (Farr Associates, 2000).

THE SIGMA SITE

Because of the swampy soil conditions, construction of the Sigma site building required a deep foundation system of 82 concrete-filled steel pipe piles and a passive methane venting system. Reinforced concrete grade beams were constructed on the concrete pile caps needed to support the steel superstructure finished with bow-truss roof beams. The building was completed with a brick and metal panel exterior and designed with clerestory windows to provide natural day lighting. Mechanically, the site building was equipped with a high-efficiency electrically powered HVAC system equipped with a critical zone reset, thermostat setback, and destratification fans. Low-voltage lighting, light timers, and occupancy sensors were also installed to reduce electrical use. In addition, reflective glazing was constructed on the south side of the building to help eliminate heat gains during the summer. The final site grading was finished with low-maintenance native plants and a grass swale to reduce runoff from the site (Sigma, 2011).

LCA METHODOLOGY

SCOPE AND BOUNDARY

The scope of the LCAs was designed to determine the CEDs and the associated GHG emissions required to perform the brownfield remediation activities and the building rehabilitation activities, and to operate the redeveloped sites. Three primary life cycle stages were analyzed: (i) brownfield assessment and remediation; (ii) building rehabilitation/construction and site redevelopment; and (iii) the energy consumed during the operation of the sites. Consistent for both sites, the boundary of the LCAs is shown in Figure 10.3.

It should be noted that the recycling/disposal (of the building) life cycle stage was not included in the scope because both buildings are still operational. The system boundary included the transportation of building materials to the sites as well as the transportation of contaminated soil away from the sites. The system boundary did not extend beyond the transportation of contaminated soil to its destination because the scope of this LCA included only the removal of the contaminated soil (excavation and trucking) from the sites.

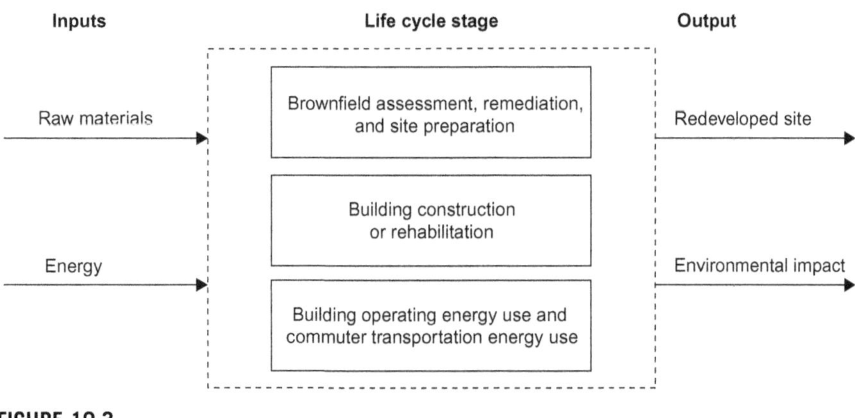

FIGURE 10.3

LCA system boundary.

TOOLS

The SimaPro software package, version 7.3.3 (Product Ecology Consultants, 2012), was used to perform the CED and GHG emission calculations for the LCAs. Based on the processes in the Ecoinvent database, the single-issue CED calculation determines the cumulative energy required to provide a good or service from the following five impact categories: (i) non-renewable fossil energy; (ii) non-renewable nuclear energy; (iii) renewable biomass energy; (iv) renewable wind, solar, and geothermal energy; and (v) renewable wind energy. The tool for the reduction and assessment of chemical and other environmental impacts (TRACI 2), a model developed by the USEPA, was used to calculate the GHG emissions (global warming potential). TRACI 2 is a midpoint-based life cycle impact assessment methodology, where the impact categories are characterized at the midpoint level for reasons including a higher level of societal consensus concerning the certainties of modeling at this point in the cause-and-effect chain (Goedkoop et al., 2010).

SimaPro contains several databases; however, only the Ecoinvent and United States Input–Output (US IO) databases were used. The Ecoinvent database includes more than 4,000 datasets based on LCA research in the following fields: energy; building products; chemicals; wood; metals; packaging and graphical paper; detergents; waste treatment services; transportation services; agricultural production systems; biofuels; electric and electronic equipment; pure chemicals; renewable materials; petrochemical solvents; and metals processing (Frischknecht et al., 2007).

The US IO database consists of a commodity matrix from 1998, supplemented with data for capital goods. The IO commodity matrix is linked to a large environmental intervention matrix. Environmental data have been compiled using several data sources: Toxic Releases Inventory 98 (TRI), Air Quality Planning and Standard (AIRS) data of the USEPA, Energy Information Administration (EIA) data of the US Department of Energy, Bureau of Economic Analysis (BEA) data of the US Department of Commerce (DOC), National Center for Food and Agricultural Policy (NCFAP), and the World Resource Institute (WRI) (Kellenberger et al., 2007). Default materials and processes included within SimaPro's databases were utilized as much as practical.

DATA

Brownfield remediation

Environmental brownfield assessment and remediation data were collected manually through the review of technical environmental reports. These reports described the activities that were completed to assess and remediate the sites. The reports that were reviewed for the sites included the aforementioned Phase I ESAs, Phase II ESAs, and the state-specific voluntary clean-up program reports. These state-specific environmental reports were provided by the site owners and by the IEPA and WDNR through the Freedom of Information Act (FOIA) requests.

Site redevelopment

The city of Chicago provided the architect's rehabilitation blueprints for the CCGT site (Farr, 2000) as well as the LEED certification documentation that was submitted to the USGBC. Sigma also provided the building's construction blueprints for the project (Redmond, 2002). The construction blueprints for each site were used to determine quantities of building materials needed for the renovation and construction of the respective site buildings. Information obtained from the review of construction blueprints included: the amount of steel framing, drywall, and drywall insulation needed to construct the interior walls; the amount of glass needed for the interior and exterior windows; the number of new steel and wooden doors and the associated door frames; the area of various floor and ceiling finishes; the amount of paint needed to paint the interior walls; the external building finishes of brick or metal siding, external roof details; the volume of structural steel and reinforced concrete required for the building foundation and superstructure, including the concrete-filled steel pipe piles that were constructed for the Sigma site building; and other components. In addition, the origins of the building materials were provided, thus enabling the transportation of the building materials to the site to be included in the analysis.

Postdevelopment building operating energy

The owners of each site provided long-term building operational energy data. The city of Chicago provided 10 years of electricity and natural gas consumption data from July 2002, when the site became operational, through June 2012. The 10-year electricity consumption is illustrated in Figure 10.4. Superimposing the electricity consumption over the average temperature suggests that the increased electricity consumption

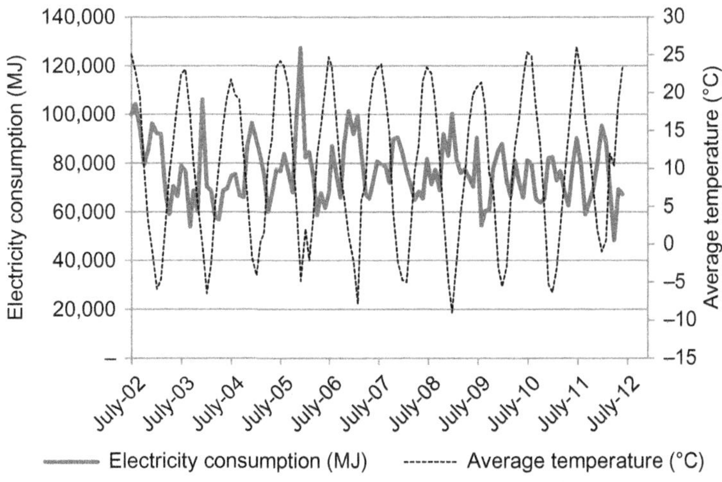

FIGURE 10.4

CCGT monthly electricity consumption (July 31, 2002 through June 30, 2012).

during the hot summer months was required to operate the heat pumps and provide cooling from the geothermal heating and cooling system. Natural gas consumption data, although not illustrated graphically herein, were provided in a similar format.

The owner of the Sigma site provided only 1 year (2010) of building operational energy consumption. Thus, it was necessary to extrapolate this year of baseline energy consumption data to the remaining years of the 10-year study period (January 2004 through December 2013). First, the annual lighting load and the computer load were estimated and subtracted from the overall 2010 energy demand to obtain the heating and cooling load. The heating and cooling load was then normalized for heating and cooling degree days and extrapolated over the remaining years of the 10-year analysis period. For example, the electrical consumption of the Sigma site during the year 2010 was 1,478,016 MJ. Based on the construction specifications, the building's floor area, and the number of employees, the annual lighting and computer loads were estimated at 93,744 and 150,782 MJ, leaving 1,233,490 MJ (1,478,016 MJ − 93,744 MJ − 150,782 MJ) available for heating and cooling. In Milwaukee in 2010, there were 6,229 heating degree days and 922 cooling degree days, for a total of 7,151 degree days (NOAA, 2013, 2014), resulting in a heating and cooling load of 172.6 MJ per degree day. This factor, coupled with the climatic data, was used to estimate the electricity consumption for the remaining years of the analysis period. Meanwhile, lighting and computer loads were assumed to be constant for each year within the analysis period.

Postdevelopment commuter transportation operating energy

Each site's long-term operational energy life cycle stage also included the commuter transportation impacts for the occupants of each site building. The commuter transportation data were generated through the administration of a commuter transportation survey to the occupants of each site, which ascertained the number of trips an employee commutes to and from each site over the course of a typical year, the distance traveled to and from each site daily, and the mode of transportation used (i.e., driving alone, carpooling, bus, light rail, commuter rail, bicycle, walk) to commute to and from the building (USGBC, 2011). Response rates of 70% and 45% were achieved for the respective CCGT site and Sigma site. A copy of the transportation survey has been included in Appendix A.

LCA RESULTS

BROWNFIELD ASSESSMENT AND REMEDIATION

The review of the environmental reports related to the CCGT site provided details involving the following primary activities:

- The excavation and removal of 382,500 m^3 of C&D debris
- The performance of the Phase I and Phase II ESAs

- The removal of one 38,000 L heating oil UST
- The excavation and disposal of 2,450 m^3 of contaminated soil, replacement with clean backfill materials, and the construction of an engineered barrier and vegetated bioswale

The complete set of data collected for the CCGT site's remedial activities are summarized in Table B.10.1.

The review of the environmental reports related to the Sigma site provided information pertaining to the following activities:

- The performance of multiple Phase I and Phase II ESAs
- The excavation and disposal of 142 m^3 of contaminated soil, replacement with clean backfill materials, and the construction of an engineered barrier

The collected data for the remedial activities of the Sigma site have been summarized in Table C.10.1.

The energies required for the brownfield assessment and remediation activities for each site are illustrated for comparison in Figure 10.5A.

The CED associated with the CCGT site's brownfield remediation activities was found to be 26.5 TJ. Ninety-four percent of the energy was required for the removal of the 382,500 m^3 of C&D debris. The site remediation and engineered barrier construction accounted for 5% of required energy, whereas the combined UST removal and site assessment activities required less than 1% of the energy for this life cycle stage.

The CED associated with the Sigma site brownfield remediation activities was 15.3 TJ. Ninety-six percent of the energy was required for the construction of the asphalt parking lot, curb, and gutter that was used as an *in situ* cap. The soil remediation and methane venting system accounted for almost 4% of required energy, whereas the combined geotechnical and ESA activities again required less than 1% of the energy for this life cycle stage.

The CED and the associated GHG emissions for the brownfield assessment and remediation life cycle stage have been compared for both sites in Figure 10.5B.

Both the CED and the associated GHG emissions for the CCGT were significantly higher than those for the Sigma site because the CCGT remediation activities included the excavation and trucking of 382,500 m^3 of illegally stockpiled C&D debris, a process that demanded a large quantity of diesel fuel for combustion in heavy duty excavation and trucking equipment.

BUILDING REHABILITATION AND CONSTRUCTION

Based on a review of the architect's blueprints and LEED documentation provided for the CCGT, the following activities were quantified:

- The rehabilitation of a 2,600 m^2 two-story brick building
- The construction of a 160 kW geothermal heating and cooling system

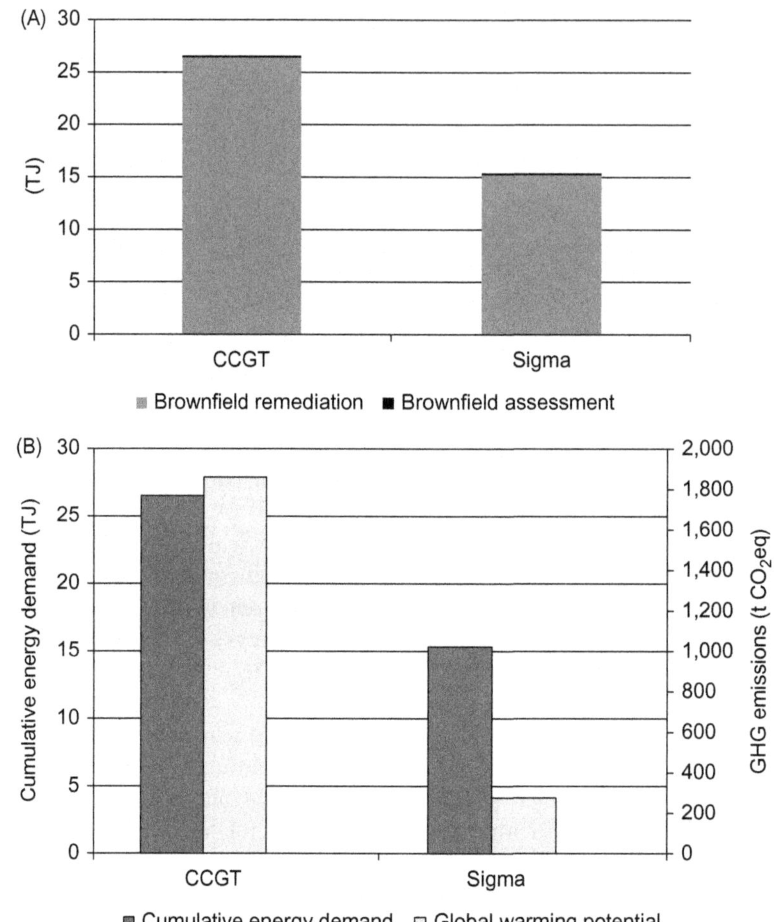

FIGURE 10.5

(A) Energies required for each site's brownfield assessment and remediation activities. (B) Required energies and associated GHG emissions for each site's brownfield assessment and remediation activities.

- The installation of a 970 m^2 photovoltaic array, which was designed to generate 490,000 MJ annually
- The construction of a 230 m^2 green roof

The collected input data have been summarized in Table B.10.2. The energies and associated GHG emissions required for the CCGT building rehabilitation activities are shown in Figure 10.6A.

The CED associated with the CCGT building rehabilitation activities was 12.0 TJ, distributed among several components (Figure 10.6A). The external

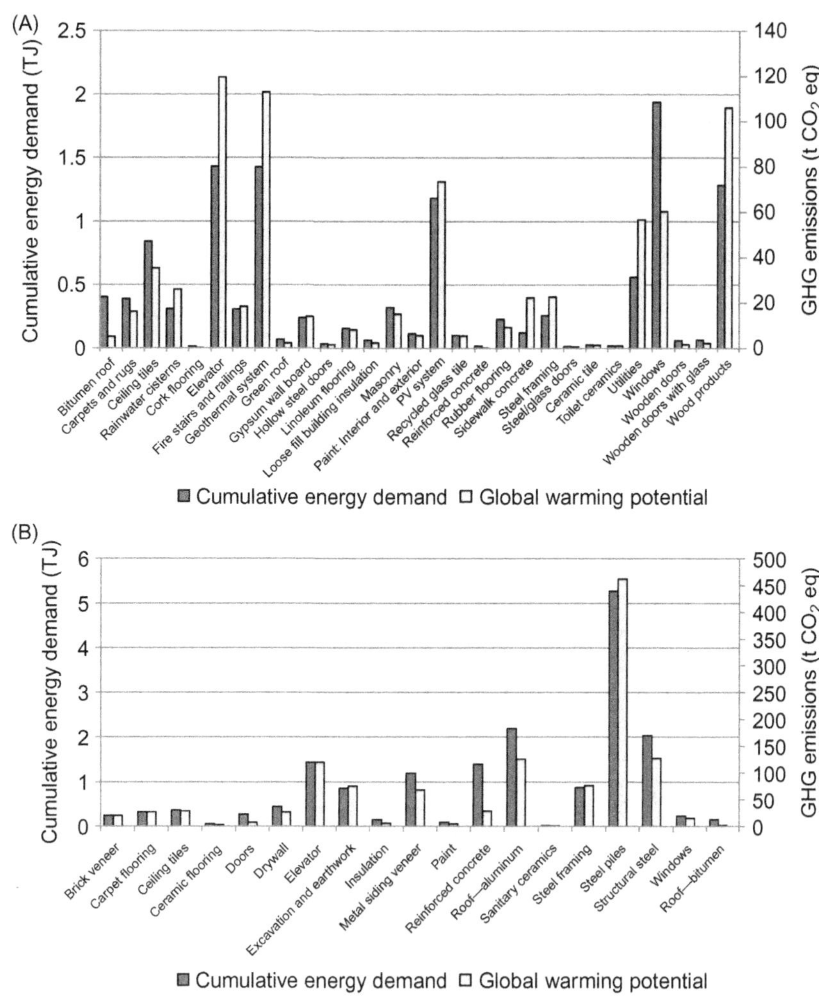

FIGURE 10.6

(A) Embodied energies and GHG emissions associated with CCGT building rehabilitation activities. (B) Embodied energies and GHG emissions associated with Sigma building construction activities.

windows, the geothermal heating and cooling system, the elevator, wood products, and the photovoltaic system accounted for approximately 60% of the building's embodied energy. The building's windows, which included wood window frames, and the building's wood products have a noticeably high CED because of the energy embodied in the wood. In other words, the CED accounts for the potential energy contained within the wood products that could be released as kinetic energy in the future (e.g., during combustion). The corresponding GHG emissions largely represent the processing and manufacturing of the glass portion of the windows. The remaining building components with the highest embodied energies included the elevator, the geothermal heating and cooling system, and the photovoltaic system. These building components also exhibited the highest "embodied" GHG emissions associated with their manufacture.

Based on a review of the contractor's construction blueprints provided for the Sigma site, the following activities were quantified:

- The new construction of a 2,600 m^2 two-story building
- The installation of 82 concrete-filled steel pipe piles for the deep foundation of the building

The collected data have been summarized in Table C.10.2. The energies and associated GHG emissions required for the Sigma site's building construction activities are shown in Figure 10.6B.

The CED associated with the Sigma building construction activities was 17.5 TJ, distributed among several components of the rehabilitation activities (Figure 10.6B). The deep steel pipe piles, structural steel, reinforced concrete, aluminum veneer, standing seam metal roof, and the elevator accounted for more than 75% of the building's embodied energy. The majority of the GHG emissions are associated with the processing and manufacture of the carbon-steel pipe piles and structural steel for building foundations and superstructures.

OPERATING ENERGY

Chicago Center for Green Technology

Based on a review of the energy consumption records and the data generated from the commuter transportation survey, the following activities were quantified for the CCGT site:

- 10-year electricity consumption
- 10-year natural gas consumption
- 10-year commuter transportation energy consumption

The operational energy data for the CCGT site are summarized in Table B.10.3.

Sigma site

Based on the provided energy data and the data generated from the commuter transportation survey, the following activities were quantified for the Sigma site:

- 10-year electricity consumption
- 10-year natural gas consumption
- 10-year commuter transportation energy consumption

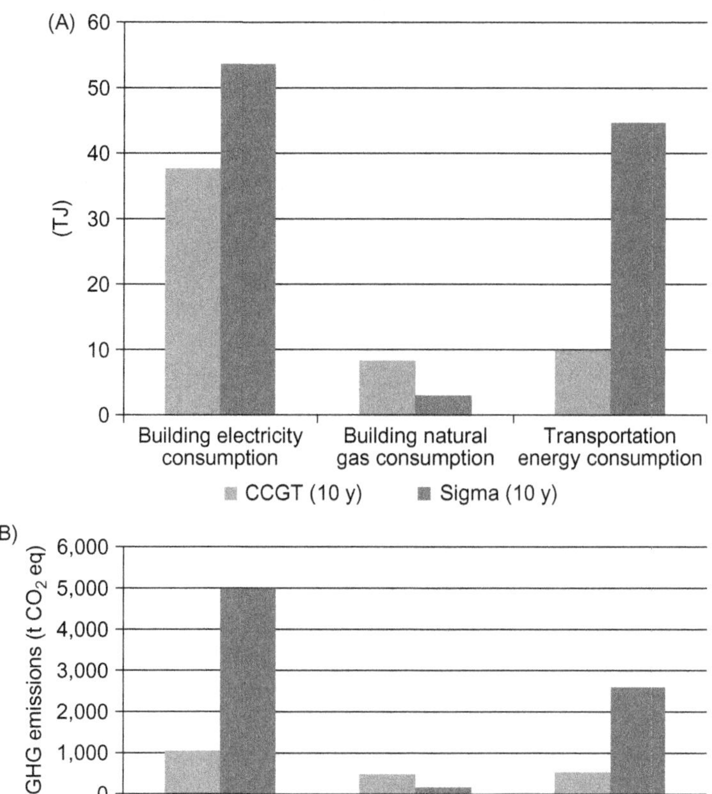

FIGURE 10.7

(A) Long-term energy use during postdevelopment site operations. (B) Long-term GHG emissions during postdevelopment site operations.

The operational energy data for the Sigma site have been summarized in Table C.10.3. The CEDs and the associated GHG emissions for each site's long-term operating energies have been compared over consumption categories (Figure 10.7A and B).

The respective CEDs associated with the CCGT and the operational activities of the Sigma site were 55.8 and 101.2 TJ over 10 y (Figure 10.7A). Each site's building electricity demand was the dominant component of their CEDs (Figure 10.7A). The Sigma site, relying on electricity for its heating and cooling requirements, consumed more energy than the CCGT, which reduced its energy demand by harnessing geothermal and solar energy on-site. Electricity consumption accounted for approximately 67% of the operational energy for CCGT and 53% of the operational energy for the Sigma site. The CCGT consumed more natural gas to supplement its geothermal heating and cooling system during the winter months, whereas the Sigma site relied on electricity to provide its heat. As a result, natural gas provided 15% of the CCGT's CED and only 3% of the Sigma site's CED. The transportation energy consumption accounted for 44% of the Sigma site's operational energy; most commuters to the Sigma site were found to make single-occupancy vehicle (SOV) trips over a round trip distance of 59 km, at least partially because public transportation options are limited in Milwaukee. Conversely, the CCGT's transportation energy accounted for only 18% of its operational energy, because its commuters took advantage of the Chicago Transit Authority's (CTA) public bus and train routes serving the area. Only 24% of the CCGT's commuters made SOV trips, and their average round trip distance was 30 km.

The GHG emissions associated with the postdevelopment use of each site have been compared in Figure 10.7B. The GHG emissions associated with Sigma site operational activities were 7,760 t of carbon dioxide equivalents (CO_2 eq), almost four-times more than the 2,070 t of CO_2 eq related to the CCGT's CED. Increased GHG emissions from the Sigma site's electricity consumption are commensurate with its electricity being generated at a neighboring coal-fired power plant. The plant is located near residential areas and is known to emit various air pollutants including, but not limited to, particulates, heavy metals, SO_x, NO_x, and CO_2 (CMD, 2013).

Notwithstanding the use of on-site photovoltaic and geothermal energy, 62% of the CCGT's purchased electricity was produced using nuclear technology, whereas only 31% was generated using coal (ICC, 2012). Chicago, where the CCGT is situated, is part of the electricity grid subregion known as Reliability First Corporation West, which includes northeastern Illinois, Indiana, and portions of Wisconsin and Michigan. In this subregion, Commonwealth Edison (Com-Ed) produced electricity from a mixture of sources that is different from the rest of the region, in that it is more dependent on nuclear energy (Klein-Banai et al., 2010). Thus, the CCGT's operational electricity was specified using a custom mixture of power sources instead of a more broad regional average.

The impact of commuter transportation and electricity source

The data generated through the administration of the commuter transportation survey to the occupants of the Sigma site, based on a 45% survey response rate, indicated that the majority (85%) of commuters drive alone and have an average round trip distance of 59 km, which comprised 96% of the CED for commuter transportation. The remainder of commuters carpooled (5%), rode the public bus (5%), and biked (5%), which accounted for the remaining 4%

of the commuter transportation CED. These data were based on a total of 102 occupants of the Sigma site who commuted an average of 5 d/week and 50 weeks/y.

In contrast, 113 full-time occupants of the CCGT site commuted an average of 4.25 d/week and 46 weeks/y. Nearly 57% of commuters to the CCGT site utilized public transportation, including trains and buses, and another 19% of commuters carpooled, biked, or walked to the CCGT site. Only 24% of commuters, based on a 70% survey response rate, drove alone and had an average round trip distance of 30 km, which comprised 58% of the commuter transportation CED. The CCGT was located in an area of metropolitan Chicago where both bus and light rail systems made public transportation more accessible for its commuters. The Sigma site was located in Milwaukee, where no intra-city light rail system exists, although an inter-city commuter rail (Amtrak) station was located near the site. As a result, the commuters for the Sigma site were not afforded the same transportation options as the CCGT's commuters.

Figure 10.8A shows that when considering only each site's building, the average annual CEDs are within the same order of magnitude. Expanding the LCA goal and scope to include the impacts of commuter transportation nearly doubles the Sigma site's annual operating energy requirement, and its overall CED is nearly twice the CED for the CCGT. Nonetheless, the results of the CED analysis reveal only one facet of the life cycle.

Figure 10.8B shows that although the Sigma site's average annual CED is approximately twice as high as the CCGT's CED (Figure 10.8A), the corresponding GHG emissions are nearly four times as high, because of the commuter transportation and the building operational energy consumption. Again considering only each site's building operational energy, despite the similar CEDs (Figure 10.8A), the GHG emissions associated with the Sigma site's building are three times as high as the GHG emissions associated with the CCGT's building. The cause of the elevated GHG emissions is that the Sigma site uses electricity as the primary source of heating and cooling, and also because the electricity is generated from a coal-fired power plant. Despite the additional electricity demand to power the heat pumps associated with the on-site geothermal heating and cooling system, the CCGT site purchased less electricity than the Sigma site; the source of the CCGT's purchased electricity, reduced through the generation of solar electricity on-site, consisted of an average mix of 62% nuclear, 31% coal, 5% natural gas, and 2% renewable energy (biomass, hydropower, and wind power). In addition, Figure 10.8B shows that the Sigma site's transportation GHG emissions are nearly five times higher than those from the CCGT because of the increased vehicle miles traveled (VMT) commensurate with commuters to the Sigma site making more SOV trips over longer distances.

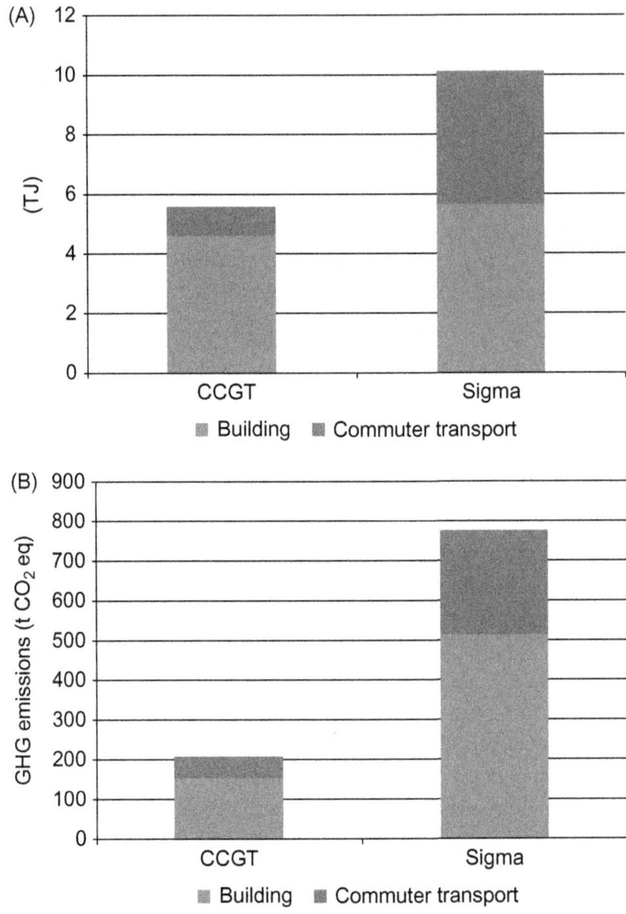

FIGURE 10.8

(A) Annual operating energy comparison between the redevelopment projects. (B) Annual GHG emissions comparison between the redevelopment projects.

COMPARISON OF LIFE CYCLE STAGES

Illustrations of the energies and associated GHG emissions required for each site's full life cycle stages are provided in Figure 10.9A and B.

The CED, including embodied energy and operational energy, for the CCGT brownfield redevelopment project totalled approximately 94.3 TJ after redevelopment and 10 y of operation; the operating energy life cycle stage contributed 59% of the life cycle energy. The CED for the Sigma brownfield redevelopment project totalled approximately 134.0 TJ after redevelopment and 10 y of operation.

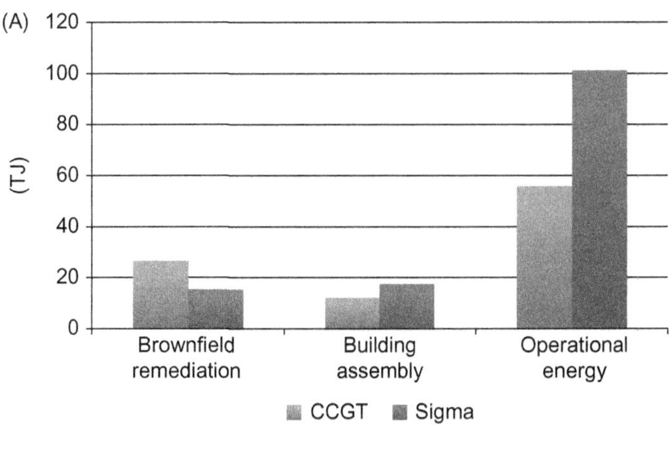

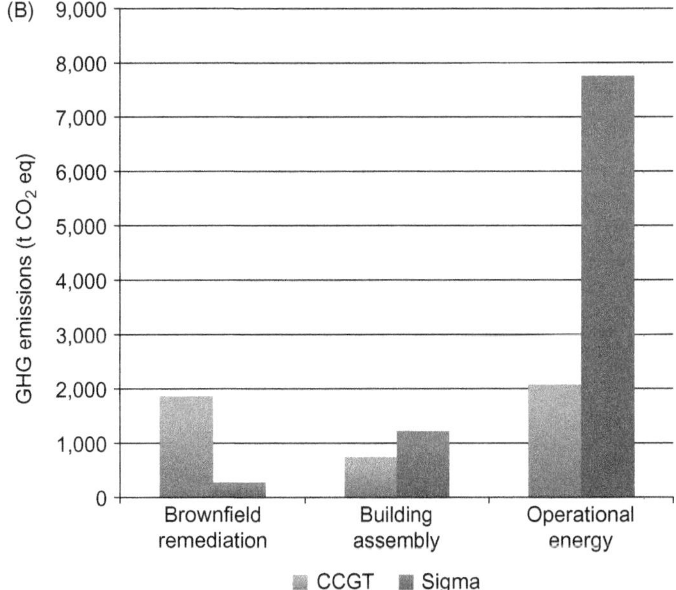

FIGURE 10.9

(A) CED comparison over life cycle stages of two brownfield redevelopments. (B) GHG emission comparison over life cycle stages of two brownfield redevelopments.

The Sigma site's operating energy, including commuter transportation, comprised 76% of the life cycle energy for the brownfield redevelopment project. The percentages of CED attributed to the operational life cycle stages will continue to increase with time and are expected to contribute 80–90% of the overall life cycle energy at the end of the building's operational life (Kotaji et al., 2003).

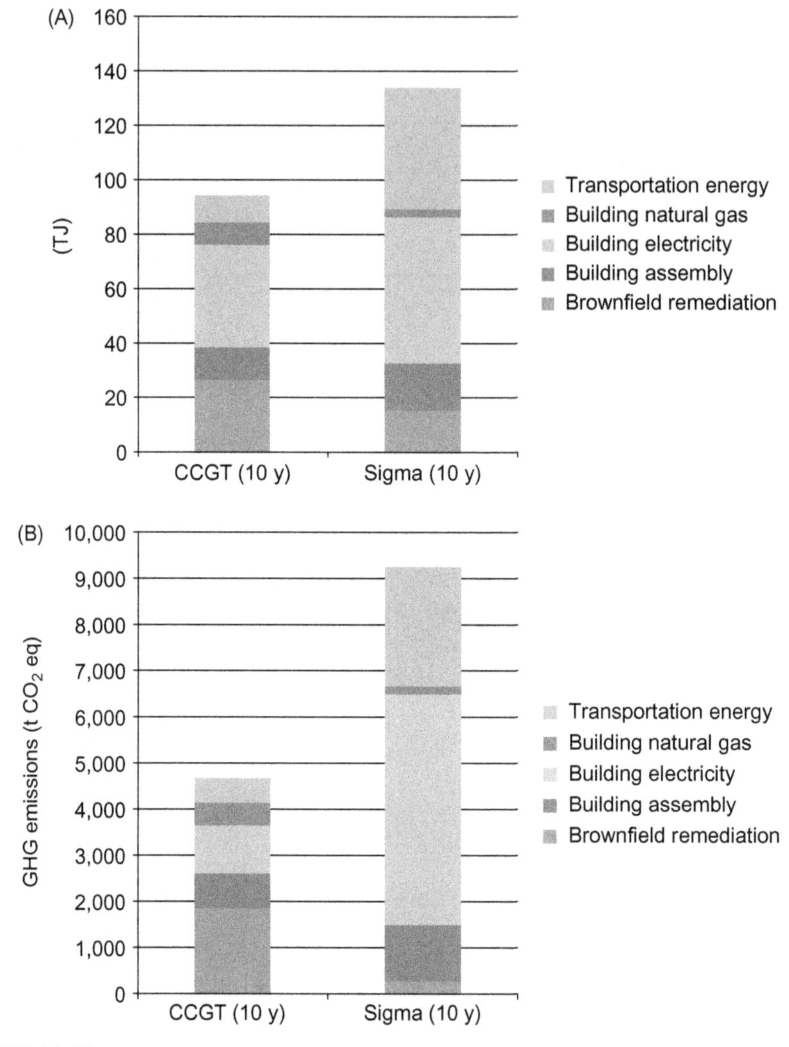

FIGURE 10.10

(A) Comparison of CED distributions between the two sites. (B) Comparison of GHG emissions between the two sites.

Perhaps the best way to compare the two sites is to aggregate the CED and associated GHG emissions into a single score for each site. Figure 10.10A and B illustrates the components that comprise the respective CEDs and the associated GHG emissions for each site.

DISCUSSION

Two significant factors affected the outcome of this study. First, the source of a building's operating energy has a significant impact on the associated GHG emissions during the operational life cycle stage, regardless of how efficiently a building operates. The second factor is that the expansion of the LCA scope and boundary to include commuter transportation as a component of long-term operational use has a significant impact on a site's postdevelopment operational energy and GHG emissions.

BUILDING OPERATING ENERGY

From a demand side, both buildings appear to perform as energy-efficient buildings. When normalizing the annual building energy consumption on an area basis ($MJ/m^2/y$) to determine the building's energy use intensity (EUI), Brecheisen and Theis (2013) found the CCGT consumed 601 $MJ/m^2/y$. The Sigma site building consumed approximately 651 $MJ/m^2/y$. According to previous studies, the median energy consumption for commercial buildings in the Midwest is 1,124 $MJ/m^2/y$, and the median energy consumption from 17 LEED-certified projects in Illinois is 1,067 $MJ/m^2/y$ (USGBC, 2009). Both the CCGT and the Sigma sites outperformed those benchmarks, suggesting that the design and construction of high-performance buildings lead to long-term savings in energy and costs.

Figure 10.10A illustrates the impact categories that comprise the CEDs for each site. Because the site buildings were both two-story buildings with the same floor area, geographically located in similar climates, it seems reasonable to expect the CEDs for these impact categories to be similar. The same holds true for the operational energies of the buildings, given the similar EUIs of 651 and 601 $MJ/m^2/y$ for the respective Sigma and CCGT sites. If transportation energy requirements are neglected, then the two site buildings required similar amounts of energy to operate over 10 years and the main difference in operating energy consumption lies in the commuter transportation impacts (Figure 10.10A).

However, when considering the supply side of the operational energy, Figure 10.10B illustrates the difference in the associated GHG emissions between the two sites, considering the Sigma site obtained its electricity from a coal-fired power plant, but coal only provided 31% of the CCGT's purchased electricity. Cleaner sources of electricity and heat are required; otherwise, as can be shown in Figure 10.10A and B, energy-efficient buildings can still have adverse impacts on the environment over the course of their life cycles.

The coal-fired Valley plant, constructed in the late 1960s, lacks modern pollution controls and a conversion from coal to natural gas is required (Content, 2014). Wisconsin energy regulators have approved the plan of We Energies to convert the coal-fired power plant in the Menomonee Valley to burn natural gas instead (Content, 2014). In an oversimplified sensitivity analysis substituting electricity generated using natural gas in lieu of coal-fired electricity, the GHG emissions associated with the Sigma site's 10-year building operational energy would be reduced 40%, from 5,160 t to 3,040 t of CO_2 eq. This supports the assertion that natural gas burns cleaner than coal and generates only half as much CO_2 per unit of energy produced as coal (Withgott and Laposata, 2012).

TRANSPORTATION OPERATING ENERGY

Commuter transportation contributes significantly to each site's operational energy consumption and associated GHG emissions. Both the transportation mode and round trip distance affect the transportation component of each site's operational energy. As expected, more energy is required for commuters who travel using SOV trips over long distances, and the associated GHG emissions are correspondingly higher. The results of this study suggest that transit-oriented development (TOD), which maximizes public access to alternative transportation, has the potential to realize reductions in commuter transportation energy consumption and GHG emissions. With respect to transportation energy inputs, the CCGT site outperformed the Sigma site, although it must be stated that the availability of public transportation options is a function of municipality and not under the control of either site.

CONCLUSIONS

The operational energy is the dominant life cycle stage for both sites after 10 y of operation and, thus, it is evident that the long-term operational energy consumption will be the most energy-intensive life cycle stage over a building life cycle of 50–100 y. Each of the two sites exhibited building operational EUIs that were only 54% and 58% of the median EUI for commercial buildings in the Midwest. It was estimated that the CCGT avoided energy consumption impacts of 35 TJ in comparison with an average commercial building (Brecheisen and Theis, 2013). Similarly, when compared with the median energy consumption for commercial buildings in the Midwest (1,124 $MJ/m^2/y$), the Sigma site was estimated to avoid building energy consumption impacts of 40 TJ through its energy-efficient building design and construction. The Sigma site's avoided energy impacts 40 TJ over 10 y were slightly greater than the CCGT's avoided energy impacts of 35 TJ over 10 y. The CCGT building energy consumption was 59% electricity and 41% natural gas, whereas the Sigma site building energy consumption was 87% electricity and 13% natural gas, suggesting that coal-fired electricity is more energy-intensive and environmentally more damaging than electricity derived from natural gas.

The EUIs for the two buildings outperformed several LEED-certified buildings in Illinois. Despite the CCGT utilizing both an on-site photovoltaic system and a geothermal heating and cooling system, the Sigma site, relying on conventional electricity from a coal-fired power plant for its heating and cooling, maintained a comparable EUI (651 MJ/m^2/y) in comparison with the CCGT (601 MJ/m^2/y). Although the Sigma site did not seek LEED certification, the building design incorporated innovative technologies in day lighting and energy efficiency. Thus, the results of the comparative LCAs presented herein indicate that the Sigma site building, despite not pursuing LEED certification, performs nearly as well as the CCGT, a LEED-certified building that was recognized for its sustainable site design with an awarded platinum status, the highest level attainable through the USGBC.

The following discussion emphasizes the importance of site selection for sustainable development. The fact that nearly 90% of Sigma site occupants commute by driving alone an average round trip distance of 59 km for 250 days/year erodes some of the efficiency gains of the Sigma site building; the CED associated with commuting equalled 79% of the CED associated with the Sigma site building and 44% of the overall operational energy. Conversely, the CED associated with the commuting of the CCGT site totalled only 21% of the CCGT site building operational energy and only 18% of the overall operational energy. The CCGT was awarded points toward its platinum LEED certification as a sustainable site based on its proximity to neighboring public rail and bus systems.

One of the great benefits of brownfield redevelopment is existing infrastructure, including public transportation alternatives, the presence or absence of which can be a great strength or weakness of a project (Lange et al., 2014). In this study, the existence of public transportation alternatives for the CCGT site were a strength of the project that resulted in significantly lower operating energy requirements and associated GHG emissions when compared with the Sigma site. Hou and Al-Tabbaa (2014) point out that the increased recognition of these secondary adverse impacts is a primary force to drive brownfield redevelopment toward sustainability and that these secondary impacts can be quantified using LCA.

There is a need to replace SOV commuter trips with alternative trips to reduce the energy consumption and associated air pollution associated with the transportation sector. Transportation planning decisions present a major challenge in moving toward sustainability (Eckelman 2013). It has been shown that cities with higher population densities have lower per capita transportation emissions (Kennedy et al., 2009). Mashayekh et al. (2012) also found that travel demand reductions from multiple studies of smart growth and brownfield development strategies can result in up to a 75% reduction in life cycle GHG and air pollutant emissions, whereas a related study in Toronto regarding life cycle energy use and GHG emissions for high-density and low-density development strategies found an approximate 60% difference in GHG emissions, largely because of transportation (Norman et al., 2006).

This study focused only on CED and GHG emissions as the sustainability metrics for the basis of comparing two sites using common indices. The evaluation of sustainability metrics based on LCA has the potential to offer deeper insight into the trade-offs associated with various site-specific brownfield redevelopment practices.

This study concludes that limiting factors such as access to alternative public transportation modes and clean energy sources have an overarching and synergistic effect in unlocking sustainable development. This study shows that two energy-efficient buildings successfully reduced their energy consumption using sustainable building design features. However, looking at the life cycle impacts from a broader view, gains in building energy efficiencies can be quickly eroded in the absence of clean energy and alternative transportation modes. The conversion of the Valley plant from coal to natural gas will significantly reduce the Sigma site building's life cycle GHG emissions.

ACKNOWLEDGMENTS

We acknowledge the USEPA, who made this project possible through the issuance of a K6 grant (TR-83418401) to investigate the best management practices and benefits of sustainable redevelopment of brownfield sites. The contents of this paper are solely the responsibility of the authors and do not necessarily represent the official views of the USEPA. We thank Jenny Babcock, Bryan Glosik, Emma Peng, Steve Pincuspy, and Kelly Reiss of WRD Environmental; Stacey Munroe of the Chicago Department of General Services; Kunal Dasai of the University of Illinois at Chicago; Ben Gramling of the Sixteenth Street Community Health Center; and Tom Lamb, Robert Peschel, and Steven Syburg from the Sigma Group.

Data for this LCA were obtained from project reports, construction blueprints, and the site owner. Please contact Thomas Brecheisen, PhD candidate, at the Institute for Environmental Science and Policy, University of Illinois at Chicago; tbrech2@uic.edu.

APPENDIX A COMMUTER TRANSPORTATION SURVEY

Commuter Transportation Survey LEED EBOM SS Credit 4/USEPA K6

The University of Illinois at Chicago (UIC) is conducting a survey to determine how employees and occupants of the CCGT commute to and from work. Thus, we are seeking your input to assess the current commuting habits of employees at CCGT and to evaluate awareness with regard to various forms of alternative transportation. The results of this survey will further be used to earn LEED certification for the CCGT. Please fill out the survey below and return it to Tom Brecheisen from UIC. Thank you in advance for your cooperation and assistance with this program.

Personal Commuting Information

1 I live in the following zip code: ___________________________

2 The closest major intersection to my home is [NOT the exact address]:

____

3 My average ROUND TRIP commute distance is: _______________ miles.

4 Typically, I commute to work _______________ weeks/year.

5 Typically, I commute to work _______________ days/week.

Weekly Commuting Schedule

6 Please indicate the mode(s) of transportation you used to commute to and from work during a typical week.

____

7 Please specify the typical 7-day survey period here (Ex: July 24, 2011 through July 30, 2011).

____

8 If you used more than one mode of transportation on a given day, please indicate the approximate distance [in miles] for each mode of transportation used ROUND-TRIP on your daily commute.

	Sun	Mon	Tue	Wed	Thu	Fri	Sat
Drove Alone							
Carpool/Vanpool							
Walk							
Bicycle							
Motorcycle/Moped							
Bus							
Train							
Taxi							
Telecommute							
Sick/Vacation Day							
Other							

Vehicle Information [fill out if you drive alone OR carpool for any portion of your commute]

9. If you drive to work, please indicate the make, model, and year of your vehicle below:
   Make:_______________ Model:_______________ Year: _______________

10. Please indicate whether your vehicle uses any of the following alternative fuels:
    - ☐ Electricity (including gasoline-electric or diesel-electric hybrids)
    - ☐ Hydrogen
    - ☐ Propane or compressed natural gas

☐ Liquid Natural Gas
☐ Methanol
☐ Ethanol

11. If you carpool or vanpool, please indicate the total number of people typically commuting with you:
☐ 1 ☐ 2 ☐ 3 ☐ 4 ☐ 5 ☐ 6 ☐ 7 ☐ 8 ☐ 9

Variation in Commuting Behavior

12. Do you usually travel home using the same mode of transportation used to get to work?
☐ Yes
☐ No
If "no," please briefly explain your mode of transportation used to return home from work below:
____
____
____
____
____
____

13. Does your typical commuting pattern change significantly depending on the time of year? If so, please explain below (i.e., bike in the summer instead of bus).
____
____
____
____
____
____

Commuter Amenities

Please indicate whether you utilize any of the following amenities at your workplace.

Answer Options	I use regularly	I use occasionally	I do not use	My workplace does not offer	I don't know if my workplace offers
Employer-provided vehicle for drips during the workday					
Reimbursable taxi or transit trips during the workday					
An immediate ride home in case of emergency (guaranteed ride home)					

Answer Options	**I use regularly**	**I use occasionally**	**I do not use**	**My workplace does not offer**	**I don't know if my workplace offers**
Employer subsidy or coordination for carpools or vanpools					
Priority, reserved, or discounted parking for carpools or vanpools					
Priority, reserved, or discounted parking for hybrid vehicles					
Parking cash outs					
Pre-tax transit benefits					
Secure and/or indoor parking for bicycles					
Changing room with lockers (or similar storage) and/or showers for bicyclists and walkers					
On-site childcare, banking, dry cleaning, fitness center, or other services					
On-site food service or other kitchen facilities					
Childcare, banking, dry cleaning, fitness center, or other services within 5 min walking distance from building					
Food service options within 5 min walking distance from building					
Telecommuting, compressed workweek, or flex time					

14. What is the one thing you like most about your commute?

15. What is the one thing you would most like to see improved about your commute?

APPENDIX B CCGT LCA INPUT DATA

Table B.10.1 Summary of CCGT Brownfield Assessment and Remediation Data

Material or Process	Quantity	Units	Description
1. Construction and Demolition Debris Removal			
Excavation hydraulic digger	382,500	m^3	C&D Removal
Excavation skid-steer loader	382,500	m^3	C&D Removal
Transport combination truck diesel powered	15,000,000	tkm	C&D Removal
2. Site Assessments			
Diesel combusted in industrial equipment	76	L	Environmental Drill Rig
Passenger car	242	km	Engineer Oversight Transportation
Computer with monitor	60	h	Site Assessment Reporting
3. Underground Storage Tank Removal			
Diesel combusted in industrial equipment	57	L	Vacuum Truck
40t semi	4,109	tkm	UST Liquids Removal
Excavator	589,550	kg	UST Excavation
40t semi	77	tkm	UST Removal Transportation
Passenger car	105	km	Engineer Oversight Transportation
Computer with monitor	40	hr	UST Removal Reporting
4. Site Remediation and Engineered Barrier Construction			
Textile, woven cotton	500	kg	Geotextile for plant storage area
Gravel, crushed, at mine	695,111	kg	Gravel for plant storage area
Clay, at mine	1,540,909	kg	Clay for demonstration Garden
Gravel, crushed, at mine	915,825	kg	Gravel for parking lot subbase
Clay, at mine	2,121,212	kg	Clay for stormwater retention swale
Chemi-thermomechanical pulp, at plant	10,000	kg	Parking lot emulsion polymer
Truck 40t	57,477	tkm	Transportation of contaminated soil to landfill
Excavation, hydraulic digger	2,450	m^3	Excavation of contaminated soil
Truck 40t	13,429	tkm	Transportation of gravel backfill
Truck 40t	49,617	tkm	Transportation of clay backfill
Excavation, hydraulic digger	4,500	m^3	Site grading

Table B.10.1 Summary of CCGT Brownfield Assessment and Remediation Data *Continued*

Material or Process	Quantity	Units	Description
Truck 40t	106,003	tkm	Transportation of site grading material
Truck 28t	12,360	tkm	Transportation of parking lot material
Excavator, technology mix, construction	383,838	kg	Construction of parking lot
Computer with monitor	500	h	Remediation Reporting and Project Management

Note: "tkm" denotes "tonne-kilometer," which is the work required to transport one tonne of material a distance of 1 km.

Table B.10.2 Summary of CCGT Building Rehabilitation Input Data

Material or Process	Quantity	Units[a]	Description
Bitumen adhesive compound, hot	7,000	kg	Bitumen Roof
Truck 40t	8,126	tkm	Bitumen Roof Transportation
Carpets and rugs	15,625	USD	Carpet Flooring
Truck 28t	507	tkm	Carpet Flooring Transportation
Mineral wool	27,851	USD	Ceiling Tiles
Truck 28t	9,130	tkm	Ceiling Tile Transportation
Galvanized steel sheet, at plant	8,082	kg	Rainwater Cisterns
Zinc coating, pieces	107	m^2	Rainwater Cisterns
Truck 28t	22,563	tkm	Rainwater Cisterns
Wood, cork oak	1	m^3	Cork Flooring
Truck 28t	250	tkm	Cork Flooring
Elevator	79,500	USD	Elevator
Architectural and ornamental metal work	21,015	USD	Fire Stairs and Railings
Borehole heat exchanger 150 m	12	p	Geothermal heating and cooling system
Heat distribution, hydronic radiant floor heating, 150 m^2	18	p	Geothermal heating and cooling system
Heat pump 30 kW	6	p	Geothermal heating and cooling system
Textile, woven cotton	40	kg	Green Roof
High-density polyethylene resin	344	kg	Green Roof
Low-density polyethylene resin	344	kg	Green Roof
Grass seed IP, at farm	4	kg	Green Roof
Excavation, hydraulic digger	37	m^3	Green Roof
Truck 28t	1,032	tkm	Green Roof Transportation

(Continued)

Table B.10.2 Summary of CCGT Building Rehabilitation Input Data
Continued

Material or Process	Quantity	Units[a]	Description
Gypsum plaster board	38,080	kg	Gypsum Wall Board
Truck 40t	3,020	tkm	Gypsum Wall Board Transportation
Steel hot rolled section	971	kg	Hollow Steel Doors
Zinc, from combined metal production	123	g	Hollow Steel Doors
Zinc coating, pieces	116	m^2	Hollow Steel Doors
Cellulose fibre, inclusive blowing in	25	kg	Hollow Steel Doors
Cold rolled sheet, steel, at plant	280	kg	Hollow Steel Doors
Truck 28t	603	tkm	Hollow Steel Doors Transportation
Hard surface floor coverings, n.e.c.	7,400	USD	Linoleum Flooring
Truck 28t	956	tkm	Linoleum Flooring Transportation
Cellulose fibre, inclusive blowing in	6,623	kg	Loose Fill Building Insulation
Truck 28t	11	tkm	Loose Fill Building Insulation Transportation
Cut stone and stone products	26,351	USD	Masonry
Alkyd paint, white, 60% in H_2O	1,873	kg	Paint: Interior and Exterior
Truck 28t	2,114	tkm	Paint: Interior and Exterior Transportation
Photovoltaic panel, a-Si, at plant	970	m^2	Photovoltaic System
Truck 40t	13,556	tkm	Photovoltaic System Transportation
Packaging glass, white, at plant	3,391	kg	Recycled Glass Tile
Ceramic tiles, at regional storage	2,456	kg	Recycled Glass Tile
Truck 28t	2,390	tkm	Recycled Glass Tile Transportation
Pre-cast concrete, min. reinf.	27,195	kg	Reinforced Concrete
Polybutadiene	2,227	kg	Rubber Flooring
Truck 28t	2,180	tkm	Rubber Flooring Transportation
Concrete, normal	85	m^3	Sidewalk Concrete
Cold rolled sheet, steel	8,628	kg	Steel Framing
Truck 40t	430	tkm	Steel Framing Transportation
Steel hot rolled section	295	kg	Steel/Glass Doors
Zinc, from combined metal production	38	g	Steel/Glass Doors
Zinc coating, pieces	35	m^2	Steel/Glass Doors

Table B.10.2 Summary of CCGT Building Rehabilitation Input Data *Continued*

Material or Process	Quantity	Units[a]	Description
Glazing, double (2-IV), $U < 1.1$ W/m^{2}K, laminated safety glass	8	m^2	Steel/Glass Doors
Cold rolled sheet, steel, at plant	85	kg	Steel/Glass Doors
Truck 28t	176	tkm	Steel/Glass Doors Transportation
Ceramic tiles	1,775	kg	Ceramic Tiles
Truck 28t	309	tkm	Ceramic Tile Transportation
Sanitary ceramics, at regional storage	460	kg	Sanitary Ceramics
Truck 28t	80	tkm	Sanitary Ceramic Transportation
Other new construction	68,487	USD	Utilities
Window frame, wood	370.407	m^2	Windows
Glazing, double (2-IV)	297.135	m^2	Windows
Door, inner, wood	30	m^2	Wooden Doors
Cold rolled sheet, steel	158	kg	Wooden Doors
Truck 28t	1,130	tkm	Wooden Doors Transportation
Door, inner, glass-wood, at plant	32	m^2	Wooden Doors with Glass
Cold rolled sheet, steel, at plant	171	kg	Wooden Doors with Glass
Truck 28t	1,120	tkm	Wooden Doors with Glass Transportation
Wood products	55,139	USD	Ornamental Woodwork
Woodworking machinery	60,377	USD	Ornamental Woodwork

[a]*Note: "p" denotes "piece;" "USD" denotes "US dollars."*

Table B.10.3 Summary of CCGT Long-Term Energy Consumption Input Data

Material or Process	Quantity	Units	Description
Electricity, nuclear, at power plant	5,752,365	MJ	Electricity Consumption (10 y)
Electricity, hard coal, at power plant	2,832,354	MJ	Electricity Consumption (10 y)
Electricity, natural gas, at power plant	457,708	MJ	Electricity Consumption (10 y)
Electricity, biomass, at power plant	81,972	MJ	Electricity Consumption (10 y)

(Continued)

Table B.10.3 Summary of CCGT Long-Term Energy Consumption Input Data *Continued*

Material or Process	Quantity	Units	Description
Electricity, hydropower, at power plant	45,140	MJ	Electricity Consumption (10 y)
Electricity, at wind power plant	35,204	MJ	Electricity Consumption (10 y)
Heat, natural gas, at industrial furnace	6,415,398	MJ	Natural Gas Consumption (10 y)
Transport, bicycle	83,040	person km	Commuter Transportation (10 y)
Transport, passenger car, petrol, fleet average 2010	1,853,920	person km	Commuter Transportation (10 y)
Transport, regular bus	1,644,680	person km	Commuter Transportation (10 y)
Transport, metropolitan train	1,306,570	person km	Commuter Transportation (10 y)
Walking	86,850	km	Commuter Transportation (10 y)

APPENDIX C SIGMA SITE LCA INPUT DATA

Table C.10.1 Summary of Sigma Brownfield Assessment and Remediation Data

Material or Process	Quantity	Units	Description
1. Geotechnical Assessments			
Diesel, combusted in industrial equipment	225	L	Geotechnical Drill Rig
Use, computer, desktop with CRT monitor, active mode	160	h	Geotechnical Report Preparation
Passenger car	241	km	Geotechnical Personnel Travel
2. Environmental Site Assessments			
Diesel, combusted in industrial equipment	751	L	Environmental Drill Rig
Use, computer, desktop with CRT monitor, active mode	300	hr	Environmental Report Preparation
Passenger car	282	km	Environmental Personnel Travel

Table C.10.1 Summary of Sigma Brownfield Assessment and Remediation Data *Continued*

Material or Process	Quantity	Units	Description
3. Soil Removal and Disposal			
Excavation, hydraulic digger	141	m^3	Contaminated Soil Removal
Truck 40t	8,300	tkm	Transportation of Contaminated Soil to Onyx Landfill
Excavation, hydraulic digger	498	m^3	Placement of Clean Backfill
Truck 40t	30,380	tkm	Transportation of Clean Backfill
Textile, woven cotton, at plant	500	kg	Geotextile Barrier for Vegetated Areas
Clay, at mine	826	tonne	Six-inch Clay Cap for Vegetated Area
4. Passive Methane Venting System			
Construction machinery and equipment	19,983	USD	Passive Methane Venting System Construction
5. Engineered Barrier Construction			
Asphalt Paving Mixtures and Blocks	90,698	USD	Asphalt Parking Lot, Concrete Curb and Gutter

Note: "tkm" denotes "tonne-kilometer," which is the work required to transport one tonne of material a distance of 1 km.

Table C.10.2 Summary of Sigma Site Building Construction Input Data

Material or Process	Quantity	Units	Description
Brick, at plant	81	t	Brick Vaneer
Truck 28t	2,028	tkm	Brick Vaneer Transportation
Textile, woven cotton, at plant	987	km	Carpet Flooring
Truck 28t	1,328	tkm	Carpet Flooring Transportation
Mineral Wool	13,368	USD	Ceiling Tiles
Ceramic Tiles, at Regional Storage	3,554	km	Ceramic Floor Tiles
Truck 28t	89	tkm	Ceramic Floor Tile Transportation
Door, inner, wood, at plant	123	m^2	Interior Doors
Door, outer, wood-glass, at plant	22	m^2	Exterior Door
Truck 28t	2,000	tkm	Interior and Exterior Door Transportation

(Continued)

Table C.10.2 Summary of Sigma Site Building Construction Input Data *Continued*

Material or Process	Quantity	Units	Description
Gypsum plaster board, at plant	72,825	kg	Drywall Plasterboard
Truck 28t	1,821	tkm	Drywall Plasterboard Transportation
Elevators and Moring Stairways	79,500	USD	Elevator
Construction Machinery and Equipment	66,853	USD	Excavation and Earthwork
Cellulose fiber, inclusive blowing in, at plant	14,682	kg	Insulation
Truck 28t	367	tkm	Insulation Transportation
Cladding, crossbar-pole, aluminum, at plant	448	m^2	Metal Siding Veneer
Truck 28t	43	tkm	Metal Siding Veneer Transportation
Alkyd paint, white, 60% in H_2O, at plant	1,513	km	Paint
Truck 28t	38	tkm	Paint Transportation
Pre-cast concrete, reinforced mix	1,537	t	Reinforced Concrete
Truck 28t	107,586	tkm	Reinforced Concrete Transportation
Bitumen adhesive comound, hot, at plant	2,715	kg	Roof-Bitumen
Truck 28t	90	tkm	Roof-Bitumen Transportation
Cladding, crossbar-pole, aluminium, at plant	828	m^2	Roof-Aluminum
Truck 28t	129	tkm	Roof-Aluminum Transportation
Sanitary cermaics, at regional storage	344	kg	Sanitary Ceramics
Cold rolled sheet, steel, at plant	29,318	kg	Steel Framing
Truck 40t	890	tkm	Steel Framing Transportation
Cold rolled sheet, steel, at plant	176	t	Steel Pipe Piles
Truck 40t	12,958	tkm	Steel Pipe Piles Transportation
Reinforcing steel, at plant	88	t	Structural Steel
Truck 40t	2,641	tkm	Structural Steel Transportation
Glazing, double (2-IV), $U < 1.1$ W/m^2 K, at plant	474	m^2	Windows
Truck 28t	1,500	tkm	Windows Transportation

Note: "p" denotes "piece"; "USD" denotes "US dollars."

Table C.10.3 Summary of Sigma Site Long-Term Energy Consumption Input Data

Material or Process	Quantity	Units	Description
Electricity, hard coal, at power plant	15,098,933	MJ	Electricity Consumption (10 y)
Heat, natural gas, at industrial furnace	2,278,483	MJ	Natural Gas Consumption (10 y)
Transport, passenger car, petrol, fleet average 2010	13,286,318	person km	Commuter Transportation (10 y)
Transport, regular bus	502,812	person km	Commuter Transportation (10 y)
Transport, bicycle	241,350	person km	Commuter Transportation (10 y)

REFERENCES

Angela, A., Hutzler, N., 2005. Operational life-cycle assessment and life-cycle cost analysis for water use in multi-occupant buildings. J. Arch. Eng. 11 (3), 99–109.

Adalberth, K., Almgren, A., Peterson, E.H., 2001. Life-cycle assessment of four multifamily buildings. Int. J. Low Energy Sustainable Build. 2, 1–21.

Asif, M., Muneer, T., Kelley, R., 2007. Life cycle assessment: a case study of a dwelling home in Scotland. Build. Environ. 42 (3), 1391–1394, School of Engineering, Napier University, Edinburgh, UK.

Bartsch, C., 1999. National lessons and trends. In: Rafson, H.J., Rafson, R.N. (Eds.), Brownfields—Redeveloping Environmentally Distressed Properties. McGraw-Hill, New York, NY, USA.

Bastos, J., Batterman, S.A., Freire, F., 2013. Life-cycle energy and greenhouse gas analysis of three building types in a residential area in Lisbon. Energy Build. 69 (2014), 344–353, University of Coimbra, Polo II Campus, Department of Mechanical Engineering, Rua Luis Reis Santos, Coimbra, Portugal.

Bishop, P., 2000. Pollution Prevention Fundamentals and Practice. Waveland Press, Inc., Long Grove, IL, USA.

Brecheisen, T., Theis, T., 2013. The Chicago Center for Green Technology: life-cycle assessment of a brownfield redevelopment project. Environ. Res. Lett. 8, 015038.

Bribian, I.Z., Capilla, A.Z., Uson, A.A., 2011. Life cycle assessment of building materials: comparative analysis of energy and environmental impacts and evaluation of the eco-efficiency improvement potential. Build. Environ. 46, 1133–1140, Centre of Research for Energy Resources and Consumption, Campus Rio Ebro. University of Zaragoza, Mariano Esquillor Gomez, Zaragoza, Spain.

Building Green, Inc, 2010. Case Studies: Chicago Center for Green Technology. Brattleboro, Vermont. <www.buildinggreen.com/hpb/overview.cfm?projectID = 97> (accessed 12.08.2010).

Cabeza, L.F., Rincon, L., Vilarino, V., Perez, G., Castell, A., 2013. Life cycle assessment (LCA) and life cycle energy analysis (LCEA) of buildings and the building sector: a review. Renewable Sustainable Energy Rev. 29 (2014), 394–416, GREA Innovacio Concurrent, Universitat de Lleida, Edifici CREA, Pere de Cabrera s/n, 25001 Lleida, Spain.

Center for Media and Democracy (CMD), 2013. Sourcewatch "Valley Power Plant." Madison, WI. <www.sourcewatch.org/index.php?title = Valley_Power_Plant> (accessed 29.03.2014).

Consoli, F., Allen, D., Boustead, I., Fava, J., Franklin, W., Jensen, A.A. (Eds.), 1993. Guidelines for Life Cycle Assessment. A Code of Practice. SETAC Press, Pensacola, FL, USA.

Content, T., 2014. Regulators OK We Energies conversion of Menomonee Valley coal plant Milwaukee Journal Sentinel Milwaukee, WI, USA. <www.jsonline.com/business/regulators-ok-we-energies-conversion-of-menomonee-valley-coal-plant-b99195240z1-242865891.html> (accessed 30.03.2014).

De Sousa, C., 2012. Milwaukee's Menomonee Valley: A Sustainable Re-Industrialization Best Practice. Chicago, IL, USA. <www.uic.edu/orgs/brownfields/research-results> (accessed 02.08.2013).

De Sousa, C., D'Souza L.A., 2012. The Chicago Center for Green Technology: A Sustainable Brownfield Revitalization Best Practice. <www.uic.orgs/brownfields/research-results> (accessed 30.11.2012).

Eckelman, M.J., 2013. Life cycle assessment in support of sustainable transportation. Environ. Res. Lett. 8 (2013), 021004.

ESE (Environmental Science & Engineering, Inc.), 2000a. Additional Site Investigation Report, 445 N. Sacramento Boulevard Site-Front, Chicago, IL, USA. ESE Project No. 559-9198.5104. Chicago, IL.

ESE (Environmental Science & Engineering, Inc.), 2000b. Remedial Objectives Report and Remedial Action Plan, 445 N. Sacramento Boulevard Site-Front, Chicago, IL, USA. ESE Project No. 559-9198. Chicago, IL.

Farr Associates, 2000. A-0.1 site plan, A-1.1, first floor plan, A-1.2, second floor plan, A-1.3 roof plan, A-1.4 roof mounted photovoltaic panels plan, A-2.1 building elevations existing, C-3.1 Phase 1 grading plan, C-3.2 Phase 2 grading plan. Chicago, IL, USA.

Feist, W., 1996. Life-cycle energy balances compared: low-energy house, passive house, self-sufficient house. In: Proceedings of the International Symposium of CIP W67, Vienna, Austria, pp. 183–190.

Frischknecht, R., Jungbluth, N., Althaus, H.-J., Doka, G., Heck, T., Hellweg, S., Hischier, R., Nemecek, T., Rebitzer, G., Spielmann, M., Wernet, G., 2007. Overview and Methodology. Ecoinvent Report No. 1. Swiss Centre for Life Cycle Inventories. Dubendorf, Switzerland, 2007.

Goedkoop, M., De Schryver, A., Oele, M., 2008. Introduction to LCA with SimaPro 7. PRé Consultants. The Netherlands.

Goedkoop, M., Oele, M., De Schryver, A., Vieria, M., Hegger, S., 2010. SimaPro 7 Database Manual. PRé Consutants. The Netherlands.

Harding ESE, Inc., 2002. Remedial Action Completion Report, Sacramento Crushing Site, 445 North Sacramento-Front, Chicago, IL. Harding ESE Project No. 559198, Chicago, IL.

Hou, D., Al-Tabbaa, A., 2014. Sustainability: a new imperative in contaminated land remediation. Environ. Sci. Policy 39, 25–34.

IBC Engineering Services, Inc., 2010. Chicago Center for Green Technology, An Engineering Perspective. <www.ibcengineering.com/featurs/ccgt/p2.html> (accessed 12.08.2010).

Illinois Commerce Commission (ICC), 2012. Environmental Disclosure Statements. Springfield, IL. <www.icc.illinois.gov/electricity/environmentaldisclosure.aspx> (accessed 21.03.2012).

Kellenberger, D., Althaus, H.-J., Jungbluth, N., Kunniger, T., Lehmann, M., Thalmann, P., 2007. Life Cycle Inventories of Building Products. Final Report Ecoinvent Data v2.0 No. 7. EMPA Dubendorf, Switzerland, Centre for Life Cycle Inventories, Dubendorf, Switzerland.

Kennedy, C., Steinberger, J., Gasson, B., Hansen, Y., Hillman, T., Havranek, M., Pataki, D., Phdungsilp, A., Ramaswami, A., Mendez, G.V., 2009. Greenhouse gas emissions from global cities. Environ. Sci. Technol. 43, 7297–7302.

Khasreen, M.M., Banfill, P., Menzies, G.F., 2009. Life-cycle assessment and the environmental impact of buildings: a review. Sustainability 1, 674–701.

Klein-Banai, C., Theis, T.L., Brecheisen, T.A., Banai, A., 2010. A greenhouse gas inventory as a measure of sustainability for an Urban Public Research University. Environ. Pract. 12 (1), 35–47.

Lange, D., Wang, D., Zhuang, Z., Fontana, W., 2014. Brownfield development selection using multiattribute decision making. J. Urban Plann. Dev. 140 (2), 1–6 (June 2014), 04013009 American Society of Civil Engineers, New York, USA.

Mashayekh, Y., Jaramillo, P., Samaras, C., Hendrickson, C.T., Blackhurst, M., MacLean, H.L., Matthews, H.S., 2012. Potentials for sustainable transportation in cities to alleviate climate change impacts. Environ. Sci. Technol. 46, 2529–2537.

National Oceanic and Atmospheric Administration, 2013. National Climatic Data Center. Monthly Summaries of GHCN-Daily Order No. 182184. <www1.ncdc.noaa.gov/pub/orders/cdo/182184.pdf> (accessed 26.07.2013).

National Oceanic and Atmospheric Administration, 2014. National Climatic Data Center. Monthly Summaries of GHCN-Daily Order No. 309691. <www1.ncdc.noaa.gov/pub/orders/cdo/309691.pdf> (accessed 25.03.2014).

Norman, J., MacLean, H.L., Kennedy, C.A., 2006. Comparing high and low residential density: life-cycle analysis of energy use and greenhouse gas emission. J. Urban Plann. Dev. 132, 10–21.

Patrick Engineering, Inc., 1999a. Phase I Environmental Site Assessment for the Sacramento Crushing Yard, 445 N. Sacramento Blvd. located in Chicago, IL, USA. PEI Project No. 8008-A0-13. Chicago, IL.

Patrick Engineering, Inc., 1999b. Report of Underground Storage Tank Removal Activities at 445 N. Sacramento Blvd., Chicago, IL. PEI Project No. 8008.A0-12. Chicago, IL, USA.

Peuportier, B.L.P., 2001. Life cycle assessment applied to the comparative evaluation of single family houses in the French context. Energy Build. 33, 443–450, Ecole des Mines de Paris, 60 B. St. Michel, 75272. Paris Cedex 06, France.

Product Ecology Consultants, February 2012. SimaPro 7.3 Installation Manual. Amersfoort, The Netherlands.

The Redmond Company, 2002. A1.0, first floor plan, A1.1, second floor plan, A3.0, roof plan, A3.1, roof details, A4.0, elevations, A4.1, elevations, A5.0, building sections, A5.1, building sections, A6.0, wall sections and details, A6.1, wall sections and details, A6.2, wall sections and details, A6.3, wall sections and details, A6.4, wall sections and details, A6.5, wall sections and details, A7.0, floor finishes and enlarged plans, A7.1, floor finishes and enlarged plans, A8.0, casework/interior elevations, A8.1, casework/interior elevations, A8.2, casework/interior elevations, A9.0, door & room finish schedules, A9.1, wall types & door schedule results, C1, site layout and paving plan, C2, site preparation and erosion control plan, C3, site grading and drainage plan, C4, site utility plan, C5, site details trash enclosure, E1.0, site lighting plan, E1.1, first floor lighting plan, E1.2 first floor power plan, E2.1, second floor lighting plan, E2.2, second floor power plan, L1.0, landscape plan, M1, first floor HVAC plan, M2 second floor HVAC plan, M3, schedules, M4, schedules, M5, schedules, S2.0, foundation plan, S2.1, second floor framing plan, S2.2, roof framing plan, S3.0, foundation details, S3.1, floor framing details, S3.2, roof framing details. Job No. 01295 Waukesha, WI, USA.

Sartori, I., Hestnes, A.G., 2007. Energy use in the life cycle of conventional and low-energy buildings: a review article. Energy Build. 30 (2007), 249–257, Department of Architectural Design, History and Technology. Norwegian University of Science and Technology (NTNU). Trondheim, Norway.

Scheuer, C., Keoleian, G.A., Reppe, P., 2003. Life cycle energy and environmental performance of a new University building: modeling challenges and design implications. Energy Build. 35 (10), 1049–1064.

Sharma, A., Saxena, A., Sethi, M., Shree, V., Vurun, 2011. Life cycle assessment of buildings: a review. Renewable Sustainable Energy Rev. 15, 871–875, Department of Mechanical Engineering. National Institute of Technology, Hamirpur, India.

Sigma, 1994. Report of a Phase II Environmental Assessment at 1300 W. Canal Street, Milwaukee, WI, USA.

Sigma, 1999a. 1999 Summary Report for Remedial Investigation. Milwaukee, WI, USA.

Sigma, 1999b. 1999 Remedial Action Plan. Milwaukee, WI, USA.

Sigma, 2000. Subsurface Methane Study & Pilot Test Results. Milwaukee, WI, USA.

Sigma, 2006. VPLE Case Summary & Site Closure Request. Milwaukee, WI, USA.

Sigma, 2004a. Locating in Milwaukee's Menomonee River Valley. An Impact Report. Milwaukee, WI, USA.

Sigma, 2004b. Construction Cost Summary. Milwaukee, WI (accessed 19.08.2011).

Sigma, 2011. Green Building Elements. Milwaukee, WI (accessed 19.08.2011).

Singh, K., Associates (Singh), 1991. Soil Contamination Assessment & Foundation Investigation. Waupaca, WI, USA.

United States Energy Information Administration (USEIA), 2013. Short Term Energy Outlook. Washington, DC. <www.eia.gov/forecasts/steo/pdf/steo_full.pdf>.

United States Environmental Protection Agency. 2002. Small Business Liability Relief and Brownfields Revitalization Act. Public Law 107-118 (H.R. 2869). <epa.gov/brownfields/overview/glossary.htm> (accessed 19.09.2012).

United States Green Building Council, 2009. Regional Green Building Case Study Project: A post-occupancy study of LEED projects in Illinois. Year One Report. US Green Building Council—Chicago Chapter. Chicago, IL, USA.

United States Green Building Council and Center for Neighborhood Technology, 2011. Regional Green Building Case Study: Year Two Report. CNT Energy and the U.S. Green Building Council—Illinois Chapter. Chicago, IL, USA.

Wedding, G.C., Crawford-Brown, D., 2007. Measuring site-level success in Brownfield redevelopments: a focus on sustainability and Green Building. J. Environ. Manage. 85, 483–495, Department of Environmental Sciences and Engineering and Carolina Environmental Program. University of North Carolina at Chapel Hill. Chapel Hill, NC, USA.

Winther, B.N., Hestnes, A.G., 1999. Solar versus Green: the analysis of a Norwegian row house. Solar Energy 66 (6), 387–393.

Wisconsin Department of Natural Resources (WDNR), 2009. Sigma Environmental PUB-RR-808 Remediation and Redevelopment Program, Madison, WI, USA.

Withgott, J., Laposata, M., 2012. Essential Environment the Science Behind the Stories, fourth ed. Pearson Education, Inc., Upper Saddle River, NJ, USA.

Yohanis, Y.G., Norton, B., 2002. Life-cycle Operational and Embodied Energy for a Generic Single-storey Office Building in the UK. Energy 27 (2002), 77–92, Centre for Sustainable Technologies, School of the Built Environment, University of Ulster. Northern Ireland, UK.

Zvenyach, L., Littman, T.H., 2006. From brownfield to sustainability showcase: Chicago Center for Green Technology. ASHRAE J. 48, 20–30.

CHAPTER

The Environmental Performance Strategy Map: an integrated life cycle assessment approach to support the strategic decision-making process

11

Luca De Benedetto and Jiří Jaromír Klemeš

Centre for Process Integration and Intensification—CPI[2], Research Institute of Chemical and Process Engineering—MÜKKI, Faculty of Information Technology, University of Pannonia, Veszprém, Hungary

INTRODUCTION

BACKGROUND

Life Cycle Assessment (LCA) is a tool for analyzing environmental impacts on a wide perspective with reference to a product system or economic activity (Figure 11.1). The concept varies depending on the adoption pattern and on the precision that needs to be achieved. Because of the constraints on resource or data availability, industrial companies perform analyses based on a more simplified approach (Life Cycle Approach) or simply apply the general principles to certain aspects of the production system (Life Cycle Thinking), even though they are usually all referred as LCA activities.

Life Cycle Thinking requires considering all environmental and toxicological impacts associated with the life cycle of a product. Life Cycle Thinking is a key element in different policies from the United States to the European Union. Because lowering our overall environmental impact is critical, the research for ways to gauge the sustainability of our actions has intensified. The importance of specific metrics to support policy making and the decision-making process is paramount. The challenge is to develop indicators that are not too generic or too broad. At the same time, it should be possible to aggregate them into a meaningful single indicator of performance to be used for strategic decision-making. This

Assessing and Measuring Environmental Impact and Sustainability.

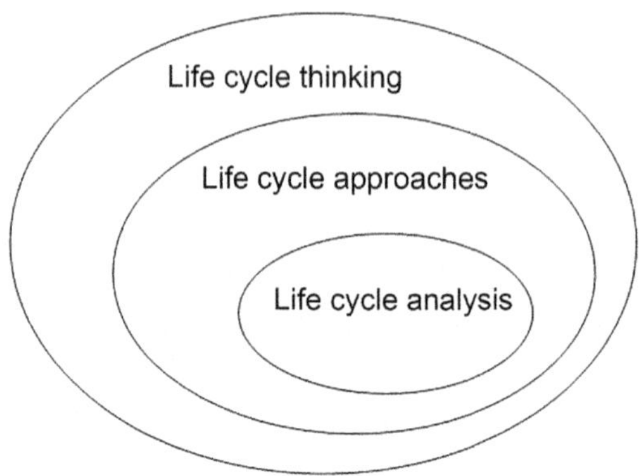

FIGURE 11.1

Basic levels of LCA.

Adapted from Frankl et al. (2000).

is particularly useful because it can provide the practitioner with a tool that, balancing different aspects, qualitatively points at the option that performs best from an environmental perspective.

LCA: HISTORY

LCA involves the evaluation of specific elements of a product system to determine its environmental impact. Also called Life Cycle Analysis or cradle-to-grave approach, LCA comprises a conceptual framework and a set of tools that have been studied and developed in the past 45 years. The core of the concept is the assessment of the impacts at each stage of the product life cycle. The term "product life cycle" is not used here with reference to a product's sales and profits course over time, but rather with reference to the notion of production, manufacturing, distribution, use, and disposal, including all necessary transportation steps. The proposed view is therefore a holistic one that includes the entire life span of a product from the extraction of the raw materials to its disposal.

The first studies of LCA date from the late 1960 and early 1970s. In 1969, for example, the Coca-Cola Company funded a study to compare resource consumption and environmental releases associated with beverage containers (Udo de Haes and Heijungs, 2007). Similar studies were then started in the United Kingdom, Switzerland, and Sweden. In these early studies, LCA was closely linked with energy analysis. Also, because of the energy crisis of the early 1970s, waste and outputs were initially not considered and attention was concentrated on calculating the total energy used in production of various household goods. For example, Boustead (1996) in the United Kingdom studied various types of beverage containers, including glass, plastic, steel, and aluminum. This demonstrated the high embodied energy value of aluminum in contrast to glass. Glass scores even better if it is reused or recycled. Nevertheless, the winners in this battle were the reusable plastic bottles, because glass and aluminum were transported great distances to be recycled.

After the oil crisis subsided, the energy issues and the use of LCA in this application lost prominence. It was only in the late 1980s and early 1990s that a new interest in the tool was found and coupled with efforts to bring standardization to its use. In 1989, Society of Environmental Toxicology and Chemistry (SETAC) started working on defining a common terminology and a methodology framework. A first result of this work was the definition of the *functional unit*. This is a quantified description of the product systems to which impacts are attributed. This unit sets the scale for comparison of two or more products, and one of its main purposes is to provide a reference to which the input and output data are normalized. Three aspects have to be taken into account when defining the functional unit (Lindfors et al., 1995):

1. Efficiency of the product
2. Durability of the product
3. Performance quality standards.

When performing an assessment of more complicated systems (e.g., multi-functional systems like waste treatment systems), special attention has to be given to by-products.

This standardization work was then picked up by the International Standard Organization (ISO) in 1994 with the first of its 14040 series. The rigid context of the ISO offered coherence to the different methodologies and approaches in LCA without imposing one. The ISO work has resulted in the definition of specific steps that allow the separation of the subjective and objective phases within the proposed method. The principles and framework for LCA in these documents include goal and scope definition, life cycle inventory analysis (LCI), life cycle impact assessment (LCIA), and life cycle interpretation phase and reporting.

These phases are the codification of the same steps individuated by SETAC in the previous years, with the exception that Life Cycle Improvement has been considered an activity that should permeate all other phases and not one of its own. The interpretation phase was, instead, added. The interest in this topic is witnessed by newer versions of the aforementioned series, the latest of which was published in 2006 14040: (ISO, 2006) effectively replaces 14040: (ISO, 1997), 14041: (ISO,1998), 14042: (ISO, 2000a), and 14043: (ISO, 2000b).

Although in past years the pace of development of LCA has been slowing, the methodology is beginning to consolidate. Nevertheless, the usefulness of the technique is still very much under debate. In recent times, the need to extend the complement ISO LCA recommendation with other tools has also been expressed (Jeswani et al., 2010).

A survey by the European Environment Agency (Jensen et al., 1996) indicated the following social impacts of LCA:

- LCA is now seen as necessary by all stakeholders as an integral part of the environmental management tool kit

- Use of this tool is also seen as important in the process of corporate strategy formulation
- LCA remains at an early stage of development and further research is needed
- Level of knowledge of LCA remains troublingly low in the general public
- Level of progress in LCA adoption varies between countries
- Quality control mechanism remains relatively weak
- Involvement of external stakeholders in defining study boundaries is seen as increasingly important.

LCA: AN OVERVIEW OF THE GENERAL FRAMEWORK

The methodology for Life Cycle Analysis includes the phases described in Figure 11.2. It has to be noted that ISO 14040 does not describe the technique in detail, nor does it specify which methodology should be used for each phase. It instead provides a framework in which these elements can be developed.

Goal and scope definition

This is the first subjective phase of the application of LCA. At this stage, it is necessary to identify the aim of the analysis and to select the system boundaries to ensure that no relevant part of the system to be investigated is actually left out. The definition of the goal and the scope are critical elements because the results will depend greatly on them.

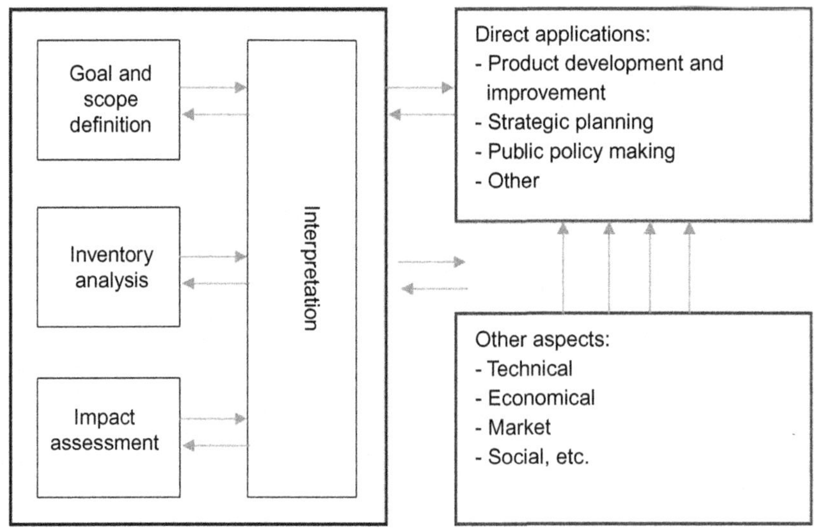

FIGURE 11.2

Phases and application of LCA.

Adapted from ISO 14040:1997.

The *goal* needs to state clearly and without ambiguity which is the application, what are the reasons, why the study is performed, and who the recipients of the results of the study are. A goal identified in such a way will allow the practitioner to make the correct choices throughout the study. Obviously, the goal could also be adjusted depending on specific and relevant findings of some later steps of the analysis.

The *scope* sets the borders of the assessment. Different elements should be considered to specify the scope correctly: product group; functional unit; the system and its boundaries; impact assessment boundaries; the data quality requirements; and limitations. The definition of the system boundaries (inputs and outputs) are critical to determine the amount of work to be performed. The impact categories need to be chosen from a list of standard ones: assessing the boundaries will also limit the categories to be considered during the study. The work provided in the ISO 14040 actually lists the standard impact categories. The national agencies for environment usually rework this list according to their national characteristics and release an official one. Regarding the goal, even the scope can be adjusted during the iterative process of the analysis.

Inventory analysis

The Aim of this second phase is to perform mass and energy balances to quantify all the material and energy inputs, waste, and emissions from the system (i.e., the environmental burdens). The following main issues (as defined by ISO 14040:1997) should be considered during this phase: data collection; refining system boundaries; calculation; validation of data; relating data to the specific system; and allocation.

Data can be specific (to the process, the company, and the geographical area) or more generic (for instance, extracted from trade organizations or governmental institutions). According to the availability, it is possible to rely on quantitative or qualitative data to perform this phase. Generally, the results obtainable with LCA are very sensible to the set of data used, so it is important to understand the intrinsic criticality of this phase. More generic and qualitative data could be used for a first simplified analysis to be reiteratively repeated using more specific and system-related data.

The data initially collected can be used to review the *system boundaries*, as defined in the previous phase. We should also consider that if the system is very complex, then it might be necessary either to review the system boundary to include more data or to allocate the relevant environmental burdens to the system, ensuring that the approximation of the input–output relationship and the main characteristics are possible. *Allocation* might actually be necessary in case of multi-input or multi-output systems, in which a direct relation between inputs and emissions is not evident.

The inventory should then be interpreted considering all the specified uncertainties and lack of data. In particular, *validation* should also be considered and performed during the whole process of data collection to reduce eventual discrepancies and data quality issues later.

Impact assessment

In the third phase, and as a natural sequence of the work performed in the inventory phase, it is necessary to aggregate the environmental impacts quantified in the Inventory Analysis into a limited set of recognizable impact categories (e.g., global warming, ozone depletion, acidification, etc.). This phase comprises classification, characterization, normalization, and weighting.

According to ISO 14042:2000, there are three groups of categories to be considered: Resource use; Human Health consequences; and Ecological consequences. These broad groups should include all categories such as climate change, stratospheric ozone depletion, photochemical oxidant formation (smog), eutrophication, acidification, water use, and noise. It has to be noted that, currently, there is no consensus regarding a single list of impact categories. The impact categories therefore should be selected from a list of examples and should be relevant to the system undergoing investigation.

The second step in this phase is mainly a quantitative step: *characterization*. In this step, it is necessary to assign the relative contribution of each input and output to the selected impact categories. Pennington et al. (2004) proposed a generic equation to calculate, from the inventory data, indicators for each impact category using generic characterization factors.

$$\text{Category Indicator} = \sum_{s} \text{Characterization Factor}(s) \times \text{Emission Inventory}(s) \qquad (11.1)$$

where "s" indicates the given element (e.g., chemical compound) and the "Factors" can be found in literature as databases or are available in various LCA support tools. The following equation takes into account some of the potential variables of non-generic characterization factors in the context of human health and natural environment (Pennington et al., 2004):

$$\text{Characterization Factor}(s,i,t) = \sum_{j} \frac{\text{Effect}(s,j,t)}{\text{Emission}(s,i)} = \sum_{j} \left(\frac{\text{Fate}(s,j,t)}{\text{Emission}(s,i)}\right) \cdot \left(\frac{\text{Exposure}(s,j,t)}{\text{Fate}(s,j,t)}\right) \cdot \left(\frac{\text{Effect}(s,j,t)}{\text{Exposure}(s,j,t)}\right) \qquad (11.2)$$

Subscript s again denotes the chemical, i is the location of the emission, j is the related location of exposure of the receptor, and t is the time period during which the potential contribution to the impact is taken into account. "Fate(s,j,t)" is the probability that the given emission or exposure occurs, depending on the chemical, the location, and the related time interval.

The next step in this phase is the *normalization*. This activity was described by Stranddorf et al. (2003) as necessary to calculate the magnitude of the category indicator results relative to reference values when the different impact potentials and consumption of resources are expressed on a common scale. The goal of normalization is to set a common reference enabling comparison of different environmental impacts.

Because quantitative results of the aforementioned characterization of impact categories are not always comparable, an additional step is necessary: *weighting*. This activity aims at comparing the impact categories against each other. This would allow ranking and possibly defining the relative importance of these different results. Weighting can be a quantitative or qualitative activity, not always based on science, but often based on social or political considerations. Different weighting methods have been developed (Lindeijer, 1996).

Interpretation

This is the last phase as indicated by ISO 14040:1997. Interpretation is a systematic procedure to evaluate information from the conclusions of the inventory analysis and impact assessment of a product system and to present them to meet the requirements of the analysis as described in the goal and scope of the study. The following tasks should be accomplished in this phase (Jensen et al., 1997).

1. Identify the significant environmental issues
2. Evaluate the methodology and results for completeness, sensitivity, and consistency
3. Check that conclusions are consistent with the requirements of the goal and scope of the study, including, in particular, data quality requirements, predefined assumptions and values, and application-oriented requirements
4. If so, report as final conclusions. If not, return to step 1 or 2.

LIMITATIONS OF LCA APPROACHES

Even though LCA is a powerful tool to assess the environmental impact of product/services, some important limitations have been evidenced in the past years. The main limitations are all related to the LCA methodological approach, especially data quality and collection (Lee et al., 2007), definition of system and time boundaries (Stoeglehner and Narodoslawsky, 2008), and multi-functionality and allocation (Jeswani et al., 2010).

LCA is a methodology that is very data-dependent. The *quality* and *availability of data* influence the results significantly. Some of the steps of LCA can be reiterated to better tune the analysis to the systems undergoing investigation. Therefore, starting with easily accessible data and eventually refining the data quality with reference to the results are suggested. In some cases it is unavoidable to introduce simplification and limitation of assumptions because of uncertainty of specific data. For instance, toxicological categories as well as some energy production impact categories are deeply affected by lack of data (Lee et al., 2007). There can be different types of data uncertainty (Schmidt et al., 2007), and it must be noted that the methodological uncertainties are sometimes larger than the data uncertainties.

The *time aspect* is often critical in including or excluding some effects of the systems undergoing analysis. LCA should consider environmental impacts on the longest possible timeframe, possibly an infinite one. Nevertheless, most of the studies use shorter time periods, leading to contestable conclusions (for instance, in Municipal Solid Waste treatments, landfills act in a limited period of time as carbon sinks and therefore become a more favorable solution than incineration).

The holistic approach of LCA is one of its main strengths, but it is also a cause of complexity during the actual execution of the analysis. Having to collect and analyze data from so many different elements can be cumbersome. This is the main reason why some assumptions are made and the *system boundaries* are modified to omit some elements. In particular, the upstream elements of the supply chain are usually not included in the analysis because of the inherent difficulty in gathering complete information for elements outside the specific product system.

Results of LCA are often used for process optimization. The applicability of these results depends greatly on the *model of the process* that has been adopted at the beginning of the study. This model is frequently simplified to be able to take into consideration all possible inputs and outputs, and in most cases it does not include *health and safety* elements. This is very reductive because not all results from LCA can be applied directly in process improvements; choices that reduce the environmental impact might not always be applicable for human or industrial constraints or, in some cases, can prove to be dangerous. Therefore, it is necessary to take into due account the human factor and to integrate the work environment in the holistic approach of LCA.

Uncertainty estimation is also another challenge in using LCA. A possible approach to overcome this challenge as well as the challenge of accounting for economic value is given by Hybrid LCA. This combines a bottom-up construction of the supply chain based on facility-level data for material/energy use with a top-down economic input–output (EIO) model to account for processes for which direct data are usually not unavailable (Deng et al., 2011). In this approach, the environmental burdens are calculated for the process that generated a specific product, whereas the economic impacts are derived from the decomposition/recycling of the final product at the end of the life cycle. In the context of waste management, different approaches have been worked out, but there is still the need to develop a method for uncertainty estimation and accounting of waste LCA (Clavreul et al., 2012).

The great amount of detailed information required in completing an LCA, which also takes into consideration cost, might discourage some practitioners from using this approach as a decision-making support tool.

THE HUMAN FACTOR: WORK ENVIRONMENT IN LCA

As noted previously, one of the main limitations of the LCA methodology, as described by ISO 14040, is the lack of inclusion of work environment issues.

This does not mean that safety and health analysis of processes is not performed by the company. Most frequently, these issues are addressed after the fact to analyze the suggestions indicated by the application of an environmentally oriented LCA. There is still a tendency for companies to treat safety, health, and environment (SHE) as separate issues (Crawley and Ashton 2002). This adds complexity to environmental management systems and makes the companies lose out on possible synergies between environmental and safety issues. From an economical point of view, this practice is not optimal because design could be taken too far before it is found to be too dangerous from a work environment perspective.

Historically, the first efforts including the human factor in LCA have been made in Scandinavia. The Nordic countries therefore have produced different approaches for a Work Environment LCA (WE-LCA). Antonsson and Carlsson (1995) proposed a method based on five quantitative and two qualitative impact categories. The WE-LCA is performed in a way similar way to that for the external environment, with the four steps of goal and scope definition, inventory analysis, impact assessment, and interpretation. The method requires the use of an inventory of effects, instead of emissions, followed by the impact assessment. The quantitative impact categories are:

1. Deaths caused by work-related accidents
2. Workdays lost because of work-related accidents
3. Workdays lost because of illness
4. Hearing loss
5. Allergies.

The qualitative impact categories are:

- Carcinogenic impact
- Impact on reproduction.

Data for the quantitative categories can be collected from single companies or trade statistics organizations. It must be noted that the final result will depend greatly on the quality and precision of this set of data. Moreover, the level of detail must be balanced against the goals of the analysis. Another source of uncertainty in the method is the fact that not all impact categories can be estimated quantitatively. Work environment issues have been omitted not only from LCA methodology but also from environmental technology databases and reports of Best Available Technologies. In a project conducted for the European Commission on a selection of cases of the International Cleaner Production Information Clearinghouse (ICPIC) system, Ashford (1997) evidenced the following: the most striking feature of the case studies is their complete lack of information regarding the interactions of humans with the production processes, materials, or products. A serious lack of integration of safety concerns with legislation, regulations, and policies addressing environment and, more generally, industrial ecology is also evident.

The roles of personal risk perception and involvement in occupation health and safety issues in environmental management systems have been interestingly analyzed by Honkasalo (2000). In particular, risks in industrial environments seem to be perceived differently depending on the level of involvement of the perceiver. Risks caused by others, for instance, global environmental issues, are not tolerated easily, because the perceiver feels that they cannot be affected. Risks taken voluntarily, for instance, safety risks, are more accepted.

In 2004, the Danish Environmental Agency formed (Schmidt et al., 2004) guidelines regarding how to calculate the potential Work Environment impacts per functional unit by adding the impacts from a number of processes and activities. This method, based on the collection of goods statistics, aims at calculating the number of reported accidents per produced weight unit on the sector level. The steps are the same as the ones for the environmentally oriented LCA. The first step is the inventory procedure, whereby material flows are calculated for the product (with reference to a set of data obtained from the governmental statistical office). The material flows are then aggregated on relevant processes in the provided database and for each process the weight is multiplied with the impacts per weight unit for each of the affected categories. It must be noted how the method introduces a great source of uncertainty. One of the main challenges is to match the actual activities with data sets in the database (several thousand product groups must be related to a small number of sectors (<300). Aim of the normalization activity in the impact assessment step is to relate the total number of accidents and work-related diseases with the Danish population (the same could be applied to other countries if similar databases are available). In Table 11.1, a list of impact categories and normalization factors are illustrated.

Interpretation can be performed after the inventory or after the normalization. In the first case it is possible to establish an overview of how much each of the activities contributes to the single effect categories. After the normalization, it is possible to depict which are the most important impact categories in the life cycle of a product. The method described is a comprehensive approach to determine the

Table 11.1 Impact Categories and Normalization Factors

Basic for Normalization	Person Equivalents, PE	Worker Equivalents
Effect Category	**Danish Population**	**Danish Work Force**
Fatal accidents	1.54×10^{-5}	3.06×10^{-5}
Accidents	9.69×10^{-3}	1.92×10^{-2}
Cancer	3.54×10^{-5}	7.02×10^{-5}
Psycho-social damages	1.40×10^{-4}	2.77×10^{-4}
CNS function disorders	6.37×10^{-5}	1.26×10^{-4}
Hearing damages	4.56×10^{-4}	9.06×10^{-4}
Airway diseases, non-allergic	1.00×10^{-4}	1.99×10^{-4}
Airway diseases, allergic	7.93×10^{-5}	1.57×10^{-4}
Skin diseases	3.12×10^{-4}	6.19×10^{-4}
Musculoskeletal disorders	1.44×10^{-3}	2.85×10^{-3}

From Schmidt et al. (2004).

impact of work environment issues with an LCA approach. The established database associated with it covers approximately 80 economic sectors and provides an important tool for this kind of analysis in Denmark. It would be most useful if other countries collected the same kind of information so that similar analyses could be performed with increased reliability in those countries.

FROM ENVIRONMENTAL ASSESSMENT TO STRATEGIC ENVIRONMENTAL MAPS

INTRODUCTION

The ecological footprint is a way to compare human demand with our planet capacity to regenerate it, and it is a measure of our burden on the ecosystem (Stoeglehner, 2003). Usually, it represents the amount of biologically productive land and sea area needed to regenerate the resources consumed and to absorb the corresponding waste. Different footprints have been developed to consider the impact of different resources. In a broader view, the ecological footprint is related to the method of LCA, which is typically used for products and services but is also applicable for production plants and regions. One of the advantages of LCA is that it more aptly covers the whole range of impacts, and it may also provide an account of the upstream impacts. Recent studies have tried to combine the different footprints from the environmental assessment into a "family" to provide support in the decision-making process (Galli et al., 2012).

Nevertheless, one of the most important limitations in the application of LCA as an input for strategic decision-making from an environmental perspective is the limited inclusion of cost and investment considerations. In this Chapter we propose a new approach to integrate financial, environmental, resource, and toxicological considerations into a single analysis. The core of this new concept is to calculate some specific sustainability indicators based on LCA. This will help one define the relevant contributions to support strategic decision making. The cradle-to-grave approach will assure that all environmental and human consequences are taken into account. These must be further balanced against financial and resource consumption considerations.

When evaluating different options from a strategic perspective, they are usually evaluated against the following categories (Čucek et al., 2012a):

- Carbon footprint
- Water footprint
- Energy footprint—Land, Renewables, Non-Renewables
- Emission footprint—emissions in Air, in Water, and in Soil Waste
- Work environment footprint—work environment and toxicological impacts.

Cost should also be considered as an additional category possibly representing the crucial relation that it has with all other categories (Čucek et al., 2012b).

To represent these relations and to compare options from an environmental and, more generally, business perspective, we introduced a new graphical representation: the Environmental Performance Strategy Map (EPSM). The objective of this representation is to build on the strength of the Ecological Footprint and Life Cycle Analyses to provide a single indicator for each option. The practitioner can make use of this indicator to direct the decision-making process toward the best option from a sustainability and environmental perspective.

The first step in building the EPSM correctly is to calculate the impact of the option undergoing analysis for all aforementioned footprints. The combination of these elements and the cost perspective will provide a single indicator to assign to each option. The comparison between different options with different characteristics and ratio of advantages and disadvantages will also be facilitated by a graphical representation. The best option from an environmental and financial perspective will be selected based on this approach.

WHAT FOOTPRINTS?

Different methods have been developed in past years to correlate environmental sustainability of specific activities with land and water areas required to supply this activity with resources and to absorb its wastes (Monfreda et al., 2004). This is usually referred to as Ecological Footprint.

Some initial objections to the original method regarding the way energy have been accounted for (Ferng, 2005) and the difficulty in using the tool in the decision-making process (Ayres, 2000) have been overcome by the development of specific indicators (SPI) (Krotscheck and Narodoslawsky, 1996) and DAI (Eder and Narodoslawsky, 1999). In particular, the Sustainable Process Index (SPI) considers the area as a basic measure: the more area a process requires, the more its burden from an ecological point of view. The SPI method is based on the comparison of natural flows with the mass and energy flows generated by a technological process. The calculation of an SPI centers on the computation of the total area required (A_{tot}):

$$A_{tot} = A_R + A_E + A_I + A_S + A_P \tag{11.3}$$

where A_R is the area required to produce the raw materials (given as the sum of the areas to provide renewable raw materials, fossil raw materials, and non-renewable raw materials), A_E is the area needed to produce process energy, A_I is the area required for the process installations (equipment/plant), A_S is the area required for support staff, and A_P is the area required for the accommodation of products and by-products (Krotscheck and Narodoslawsky, 1996).

A model that proposes the combination of ecological footprinting with economic considerations is proposed in the Ecological Value-Added system (Kratena, 2004). This is based on an input–output system and on the ecosystem pricing concept introduced via energy values and the ecological footprint. The balance between carbon sinks and emissions defines the sustainability target for this model. To provide a more comprehensive analysis of the interaction of the environmental burdens and financial costs, the EPSM is based on the combination of the following five footprints (Čucek et al., 2012c).

Carbon footprint

With environmental issues high on the business and political agenda, different definitions of the individual contribution to CO_2 emissions have been proposed in past years (Wiedmann and Lenzen, 2007). Usually, they are referred to as carbon footprint. In response to this public attention, different tools have been proposed to calculate the value of the carbon footprint in relation to a product or process (Padgett et al., 2008). Even though these tools are useful in increasing public awareness, they often lack transparency and might provide conflicting results.

For the purpose of building the EPSM, we refer to a land-based definition indicating that the carbon footprint estimates the land area required to sequester atmospheric fossil CO_2 emissions through afforestation (Monfreda et al., 2004). This area is calculated as (Hujbregts et al., 2008)

$$CF = M_{CO_2} \times \frac{1 - F_{CO_2}}{S_{CO_2}} \times EF \tag{11.4}$$

where CF is the footprint of indirect land occupation by fossil fuel and cement-related CO_2 emissions, M_{CO2} is the product-specific emission of CO_2 (kg CO_2), F_{CO2} is the fraction of CO_2 absorbed by the oceans, S_{CO2} is the sequestration rate of CO_2 by biomass (kg CO_2 m^{-2} y^{-1}), and EF is the equivalence factor for forests. This footprint unit of measure is expressed in m^2.

Water footprint

The concept of water footprint is a relatively new one; it is related to the concept of virtual water (Hoekstra and Hung, 2002) and later (Hoekstra, 2007). Virtual water is the amount of water required to produce a service or a product. Analogous to ecological and carbon footprints, this indicator is designed to summarize the contribution of a product or activity to the deterioration of the environment. The focus is on the consumption of a limited resource, water. Although the ecological footprint is designed to calculate the area needed to sustain specific human activities, the water footprint examines the volume of water. With two different methods (top-down or bottom-up), the water footprint measures the amount of water related to human consumption and takes into consideration blue and green water, as well as the production of polluted gray water (Hoekstra and Chapagain, 2007). For instance, in the case of crops, we can define the green virtual water content as a ratio between the effective rainfall and the crop yield. Analogously, the blue virtual water content is the ratio between the effective amount of irrigated water and the crop yield. The total virtual water content is given by the sum of these two elements.

In this study the EPSM is used to represent an overall indication of the comparative sustainability of different options from a strategic decision-making point of view. The water footprint of an activity therefore consists of two components: the direct water used (for producing/manufacturing or for supporting activities) and the indirect water use (that propagates throughout the supply chain). This footprint unit of measure is m^3.

Energy footprint

The energy supply footprint (Stoeglehner and Narodoslawsky, 2008) takes into account different energy supplies as related to different demand categories, such as heating and hot water production, process energy, electricity, and traffic. The footprint is calculated by multiplying the final energy use of different energy carriers with their land need indices and adding these results to the footprint of the whole energy supply. This footprint unit of measure is m^2. It is important to notice that the Energy footprint, as defined elsewhere (Stoeglehner and Narodoslawsky, 2008), includes some CO_2 contributions from burning processes. However, it does not include all other CO_2 contributions, and that is why it is important to make use of the Carbon footprint as defined previously.

$$\text{Energy Usage(MJ)} \times \text{Replenishment Rate}\left(\frac{m^2}{MJ}\right) = \text{Energy Footprint}(m^2) \qquad (11.5)$$

Emissions footprint

To identify the real environmental burden, we define the Emissions footprint as the quantity of emissions of the process undergoing investigation in water, soil, and air converted to area requirements. The conversion of emissions is calculated according to the principle that anthropogenic mass flows must not alter the quality of local compartments (Sandholzer and Narodoslawsky, 2007). Maximum flows are defined based on the natural existing quality of the compartment and their replenishment rate per unit area. For emissions to soil, the replenishment rate is given by the decomposition of biomass to humus (measured by the production of compost by biomass). For ground water, this is the seepage rate (given by local precipitations). Emissions to the compartment air are treated slightly differently, because there is no natural replenishment rate for this compartment. Here, the natural exchange of substances between forests and air per unit area (which is known for most airborne substances) is taken as a base of comparison between natural and anthropogenic flows (Sandholzer and Narodoslawsky, 2007). Different emissions to air are not weighted, because only the largest dissipation areas are to be considered. Lower area consumptions emissions may be dissipated without violating the principle that anthropogenic mass flows must not alter the quality of local compartments. This footprint unit of measure is m^2.

Work environment footprint

For the purpose of building the EPSM, the Work environment footprint is the Work Environment LCA as proposed previously (Schmidt et al., 2004). This method, based on the collection of goods statistics, is designed to calculate the number of reported lost days of work per produced weight unit on the sector level. The following impact categories are included in the assessment (Schmidt et al., 2004):

- Fatal accidents
- Total number of accidents
- CNS function disorder
- Hearing damage
- Cancer
- Musculoskeletal disorders
- Airway diseases (allergic and non-allergic)
- Skin diseases
- Psycho-social diseases.

The calculation used in this study, based on the previously mentioned theory, is modified and calculated as number of lost working days per employee. This is to increase its generality and use of readily available statistical sources. In case of fatalities, the number of lost working days is calculated until replacement of the workforce.

BUILDING THE MAP

Once the contribution of each option to the specific footprints has been calculated, it is possible to build the EPSM. The basic concept is to map the footprint on a specific spider-web plot to identify a meaningful combination (Figure 11.3). To obtain comparable measures, the results of each footprint are normalized, resulting in a scale from 0 to 100. Therefore, we apply a deviation-from-target methodology wherein, for each of the footprints, it is possible to define a maximum target and express each value recorded as a percentage. The aim is to lower as much as possible, in percentage points, the contributions of each footprint to the overall combined Indicator. The targets specified in Table 11.2 either are based on maximum available resources or are drawn from scientific consensus or regulatory requirements.

Each option has an area assigned that represents a combination of all footprints. To specify the cost and financial impacts, we introduce an additional dimension (Figure 11.3). This is the cost of the option undergoing analysis. The cost is considered as additional dimension because it is not used for comparative reasons. The indicator takes into account the total financial investment required for each options. When geometrically combining these five footprints in one plane with the orthogonal dimension of the cost, we obtain a pyramid.

The volume of each pyramid represents the overall environmental and financial impact of the option undergoing consideration. We can define this index as the Strategic Environmental Performance Indicator (SEPI). Finally, it is possible to plot all options under consideration in a specific EPSM. The map thus enables comparison of different options for strategic decision-making purposes based on a SEPI.

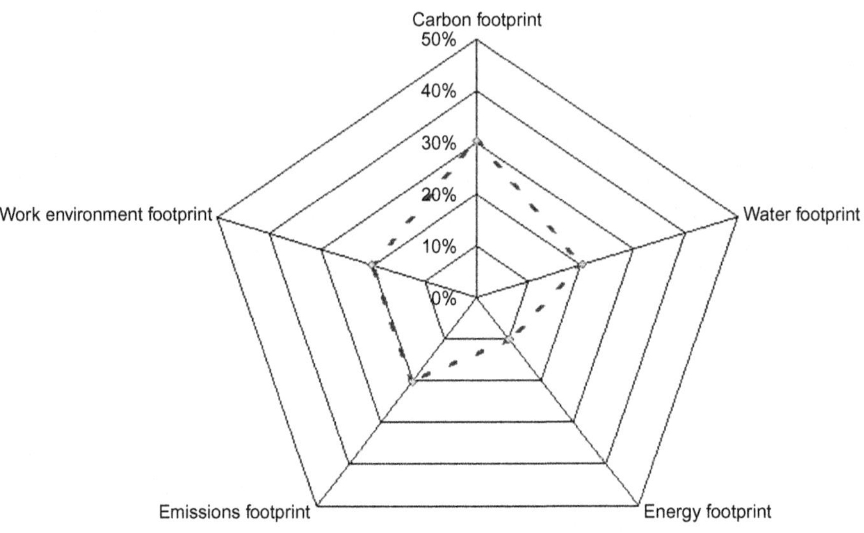

FIGURE 11.3

Plotting the footprints in the EPSM.

Table 11.2 Normalization Target Factors

Footprints	Target	Source
Carbon Footprint	Max area available (m^2)	Problem definition
Water Footprint	Average Water required for a specific category of product or service (m^3)	UN statistical office (unstats.un.org/unsd/ENVIRONMENT/waterresources.htm)
Energy Footprint	Max area available (m^2)	Problem definition
Emission Footprint	Max area available (m^2)	Problem definition
Work Environment Footprint	Total number of accidents and work-related diseases per product group in the specific country/area (accidents/person)	National Statistical office (e.g., epp.eurostat.ec.europa.eu/portal/page?_pageid = 1073,46587259&_dad = portal&_schema = PORTAL&p_product_code = KS-BP-02-002-3A)
Cost	Max budget available for EHS management (€)	Problem definition

Table 11.3 Calculation of Footprints and Deviation from the Targets

Total Consumptions	Units	Values
Water consumption	m^3	49,749,000
Energy consumption		
Natural gas	MJ	809,000
Electricity	MJ	165,000
Heating oil	MJ	5,100
Emissions		
Water	t	765
Air	t	94
Soil	t	94,550
CO_2	t	52,500
Work environment		
Absence from work due to accidents		4
Absence from work due to accidents (per worker)		0.004

To illustrate the use of the EPSM, a demonstrative example of a plant in a Nordic country producing fertilizers and pesticides is presented. This example is part of a case study defined to validate the applicability of the concept. For confidentiality reasons, names and sensitive data are hidden. The company develops and markets plant protection products for controlling weeds and fungal diseases. It employees 850 people, and it is located in a municipality in the northern part of Jutland (Denmark). The company budget for EHS management is 500,000€. Table 11.3 characterizes the production process.

Applying the method described in the previous paragraphs, the different footprints are calculated (Table 11.2). To calculate the percentages of deviation from targets, the values from Table 11.3 were used. These values are derived from Table 11.4. The deviation-from-target values are used for the EPSM.

The SEPI valuc from the area of the polygon is equal to 819.50. In this case for the Cost we will only consider the value proposed for the investment and not a specific target. This assumption can be made because there are no other additional options to be considered. Thus, the volume of the pyramid we obtained (Figure 11.4) represents the space of all possible solutions.

THE SEPI AND POLICY MAKING

Considering the inter-relations and the complexity of Environmental issues, decision-making in this field is very difficult. This is particularly true if it is not supported by analytical tools and reliable metrics. The SEPI, as proposed in this study, does not aim at being the single metric that policy-makers should rely on.

Table 11.4 Maximum Target Values in the Given Geographical Area (North Jutland, Denmark)

Targets	Units	Values
Total budget available	€	500,000
Total site	m^2	100,000
Total area available (municipality)	m^2	468,000,000
Total water resources (municipality)	m^3	244,897,959
Average accidents per worker (national average)		0.022

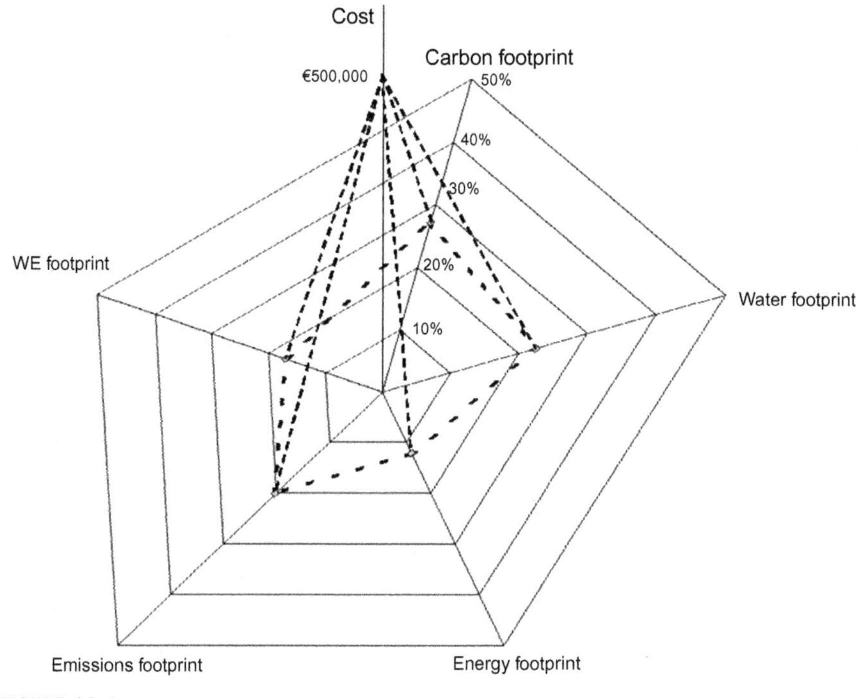

FIGURE 11.4

The additional dimension in the EPSM.

Encompassing environmental and financial issues, it is a useful tool that can help in the decision-making process and could be adopted to compare different options and their comparative impacts on societies and on the ecosystems on which they are totally dependent. In particular, with its "deviation-from-target approach," SEPI provides a way to measure the effectiveness of environmental policies against performance targets. In this sense, the targets proposed in Table 11.2 can easily be adjusted to reflect local communities, counties, or national reference targets. SEPI can be a valuable tool to investigate different options to a given environmental problem. For instance, if we mandate the reduction of the water consumption by 20%, this is reflected in a decrease in SEPI by 7%. In this way, we can simulate the different options and propose the one with the highest possibility of reduction of the environmental burden.

CONCLUSIONS

The limitations in the use of LCA as a tool for strategic decision-making allowed us to introduce the EPSM as a possible solution. This particular graphical representation is designed to provide a single indicator—the SEPI—to overcome the use of footprints as mere communication and awareness tools. The introduction of the financial aspect complements the environmental and work environment considerations to provide a more holistic answer to the sustainability of specific options. In particular, SEPI can be successfully applied to provide an overall indicator of the environmental performance of existing applications or can be used as a supporting tool in comparing competing options in a strategic decision-making process. This new approach has been demonstrated with a specific case to illustrate the main steps needed to find the right balance between cost and environmental impacts. This offers potential balancing and minimizing environmental impacts, such as energy/carbon footprint, water, emissions, and working environment, and quantifying them into one indicator.

The SEPI is an important step in the debate on defining the appropriate metrics and methodologies for evaluating environmental performance. It is important to point out the difficulties in converting some of the footprints to area requirement. In general, although it is relatively easy to express as area processes that are area-based, such as an agricultural process, converting to area processes that are not primarily area-based, such as a chemical process, can prove problematic. Another important development could be addressing the long-term human health and ecosystem degradation costs. By identifying best practices and costs associated with those, a target value could be set. This should be the basis for a comparison with current cost and practices.

THE ENVIRONMENTAL BILL OF MATERIALS AND TECHNOLOGY ROUTING

The term "Bill of Materials" (BOM) is used to indicate basic materials, components, parts, and the quantities of each needed to manufacture a product or service (Reid et al., 2002). The idea is to associate contributions to the aforementioned footprints to each component of a manufactured good or service, which we can define as Environmental Performance Points (EPPs). To calculate these

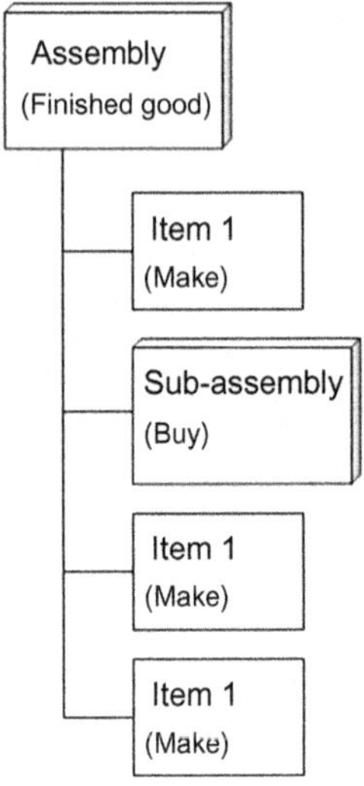

FIGURE 11.5

Example of a simplified Bill of Materials.

contributions, it is necessary to isolate the environmental burden of each item of the BOM. Let us consider the case of an assembled item with a simple BOM where a subassembly and few other items appear (Figure 11.5).

In this case it is important to specify if items and subcomponents are made in-house or purchased from outside (Make or Buy items). Make items are typically known and provide fewer issues in calculating the contribution to the environmental burden. This might prove more difficult for the Buy items. This distinction is also important to draw the boundaries of the system we want to investigate. If information regarding Subassembly (Buy) is available, then it is possible to decide not to explode it in its subcomponents or basic materials.

With regard to previous work on this subject (De Benedetto and Klemeš, 2009), the contributions of each item to the footprints specified in the previous paragraph is recorded together with the related EPP. The definition of this new BOM as an Environmental Bill of Materials (Env-BOM) is therefore suggested (Table 11.5).

When all components are recorded, it is possible to consider the operations and the technology process required to transform the materials into the final product. All operations and processing steps should be considered. This work defines the collection of all operations Technology Routing (Table 11.6). As performed for the items in the BOM, it is necessary to define the contributions in terms of the footprints for each step of the process. It is important at this stage to review the scope of the analysis to avoid filling the Technology Routing with too many steps that might not have significant impact on the definition of the EPSM.

Table 11.5 Example of Environmental BOM Form

Env-BOM				Environmental Performance Points				
Assembly	**Type**	**UOM**	**Qty**	**Carbon**	**Water**	**Energy**	**Emissions**	**Work Environment**
Item 1	Make	Each	1					
Sub-assembly	Buy	Each	1					
Item 2	Buy	Each	1					
Item 3	Make	Each	1					

Table 11.6 Example of Technology Routing Table Form

Technology Routing		Environmental Performance Points				
Operation	**Quantity**	**Carbon**	**Water**	**Energy**	**Emissions**	**Work Environment**
Manual assembly	1					
Painting	1					
Drying	1					
Quality Control	1					
Packaging/shipping	1					

When all EPPs are defined, it is possible to sum them all to provide the final figures to build the EPSM.

HOW TO USE THE ENV-BOM AND TECHNOLOGY ROUTING

To illustrate the use of the Environmental BOM and the Technology Routing, a demonstrative example of a plant in a Central European country producing fertilizers and pesticides is presented. For confidentiality reasons, names and sensitive data have not been presented. The company develops and markets Ammonium Nitrate (AN)-based fertilizers. It employees 800 people, and it is located in a municipality in the Western Trans-Danubian region in Hungary. The capacity of the plant is 800–900 t/d.

Short process description

The whole production process is divided into two parts: (i) the production of 87% AN solution from raw materials and (ii) the production of solid AN prills from 87% AN solution. The production of AN started from 58% nitric acid solution and Ammonia (gas form, 3.5 bar pressure). The neutralization reaction takes place in four reactors. The pH is set to 5.5 in the last reactor. Because of the reaction, heat water vapor leaves and the solution reaches a temperature of 140–145 °C and 80–85% concentration. The water vapor leaves the reactor through a vapor line (there is acidic washing to retain the Ammonia) and condenses through a drop collector. After condensation, it is pumped into a wastewater plant.

The AN solution coming from the reactors is concentrated in a three-stage (one atmospheric and two vacuum) distillation process. All the distillers are heated with 10-bar saturated water steam. The first stage has an auto-circulation system and consists of a distiller and a separator. It works on atmospheric pressure and on 155 °C. The solution is concentrated up to 95%. The vapors from the separator leave to the atmosphere.

The second stage is similar to the first one. It works on 120 mmHg pressure and 160 °C. This stage is sucked through the separator by a vacuum pump. The vapor generated goes to a condenser, and from there to the wastewater plant.

The solution from the separator gets to the third (end) stage. The distiller contains a double tube system and works on 50 mmHg and 178 °C. The separator is sucked by jet pump. The vapors are condensed and driven to the wastewater treatment plant.

The final solution has less than 0.5% water and the pH is 5.5 (corrected with Ammonia). The solution gets to a scatter basket through a vacuum seal. The basket is at the top of the prilling tower. The dispersed solution droplets cool and solidify as they fall down in the tower on a fluid bed cooler. At the bottom of the tower, big fans blow air in the tower to accelerate the cooling. The 30–35 °C AN granules are surface-treated by fat amine. The product is stored and distributed in bulk form or in big bags.

Definition of the environmental bill of materials

Table 11.7 shows the Bill of Materials for the main components of the AN-based fertilizer under consideration. The amounts are reduced to 1 ton of final product. All values are estimates and are used here only for demonstrative purposes.

The specific amounts to produce 1 t of 87% AN solution are shown in Table 11.8.

The following main operations define the technology routing:

1. Neutralization reaction
2. Three-stage concentration
3. Prilling
4. Packaging and shipping.

The boundaries of the study are drawn around the actual production process to keep this illustrative example simple. The final Environmental Bill of Materials should specify the contribution of each material and operations of the five main areas of environmental impact. Table 11.9 reports the Env-BOM. Emissions values are estimates and are used here only for demonstrative purposes. Table 11.9 exemplifies the different steps of the process and therefore are listed in the technology routing.

Once these contributions are defined, the process to build the EPSM is the same as outlined by De Benedetto and Klemeš (2009). It is also possible to define the EPPs as the percentage deviation from targets (Table 11.10) followed by technology routing (Table 11.10) and an example of maximum target values in the given geographical area (Table 11.11).

To calculate these deviations the values from Table 11.11 were used. The Performance Points can be directly mapped on an EPSM without further

Table 11.7 Bill of Materials of 1 Ton of AN Fertilizer

Designation	Unit	Specific Amount
Raw dolomite aggregate	t	0.028
Fat amine	kg	0.589
Ammonia	t	0.003
AN solution 87%	t	1.116
Instrument air	Nm^3	4.481

Table 11.8 Bill of Materials of 1 Ton of AN Solution Used to Produce AN Fertilizer

Designation	Unit	Specific Amount
Ammonia	t	0.185
Nitric acid 56–58%	t	1.181

Table 11.9 Environmental Bill of Materials and Technology Routing

ENV-BOM	UOM	Quantity	Water (m^3)	Energy Consumption (MJ)	Emissions Water (kg)	Emissions Air (kg)	Emissions Soil (kg)	CO_2 (kg)	Work Environment (*n*)
Raw dolomite aggregate	t	0.028		50				10	
Fat amine	kg	0.589		22.5					
Ammonia	t	0.209	500	200		1.8	0.1	100	
Instrument air	Nm^3	4.481		30					
Nitric acid 56–58%	t	1.318	150	70	0.2	1.3		310	
Technology Routing									
Neutralization reaction			100	225	0.8	1.18			
Three-stage concentration			150	230	0.3			80	4
Prilling				50			0.31		
Packaging and shipping				30			0.5	90	3.8
Total			900	907.5	1.3	4.28	0.91	590	7.8

Table 11.10 Normalized Env-BOM with EPPs and Technology Routing

Env-BOM	UOM	Quantity	Water (EPP)	Energy Consumption (EPP)	Emissions Water (EPP)	Emissions Air (EPP)	Emissions Soil (EPP)	CO_2 (EPP)	Work Environment (EPP)
Raw dolomite aggregate	t	0.028	0%	0.79%	0%	0%	0%	0.02%	0%
Fat amine	kg	0.589	0%	0.36%	0%	0%	0%	0%	0%
Ammonia	t	0.209	18.52%	3.17%	0%	10.91%	0%	0.21%	0%
Instrument air	Nm^3	4.481	0%	0.48%	0%	0%	0%	0%	0%
Nitric acid 56–58%	t	1.318	5.56%	1.11%	0%	7.88%	0%	0.66%	0%
Technology Routing									
Neutralization reaction			3.70%	3.57%	0%	7.15%	0%	0%	0%
Three-stage concentration			5.56%	3.65%	0%	0%	0%	0.17%	1.93%
Prilling			0%	0.79%	0%	0%	0%	0%	0%
Packaging and shipping			0%	0.48%	0%	0%	0%	0.19%	1.84%
Total			33.33%	14.4 %	0%	25.93%	0%	1.27%	3.77%

Table 11.11 Maximum Target Values in the Given Geographical Area—Trans-Danubia Region

Resources	Units	Values
Total site area	m^2	100,000
Total area available (municipality)	m^2	126,000,000
Total water resources (municipality)	m^3	647,976,352
Average days of absence per worker (national average 2004)		0.24355

From De Benedetto and Klemeš (2009).

processing to obtain the SEPI for the given combination of materials and process. At this stage the industrial operator can investigate alternatives in materials or process that could lower any of the percentages indicated. That would be reflected immediately on the Map and ultimately the SEPI. In this case, it is easy to note the Ammonia is one of the main contributors to the emissions category.

Applying a different production technique, such as Reduced Primary Reforming (EIPPCB 2007), could abate the NO_x emissions to 0.3 kg/t, more than halving the current level of emissions to air (as described in Table 11.9, where the current value is 1.3 kg). Inputting this new value in the calculations would result in reduction of 6% in the final EPP, and therefore it would lower the final SEPI for the whole process. Another possibility would be lowering the water requirements for the process using water integration techniques (Ku-Pineda and Tan, 2006).

CONCLUSIONS

The SEPI and the associated Strategy Map provide an overall indicator of the environmental performance of existing applications or can be used as a supporting tool in comparing competing options in a strategic decision-making process. One of its main characteristics is that it aggregates all contributions at a very high level. In case, this is not desirable and a more granular approach is required; we can work with two new elements: the Environmental Bill of Materials and Technology Routing. These elements allow the decomposition of each environmental burden at materials and process steps level. It has been demonstrated that the use of this new technique could isolate the materials or operations steps responsible for excessive resource consumption or emissions. A possible reduction of 6% in the emissions to air has been evidenced in the associated case study. This would have not been evidenced directly using the EPSM, because this approach would consider only process numbers. Although this approach still offers potential balancing and minimizing environmental impacts as energy/carbon footprint, water, emissions, and working environment, and quantifying them into one indicator, it also guides the practitioner directly toward those elements of the bill of materials or those steps of the process that contribute the most to the final environmental burden.

UNCERTAINTY ESTIMATION IN THE DEFINITION OF THE EPSM

Environmental impact assessment involves the evaluation of effects of diverse actions on a number of different environmental impacts. The method proposed in this chapter is based on ecological footprinting (discussed in several other chapters in details), and LCA requires the calculation of some specific sustainability indicators.

To represent these relations and to compare options from an environmental perspective, a new graphical representation has been introduced: the EPSM. This leads to the calculation of a single indicator of the environmental viability of a given scenario (SEPI).

This methodology has been proved for deterministic values of impact categories. However, most of the time, these exact values are not available to practitioners. Considering the uncertainty and inaccuracy inherent in the process of allocating values to environmental impacts, fuzzy logic is a suitable and useful tool to perform these evaluations.

INTRODUCTION TO FUZZY LOGIC

Fuzzy logic is based on the concept of a fuzzy set. A *fuzzy set* is a set without a sharp, clearly defined boundary. It can contain elements that belong fully to it or that are characterized by only a partial degree of membership. Fuzzy inference is the process of formulating the mapping from a given input to an output using fuzzy logic. The process of fuzzy inference involves membership functions (MFs) and fuzzy rules.

A *membership function* (MF) is a curve that defines how each point in the input space is mapped to a membership value (or degree of membership) between 0 and 1. Fuzzy sets and fuzzy operators are the main elements of fuzzy logic. These if–then rule statements are used to formulate the conditional statements used in fuzzy logic.

A single fuzzy if–then rule assumes the form if x is A, then y is B, where A and B are linguistic values defined by fuzzy sets on the ranges X and Y. The if part of the rule "x is A" is called the *antecedent* or premise, whereas the then part of the rule "y is B" is called the *consequent* or conclusion. Fuzzy inference process comprises of five parts: *fuzzification* of the input variables; application of the fuzzy operator (AND or OR) in the antecedent; implication from the antecedent to the consequent; aggregation of the consequents across the rules; and *defuzzification.*

CASE STUDY

As an application, the operation of a fertilizer production plant used for the previous case study is considered here as well. The process is divided in four main technology steps that cause impact:

1. Neutralization reaction
2. Three-stage concentration
3. Prilling
4. Packaging and shipping

To apply the fuzzy logic inference system, the focus of this example is on the calculation of the first contribution to the EPSM: the Carbon Footprint.

The following tasks will be accomplished:

- Definition of the fuzzy variables (input and output variables)
- Definition of the MF for all variables
- Definition of the fuzzy rule set
- Assuming that MF of all impact categories are similar to triangular fuzzy numbers, a total positive or negative fuzzy value for the environmental impacts is going to be calculated
- Finally, the defuzzification will lead us to the punctual impact estimator and its corresponding uncertainty interval.

To perform these tasks, the Fuzzy Logic toolbox in MatLab is used.

FUZZY INFERENCE SYSTEM FOR EPSM CALCULATION

The first step in the calculation is to take the inputs and determine the degree to which they belong to each of the appropriate fuzzy sets via MF (Table 11.12).

X_{cn} represents the contribution in carbon dioxide emissions from the operation of Neutralization, X_{cc} represents the contribution in carbon dioxide emissions from Concentration, X_{cp} represents the contribution in carbon dioxide emissions from Prilling, and, finally, X_{cs} represents the contribution in carbon dioxide emissions from Shipping.

Before the rules can be evaluated, the inputs must be fuzzified according to each of the linguistic set. In this case, the linguistic set will include the following descriptions: *very low, low, high,* and *very high.* All fuzzy variables considered

Table 11.12 Input Variables Definition

Neutralization	Concentration	Prilling	Shipping
X_{cn}	X_{cc}	X_{cp}	X_{cs}

here will have a triangular distribution. The MF for each of the variables are defined in Figures 11.6–11.9.

The second step in the fuzzy inference system definition is to apply the logical operators on the *antecedents*. That is, we need to define what kind of logical operation might relate each variable and their linguistical definition. As each variable comes from a different step of the production process, the logical operator that is considered is the AND:

$$X_{c,n} \text{ AND } X_{c,c} \text{ AND } X_{c,p} \text{ AND } X_{c,s} = > \text{Carbon Footprint} \qquad (11.6)$$

The third step is to work on the second part of this relation: the *consequent*. A consequent is a fuzzy set represented by an MF, which weights appropriately the linguistic characteristics that are attributed to it. The consequent is reshaped using a function associated with the antecedent (a single number). The input for the implication process is a single number given by the antecedent, and the output is a fuzzy set. Implication is implemented for each rule. In

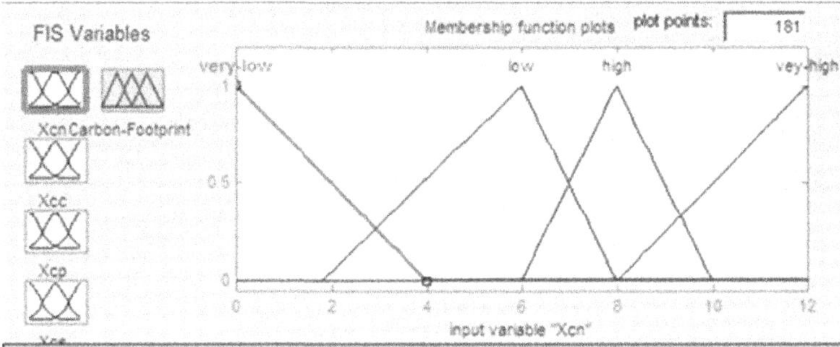

FIGURE 11.6

X_{cn} membership function.

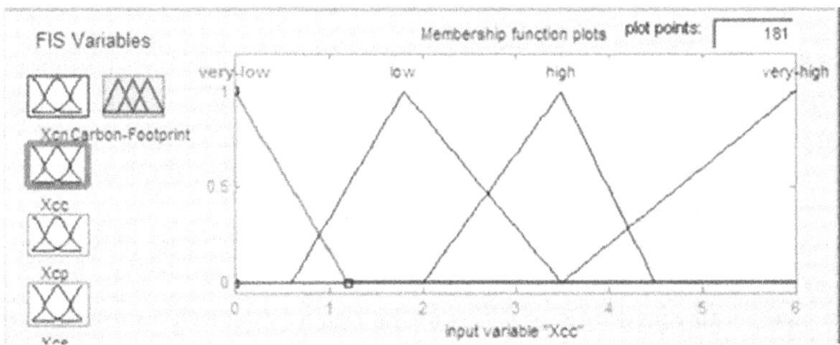

FIGURE 11.7

X_{cc} membership function.

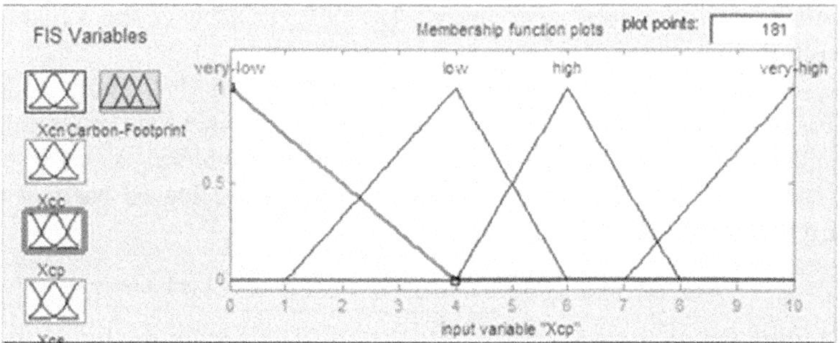

FIGURE 11.8

X_{cp} membership function.

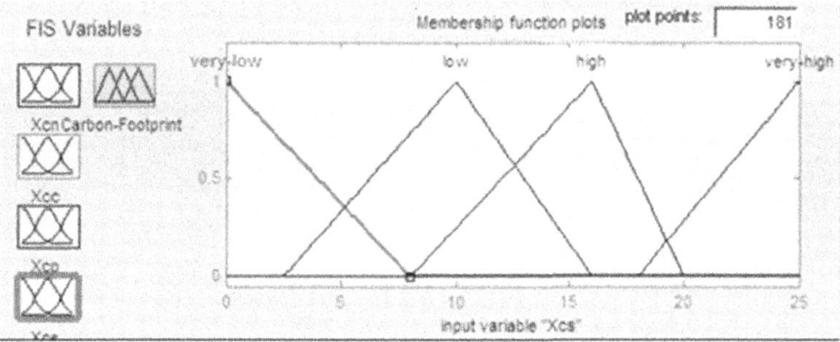

FIGURE 11.9

X_{cs} membership function.

this particular case the AND method is used: *min* (minimum), which truncates the output fuzzy set. The consequent—in this case the value for the carbon emission—is considered in this study to belong to the following classes: *unacceptable*, *neutral*, and *acceptable*.

It is now necessary to define the rules that link the input and the output variables. To do that, we define what can be the allowed combinations of the input variables and their influence on the output variable (Table 11.13).

For the purpose of this study we consider all possible combinations of the input variables with their specific linguistic identifiers.

All the rules have been coded in the MatLab toolbox to be able to complete the fuzzy inference.

The fuzzy inference diagram is the composite of all the smaller diagrams introduced so far. It also takes into consideration the relations induced by the fuzzy rules. To obtain the inference diagram we need to first consider all the

Table 11.13 Fuzzy Rules for Carbon Footprinting Calculation

	IF				THEN
	X_{cn}	X_{cc}	X_{cp}	X_{cs}	X_c
1	Very low	Low	High	Very high	Unacceptable
2	Very low	Low	Very high	High	Unacceptable
3	Very low	High	Low	Very high	Unacceptable
4	Very low	High	Very high	Low	Neutral
5	Very low	Very high	High	Low	Neutral
6	Very low	Very high	Low	High	Unacceptable
7	Low	Very low	High	Very high	Unacceptable
8	Low	Very low	Very high	High	Unacceptable
9	Low	High	Very low	Very high	Unacceptable
10	Low	High	Very high	Very low	Acceptable
11	Low	Very high	High	Very low	Acceptable
12	Low	Very high	Very low	High	Unacceptable
13	High	Low	Very low	Very high	Unacceptable
14	High	Low	Very high	Very low	Acceptable
15	High	Very low	Low	Very high	Unacceptable
16	High	Very low	Very high	Low	Neutral
17	High	Very high	Very low	Low	Acceptable
18	High	Very high	Low	Very low	Unacceptable
19	Very high	Low	High	Very low	Neutral
20	Very high	Low	Very low	High	Unacceptable
21	Very high	High	Low	Very low	Acceptable
22	Very high	High	Very low	Low	Acceptable
23	Very high	Very low	High	Low	Neutral
24	Very high	Very low	Low	High	Unacceptable

MFs. For example, if we analyze the first row of the inference diagram, we notice the MFs already introduced in Figures 11.6–11.9. When defining a value of 2.7 for X_{cn} we are in the "very low" area of the MF. For X_{cc} the value of 3.5 is recorded as "low." For X_{cp} the value of 4 is to be recorded in the area of "high" and, finally, for X_{cs} the value of 19 is "very high." When combining these membership values with the rules introduced in Table 11.12, we obtain the MF for the combined values (remember that the value is obtained with a sequence of AND logic functions).

The last step is the defuzzification. The input for the defuzzification process is a fuzzy set (the aggregate output fuzzy set) and the output is a single number. As much as fuzziness helps the rule evaluation during the intermediate steps, the final desired output for each variable is generally a single number. However, the aggregate of a fuzzy set encompasses a range of output values and therefore must be defuzzified to resolve a single output value from the set. The defuzzification method used here is the centroid calculation, which returns the center of area under the curve.

CASE STUDY RESULTS

As already presented in Figures 11.10 and 11.11, the values considered are:

$$[X_{c,n}\ X_{c,c}\ X_{c,p}\ X_{c,s}] = [2.7\ 3.5\ 4\ 19] \tag{11.7}$$

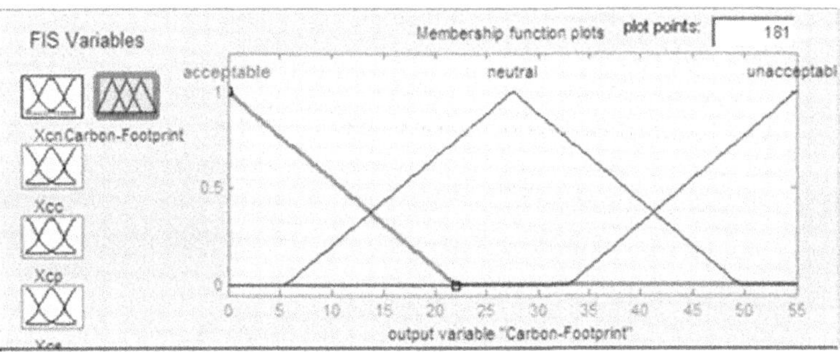

FIGURE 11.10

Membership function of the output variable.

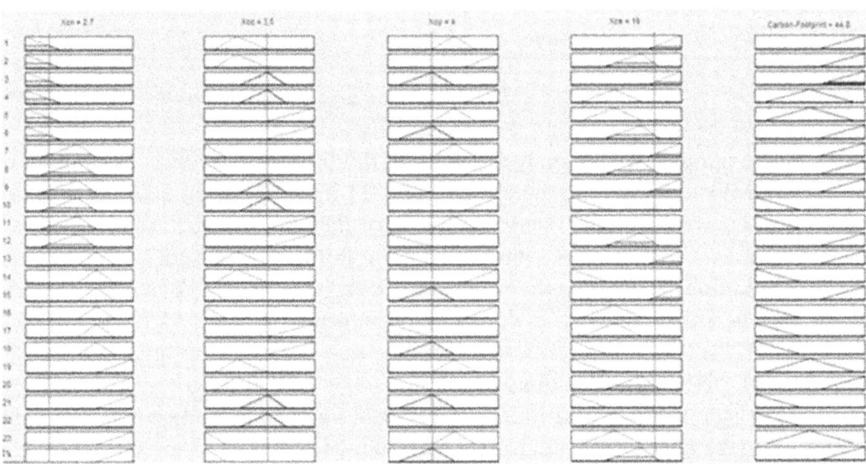

FIGURE 11.11

Fuzzy inference diagram.

Table 11.14 The Inference System Table

Rules	Carbon Footprint: Fuzzy Value
1	Unacceptable
2	Unacceptable
3	Unacceptable
4	Neutral
5	Neutral
6	Unacceptable
7	Unacceptable
8	Unacceptable
9	Unacceptable
10	Acceptable
11	Acceptable
12	Unacceptable
13	Unacceptable
14	Acceptable
15	Unacceptable
16	Acceptable
17	Acceptable
18	Acceptable
19	Neutral
20	Unacceptable
21	Acceptable
22	Acceptable
23	Neutral
24	Unacceptable

If the input is analyzed considering the linguistic description, we realize that the carbon dioxide emissions related to the neutralization step are surely low but can be identified both as *very low* and *low*. The same applies to the shipping step, whereby the emissions can be classified both as *high* and *very high*. However, the emissions level for concentration and prilling have crisp values that could easily be attributed to the class of *high* and *low*. The inference system, using these values and applying these rules, returns a fuzzy set for the output variable (Table 11.14).

By first look it is possible to realize that the Unacceptable results are in the majority. Using the centroid method to defuzzify the output, we obtain a crisp value of 44.8 for CO_2 emissions. This value can be used to calculate the carbon footprint for this particular case study.

It has to be noted that the emission level for the shipping operations are the ones that have the highest impact, given the way the rules have been made, on the final value for the output variable. This is attributable to the fact that the shipping operation also includes the transportation costs. Whether these emissions should be accounted for in the scope of the study should be the subject of further discussion.

Generally, we could also define different weights for each rule in case more refined characterization of the impacts might be needed. For the purpose of this study, all rules have the same weights.

CONCLUSIONS

The EPSM case study in this report clearly illustrates the advantages of using Fuzzy Logic as a way to deal with uncertainty in the environmental footprint area. When analyzing the environmental impacts in the process industry, the practitioner might face situations where the definition of an acceptable level of carbon dioxide emissions is not that sharp. That is attributable to the fact that most of the time, the levels of emissions might relate to the size of the production facility, the geographical position of the plant, as well as the proximity of inhabited centers or other natural resources that might become contaminated. The use of fuzzy variables to describe the contribution of each impact might better-suit the long-term strategic decision-making process. The combination of these impacts with their linguistic identifiers allowed us to define a single value for the carbon dioxide emissions. The same approach can be easily applied to all environmental contributions and the defuzzified values can be used for the basic calculations of the EPSM.

THE E^3-METHODOLOGY IN LCA EVALUATION

In the previous chapters, we have identified a new approach to evaluate overall environmental impacts, both at general and process step levels. The introduction of the fuzzy logic has also allowed us to improve the approach, including the uncertainty estimation in case of imperfect knowledge of some variables.

When looking at sustainability issues, it becomes clear that a methodology that is able to evaluate the overall environmental impact and the impact of each step in the process is also able to deal with uncertainty estimation, which is highly valuable. As already mentioned quite a few studies dealing with, e.g., bottom ash from incineration of municipal waste (Olsson et al., 2006), municipal solid waste (Liamsanguan and Gheewala, 2007), Danish case study on waste paper (Schmidt et al., 2007), LSA perspective of waste management (Fruergaard and Astrup, 2011), household waste in planning (Slagstad and Brattebø, 2012), household waste composition in management (Slagstad and Brattebø, 2013), and integrated solids state management in Asia (Othman et al., 2013), have identified in Sustainability and Waste to Energy (W2E) the potential of LCA in the decision-making process. Nevertheless, these and other studies have still fallen little short of providing an overall methodology to define a complete picture that takes into account not only the environmental burden but also the cost perspective. This perspective can be crucial when dealing with decisions at a strategic level and on the long-term horizon.

THE METHODOLOGY

To support the strategic decision-making process, we need to take into consideration the following steps:

- Problem statement
- Definition of the system boundaries and level of detail (overall impacts or process steps impact)
- Definition of the maximum targets for all main five footprints (including normalization factors)
- Definition of the financial key (either maximum available budget or expected investment request)
- Calculation of contribution to the main five footprints
- Application of fuzzy logic, in case uncertainty estimation should be accounted for
- Definition of the EPSM
- Calculation of the SEPI
- Comparison of different options—definition of the best possible solution from an overall environmental and economic perspective.

In real-life applications such as Life Cycle Analysis and Waste Management Systems, there is a need for a considerable capacity to handle large data sets. Several previous works, e.g., Krotscheck and Narodoslawsky (1996), provide a background for the future development. Some standard software programs are available to perform some of these steps, whereas others can be performed manually or with the help of basic calculation packages. To arrive at the definition of the SEPI, the different tasks need to be defined and the best available combination of software and manual extensions can be identified.

The combination process that defined, for the first time, the algorithm presented is based on the basic idea of collecting all tasks and their related business requirements. In a second step (after having verified that all tasks and, therefore, requirements are covered), it is necessary to use these requirements to specify the technical requirements for software adoption and integration. It is not the aim of this study to propose a thorough evaluation of available LCA systems or the development of specific new software. Nevertheless, after having defined the new methodology to deal with the environmental impact of strategic decision-making in the previous chapters, and before validating the end-to-end proposed approach, it is valuable to define the main steps required to evaluate the best combination of software tools to support the problem-solving techniques. This newly introduced procedure with all the steps is presented in Figure 11.12.

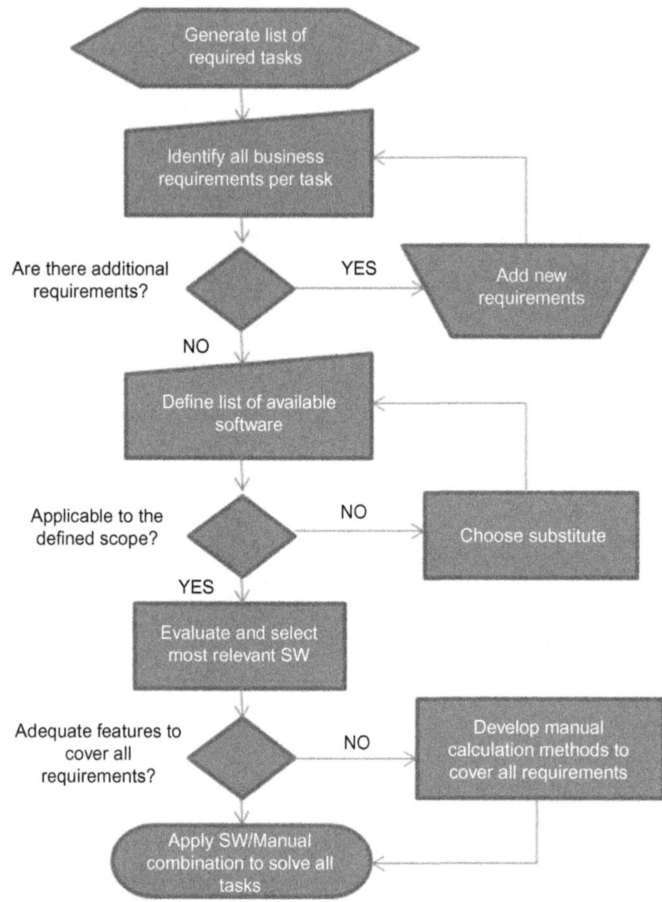

FIGURE 11.12

Evaluation software methodology.

The combined use of standard LCA programs as SIMAPro or SPIonExcel with MatLab and the calculation methods proposed for the SEPI has been the most effective. Therefore, these are the software programs that, combined with the standard Excel calculation sheets used for the SEPI definition, will be the basis of this new methodology.

THE E^3-METHODOLOGY—THE STEPS

Step 1

The first step requires the identification of the problem statement and the objectives that need to be accomplished. What exactly is it that we want to measure or optimize? Is this the impact of a specific business decision on the sustainability of a product or service? Or is it the need to identify within a current/future setup the main contributors to the environmental performance? In this phase, it also required to define the scope of the system or process we take under consideration. It is clear that only after these elements have been clarified can we proceed in the definition of the maximum targets. These need to be consistent with the system boundaries and are usually global or local industry-specific data available from different institutions or from the problem description. Finally, it is necessary to also define the financial key, because this will be included in the SEPI calculations and will provide either the baseline for the available budget or the possible deviation from target. See Chapter 2 for a detailed description of the definition of the maximum targets for the SEPI, including the financial key.

Step 2

The second step requires defining the level of uncertainty that needs to be estimated. During the problem definition at Step 1, we need to obtain exact values for the contribution of the different elements needed for the calculation of the footprints (Carbon, Water, Emissions, Energy, Work Environment), and then we can proceed directly to using one of the selected software, such as SPIonExcel. In case these values are not exact, we should proceed in applying the fuzzy logic and inferring the final results. The following sub-steps need to be accomplished (with the use of MatLab):

1. Definition of the fuzzy variables (input and output variables)
2. Definition of the MF for all variables
3. Definition of the fuzzy rule set
4. Assuming that MFs of all impact categories is similar to triangular fuzzy numbers, a total positive or negative fuzzy value for the environmental impacts is going to be calculated
5. Finally, the defuzzification will lead us to the punctual impact estimator and its corresponding uncertainty interval.

Step 3

Starting from the values defined in the problem statement (Step 1), it is possible to calculate the different impacts. The definition of the calculations is provided "What footprints?" section. It is also possible to directly use a tool, such as SPIonExcel, to derive some of these values. Refer to <spionexcel.tugraz.at/index.php> for further details regarding how to use SPIonExcel (2014) to calculate the environmental impacts.

Step 4

Whichever way we obtained the footprints, by direct calculation or fuzzy logic inference, it is now possible to calculate the deviation from targets for all of them. Using the SEPI calculation sheet, we can define the value for the sustainability indicator (see "Building the map" section).

1. The results of each footprint are normalized, resulting in a scale from 0 to 100.
2. The combination of all the footprints is represented by the area of the irregular pentagon (the "Map").
3. The required investment—compared with the available budget—provides the additional dimension necessary to define the SEPI.
4. The calculation of the volume of the irregular pyramid is performed as explained in Chapter 2.

Step 5

When the SEPI has been calculated for the different options, a direct comparison can be performed. Finally, detail analysis of the main contributors to the result and simulations of different changes can be performed with variation within the Environmental Technology Routing or the Env-BOM.

The different tasks, together with the best available supporting tool, are represented in Figure 11.13 and describe the E^3-Methodology in its entirety.

CHAPTER CONCLUSIONS

LCA is a tool for analyzing environmental impacts on a wide perspective and with a reference to a product system and economic activity. The ecological footprint is a way to compare human demand with our planet capacity to regenerate it, and it is measure of our burden on the ecosystem. One advantage of LCA, as compared with the ecological footprint, is that it more aptly covers the whole range of impacts, and it may also account for the upstream impacts. Nevertheless, one of the most important limitations in the application of LCA as an input for strategic decision-making is the limited inclusion of cost and investment considerations. When trying to apply LCA thinking to strategic decision-making in sustainability topics, the limitations become more relevant.

In this chapter, the authors proposed a methodology to deal with and solve these limitations. The methodology also offers the possibility to define different levels of granularity to study the options and define the quantitative contributions of each element to the overall environmental impact. Practical application of this theory has provided the possibility to refine the approach to an environmental end-to-end (E^3) methodology, which combines different software tools and methods to provide the single indicator of overall better financial and environmental performance. The E^3-methodology, also tested with some applications in Waste-to-Energy cases, has proven to be a comprehensive toolbox for use in strategic decision-making in this field.

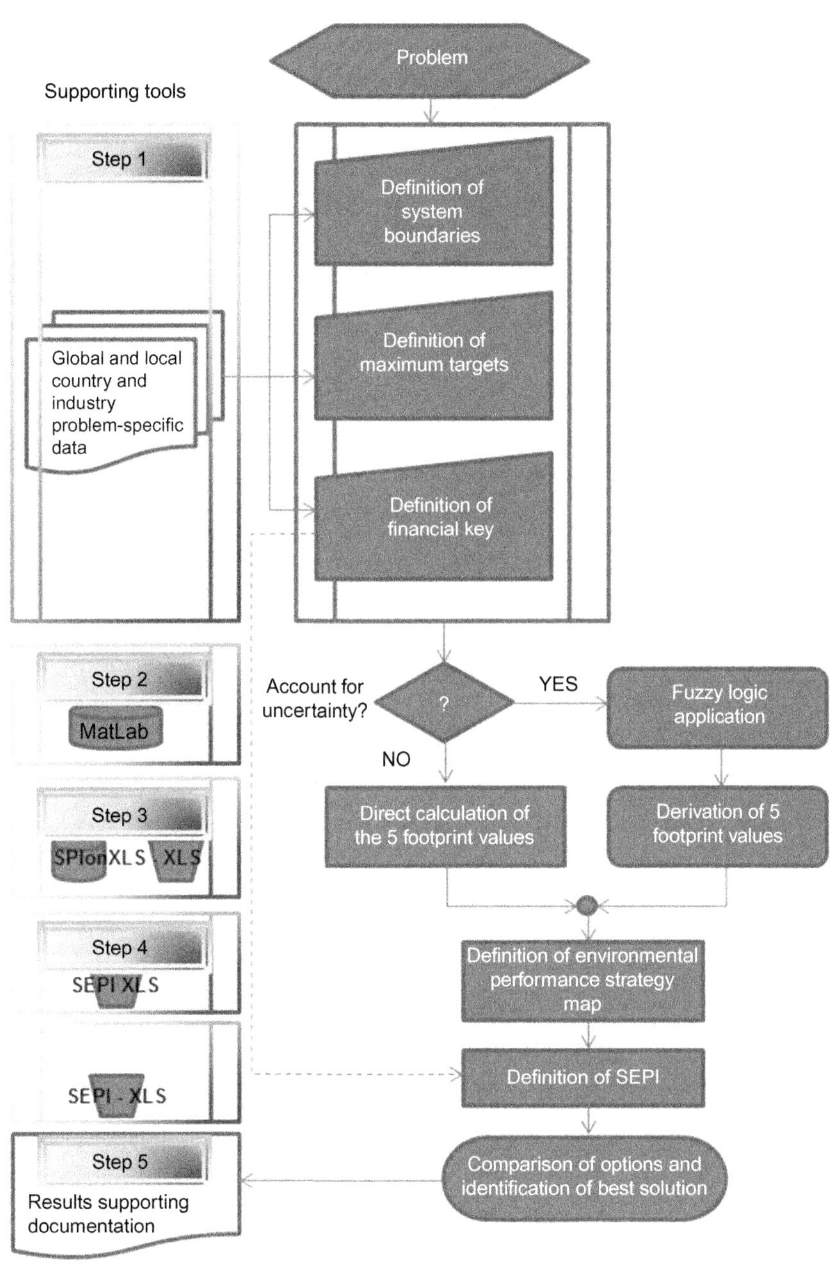

FIGURE 11.13

The E^3-Methodology to define the SEPI.

REFERENCES

Antonsson, A.B., Carlsson, H., 1995. The basis for a method to integrate work environment in life cycle assessment. J. Cleaner Prod. 3, 215–220.

Ashford, N.A., 1997. Industrial safety: the neglected issue in industrial ecology. J. Cleaner Prod. 5, 115–121.

Ayres, R.U., 2000. Commentary on the utility of the ecological footprint concept. Ecol. Econ. 32, 357–358.

Boustead, I., 1996. LCA—how it came about: the beginning in the UK. Int. J. Life-Cycle Assess. 1, 147–150.

Čucek, L., Klemeš, J.J., Kravanja, Z., 2012a. A review of footprint analysis tools for monitoring impacts on sustainability. J. Cleaner Prod. 34, 9–20.

Čucek, L., Klemeš, J.J., Kravanja, Z., 2012b. Carbon and nitrogen trade-offs in biomass energy production. Clean Technol. Environ. Policy 14, 389–397.

Čucek, L., Klemeš, J.J., Kravanja, Z., 2012c. Total FPs (footprints)-based MCO (multi-criteria optimisation) of regional biomass energy supply chains. Energy 44 (1), 135–145.

Clavreul, J., Guyonnet, D., Christensen, T.H., 2012. Quantifying uncertainty in LCA-modelling of waste management systems. Waste Manage. 32 (12), 2482–2495.

Crawley, F.K., Ashton, D., 2002. Safety, health or the environment—which comes first? J. Hazard. Mater. 93, 17–32.

De Benedetto, L., Klemeš, J., 2009. The environmental performance strategy map: an integrated LCA approach to support the Strategic Decision Making Process. J. Cleaner Prod. 17, 900–906.

Deng, L., Babbit, C.W., Williams, E., 2011. Economic-balance hybrid LCA extended with uncertainty analysis: case study of a laptop computer. J. Cleaner Prod. 19 (11), 1198–1206.

Eder, P., Narodoslawsky, M., 1999. What environmental pressures are the region's industry responsible for? A method of analysis with descriptive input, output models. Ecol. Econ. 29, 359–374.

EIPPCB, 2007. Integrated Pollution Prevention and Control (IPPC)—Reference Document on Best Available Techniques in the Large Volume Organic Chemical Industry. European IPPC Bureau, Seville, Spain.

Ferng, J.J., 2005. Local sustainable yield and embodied resources in ecological footprint analysis. Ecol. Econ. 53, 415–430.

Fruergaard, T., Astrup, T., 2011. Optimal utilization of waste-to-energy in an LCA perspective. Waste Manage. 31 (3), 572–582.

Galli, A., Widemann, T., Ercin, E., Konblauch, D., Erwing, B., Giljum, S., 2012. Integrating ecological, carbon and water footprint into a "Footprint Family" of indicators: definition and role in tracking human pressure on the planet. Ecol. Indic. 16, 100–112.

Hoekstra, A.Y., 2007. Human appropriation of natural capital: comparing ecological footprint and water footprint analysis, Value of Water Report Series 23. UNESCO-IHE. Available at: <www.waterfootprint.org/Reports/Report23-Hoekstra-2007.pdf> (accessed 12.03.2012).

Hoekstra, A.Y., Chapagain, A.K., 2007. Water footprints of nations: water use by people as function of their consumption pattern. Water Resourc. Manage. 21, 35–48.

Hoekstra, A.Y., Hung, P.Q., 2002, Virtual water trade: a quantification of virtual water flows between nations in relation to international crop trade. Value of water research series: 11. Available at: <www.waterfootprint.org/Reports/%20Report11.pdf> (accessed 09.05.2008).

Honkasalo, A., 2000. Occupational health and safety and environmental management systems. Environ. Sci. Policy 3, 39–45.

Hujbregts, M.A.J., Hellweg, S., Frischknecht, R., Hungerbühler, K., Hendriks, J., 2008. Ecological footprint accounting in the life cycle assessment of products. Ecol. Econ. 64, 798–807.

ISO, 1997. (ISO 14040). Environmental Management—Life Cycle Assessment—Principles and Framework. International Organisation for Standardisation, Geneva, Switzerland.

ISO, 1998. (ISO 14041). Environmental Management—Life-Cycle Assessment; Goal, Scope Definition and Inventory Analysis. International Organisation for Standardisation, Geneva, Switzerland.

ISO, 2000a. (ISO 14042). Environmental Management—Life-Cycle Assessment; Life-Cycle Impact Assessment. International Organisation for Standardisation, Geneva, Switzerland.

ISO, 2000b. (ISO 14043). Environmental Management—Life-Cycle Assessment; Life-Cycle Interpretation. International Organisation for Standardisation, Geneva, Switzerland.

Jensen, A.A., Møller, B.T., Søborg, L., Potting, J., 1996. Work Environmental Issues in Life Cycle Assessment, LCANET Summary Report, Copenhagen, Denmark.

Jensen, A.A., Hoffman, L., Møller, B., Schmidt, A., Christiansen, K., Elkington, J., van Dijk, F., 1997. Life Cycle Assessment—A guide to approaches, experiences and information sources Environmental Issues Series no. 6, European Environment Agency, Paris, France. Available at: <www.lca-center.dk/cms/site.asp?p = 2867> (accessed 08.09.2008).

Jeswani, H.K., Azapagic, A., Schepelmann, P., Ritthoff, M., 2010. Options for broadening and deepening the LCA approaches. J. Cleaner Prod. 18, 120–127.

Kratena, K., 2004. Ecological value added in an integrated ecosystem-economy model—an indicator for sustainability. Ecol. Econ. 48, 189–200.

Krotscheck, C., Narodoslawsky, M., 1996. The Sustainable Process Index. A new dimension in ecological avaluation. Ecol. Eng. 6, 241–258.

Ku-Pineda, V., Tan, R., 2006. Environmental performance optimization using process water integration and Sustainable Process Index. J. Cleaner Prod. 14, 1586–1592.

Lee, S.H., Choi, K., Osako, M., Dong, J., 2007. Evaluation of environmental burdens caused by changes of food waste management systems in Seoul, Korea. Sci. Total Environ. 387, 42–53.

Liamsanguan, C., Gheewala, S.H., 2007. LCA: a decision support tool for environmental assessment of MSW management systems. J. Environ. Manag. 87 (1), 132–138.

Lindeijer, E., 1996. Part VI: Normalisation and valuation. In: Udo de Haes (Ed.), Towards a Methodology for Life Cycle Impact Assessment. Society of Environmental Toxicology and Chemistry (SETAC)—Europe, Brussels, Belgium.

Lindfors, L.G., Christiansen, K., Hoffmann, L., Virtanen, Y., Juntilla, V., Hanssen, O.J., Rønning, A., Ekvall, T., Finnveden, G., 1995. Nordic Guidelines on Life Cycle Assessment. Nord 1995:20. Nordic Council of Ministers, Copenhagen, Denmark.

Monfreda, C., Wackernagel, M., Deumling, D., 2004. Establishing natural capital accounts based on detailed ecological footprint and biological capacity assessment. Land Use Policy 21, 231–246.

Olsson, S., Kärrman, E., Gustafsson, J.P., 2006. Environmental systems analysis of the use of bottom ash from incineration of municipal waste for road construction. Resour. Conserv. Recycl. 48, 26–40.

Othman, S.N., Noor, Z.Z., Abba, A.H., Yusuf, R.O., Abu Hassan, M.A., 2013. Review on life cycle assessment of integrated solid waste management in some Asian countries. J. Cleaner Prod. 41, 251–262.

Padgett, J.P., Steinemann, A.C., Clarke, J.H., Vandenbergh, J.H., 2008. A comparison of carbon calculators. Environ. Impact Assess. Rev. 28, 106–115.

Pennington, D.W., Potting, J., Finnveden, G., Lindeijer, E., Jolliet, O., Rydberg, T., 2004. Life cycle assessment—Part 2: current impact assessment practice. Environ. Int. 30, 721–734.

Reid, R.D., Sanders, N.R., 2002. Operations Management. John Wiley and Sons New York, USA, 457–458. ISBN 0-471-32011-0.

Sandholzer, D., Narodoslawsky, M., 2007. SPIonExcel—Fast and easy calculation of the Sustainable Process Index via computer. Resour. Conserv. Recycl. 50, 130–142.

Schmidt, A., Poulsen, P.B., Andreasen, J., Floee, T., Poulsen, K.E., 2004. The Working Environment in LCA. A New Approach, 72. Guidelines from the Danish Environmental Agency. Available at: <www.lca-center.dk/cms/site.asp?p = 2867> (accessed 24.04.2011).

Schmidt, J.H., Holm, P., Merrild, A., Christensen, P., 2007. Life cycle assessment of the waste hierarchy—A Danish case study on waste paper. Waste Manage. 27, 1519–1530.

Slagstad, H., Brattebø, H., 2012. LCA for household waste management when planning a new urban settlement. Waste Manage. 32 (7), 1482–1490.

Slagstad, H., Brattebø, H., 2013. Influence of assumptions about household waste composition in waste management LCAs. Waste Manage. 33 (1), 212–219.

Stoeglehner, G., 2003. Ecological footprint—a tool for assessing sustainable energy supplies. J. Cleaner Prod. 11, 267–277.

Stoeglehner, G., Narodoslawsky, M., 2008. Implementing ecological footprinting in decision making processes. Land Use Policy 25, 421–431.

Stranddorf, H.K., Hoffmann, L., Schmidt, A., 2003. LCA technical report: impact categories, normalisation and weighting in LCA. Update on selected EDIP97-data, Seritikel nr 2003. FORCE Technology, Denmark. Available at: <www.lca-center.dk/cms/site.asp?p = 2867> (accessed 24.12.2010).

Technical University of Graz, Institut für Prozess- und Partikeltechnik, Austria. 2014. <spionexcel.tugraz.at/index.php> (accessed 23.04.2014).

Udo de Haes, H.A., Heijungs, R., 2007. Life-cycle assessment for energy analysis and management. Appl. Energy 84, 817–827.

Wiedmann, T., Lenzen, M., 2007. On the conversion between local and global hectares in ecological footprint analysis. Ecol. Econ. 60, 673–677.

CHAPTER

12 Green supply chain toward sustainable industry development

Hon Loong Lam, Bing Shen How, and Boon Hooi Hong
Centre of Excellence for Green Technologies, The University of Nottingham Malaysia Campus, Selangor, Malaysia

INTRODUCTION

Green supply chain management (GSCM) exposes the applications of the most important sustainable development issues. It demonstrates how green technologies and practices can be implemented and, in line with this, the motivation of saving money and increasing efficiency.

Ahi and Searcy (2013) have identified and analyzed the published definitions of GSCM and sustainable supply chain management. Some of those definitions are as follows:

> *Application of environmental management principles to the entire set of activities across the whole customer order cycle, including design, procurement, manufacturing and assembly, packaging, logistics, and distribution*
>
> **(Handfield et al., 1997)**

> *Integrating environmental thinking into supply-chain management, including product design, material sourcing and selection, manufacturing processes, delivery of the final product to the consumers as well as end-of-life management of the product after its useful life.*
>
> **(Srivastava, 2007)**

> *The process of using environmentally friendly inputs and transforming these inputs through change agents—whose byproducts can improve or be recycled within the existing environment. This process develops outputs that can be reclaimed and re-used at the end of their life-cycle thus, creating a sustainable supply chain.*
>
> **(Penfield, 2008)**

Integrating environmental concerns into the inter-organizational practices of SCM including reverse logistics.

(Sarkis et al., 2011)

Minimizing and preferably eliminating the negative effects of the supply chain on the environment.

(Andic et al., 2012)

A graphical definition of GSC is presented in Figure 12.1. The conventional supply chain (SC) interest and method have been extended with several "green techniques" such as:

- Integration of green technologies and process optimization
- Green network analysis and synthesis
- Life cycle analysis (LCA)
- Green enterprise resources planning
- Regulatory considerations and sustainability strategies

This chapter presents the GSC concept in several subsections. In the first section, an overview and discussion of the GSCM development since its introduction in 1997 are presented. Several remarkable extensions of GSCM are highlighted. The second section presents the fundamentals of model formulation and superstructure synthesis of the GSC network for optimization purposes. The implementation of green technologies and the LCA approach have been demonstrated. The last part of this chapter presents an advanced approach for GSCM. Some useful resources are introduced for further reference and study.

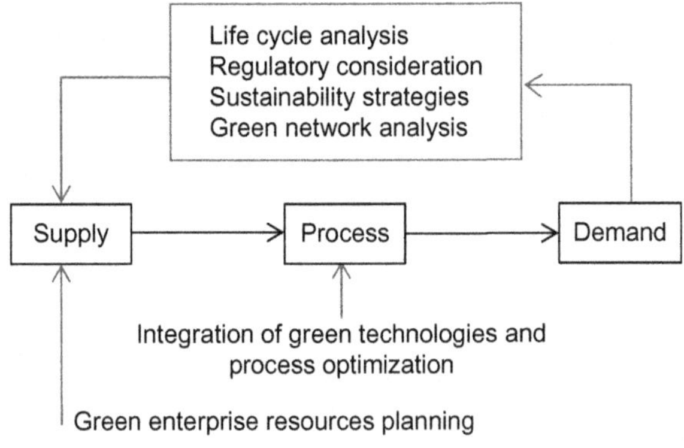

FIGURE 12.1

Graphical definition of green SC.

DEVELOPMENT OF GSCM

This section provides a summary of the important and significant milestones or elements during the history of GSCM development. It presents an overview of the "green extension" part of the SC, such as green enterprise resource planning (ERP), LCA, game theory (GT), and process optimization. These techniques became the main pillars supporting the development of GSCM.

GREEN ERP IN GSCM

GSC is an SC system that not only focuses on economic potential but also focuses on the environmental impacts and efficiency of energy used. The development of GSC will not be possible without the implementation of ERP in organizations (Kandananond, 2014). Generally, ERP is an integrated information system that is designed to automate and integrate all the business processes and operations together. To improve the effectiveness and success of the implementation of ERP in organizations, much research has been conducted during the twentieth and twenty-first centuries (Table 12.1).

However, some research has shown that, despite an ERP system being implemented in organizations, some still fail to achieve SC integration because of its complexity, nonflexibility, and inability to collaborate with others, such as reported previously in the concept of general ERP (Makey, 1998), enterprise application of integrated ERP (Linthicum, 1999) and the exploratory survey of ERP integration (Themistocleous et al., 2001) To solve this problem, Enterprise application integration technology is proposed to support the efficient incorporation of information

Table 12.1 Green ERP in GSCM Development

Year	Author	Remarks
2000	Davenport	Introduction of ERP
2000	Al-Mashari and Zairi	Re-engineer of SC by applying ERP
2002	Marinos and Zahir	Introduction of enterprise application integration (EAI) technique to integrate ERP and SC
2003	Lee et al.	Integration of enterprise with ERP and EAI
2005	Kelle and Akbulut	Information sharing, cooperation and cost optimization in SCM by implementation of ERP
2006	Koh and Saad	Integration of SCM and ERP
2007	Basoglu et al.	Organization adoption of ERP
2010	Law et al.	Introduction of full life cycle of ERP
2010	Kuhn and Sutton	Introduction of a continuous auditing system in ERP
2011	Goni et al.	Introduction of the critical success factor for ERP
2012	Lopez and Salmeron	Dynamic risk modeling in ERP
2012	Brooks et al.	Introduction of sustainable enterprise resource planning (S-ERP)

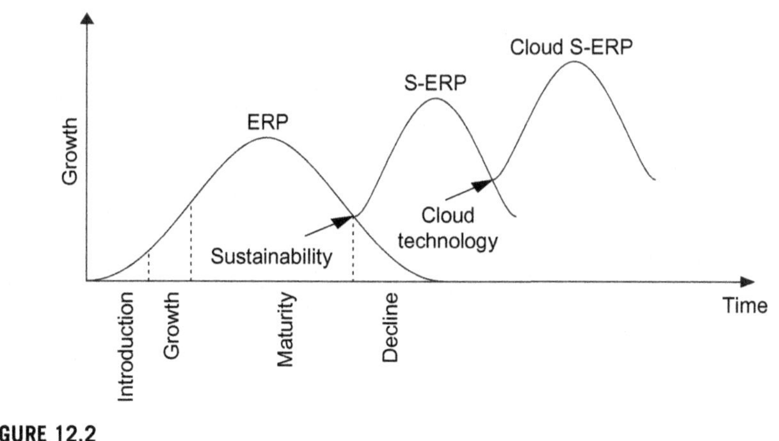

FIGURE 12.2

The life cycle of ERP and S-ERP.

From Chofreh et al. (2014).

systems, resulting in integration of the SC. However, SC development has increased environmental pressure and attendant business responsibilities. Moreover, the climate change, resource depletion, human health problem, and negative social impact are leading to a point of no return (Carvalho et al., 2013). Therefore, sustainable development is now more important than ever. However, the sustainability data are yet to be sufficiently integrated and used for decision making. An S-ERP system has been proposed to support the sustainability initiatives (Chofreh et al., 2014). S-ERP shows differences compared with the traditional ERP as its information system is driven by sustainability considerations that cover all aspects of the SC. Figure 12.2 shows an illustration of the S-ERP life cycle. To extend the development of the S-ERP system, the development of Cloud S-ERP should be researched.

LCA IN GSCM

In GSCM, the evaluation of environmental impacts of products throughout life cycle stages is very important. Among other evaluation tools, LCA is the most scientifically reliable method (Ness et al., 2007) used to assess the environmental impacts and resources used through production and disposal (Finnvedena and Moberg, 2005). ISO 14040 has been developed for LCA to provide a general framework, terminology, and principles. Moreover, these standards provide transparency and consistency in LCA studies (Cambero and Sowlati, 2014). Generally, the stages of LCA study are as follows:

- Define the goal and scope of study
- Life cycle inventory
- Life cycle impact assessment
- Interpretation (Figure 12.3)

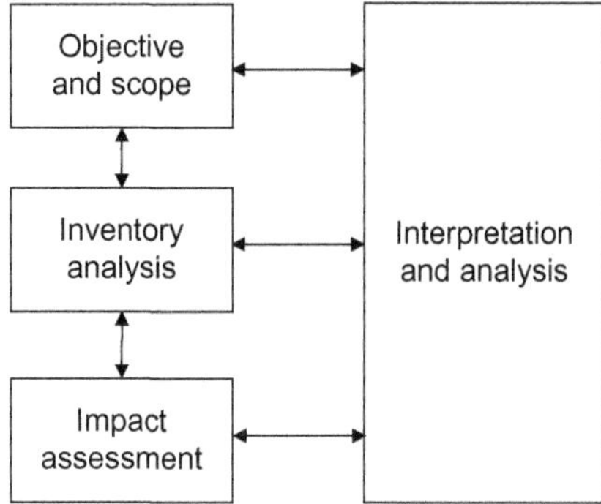

FIGURE 12.3

Stages of LCA studies.

It is worth noting that LCA results are strongly dependent on the methodological choices and parameters associated with each analyzed case. Therefore, it should not be used to provide the basis of comparative declarations of the overall environmental preferability between the products. Nevertheless, LCA results are useful to compare environmental impacts of the alternative configurations of an SC. Many studies of LCA are related to SC synthesis. However, many future research contributions are required for the success and completeness of LCA. The list of work integrating LCA into GSCM is summarized in Table 12.2.

DEVELOPMENT OF OPTIMIZATION TECHNIQUE FOR GSCM

In traditional SCM, the design of the supply chain network (SCN) is normally focusing on a single objective (either minimum cost or maximum profit). As shown in Table 12.3, several works have considered the total cost of the SC as the objective function of the studies. However, the real-life design, planning, and scheduling of tasks usually involve different objective functions that are often contradictory to each other. Many techniques and approaches have been used to solve the design problem in the SC. These techniques include mathematical modeling, agent technology, and heuristic algorithm.

In mathematical modeling, the problem is represented by an mixed-integer programming (MIP) model. For instance, a two-part model for a three-echelon SC where the MIP model is formulated to solve the supplier selection problem has been developed (Cakravastia et al., 2002). Usually, the MIP model or mixed-integer linear programming (MILP) model can be solved by the ε-constraint method (Guillen et al., 2005). The benefit of using traditional mathematic programming is that the optimum solution can be found. However, it cannot solve all the real-world SCM problems that are often fuzzy because of incomplete information and insufficient planning (Turan et al., 2012). The amount of

Table 12.2 LCA in GSCM Development

Year	Author	Remarks
2008	Lainez et al.	Application of LCA to evaluate the environmental impact while IMPACT 2002 + methodology is selected
2010	Nwe et al.	Conduction of an approach integrating LCA indicators and dynamic simulation for GSC design and action
2011	Guest et al.	Conduction of an LCA for biomass-based combined heat and power plant
2011	Kostin et al.	Integration of bio-ethanol sugar SC with economic and environmental concern where different LCA metric are used simultaneously
2012	Pucker et al.	Conduction of a greenhouse gas (GHG) and energy analysis for a biomass SC
2014	Murphy et al.	Used LCA to evaluate GHG emission and primary energy balances in Ireland

Table 12.3 Optimization Technique for GSC

Year	Author	Remarks
1998	Robinson and Satterfield	Establishment of a mixed-integer programming (MIP) model to maximize the profit in SCN
1999	Koray and Marc	Utilization of Bender's decomposition of integer programming to solve the multiproducts multi-integers problem in SCN
2000	Lee and Kim	Proposed a hybrid simulation approach to solve SCN design problem
2001	Jayaraman and Pirkul	Optimization of SC focusing on minimization of cost (single-objective optimization)
2002	Syam	Development of a model for the location problem in SCN (single-optimization problem)
2002	Syarif et al.	Used of spanning tree-based genetic algorithm (GA) to solve the multistage logistics chain problem in SCN
2002	Cakravastia et al.	Development of a two-part model for a three-echelon SC where MIP model is formulated
2003	Jayaraman and Ross	Introduction of a simulated annealing methodology for single-objective optimization
2003	Yan et al.	Development of a strategic model for single-objective optimization in SCN
2004	Chan and Chung	Introduction of a multiobjective genetic optimization procedure for the order distribution in SCM
2004	Erol and Ferrell	Development of a multiobjective optimization framework to minimize cost and maximize customer satisfaction
2004	Chan et al.	Introduction of a hybrid approach based on GA and analytic hierarchy process in SCN

Table 12.3 Optimization Technique for GSC *Continued*

Year	Author	Remarks
2004	Chen and Lee	Used multiobjective fuzzy programming approach to optimize the production and distribution planning
2005	Graves and Willems	Development of an MIP model where SC is divided into stages
2005	Guillen et al.	Development of a multiobjective stochastic MILP model for SCN
2005	Gen and Syarif	Introduction of hybrid GA for single-objective optimization
2005	Truong and Azadivar	Development of methodologies to solve the configuration problem in SC (single optimization)
2006	Amiri	Design of SCN using single-objective optimization approach
2006	Liang	Used fuzzy multiobjective linear programming method to solve transportation problem in SC
2007	Altiparmak et al.	Implication of steady-state GA to solve the multiobjectives and multiphases problem in SCN
2008	Guo and Tang	Development of a unified model to solve SCN based on JIT (Just-In-Time) system
2008	Farahani and Elahipanah	Development of a model to solve JIT distribution problem for SCM
2008	Liang	Development of a fuzzy model to solve the multiproduct and multitime period planning decision problems
2009	Peidro et al.	Used fuzzy multiobjective linear programming model for SC planning under supply, process and demand uncertainty
2010	Chang	Used GA to solve SCN design problem with adaptation of coevolution and constraints satisfaction
2010	Kannan et al.	Development of a mathematic model for battery-recycling SCN using GA and GAMS
2010	Franca et al.	Establishment of a two-objective stochastic model for SC to evaluate the relationship between profit and quality
2010	Xu and Zhai	Used fuzzy optimization approach to solve the SC coordination problem
2012	Paksoy et al.	Application of fuzzy approach to integrate the SCN of edible vegetable oils manufacturer
2013	Ramezani et al.	Development of a three-objective stochastic model for a forward/reversed logistic network
2013	Lam et al.	Development of a two-stage optimization model (macro- and microstage) for waste-to-energy SC
2013	Ng et al.	Synthesis of rubber seed SCN which support the green energy demand by using single-optimization approach

computations will increase significantly when the problem size increases (nondeterministic polynomial hard problem (NP-hard)).

Multiagent technology is another technique to solve the optimization problem in SCN design. This idea was first introduced into the design of SC by Swaminathan et al. (1998). The author suggested structuring an SC as a library of structural elements (production and transportation) and control elements (flow, inventory, demand, and supply) represented by agents that interact with each other to find the best configuration. This technology is used to interpret new messages, permit exchange between agents, and enable new policies (Ahn et al., 2003). However, several research works have pointed out that the major challenge is finding proper methodology for the coordination of the behavior of individual agents such that an optimum solution can be obtained. For example, Lim et al. (2009) presented an iterative agent bidding mechanism for responsive manufacturing, Zhang and Zhang (2007) wrote on an agent-based simulation of consumer purchase decisions, and Wang et al. (2002) demonstrated agent-based modeling and mapping of a manufacturing system.

To solve the coordination problem in the aforementioned technique, Akanle and Zhang (2008) proposed a heuristic algorithm (i.e., GA) to dynamically solve the SC design. During the past decade, GA has often been implemented to solve single-objective and multiobjective problems in production and operational management that are NP-hard. In recent years, there have been three different forms of GA:

- Traditional GA (Kannan et al., 2010)
- Steady-state genetic algorithm (ssGA) (Altiparmak et al., 2009)
- New GA for serial parallel production line (S-PPL) (Abu Qudeiri et al., 2008)

Other alternatives for the multiagent approach are summarized in Table 12.4.

A new alternative technique that has been widely used is ant colony optimization (ACO) meta-heuristics. Although optimum solutions cannot be guaranteed

Table 12.4 Multiagent Approach for GSCM

Year	Author	Remarks
1998	Swaminathan et al.	Structured an SC using multiagent technology
2000	Fox et al.	Proposed an agent building shell concept in the design of SC
2004	Silva et al.	Development of an approach which combines ACO and multiagent system
2006	Zhang et al.	Development of an agent-based architecture for SC design that has three layers
2011	Luis and Zhang	Proposed an ACO in SC design
2014	Luis and Recio	Used ACO to solve the SC configuration problem
2013	Mastrocinque et al.	Used bees algorithm (BA) to solve the SCN problem
2012	Koc	Improvement of BA using combined neighborhood size change and site abandonment (NSSA) strategy
2014	Yuce et al.	Proposed adaptive NSSA (ANSSA) strategy to solve the multiobjective optimization problem in SC

using this swarm-based optimization model, it provides a useful compromise between the amount of computation time necessary and the quality of the approximated solution space (Moncayo-Martınez and Zhang, 2011). This technique is one of the nature-inspired meta-heuristics that mimics the behavior of ant colonies and the evaporation effect of the pheromones during their (ants) search for food. As an ant exploring the forage area, it deposits a certain amount of pheromones along its trail so the concentration of the pheromones over a trail will directly affect the path selection of the other ants. In short, the higher the quantity of pheromones over a trail, the higher the probability that the ants will follow that path. However, if the pheromones over the trails are not reinforced at the same rate, they will evaporate, thus affecting the search. This searching mechanism is embedded in the ACO technique during optimization. Previously, ACO was used to solve the decision-making problem involving a single objective (Bullnheimer et al., 1999). Recently, Pareto ACO has been proposed (Doerner et al., 1992) to solve the multiobjective problem, and it has been proven to efficiently and effectively solve many real-world and theoretical problems (Moncayo-Martinez and Recio, 2014).

To improve the ACO technique mentioned, another swarm-based optimization model has been introduced (Pham and Ghanbarzadeh, 2005), called BA. In recent work conducted previously (Mastrocinque et al., 2013) BA has been proven to be a more powerful optimization tool for finding better Pareto solutions for SC design problems compared with ACO techniques. Similar to ACO, BA is an optimization algorithm that mimics the foraging behavior of honey bees to find the optimum source. The scout bees will move randomly during the process of searching for food. When they return to hive, they deposit the nectar that they have collected and share their information about the food source with other bees. As a result, some of the bees will follow the former bee to the food source. The number of the recruited bees depends on the quality of food brought by the scout bees. Once the recruited bees return to the hive, they will share their information with others. This sequence is used in the optimal solution searching process during optimization. In the past few years, several improved versions of BA have been proposed. One of the most significant improvements is the implementation of combined NSSA strategy (Koc, 2012). However, the convergence rate of an NSSA-based BA can be significantly slow if the desired locations are far from the current best sites. To overcome this challenge, an ANSSA strategy has been proposed (Yuce et al., 2014). By using this strategy, the local minima can be avoided by changing the neighborhood size adaptively (Yuce et al., 2013). Also, the testing of the ANSSA-based BA on more complex SCN optimization problems should be researched further.

GT IN GSCM

GT was first introduced in 1944 (V-Neumann and Morgenstern, 1944) and has been widely used as an approach for various research fields, such as SCM. It is a valuable tool for the identification of dominant strategies for increasing performance of each objective (Zhao et al., 2012). Many works have discussed the application of GT (normally cooperatives game) to the SC (Table 12.5). In addition, several game models have been formulated for the application to GSC, for instance, evolutionary game model (Zhu and Dou, 2007), differential game model (Chen and Sheu, 2009), bargaining model (Sheu, 2011), oligopoly game model (Nagurney and Yu, 2012), and

Table 12.5 GT in GSCM Development

Year	Author	Remarks
2004	Cachon and Netessine	Introduction of GT to solve the SC coordination problem
2004	Moyaux et al.	Analyzed information sharing and bullwhip effect in an SC through GT
2006	Yu et al.	Introduction of Stackelberg game in an SC design problem
2007	Huang et al.	Optimization of the SCN design using three-move dynamic game-theoretic approach
2007	Zhu and Dou	Development of an evolutionary game model to investigate effect of government subsidies and penalties
2008	Sobel and Turcic	Proposed a general model for SC contract negotiation with risk aversion
2008	Nagarajan and Sosic	Used cooperative bargaining model to allocate profit between SC members
2009	Chen and Sheu	Creation of a differential game model to design environmental regulation strategies
2009	Esmaeili et al.	Proposed several game model of seller–buyer relationship in an SC
2010	Zhao et al.	Used cooperative GT approach to address the SC coordination issues
2010	Zhang and Huang	Used dynamic game to integrate product development, SC configuration problem, marketing, and inventory decision
2011	Barari et al.	Development of an integrated and holistic conceptual framework in GSC by using GT
2011	Sheu	Derived bargaining game model to seek negotiation in SCM
2011	Huang et al.	Used three-level dynamic noncooperative game model to solve the coordination problem and other selection, pricing and replenishment decision in SCN
2012	Barari et al.	Proposed a dynamic evolutionary game model to solve the coordination of players in SC
2012	Nagurney and Yu	Used an oligopoly game model to design a sustainable fashion industrial SC
2012	Zhao et al.	Application of GT to select appropriate strategies in GSC with the aim of maintaining sustainability
2013	Zamarripa et al.	Used GT optimization based tool to improve the decision making of SC

dynamic evolutionary game model (Barari et al., 2012). It is worth noting that the major limitation involved in the GT is the omitted interaction between the upstream and downstream businesses. As a natural extension of research, future studies should consider coordination issues in SC (Zhao et al., 2010).

FORMULATION OF GSC MODEL

In this section, biomass network is used for the demonstration purposes because biomass utilization is one key activity in GSCM. It is a versatile source of nearly carbon-neutral energy, from which heat, electricity, and liquid biofuels can be generated. Typical examples of biomass are wood and forestry residues, energy crops, agricultural waste (e.g., wheat straw, oilseed straw, and cotton chalks), as well as biowaste from food production and wood processing. A more detailed analysis of biomass types is provided elsewhere (McKendry, 2002). It is predicted that the usage of biomass will be increased significantly in the future because of the depletion of the natural resources. Biomass is usually locally available, thus defining it as a distributed resource, and it requires extensive infrastructure networks for harvesting, transportation, storage, and processing. Therefore, the increase in biomass usage tends to increase the cost, emissions, and complexity of SCs.

As mentioned, environmental impact is an important factor in GSCM. Carbon footprint (CFP) is commonly used as an environmental impact indicator in the formulation of GSC because most industrialized countries have committed to reduce their emissions of CO_2 by an average of 5.2% during the period 2008–2010, respective of the levels of 1990 (Sayigh, 1999). The CFP is defined as the total amount of CO_2 and other GHGs emitted over the full life cycle of a process or product (POST, 2006). Energy from biomass cannot be considered truly carbon-neutral even though the direct carbon emissions from combustion have been offset by carbon assimilation during feedstock photosynthesis. The net CFP is mainly attributable to the indirect carbon emission generated along the SC itself.

With the increasing demand for energy crops, agriculture production, space for development, and land use management become environmental and societal trade-offs. The land required for growing energy crops and transforming them into marketable energy and fuels would result in negative effects on food price (increases) and deforestation (Koh and Ghazoul, 2008). This will lead to the loss of biodiversity and create a conflict between the atmospheric carbon balance and natural ecosystems (Huston and Marland, 2003).

As a result, it is vital to increase the efficiency of biomass GSC. In this section, three types of biomass GSC formulation approach are discussed and respective case studies are presented:

1. Process integration approach (Lam et al., 2008)
2. Superstructure approach (Čuček et al., 2010)
3. P-graph approach (Lam et al., 2010a,b)

PROCESS INTEGRATION APPROACH

The process integration approach utilizes the developed regional energy clustering (REC) algorithm to partition the biomass region into a number of clusters. Clusters are defined as geographic concentrations of interconnected suppliers, service providers, associated institutions, and customers in a region that competes but also cooperates (Porter, 1998). A cluster combines smaller zones to secure sufficient energy balance within the cluster (Lam et al., 2008). A zone can be a province/county, a community settlement/borough, an industrial park, or an agriculture compound from the studied region. The REC is used to manage the energy flow among the zones. The energy surpluses and deficits from various zones can be matched and combined to form energy SC clusters (Figure 12.4). Forming clusters reveal sets of zones, between which the biomass transfer is most beneficial and, therefore, minimizes the system CFP. Because at this stage only biomass exchange is considered, CFP minimization also tends to minimize the costs.

Biomass energy supply and management are then targeted using new graphical representations. Regional Energy Surplus–Deficit Curves (RESDCs) visualize the formation and sizes of introduced energy clusters. It is a pair of Cumulative Curves that represents cumulative energy surplus and deficit profiles. Regional Resource Management Composite Curve (RRMCC) is an analogy of the process integration approach, showing energy imbalances and helping trade-offs of resource management. These graphical tools provide straightforward information regarding how to manage the surplus resources (biomass and land use) in a region. The need to reduce CO_2 emission motivates the suggestion of constructing an alternative transportation route that could reduce the transportation CFP and cost. In many cases the economy has been a decisive criterion and economical payback figures can be analyzed and used as a trade-off for the CFP. Therefore, one useful

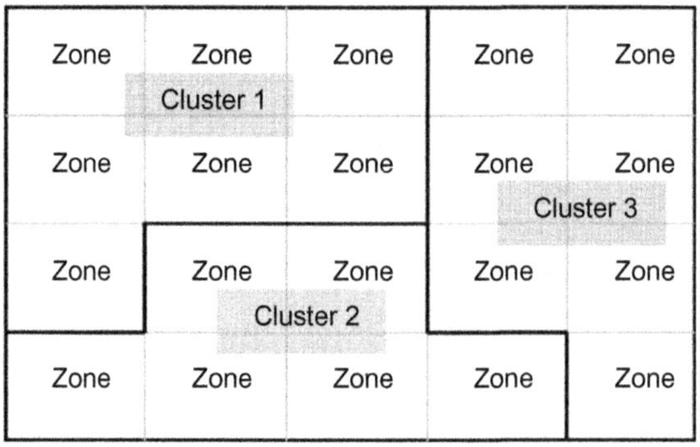

Zone	Zone	Zone	Zone	Zone
	Cluster 1			
Zone	Zone	Zone	Zone	Zone
			Cluster 3	
Zone	Zone	Zone	Zone	Zone
		Cluster 2		
Zone	Zone	Zone	Zone	Zone

FIGURE 12.4

Regional energy clusters.

From Lam et al. (2010a,b).

tool for evaluating the potential benefits of optimizing the SC is the payback analysis for infrastructure investments.

There are five main steps in the process integration approach in generating biomass GSC:

1. Identification of the region for analysis and the corresponding zones within it. The region is taken as an administrative unit. This can be a set of countries forming an administratively recognizable region (e.g., South East Asian Region). The zones are smaller administrative areas within the region, accounting for any administrative or economic boundaries (e.g., a country). However, a zone could also be defined by the economy by economic grouping and interconnection.
2. Mathematical formulation of REC algorithm
   - Energy source and demand data including the quantity of the potential biomass, the energy demand, and the collection location are tabulated. The biomass resources surpluses and deficits should be obtained from analysis of the local sources and demands inside each zone. Distribution centroid is specified by a two-dimensional (2D) Cartesian coordinate system and the centroid point for zone 1 is marked as reference point.
   - An optimum targeting result for the biomass SC is obtained using linear programming. For sources sent from zone$_i$ (source) to zone$_j$ (sink), varying $i = 1...N_{\text{zones}}$; $j = 1...N_{\text{zones}}$, $i \neq j$, the objective function is defined as follows to minimize the total CFP generated within the transfer network:

$$\text{Min CFP} = \sum_{ij} \text{CFP}_{i,j} \tag{12.1}$$

$$\text{CFP}_{i,j} = \text{FC}_{i,j} \times \text{Dist}_{i,j} \times \frac{B_{i,j}}{C} \times \text{CEF} \tag{12.2}$$

where $\text{CFP}_{i,j}$ is the CFP, $\text{FC}_{i,j}$ is the fuel consumption, $\text{Dist}_{i,j}$ is the two-way distance, $B_{i,j}$ is the biomass load, C is the truck capacity, and CEF is the carbon emission factor for diesel trucks. There are a few compulsory constraints. The total amount of biomass exported from zone$_i$ to other zones cannot exceed the available surplus AB_i, and sending flows of biomass to the same zone are forbidden:

$$\sum_{j} B_{i,j} \leq \text{AB}_i \quad \forall i \tag{12.3}$$

$$B_{i,j}||_{i=j} = 0 \tag{12.4}$$

The total bioenergy delivered to zone$_j$ cannot exceed the deficit in that zone:

$$\sum_{i} \text{TE}_{i,j} \leq D_j \quad \forall j \tag{12.5}$$

$$\text{TE}_{i,j} = \text{HV}_i \times B_{i,j} \tag{12.6}$$

where HV_i is the heating value for the particular biomass from $zone_i$ and D_j is the total deficit in $zone_j$. The biomass load in the system must be nonnegative:

$$B_{i,j} \geq 0 \quad \forall i,j \tag{12.7}$$

- Clusters are formed based on the residual bioenergy imbalance within the newly formed clusters, which is minimized (preferably zero) by using MILP.

3. RESDC plot
   - The zones within every single cluster are arranged in descending order according to their energy balance, starting with the largest surplus.
   - The clustering outcome is illustrated with a pair of monotonic Cumulative Curves. Energy Surplus Cumulative Curve is plotted with the cumulative area on the x-axis and the accumulated energy balance is on the y-axis. The cluster with the smallest imbalance will be the first cluster.
   - Plot the Energy Deficit Cumulative Curve in the same manner as plotting the Energy Surplus Cumulative Curve.
4. RRMCC can be plotted in the same way as for RESDC, but a single Energy Balance (surplus or deficit) Cumulative Curve should be used instead of a pair of Surplus Deficit Curves.
5. CFP and cost payback analysis can be defined as:

$$PB_{CFP} = \frac{CCFP}{CFP - CFP^*} \tag{12.8}$$

$$PB_{cost} = \frac{CC}{FS + CT} \tag{12.9}$$

where CCFP is the construction CFP, CFP* is the CFP of the new road, CC is the cost of new road construction, FS is the cost-saving from fuel because of the shorter distance, and CT is the carbon tax for diesel.

A case study is conducted to illustrate the process integration approach. The case study assumes energy is to be consumed by the local community in the studied region. The excess biomass is transferred from the collection point near the farm to the biomass energy conversion plant, which uses cogeneration (combined heat and power (CHP)) and a biofuel boiler as an energy generation system. Electricity produced is provided to the national grid, whereas heat generated is supplied for domestic, commercial, and industrial applications (Figure 12.5).

Table 12.6 provides the information needed for this case study. It includes information for potential biomass sources, the area and location of the collection point for a particular zone, the potential biomass heating values, and energy demand. The energy demand includes the usage for domestic direct burning, centralized heating system, and CHP. The assumptions for major parameters are as follows:

1. CEF = 2.69 (kg CO_2)/L
2. Fuel consumption by a 20-t trailer truck = 0.3 L/km
3. The road bending and level differences increase 30% of the direct distance

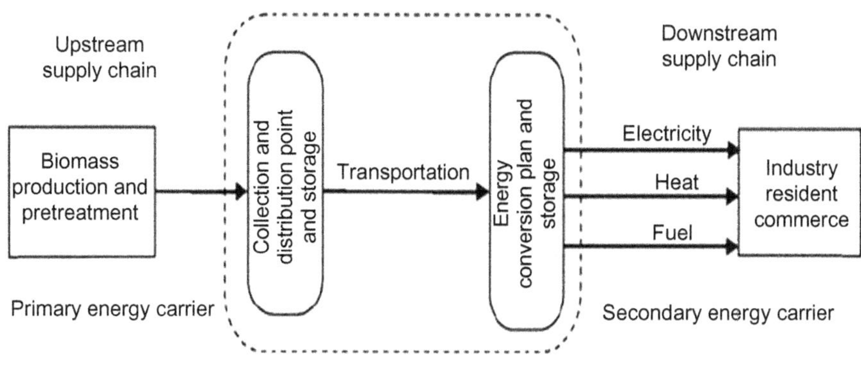

FIGURE 12.5

Biomass SC.

From Lam et al. (2010a,b).

Table 12.6 Information for the Studied Region

Z_i	Area (km^2)	Location (km, km)	Potential Biomass (t/y)	Heating Value (GJ/y)	Supply (PJ/y)	Demand (PJ/y)
1	6.12	(0,0)	2,924	17.1	0.05	2.90
2	1.60	(4.1,0.2)	135,082	17.4	2.35	0.12
3	9.58	(4.4,2.5)	56,512	13.5	0.78	0.41
4	6.35	(5.3,2.4)	45,511	15.4	0.70	0.23
5	8.38	(7.9,5.1)	56,309	19.0	1.07	0.21
6	5.57	(6.4,5.5)	15,179	14.5	0.22	2.20
7	10.63	(2.4,6.8)	103,065	19.6	2.02	0.05
8	7.83	(9.4,5.5)	58,990	13.9	0.82	0.15
9	4.12	(3.2,6.6)	63,904	20.5	1.31	0.26
10	3.15	(2.3,7.3)	44,826	17.4	0.78	3.06

Source: From Lam et al. (2010a,b).

An optimum biomass transfer network is generated using linear programming and is shown in a 2D Cartesian coordinate system (Figure 12.6). It shows the location of the zone and the bullet point represents its centroid. The biomass transfer is represented by the arrow that gives the direction and the quantity of the flow.

Three optimum clusters have been identified based on the result of biomass transfer and illustrated in the RESDC in Figure 12.7. Zones 1–4 are grouped to become cluster 1; zones 5, 6, and 8 are cluster 2; and zones 6, 9, and 10 are cluster 3. The solid line indicates the Cumulative Surplus Curve, and the dashed line represents the cumulative deficit. Figure 12.7 shows the cluster size and the total energy involved in the SC within the cluster. Because each cluster is a group

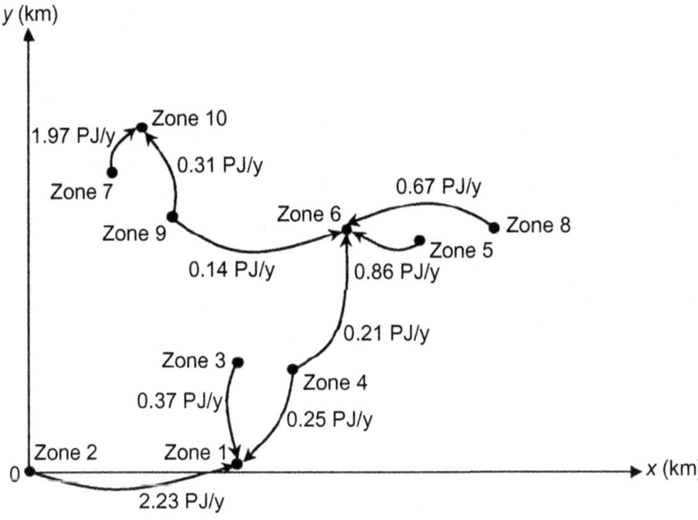

FIGURE 12.6

Biomass flow between zones.

From Lam et al. (2010a,b).

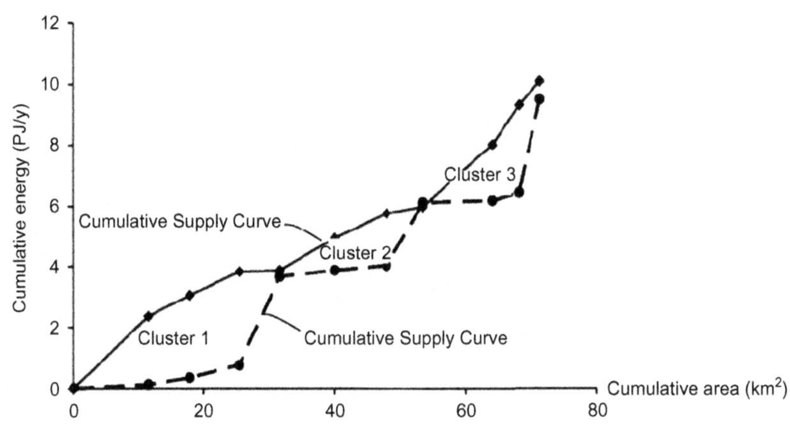

FIGURE 12.7

Regional Energy Surplus–Deficit Curve.

From Lam et al. (2010a,b).

of stringer lines, this provides an opportunity to develop efficient energy planning and management strategies within a simpler SC compared with the network of the whole region. Currently, only minimization of CFP is used as the criterion. In cases of degeneracy, additional judgment should be used, such as availability and

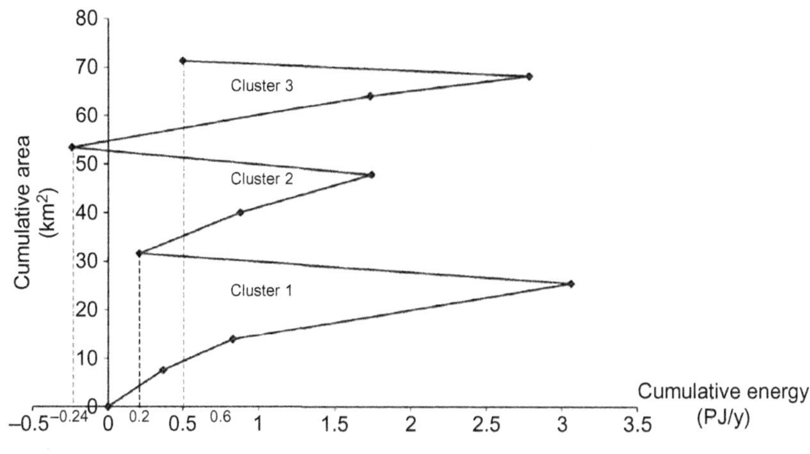

FIGURE 12.8

Regional Resource Management Composite Curve.

From Lam et al. (2010a,b).

quality of the available infrastructure, which is implicitly linked to cost and CFP of its construction.

The visual analysis of the REC result is further illustrated in the RRMCC (Figure 12.8). The zones with positive slope supply the biomass to the demanding zones, which have the negative slope. Each of the left turning points (Cluster Pinch) is the start of a new cluster. The cluster will have a surplus of biomass energy if the "turning point" for a certain cluster is on the right side of the y-axis and vice versa. Figure 12.8 shows that the studied region has a total energy surplus of 0.5 PJ/y. The energy balance for a particular cluster can be obtained by shifting the lower turning point to the y-axis which the cumulative energy starts from zero. In this case, cluster 1 has an energy surplus of 0.21 PJ/y, cluster 2 has an energy deficit of 0.44 PJ/y, and cluster 3 has an energy surplus of 0.74 PJ/y.

The slope of RRMCC represents the land area (km^2) required per unit of energy (PJ/y). This relationship gives the planner the option to assess the priority: either sell the surplus energy on the fuel market or use the land for other purposes. This case demonstrates how to apply RRMCC to maximize the free land and biofuel, and how to recover the deficit in cluster 2. The lower turning point of cluster 2 is shifted left and touches the y-axis (Figure 12.9). The curve with the positive slope on the left side of the y-axis represents the surplus from cluster 1. The shaded triangle represents the deficit from cluster 2. The opening gap on the top represents the net surplus for cluster 3 after supplying cluster 2 with required energy. As a result, the region still has a total of 0.5 PJ/y, which is 0.2 from cluster 1 and 0.3 from cluster 3. The details of RRMCC is reported elsewhere (Lam et al., 2011a,b).

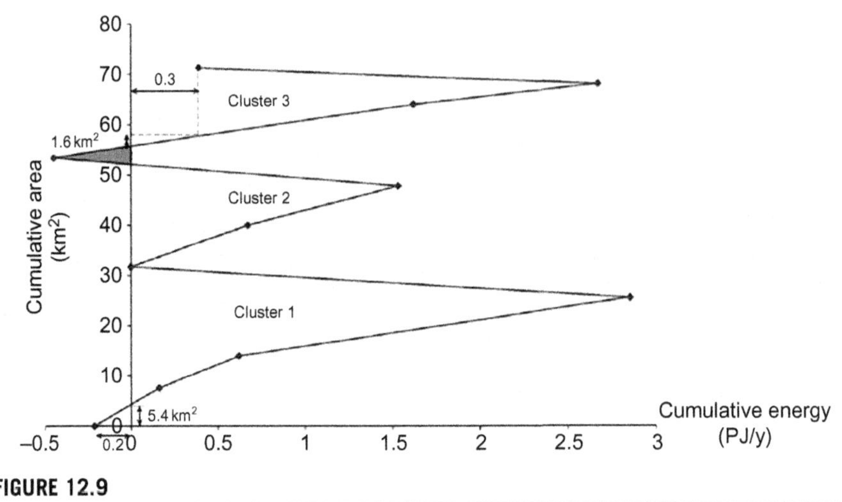

FIGURE 12.9

RRMCC for regional energy planning.

From Lam et al. (2010a,b).

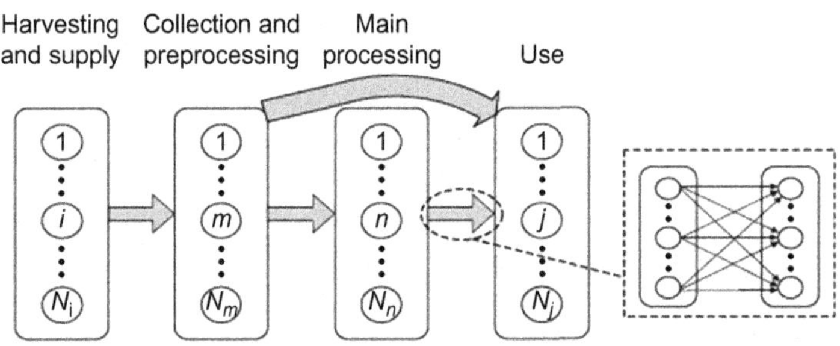

FIGURE 12.10

Biomass GSC superstructure.

From Čuček et al. (2010).

SUPERSTRUCTURAL APPROACH

The process integration approach only discusses the biomass transportation network. For an optimum GSC, the preprocessing such as drying and densification of biomass and regional energy generation are needed to be taken into account. The superstructural approach, which uses an improved algorithm, uses a four-layer structure that includes supply, preprocessing, processing, and consumption (Figure 12.10). Transportation is another component taking place between layers. Certain amounts of intermediate products are sent to the customer directly. Most of them are food crops for direct consumption. This approach is capable of accounting for different biomass types, for optimizing the locations, types, and capacities of the processing plants, and for connecting logistics network.

The model is generated within a mixed-integer nonlinear programming (MINLP) form for efficient bioenergy network optimization on a regional scale. It consists of mass balances, production and conversion constraints, cost functions, and objective function. It follows the four-layer nature of the network's superstructure, starting from harvesting and supply (layer 1 (L1)), collection and preprocessing (layer 2 (L2)), main processing (layer 3 (L3)), and use (layer 4 (L4)). The main processing steps are first represented with detailed models and are optimized by the process synthesizer to generate surrogate models based on conversion factors, investment and operating cost correlations, and CFP. These surrogate models were then inserted as block boxes into the MINLP model for the synthesis of regional networks. The three-step model formulation is as follows:

1. The first step of model formulation is based on mass balances, production, and conversion constraints. The biomass type production rate pi at supply zone$_i$ subject to hectare yield of pi and available area at zone$_I$ is expressed as follows:

$$q_{i,pi}^{m,L1} = HY_{pi} \cdot A_{i,pi}^{c} \quad \forall pi \in PI, \forall i \in I \tag{12.10}$$

Biomass pi produced at zone$_i$ is transported to preprocessing hubs m:

$$q_{i,pi}^{m,L1} = \sum_{m \in M} q_{i,m,pi}^{m,L1,L2} \quad \forall pi \in PI, \forall i \subset I \tag{12.11}$$

Total available area for biomass competing for food and energy production (pic) must be within the total competing area for food and energy at zone$_i$

$$\sum_{\text{pic} \in \text{PIC}} A_{i,\text{pic}}^{c} \leq A_{i}^{\text{UP}} \tag{12.12}$$

The collection and intermediate process hubs m have to operate within the minimum and maximum product mass flows:

$$q^{m,L2,L0} \cdot y_{m}^{L2} \leq \sum_{i \in I} \sum_{pi \in PI} q_{i,m,pi}^{m,L1,L2} \leq q^{m,L2,UP} \cdot y_{m}^{L2} \quad \forall m \in M \tag{12.13}$$

$$q_{pi}^{m,L1,L2,L0} \cdot y_{m}^{L2} \leq \sum_{i \in I} q_{i,m,pi}^{m,L1,L2} \leq q^{m,L1,L2,UP} \cdot y_{m}^{L2} \quad \forall m \in M, \forall pi \in PI \tag{12.14}$$

Pretreated intermediate product pi can be transferred from the collection and intermediate process hub m can be transferred to process plant n or directly to customer j:

$$\sum_{i \in I} q_{i,m,pi}^{m,L1,L2} \cdot f_{pi}^{\text{conv},L2} = \sum_{n \in N} q_{m,n,pi}^{m,L2,L3} + \sum_{j \in J} \sum_{pd \in PD \subseteq PI} q_{m,j,pd}^{m,L2,L4} \quad \forall m \in M, \forall pi \in PI \tag{12.15}$$

The intermediate product pi is sent to the selected technology t:

$$\sum_{m \in M} q_{m,n,pi}^{m,L1,L2} = \sum_{(pi,t) \in PT} q_{n,pi,t}^{m,T,L2,L3} \quad \forall n \in N, \forall pi \in PI \tag{12.16}$$

The inlet flow to certain technology must not exceed its maximum capacity:

$$\sum_{(pi,t)\in PT} q_{n,pi,t}^{m,T,L2,L3} \le q_t^{m,L3,UP} \cdot y_{n,t}^{L3} \quad \forall n \in N, \forall pi \in PI \tag{12.17}$$

Intermediate product *pi* is converted into the product *pp* using corresponding conversion factor:

$$q_{n,pi,t}^{m,T,L2,L3} \cdot f_{pi,pp,t}^{\text{conv},L3} = q_{n,pi,pp,t}^{m,T,L2,L3} \quad \forall (n \in N, pi \in PI, pp \in PP, t \in T, (pi,pp) \in \text{PIP} \tag{12.18}$$

All of the produced product's *pp* is sent to customers:

$$\sum_{(pi,t)\in PT} \sum_{(pi,pp)\in \text{PIP}} q_{n,pi,pp,t}^{m,T,L2,L3} = \sum_{j\in J} q_{n,j,pp}^{m,L3,L4} \quad \forall n \in N, \forall pp \in PP \tag{12.19}$$

All local demand is satisfied. The sum of the produced products from plants *pp* and the directly used products *pd* must be greater than the demand for products *p*:

$$\text{Dem}_{j^\circ,p} \le \sum_{n\in N} \sum_{pp\in PP\subseteq P} q_{n,j^\circ,pp}^{m,L3,L4} + \sum_{m\in M} \sum_{pd\in PD\subseteq P} q_{m,j^\circ,pd}^{m,L2,L4} \quad \forall j^\circ \in J, \forall p \in P \tag{12.20}$$

2. This model considers the cost functions for transportation, operating cost, and equipment cost. The transportation cost depends on the density of biomass, distances, mode of transport, rate of biomass supply, and road conditions. Total transportation cost is described by:

$$c^{\text{tr}} = \begin{pmatrix} \sum_{i\in I} \sum_{m\in M} \sum_{pi\in PI} D_{i,m}^{L1,L2} \cdot f_{i,m}^{\text{road},L1,L2} \cdot c_{pi}^{\text{tr},L1,L2} \cdot q_{i,m,pi}^{m,L1,L2} \\ + \sum_{m\in M} \sum_{n\in N} \sum_{pi\in PI} D_{m,n}^{L2,L3} \cdot f_{m,n}^{\text{road},L2,L3} \cdot c_{pi}^{\text{tr},L2,L3} \cdot q_{m,n,pi}^{m,L2,L3} \\ + \sum_{m\in M} \sum_{j\in J} \sum_{pd\in PD} D_{m,j}^{L2,L4} \cdot f_{m,j}^{\text{road},L2,L4} \cdot c_{pd}^{\text{tr},L2,L4} \cdot q_{m,j,pd}^{m,L2,L4} \\ + \sum_{n\in N} \sum_{j\in J} \sum_{pp\in PP} D_{n,j}^{L3,L4} \cdot f_{n,j}^{\text{road},L3,L4} \cdot c_{pp}^{\text{tr},L3,L4} \cdot q_{n,j,pp}^{m,L3,L4} \end{pmatrix} \cdot f^{\text{yb}} \tag{12.21}$$

The operating cost for the collection and intermediate process centers, which provide collecting, drying, and compacting, and for the process plants is expressed as:

$$c^{\text{op}} = \left(\sum_{i\in I} \sum_{m\in M} \sum_{pi\in PI} C_{pi}^{\text{op},L2} \cdot q_{i,m,pi}^{m,L1,L2} + \sum_{n\in N} \sum_{pi\in PI} \sum_{t\in T} \sum_{(pi,t)\in PT} c_{pi,t}^{\text{op},L3} \cdot q_{n,pi,t}^{m,T,L2,L3} \right) \cdot f^{\text{yb}} \tag{12.22}$$

The process plant's equipment cost is assumed to change nonlinearly with the selected size variable. Investment cost for the selected *m* centers and *n* plants is defined by:

$$c^{\text{inv}} = \left(\sum_{m\in M} c_m^{\text{fix,inv},L2} + \sum_{nN} \sum_{t\in T} \left(c_t^{\text{fix,inv},L3} \cdot y_{n,t}^{L3} + \sum_{(pi,t)\in PT} \left(c_t^{\text{var,inv},L3} \cdot q_{n,pi,t}^{m,T,L2,L3} \right)^{c_t^{\text{exp,inv},L3}} \right) \right) \cdot f^{\text{yb}} \tag{12.23}$$

The objective function maximizes the profit before taxation (P):

$$\begin{aligned} P = & \sum_{n\in N}\sum_{j^\circ\in J^\circ}\sum_{pp\in PP} q_{n,j^\circ,pp}^{m,L3,L4}\cdot c_{pp}^{\text{price}} \\ & + \sum_{m\in M}\sum_{j^\circ\in J^\circ}\sum_{pd\in PD} q_{m,j^\circ,pd}^{m,L2,L4}\cdot c_{pd}^{\text{price}} \\ & + \sum_{n\in N}\sum_{j^\circ\in J^\circ}\sum_{pp\in PP} q_{n,j^\circ,pp}^{m,L3,L4}\cdot 0.9\cdot c_{pp}^{\text{price}} \\ & + \sum_{m\in M}\sum_{j^\circ\in J^\circ}\sum_{pd\in PD} q_{m,j^\circ,pd}^{m,L2,L4}\cdot 0.9\cdot c_{pd}^{\text{price}} \\ & - \sum_{i\in I}\sum_{pi\in PI} q_{i,pi}^{m,L1}\cdot c_{pi} - c^{\text{tr}} - c^{\text{op}} - c^{\text{inv}} \end{aligned} \tag{12.24}$$

The income represents the revenue from selling products and from the tax imposed on waste. The expenses represent the raw materials cost (c_{pi}), the transport cost (c^{tr}), operating cost (c^{op}), and annualized network investments (c^{inv}).

3. The CFP (De Benedetto and Klemeš, 2010) is evaluated for the preprocessing, processing, and transportation activities. It only includes the net emissions caused by those operations that consume fossil fuels. The CFP per unit of the SCN total area is defined as:

$$\text{CFP} = \left(\begin{aligned} & \sum_{i\in I}\sum_{m\in M}\sum_{pi\in PI} D_{i,m}^{L1,L2}\cdot f_{i,m}^{\text{road},L1,L2}\cdot ei_{pi,e}^{\text{tr},L1,L2}\cdot q_{i,m,pi}^{m,L1,L2} \\ & + \sum_{m\in M}\sum_{n\in N}\sum_{pi\in PI} D_{m,n}^{L2,L3}\cdot f_{m,n}^{\text{road},L2,L3}\cdot ei_{pi,e}^{\text{tr},L2,L3}\cdot q_{m,n,pi}^{m,L2,L3} \\ & + \sum_{m\in M}\sum_{j\in J}\sum_{pd\in PD} D_{m,j}^{L2,L4}\cdot f_{m,j}^{\text{road},L2,L4}\cdot ei_{pd,e}^{\text{tr},L2,L4}\cdot q_{m,j,pd}^{m,L2,L4} \\ & + \sum_{n\in N}\sum_{j\in J}\sum_{pp\in PP} D_{n,j}^{L3,L4}\cdot f_{n,j}^{\text{road},L3,L4}\cdot ei_{pp,e}^{\text{tr},L3,L4}\cdot q_{n,j,pp}^{m,L3,L4} \\ & + \sum_{i\in I}\sum_{m\in M}\sum_{pi\in PI}\Bigg(\bigg(ei_{pi,e}^{\text{op},L2}\cdot q_{i,m,pi}^{m,L1,L2} + \sum_{n\in N}\sum_{pi\in PI}\sum_{t\in T} \\ & \qquad \sum_{(pi,t)\in PT} c_{pi,t,ei}^{\text{op},L3}\cdot q_{n,pi,t}^{m,T,L2,L3}\bigg)\Bigg)/A \quad \forall e\in E \end{aligned} \right) \tag{12.25}$$

A case study is performed to demonstrate this model. The geographical features of the studied region are illustrated in Figure 12.11, where set $I = \{i_1 \ldots i_{24}\}$ is used for the supply zones, set $M = \{m_1 \ldots m_{14}\}$ is used for the collection and preprocessing centers, set $N = \{n_1 \ldots n_{10}\}$ is used for the process plants, and set $J = \{j_1 \ldots j_7\}$ is used for demand locations, with subsets $J^\circ = \{j_1 \ldots j_5\}$ for locations at the local level and $J_e = \{j_6 \ldots j_7\}$ for locations for export.

A few processing technologies are considered: (i) the dry-grind process for starchy crop-based ethanol plants from corn, wheat, and potatoes; (ii) diluted acid from corn stover; (iii) alkaline pretreatment for ethanol plants from wheat straw; (iv) gasification/fermentation process from wood chips; (v) anaerobic codigestion of biomass waste; (vi) incinerations of municipal solid waste (MSW), corn stover, wheat straw, hay, miscanthus, poplar, and wood waste; and (vii) the sawing of timber for manufacturing boards. Food demand is satisfied directly from the production of starchy crops (i.e., corn, wheat, and potatoes). Besides the base case, which includes the pretreatment of biomass at collection centers, four scenarios are considered in the synthesis of a self-sufficient regional network: (i) without

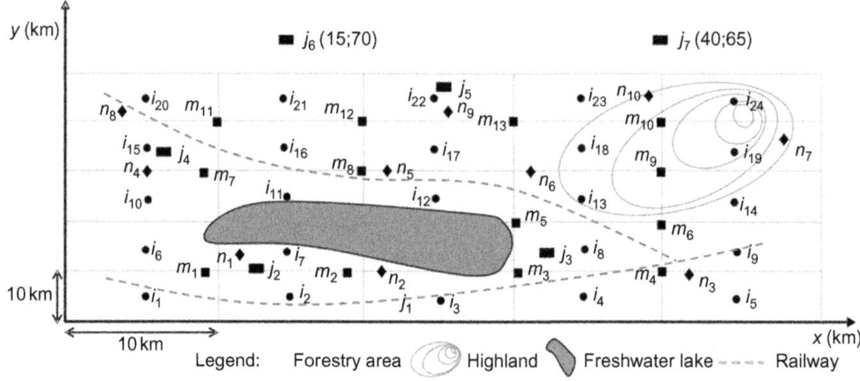

FIGURE 12.11

Regional geographical features.

pretreatment of biomass; (ii) tripled expense of transportation; (iii) 10-times larger area; and (iv) the combination of expensive transportation and a larger area. The objective is to maximize the profit while satisfying regional demand. Product surpluses can be exported. The resulting locations of the selected technologies and product surpluses are given in Table 12.7, and the effects of different scenarios can be seen in Table 12.8.

As observed in Table 12.7 and Table 12.8, for a relatively small area (2,400 km^2), only incineration is distributed in up to five plants, whereas all other technologies are centralized. When compared with the base case, if the compressing of low-density biomass is excluded, then transportation costs and CFP will significantly increase, the production costs will decrease, and one additional plant location will be selected. If transportation cost increases, then more sawing operations are selected as compared with MSW incineration, which would considerably reduce the CFP. Highly distributed energy production will be obtained if the area is increased 10 times. If the enlarged area is combined with three times more expensive transportation, then an additional AD plant will be selected and, as expected, MSW incineration will not be economically viable. Also, centralized production is preferable for ethanol production when only starchy crops are converted into ethanol; lignocellulosic biomass should be incinerated. CFP shows typical behavior indicating that, by increasing both the total area and transportation cost, much less CO_2 would release per unit of area.

The amount of the resulting biomass would satisfy the entire demand for energy and food within the region. Approximately 60% of ethanol (demand for bioethanol is set at 20% of petrol consumption), 90% of electricity, and 80% of heat can be exported. Although the demand for food will be satisfied, only corn is selected for bioethanol production. The remaining area (40% of the total area) will be better planted with miscanthus for heat and electricity generation.

Table 12.7 Plant Location, Yearly Amount of Energy, and Food for Export

	Plant Location						Export			
Scenarios	**DG**	**IN**	**AD**	**GF**	**SAW**	**MSW**	**Heat (GWh/y)**	**Electricity (GWh/y)**	**Ethanol (kt/y)**	**Food (kt/y)**
Base case	n_9	$n_{1,2,4,9}$	n_5	n_1	n_9	n_3	1,500.3	1,754.6	14.7	–
Excluded preprocessing	n_9	$n_{1,3,4,5,9}$	n_2	n_1	n_9	n_2	1,500.3	1,754.6	14.7	–
3*(Transportation costs)	n_9	$n_{1,2,3,4,9}$	n_5	n_1	$n_{1,3,9}$	–	1,463.5	1,713.2	14.7	–
10*(Area)	n_{10}	$n_{1,2,3,4,8,9}$	n_5	$n_{1,3,9,10}$	$n_{1,2,3,4,9,10}$	$n_{2,3,5}$	17,319.3	20,065.9	63.6	–
3*(Transportation costs) and 10*(Area)	n_4	$n_{1,2,3,4,5,8,9,10}$	$n_{2,5}$	$n_{1,3,4,6,9,10}$	$n_{1,2,3,4,9}$	–	16,911.4	19,607.0	63.6	–

**Dry-grind process (DG), incineration (IN), anaerobic digestion (AD), gasification/fermentation (GF), sawing (SAW), MSW incineration (MSW).*
Source: From Čuček et al. (2010).

Table 12.8 Profit, Cost, and CFP for Different Scenarios

Scenarios	Profit (M€/y)	Transportation (M€/y)	Production (M€/y)*	CFP (t/(ykm^2))
Base case	164.12	11.37	40.33	13.2
Excluded preprocessing	160.11	19.91	35.71	27.9
3*(Transportation costs)	142.87	26.26	38.07	4.5
10*(Area)	1,756.4	352.66	296.47	12.4
3*(Transportation costs) and 10*(Area)	1,286.3	595.04	293.84	1.9

**Sum of investment and operating costs (excluded raw material costs).*
Source: From Čuček et al. (2010).

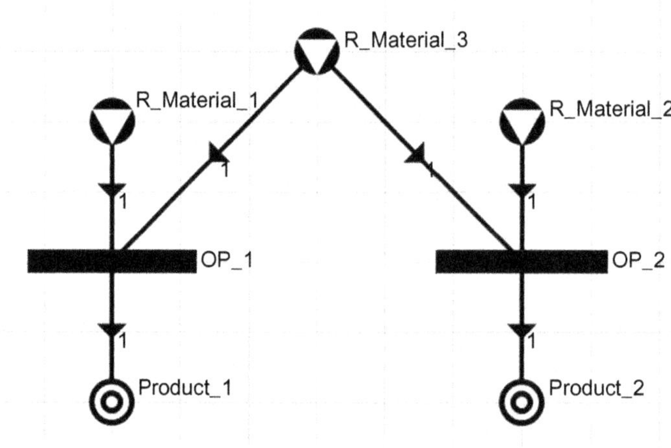

FIGURE 12.12

P-graph illustration.

From Lam et al. (2010a,b).

P-GRAPH APPROACH

The P-graph approach was initially introduced by Friedler et al. (1992) and has been implemented in the systematic optimal design, including industrial processes synthesis and supply network synthesis. As shown in Figure 12.12, P-graphs are bipartite graphs, with each comprising nodes for a set of materials, a set of operating units, and arcs linking them. Note that there are two basic algorithms, solution structure generator (SSG) and accelerated branch-and-bound (ABB), used in the P-graph approach to determine all the feasible solution structures and to exploit

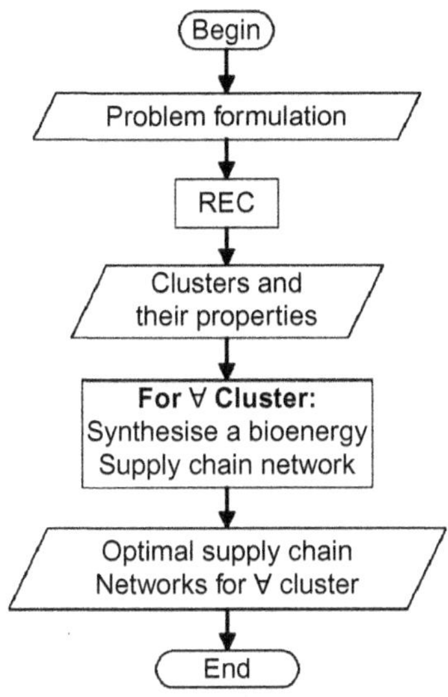

FIGURE 12.13

Methodology.

From Lam et al. (2010a,b).

the structural features of the process to be synthesized. It is a more efficient graph-theoretic method that is used to obtain the optimal (or near-optimal) network (Lam, 2013). Furthermore, one positive attribute of this approach is that it can generate several stand-by plans (near-optimal solutions) simultaneously for special purposes (different choice of strategy). Several works have studied the implementation of the P-graph approach to SC design (Barany et al., 2011), vehicle assignment problems by process-network synthesis to minimize environmental impact of transportation (Vance et al., 2013), and synthesis of sustainable energy SC (Lam et al., 2012), and open-structure biomass networks (Lam et al., 2012). These works have demonstrated the further extension of P-graph via several case studies (i.e., effective SC system, carbon emission reduction systems, and cleaner production process synthesis). It is worth noting that the results proved that the P-graph solution will provide better insights and near-optimal solutions, which are preferable for decision makers.

The P-graph approach utilizes a two-level methodology for the synthesis of regional biomass energy SCs. The SCs use various fuels as inputs (biomass and fossil) and deliver energy products—heat, power, pellets, and ethanol—to the final users. All unit operations are characterized by relevant performance specifications, reflecting their energy conversion efficiencies and linking the product generation rates to the fuel consumption rates.

The methodology is illustrated in Figure 12.13. In the text, the steps of the REC procedure are denoted by the prefix "1," indicating the first (higher) level.

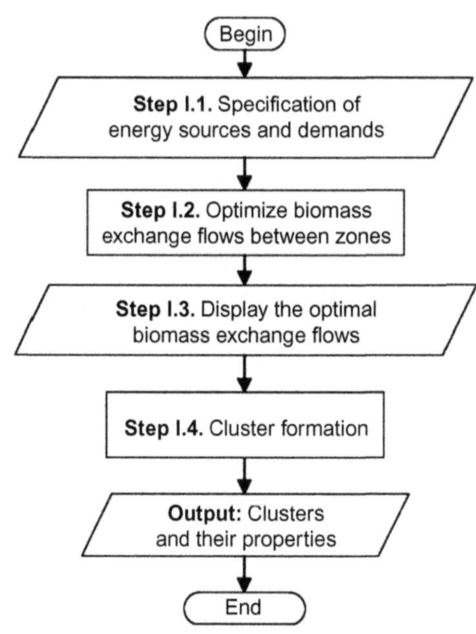

FIGURE 12.14

Synthesis of REC.

From Lam et al. (2010a,b).

The steps of the second (lower) level intracluster SC synthesis are prefixed with "2." Similar to the superstructural approach, P-graph covers not only harvesting, transportation, and storage of biomass network (in first level) but also the energy conversion process into the SC (in the second level).

The first level model can be generated using the REC algorithm in Figure 12.14, as discussed previously. As an improved model (compared with the superstructural approach), this approach utilizes P-graph instead of MINLP and superstructure to generate the model for the second level (Figure 12.15). Superstructural approach presents the selection of the operating units by integer variables, and it is more difficult and tedious to handle larger size problems because:

- The size of the algebraic optimization problems grows, where the solver needs to clearly examine infeasible combinations of integer variable values.
- The huge number of operating unit options makes it rather difficult to build the necessary problem superstructures heuristically and even automatically without rigorous combinatorial tools.
- When a superstructure is created heuristically, certain low-cost options would be missed, as would the opportunities for optimal solutions.

For handling process synthesis problems of practical complexity, the process network synthesis methodology based on the P-graph could be efficiently applied.

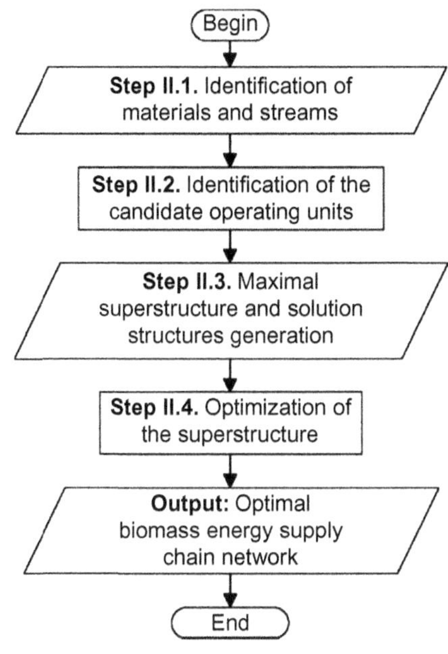

FIGURE 12.15

Synthesis of intracluster SC.

From Lam et al. (2010a,b).

Because step I has been discussed previously, the main focus of this section is on step II, which is the intracluster synthesis procedure using P-graph. To apply the P-graph approach, certain types of information need to be obtained, evaluated, and specified. Most of the necessary information has already been generated during the clustering stage. The procedure for the SCN synthesis inside each cluster follows the algorithm illustrated in Figure 12.15 and as follows:

1. Identification of the involved materials and streams. This is a preprocessing step to prepare input information for step III. The materials involve raw biomass, products (heat, power, pellets, and ethanol), and intermediates (pellets, syngas, biogas, etc.). The material prices follow a sign convention. Inputs are assigned positive prices if the plant has to pay for them and negative ones if it receives payment. Similarly, all outputs generating revenues are assigned positive prices and those generating costs receive negative prices.
2. Identification of the candidate operating units, their capital cost, and the performance of the unit by using a qualified assessment using the general workflow from Figure 12.16 as a guide. On the left side of the figure, the general sequence for harvesting, conversion, and utilization of resources is provided. In parallel, the workflow used for identifying the possible streams/ materials and the candidate operating units is shown. This is also a preprocessing

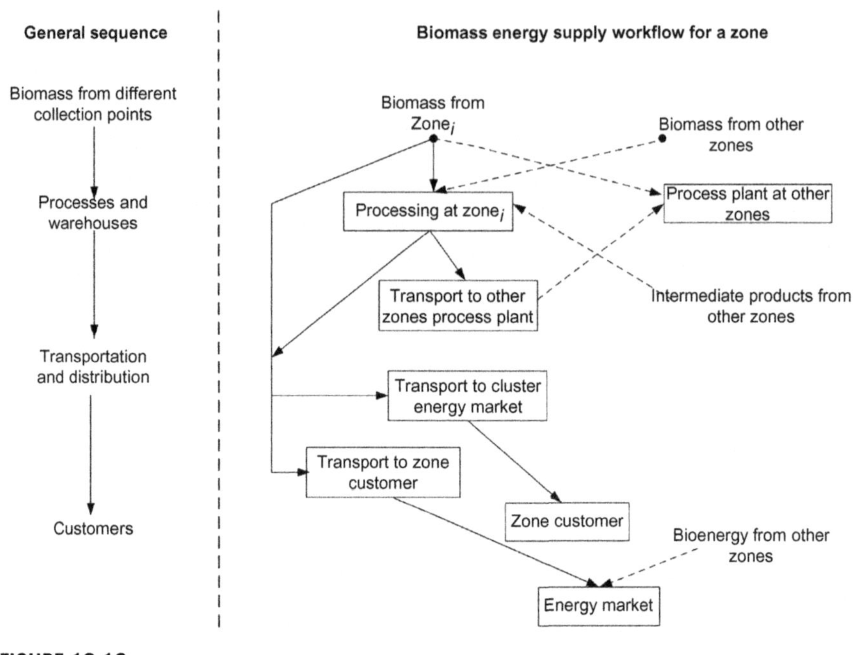

FIGURE 12.16

Workflow for energy supply of biomass.

From Lam et al. (2010a,b).

step preparing the input information for step III. The operating units are both conversion and transportation types. The capital costs of all operating units have been assumed to change linearly, adhering to the form given:

$$CC = A_{CC} + B_{CC} \cdot U_{Cap} \tag{12.26}$$

where the operating unit capacity U_{Cap} is measured by its throughput of a key inlet stream.

3. Generation of the maximal superstructure and all the combinatorial feasible individual networks between the involved materials and streams with the candidate bioenergy conversion units. This step is performed internally by the P-graph algorithms Maximal Structural Generator (MSG) and SSG.
4. Optimization of the generated superstructure. This results in the selection of the optimal network by using the ABB algorithm (Friedler et al., 1996). The latter selects the bioenergy conversion units and the quantities of the streams that formed the optimal network. More detailed explanation of the ABB algorithm has been provided by Friedler et al. (1996) and implemented into an integrated synthesis of process and heat exchanger networks by Nagy et al. (2001).

A case study is performed to synthesize the intracluster SC using the P-graph framework. Once the clusters are obtained by the REC algorithm, a biomass SC

Table 12.9 Cluster Data for Case Study

Zone	Area (km^2)	Location (km, km)	Biomass Potential (t/y)	Heating Value (MJ/kg)	Energy Supply (TJ/y)	Energy Demand (TJ/y)
1	11.60	(0, 0)	2,700	18.0	24.0	1.7
2	6.12	(4.4, 2.5)	1,100	18.5	10.2	4.5
3	9.58	(5.3, 2.4)	1,760	17.3	15.3	11.8
4	6.35	(4.1, 0.2)	940	6.4	3.0	33.0

Source: From Lam et al. (2010a,b).

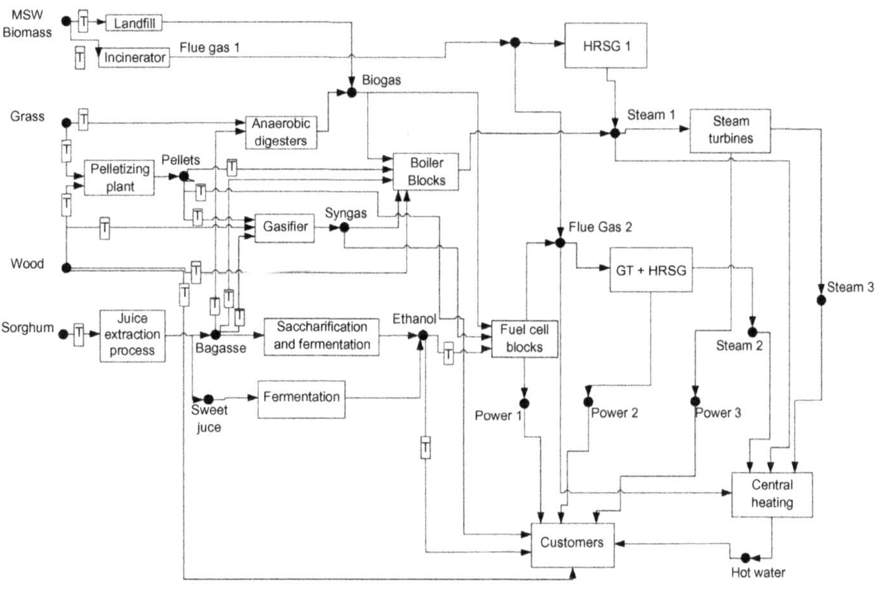

FIGURE 12.17

Schematic diagram of the intracluster SC.

From Lam et al. (2010a,b).

can be synthesized inside each cluster. The data for a particular cluster with four zones, given in Table 12.9, are used to conduct this case study.

The biomass types from zones 1–4 are wood, sweet sorghum, grass silage, and MSW, respectively. The synthesis accounts for the locations of the energy carrier conversion operations. Figure 12.17 shows the brief schematic structure for the feasible process combinations that may form the SC. The bullet point represents a material such as raw biomass, intermediate energy carriers, or products. The labeled boxes represent the operating units, and the symbol "T" with a frame represents the transportation activities. The heat-to-power ratio of the customer demands is assumed as 2:1, depending on the studied region. As a result, customer energy demands by zones become 1.13 TJ/y heating and 0.57 TJ/y power for zone 1; 3 TJ/y heating and 1.5 TJ/y power for zone 2; 7.87 TJ/y heating and 3.93 TJ/y power for zone 3; and 22.01 TJ/y heating with 10.99 TJ/y power for zone 4. Forestry wood, energy crops (grass and sweet sorghum), and MSW are the input raw materials. They are converted into other energy carriers, have higher energy densities, and are suitable for use in power generation facilities. CHP systems combining fuel cells, biofuel boilers, steam turbines, and gas turbines are defined as options to be used for the regional energy conversion system. The electricity produced is supplied to the cluster customers. The generated heat is used for domestic, commercial, and industrial applications, mainly for space heating and as hot utility.

In the first step, materials and streams have been identified (shown in Table 12.10). The second step identifies the candidate operating units (listed in Table 12.11). For each candidate operating unit, the streams/materials accepted as inputs, the outputs, the estimated performance, and capital cost coefficients are

Table 12.10 Material and Steam Properties

Symbol	Description	Price
A	Forestry wood	25 €/t
B	Sweet sorghum	20 €/t
C	Grass silage	15 €/t
D	Biomass MSW	−10 €/t
AZ 1, 2, 3, and 4	Biomass A transported to zones 1, 2, 3, and 4	30 €/t
Ptr	Petrol	1 €/L
CO_2	CO_2 emission	–
PA	Pellet from biomass A	–
PAZ 1, 2, 3, and 4	Pellet A transported to zones 1, 2, 3, and 4	–
PC	Pellet from biomass C	–
PCZ 1, 2, 3, and 4	Pellet C transported to zones 1, 2, 3, and 4	–
SG 1, 2, 3, and 4	Syngas generated from zones 1, 2, 3, and 4	–
BG 1, 2, 3, and 4	Biogas generated from zones 1, 2, 3, and 4	–
Steam 1, 2, 3, and 4	Steam generated from zones 1, 2, 3, and 4	–
Ba	Sorghum bagasse after juice extraction	–
Juice	Sorghum juice after juice extraction	–
ETN	Ethanol	–
EEM	Ethanol for energy market	350 €/t
PAEM	Pellet A for energy market	70 €/t
PCEM	Pellet C for energy market	60 €/t
Q 1, 2, 3, and 4	Heat generated for zones 1, 2, 3, and 4	–
P 1, 2, 3, and 4	Power generated for zones 1, 2, 3, and 4	–

Source: From Lam et al. (2010a,b).

Table 12.11 Unit Operation Properties

Symbol	Description	Input	Capital Cost		Performance
			A_{CC} (€)	B_{CC} (€/MW)	
T	Transport	Ptr	35,000	0.01	CO_2: 2.69 kg CO_2/L
PP	Pellet plant	A, C	150,000	3.6	Pellet: 0.98 t/t CO_2: 0.15 kg CO_2/t
JEP	Juice extraction plant	B	50,000	0.09	Juice: 0.69 t/t Bagasse: 0.31 t/t CO_2: 0.08 kg CO_2/t
FP	Fermentation plant	Juice	3,000	0.05	ETN: 0.07 t/t CO_2: 0.06 kg CO_2/t
SFP	Saccharification–fermentation plant	Ba	63,000	0.1	ETN: 0.13 t/t CO_2: 0.12 kg CO_2/t
G	Gasifier	A, PA, Ba, C, PC	42,000	0.08	SG: 0.70 MJ/MJ CO_2: 0.007 kg CO_2/MJ
B	Boiler	A, PA, Ba, C, PC	1,646	0.11	Stm: 0.80 MJ/MJ CO_2: 0.058 kg CO_2/MJ
FCGT	Fuel cell with gas turbine	SG, BG	9,200,000	0.05	P: 0.7 MJ/MJ Q: 0.2 MJ/MJ CO_2: 0.055 kg CO_2/MJ
AD	Anaerobic digester	C, Ba	4,939	0.327	BG: 0.58 MJ/MJ CO_2: 0.007 kg CO_2/MJ
HC	Home used biomass convertor (boiler, stove, etc.)	A, C, PA, PC	1,000	0.1	Q: 0.7 MJ/MJ CO_2: 0.060 kg CO_2/MJ
INC HRSG	Incinerator and heat recovery steam generator	D	52,000	0.08	Steam: 0.7 MJ/MJ CO_2: 0.055 kg CO_2/MJ
LF	Landfill	D	10,000	0.5	BG: 0.5 MJ/MJ CO_2: 0.028 kg CO_2/MJ
ST	Steam turbine	Stm	3,000,000	0.02	P: 0.24 MJ/MJ Q: 0.56 MJ/MJ

Source: From Lam et al. (2010a,b).

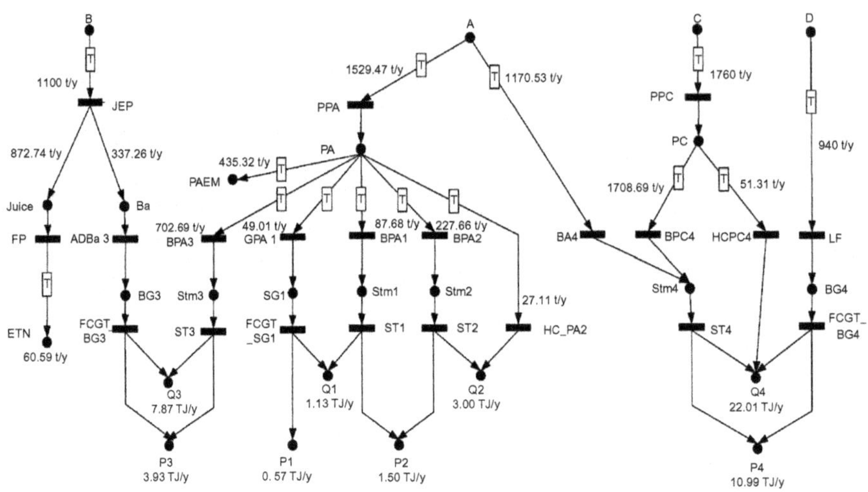

FIGURE 12.18

Optimized superstructure.

From Lam et al. (2010a,b).

available. The unit performance and economic data have been estimated to provide the basis for appropriate economic evaluation of the designs. By combining the information from Tables 12.10 and 12.11, the P-graph MSG algorithm has built the problem superstructure (maximal structure), which is shown in Figure 12.17. Finally, the software tool combinatorial process network synthesis editor (Friedler et al., 2009) is used to obtain the optimum solution for minimum production cost. The optimum solution provides the selected pathways, including input biomass quantities, type of energy carriers (input and intermediate materials), operating units, and final products for customers.

The result from the P-graph ABB optimization algorithm is illustrated in Figure 12.18. First, 1529.47 t/y of wood from zone 1 (biomass A) are pelletized, transported, and used as fuel feed for the gasifier in zone 1 as well as for the boilers in zones 1–3. Pellets A (wood origin; 22.11 t/y) are used to generate heat in zone 2 by the home bioenergy conversion unit. Another product stream, 1170.53 t/y of wood, is sent to zone 4 for direct use by steam boilers. Biomass B (sorghum) from zone 2 is converted to 60.59 t/y bioethanol for the energy market. The bagasse after the juice extraction from the sorghum (originating from zone 2) is sent to anaerobic digestion in zone 3, which produces biogas to be used as the fuel for a combined fuel cell and gas turbine unit. All of the grass feedstock (biomass C) collected in zone 3 is pelletized. Grass pellets (1708.69 t/y) are sent to zone 4 as the fuel for its pellets boiler and the rest are used as direct burning fuel for the house bioenergy conversion units in zone 4. The waste-based resource, 940 t/y of MSW fraction (biomass D), is sent to a landfill in zone 4 to produce biogas. These intermediate energy carriers (pellets, steam, biogas, and syngas) are converted to heat and power for meeting the energy demands described in the beginning of this section. For example, steam turbine and combined fuel cell and gas turbine in zone 3 produce 7.87 TJ/y of heat and 3.93 TJ/y of power to fulfill the local demands. The surplus of biomass is converted into ethanol (60.59 t/y) and wood pellet (435.32 t/y), which can be transported to another cluster or traded in the market.

The total annualized cost for the system is 1,852,723 €/y (life span: 15 y), which is equivalent to 0.036 €/MJ. Assuming the retail price for district heating is 0.03 €/MJ, the cost for power generation is 0.048 €/MJ. This compares favorably with the recent average electricity wholesale price. The recent wholesale prices are 0.013–0.045 €/MJ (EIA, 2014) in the United States in 2014, and 0.020–0.044 €/MJ (EUROSTAT, 2014) in the European Unit in 2014. The demonstration case study results show that the SC using biomass can achieve reasonable production cost levels, remaining within the profit margins for retail prices for most of Europe. CO_2 emission for energy generation is 0.123 kg CO_2/MJ, which is lower than the generation of CO_2 emissions for coal-, oil-, and LNG-fired power, which are 0.27, 0.21, and 0.17 kg CO_2/MJ (Hiroki, 2005).

FURTHER READING

To enhance the sustainability function of the GSCM, the following materials are recommended for further reading:

- Model-size reduction techniques for large-scale biomass production and supply networks (Lam et al., 2011a,b)

  This article introduces model-size reduction techniques for the analysis of large-scale renewable production and supply networks. The proposed model-size reduction techniques brought computational time improvements of several magnitudes compared with the high-performance linear system solution techniques but still with a little loss in accuracy. The techniques introduced here are: (i) reducing the connectivity in a biomass SCN; (ii) eliminating unnecessary variables and constraints; and (iii) merging the collection centers.
- Green strategy for sustainable waste-to-energy SC (Lam et al., 2013)

  A novel two-stage optimization model (macro-stage and micro-stage) is developed as a green strategy for systematic design of the waste-to-energy SC. The presented green strategy includes efficient resources management and reduction of CFP of a waste-to-energy SC. Micro-stage involves the optimization and allocation of waste (e.g., biomass, industrial waste, etc.), as well as the design of integrated processing hub. Macro-stage handles the synthesis and optimization of the waste-to-energy supply network.
- Extended P-graph applications in SC and process network synthesis (Lam, 2013)

  The article demonstrates the extension of P-graph via several case studies such as effective SC systems, carbon emission reduction systems, and cleaner production process synthesis. It overviews the application of the P-graph in new PNS areas regarding synthesis, optimization, planning, and management. The article also highlights the advantages of the P-graph in process network synthesis.

- Biomass demand-resources value targeting (Lim and Lam, 2014)

  To achieve a better-insight biomass SC, a new demand-resources value targeting (DRVT) approach is introduced. This approach investigates the value of each biomass available to fully utilize the biomass in respective applications. With systematic biomass value classification, integration of SC based on biomass value from biomass resources-to-downstream product can be developed. The DRVT model allows a better understanding of biomass and their potential downstream application.
- A supply network optimization with functional clustering of industrial resources (Ng and Lam, 2014)

  This article presents a functional clustering approach integrated in an industrial resources optimization. Under this approach, production facilities are identified and a processing hub allocation is determined. This method functionally clusters the industrial facilities based on their material interactions. Each cluster formed consists of a centralized processing hub that acts as the backbone/seed of a functional cluster. Strategic locations of centralized processing hubs are determined and functional clusters are formed by optimization modeling. The optimization result favors centralized processing hub formation. Lowered machinery capital investment and transportation cost are achieved in the functional clustered model.
- An algebraic approach for supply network synthesis (Ng et al., 2014)

  Supermatrix scoring concept and algebraic analysis with graphical visualization is proposed for GSC synthesis and analysis. The introduction of quantitative cum graphical analysis of supply network allows the performance overview of each possible choice of supply network in one picture. Also, the algebraic method allows the concurrent set-up of material allocation and ranking of possible supply network choices. This outweighs the mathematical modeling technique that allows for optimal supply network synthesis but gives no insight for the performance of next network choices; the supermatrix scoring technique ranks the general supply network choices in a whole picture yet is incapable of capturing the optimal selection that involves only part of the players, especially in the supply network problem.

CONCLUSIONS

Despite the pressure from global warming and environmental issues, there are many potential benefits of making the SC greener. Whether the SC structure is reorganized or whether a new network is synthesized, the greener policies and approaches are always cost-saving and are better for branding.

This chapter provides a general overview of the development of GSCM and several methods of constructing the GSC model. The generic model can be easily expanded and incorporated with other green strategies, such as LCA and GT. There are still many potential subjects to be developed in this area:

- Debottlenecking analysis of the current SCN.
- Biomass network synthesis for rural areas with low transportation facilities but huge amounts of unutilized biomass.
- GSC index as a standard and design guideline to determine the sustainability level of the SC or network.

ACKNOWLEDGMENT

Financial support from The University of Nottingham Early Career Research and Knowledge Transfer Award (A2RHL6) and the BiomassPlus Scholarship from Crops for Future Research Centre and Long-Term Research Grant Scheme (LRGS) "Unlocking the Potential of Fine Chemicals and Value Added Byproducts" are very much appreciated.

REFERENCES

Abu Qudeiri, J., Yamamoto, H., Ramli, R., Jamali, A., 2008. Genetic algorithm for buffer size and work station capacity in serial-parallel production lines. Artif. Life Robot 12 (1), 102–106.

Ahi, P., Searcy, C., 2013. A comparative literature analysis of definitions for green and sustainable supply chain management. J. Cleaner Prod. 52, 329–341.

Ahn, H., Lee, H., Park, S., 2003. A flexible agent system for change adaptation in supply chain. Expert Syst. Appl. 25 (4), 603–618.

Akanle, O., Zhang, D., 2008. Agent-based model for optimising supply chain configurations. Prod. Econ. 115 (2), 444–460.

Al-Mashari, M., Zairi, M., 2000. Supply-chain re-engineering using enterprise resource planning (ERP) systems: an analysis of a SAP R/3 implementation case. Phys. Distrib. Logist. Manage. 30 (3–4), 296–313.

Altiparmak, F., Gen, M., Lin, L., Karaoglan, I., 2009. A steady-state genetic algorithm for multi-product supply chain network design. Comput. Indus. Eng. 56 (2), 521–537.

Amiri, A., 2006. Designing a distribution network in a supply chain system: formulation and efficient solution procedure. Oper. Res. 171 (2), 567–576.

Andic, E., Yurt, O., Baltacioglu, T., 2012. Green supply chains: efforts and potential applications for the Turkish market. Resour. Conserv. Recycl. 58, 50–68.

Barany, M., Bertok, B., Kovacs, Z., Friedler, F., Fan, L.T., 2011. Solving vehicle assignment problems by process-network synthesis to minimize cost and environmental impact of transportation. Clean Technol. Environ. Policy 13 (4), 637–642.

Barari, S., Agarwal, G., Zhang, W.J.., Mahanty, B., Tiwari, M.K., 2011. Development of an integrated and holistic conceptual game approach. Expert Syst. Appl. Int. J. 39 (3), 2965–2976.

Barari, S., Agarwal, G., Zhang, W.J., Mahanty, B., Tiwari, M.K., 2012. A decision framework for the analysis of green supply chain contracts: an evolutionary game approach. Expert Syst. Appl. 39 (3), 2965–2976.

Basoglu, N., Daim, T., Kerimoglu, O., 2007. Organizational adoption of enterprise resource planning systems: a conceptual framework. High Technol. Manage. Res. 18, 73–97.

Brooks, S., Wang, X., Sarker, S., 2012. Unpacking Green IS: A Review of the Existing Literature and Directions for the Future. Springer-Verlag/Heidelberg, Germany/Berlin.

Bullnheimer, B., Hartl, R., Strauss, C., 1999. An improved ant system algorithm for the vehicle routing problem. Ann. Oper. Res. 89 (0), 319–328.

Cachon, G.P., Netessine, S., 2004. Game theory in supply chain analysis. Handbook of Quantitative Supply Chain Analysis: Modeling in the eBusiness Era. Kluwer, Boston.

Cakravastia, A., Toha, I., Nakamura, N., 2002. A two-stage model for the design of the supply chain networks. Prod. Econ. 80 (3), 231–248.

Cambero, C., Sowlati, T., 2014. Assessment and optimisation of forest biomass supply chains from economic, social and environmental perspectives—a review of literature. Renew. Sustainable Energy Rev. 36, 62–73.

Carvalho, A., Matos, H., Gani, R., 2013. SustainPro—a tool for systematic process analysis, generation and evaluation of sustainable design alternatives. Comput. Chem. Eng. 50, 8–27.

Chan, F.T.S., Chung, S.H., 2004. A multi-criterion genetic algorithm for order distribution in a demand driven supply chain. Int. J. Comput. Integr. Manuf. 17 (4), 339–351.

Chan, F.T.S., Chung, S.H., Wadhwa, S., 2004. A heuristic methodology for order distribution in a demand driven collaborative supply chain. Int. J. Prod. Res. 42 (1), 1–19.

Chang, Y.H., 2010. Adopting co-evolution and constraint-satisfaction concept on genetic algorithms to solve supply chain network design problems. Expert Syst. Appl. 37 (10), 6919–6930.

Chen, C.L., Lee, W.C., 2004. Multi-objective optimization of multi-echelon supply chain networks with uncertain product demands and prices. Comput. Chem. Eng. 28, 1131–1144.

Chen, Y.J., Sheu, J.B., 2009. Environmental-regulation pricing strategies for green supply chain management. Transp. Res. Part E Logistics Transp. Rev. 45 (5), 667–677.

Chofreh, A.G., Goni, F.A., Shaharoun, A.M., Ismsil, S., Klemeš, J.J., 2014. Sustainable enterprise resource planning: imperatives and research directions. Cleaner Prod. 71, 139–147.

Cucek, L., Lam, H.L., Klemeš, J.J., Varbanov, P.S., 2010. Synthesis networks for the production and supply of renewable energy from biomass. Chem. Eng. Trans. 21, 1189–1194.

Davenport, T.H., 2000. Mission Critical: Realizing the Promise of Enterprise Systems. Harvard Business Scholl Press, Boston, MA.

De Benedetto, L., Klemeš, J., 2010. The environmental bill of material and technology routing: an integrated LCA approach. Clean Technol. Environ. Policy 12 (2), 191–196.

Doerner, K., Gutjahr, W.J., Hartl, R.F., Strauss, C., Stummer, C., 1992. Nature-inspired metaheuristics for multi objective activity crashing. Omega 36 (6), 1019–1037.

EIA, 2014. Average retail price of electricity to ultimate customers by end-use sector, by state. <www.eia.doe.gov/cneaf/electricity/epm/table5_6_b.html> (accessed 17.08.2014).

Erol, I., Ferrell, J.W.G., 2004. A methodology to support decision making across the supply chain of an industrial distributor. Int. J. Prod. Econ. 89, 119–129.

Esmaeili, M., Aryanezhad, M.B., Zeephongsekul, P., 2009. A game theory approach in seller-buyer supply chain. Eur. J. Oper. Res. 195 (2), 442–448.

EUROSTAT, 2014. Electricity price for first semester 2014. <epp.eurostat.ec.europa> (accessed 17.08.2014).

Farahani, R.Z., Elahipanah, M., 2008. A genetic algorithm to optimize the total cost and service level for just-in-time distribution in a supply chain. Int. J. Prod. Econ. 111, 229–243.

Finnvedena, G., Moberg, A., 2005. Environmental systems analysis tools—an overview. J. Cleaner Prod. 13, 1165–1173.

Fox, M., Barbuceanu, M., Teigen, R., 2000. Agent oriented supply-chain management. Int. J. Flexible Manuf. Syst. 12, 1572–9370.

Franca, B.R., Jones, E.C., Richards, C.N., Carlson, J.P., 2010. Multi-objective stochastic supply chain modeling to evaluate trade-offs between profit and quality. Int. J. Prod. Econ. 127, 292–299.

Friedler, F., Tarjan, K., Huang, Y.W., Fan, L.T., 1992. Cominatorial algorithms for process synthesis. Comput. Chem. Eng. 16, 313–320.

Friedler, F., Varga, J.B., Feher, E., Fan, L.T., 1996. Combinatorially accelerated branch-and-bound method for solving the MIP model of Process Network Synthesis. In: Floudas, C.A., Pardalos, P.M. (Eds.), State of the Art in Global Optimization, Vol. 7. Kluwer Academic Publishers, Boston, MA, pp. 609–626.

Friedler, F., Fan, L.T., Bertok, B., 2009. PNS Editor. <www.p-graph.com> (accessed 16.08.2014).

Gen, M., Syarif, A., 2005. Hybrid genetic algorithm for multi-time period production/distribution planning. Comput. Ind. Eng. 48 (4), 799–809.

Goni, F.A., Chofreh, A.G., Sahran, S., 2011. Critical success factors for enterprise resource planning system implementation: a case study in Malaysian SME. Adv. Sci. Eng. Inform. Technol. 1 (2), 200–205.

Graves, S., Willems, S., 2005. Optimizing the supply chain configuration for new products. Manage. Sci. 51, 1165–1180.

Guest, G., Bright, R.M., Cherubini, F., Michelsen, O., Strømman, A.H., 2011. Life cycle assessment of biomass-based combined heat and power plants. J. Ind. Ecol. 15 (6), 908–921.

Guillen, G., Mele, F.D., Bagajewicz, M.J., Espuna, A., Puigjaner, L., 2005. Multiobjective supply chain design under uncertainty. Chem. Eng. Sci. 60 (6), 1535–1553.

Guo, R., Tang, Q., 2008. An optimized supply chain planning model for manufacture company based on JIT. Bus. Manage. 3 (1), 129–133.

Handfield, R., Walton, S., Seegers, L., Melnyk, S., 1997. 'Green' value chain practices in the furniture industry. J. Oper. Manage. 15 (4), 293–315.

Hiroki, H., 2005. Life cycle GHG emission analysis of power generation systems: Japanese case. Energy 30, 2042–2056.

Huang, G.Q., Xin, Y.Z., Lo Victor, H.Y., 2007. Integrated configuration of platform products and supply chains for mass customization: a game theoretic approach. IEEE Trans. Eng. Manage. 54 (1), 156–171.

Huang, Y., Huang, G.Q., Newman, S.T., 2011. Coordinating pricing and inventory decisions in a multi-level supply chain:A game-theoretic approach. Transport. Res. Part E 47, 115–129.

Huston, M., Marland, G., 2003. Carbon management and diversity. J. Environ. Manage. 67, 77–86.

Jayaraman, V., Pirkul, H., 2001. Planning and coordination of production and distribution facilities for multiple commodities. Oper. Res. 133 (2), 394–408.

Jayaraman, V., Ross, A., 2003. A simulated annealing methodology to distribution network design and management. Oper. Res. 144 (3), 629–645.

Kandananond, K., 2014. A roadmap to green supply chain system through enterprise resource planning (ERP) implementation. Procedia Eng. 69, 377–382.

Kannan, G., Sasikumar, P., Devika, K., 2010. A genetic algorithm approach for solving a closed loop supply chain model: a case of battery recycling. Appl. Math. Model 34 (3), 655–670.

Kelle, P., Akbulut, A., 2005. The role of ERP tools in supply chain information sharing cooperation and cost optimization. Int. J. Prod. Econ. 93, 41–52.

Koc, E., 2012. The Bees Algorithm Theory, Improvements and Applications. Cardiff University, Wales, UK.

Koh, L., Ghazoul, J., 2008. Biofuel, biodiversity and people: understanding the conflict and finding opportunities. Biol. Conserv. 141, 2450–2460.

Koh, S., Saad, S.M., 2006. Managing uncertainty in ERP-controlled manufacturing environments in SMEs. Int. J. Prod. Econ. 101, 109–127.

Koray, D., Marc, G., 1999. A primal decomposition method for the integrated design of multiperiod production distribution systems. IIE Trans. 31 (11), 1027–1036.

Kostin, A., Mele, F., Guillen, G., 2011. Multi-objective optimization of integrated bioethanol-sugar supply chains considering different LCA metrics simultaneously. Comput. Aided Chem. Eng. 29, 1276–1280.

Kuhn, J.R., Sutton, S.G., 2010. Continuous auditing in ERP system environments: the current state and future directions. Inform. Syst. 24 (1), 91–112.

Lainez, J.M., Bojarski, A., Espuna, A., Puigjaner, L., 2008. Mapping Environmental Issues within Supply Chains: A LCA Based Approach. Elsevier B.V./Ltd, Barcelona, Spain.

Lam, H.L., 2013. Extended P-graph applications in supply chain and Process Network Synthesis. Curr. Option Chem. Eng. 2 (4), 475–486.

Lam, H.L., Varbanov, P., Klemeš, J., 2008. Development of a graphical analysis method for renewable energy supply chain. In: Varbanow, P., Klemes, J., Bulatov, I. (Eds.), Energy for Sustainable Future. UoP Press, Veszprem, pp. 209–218.

Lam, H.L., Varbanov, P., Klemeš, J., 2010a. Minimising carbon footprint of regional biomass supply chains. Resour. Conserv. Recycl. 54, 303–309.

Lam, H.L., Varbanov, P., Klemeš, J., 2010b. Optimisation of regional energy supply chains including renewables: P-graph approach. Comp. Chem. Eng. 34 (5), 782–792.

Lam, H.L., Klemeš, J., Kravanja, Z., 2011a. Model-size reduction techniques for large-scale biomass production and supply networks. Energy 36 (8), 4599–4608.

Lam, H.L., Varbanov, P., Klemeš, J., 2011b. Regional renewable energy and resource planning. Appl. Energy 88 (2), 545–550.

Lam, H.L., Klemeš, J., Varbanov, P., Kravanja, Z., 2012. P-graph synthesis of open-structure biomass networks. Ind. Eng. Chem. Res. 2 (4), 172–180.

Lam, H.L., Ng, W.P.Q., Ng, R.T.L., Ng, E.H., Abdul Aziz, M.K., Ng, D.K.S., 2013. Green strategy for sustainable waste-to-energy supply chain. Energy 57, 4–16.

Law, C.C.H., Chen, C.C., Wu, B.J.P., 2010. Managing the full ERP life-cycle: considerations of maintenance and support requirements and IT governance practice as integral elements of the formula for successful ERP adoption. Comput. Ind. 61 (3), 297–308.

Lee, H.L., Kim, S.H., 2000. Optimal production distribution planning in supply chain management using a hybrid simulation-analytic approach. In: Joines, J.A., Barton, R.R., Kang, K., Fishwick, P.A. (Eds.), Proceeding of the 2000 Winter Simulation Conference.

Lee, J., Siau, K., Hong, S., 2003. Enterprise integration with ERP and EAI. Commun. ACM 46 (2), 54–60.

Liang, T.F., 2006. Distribution planning decisions using interactive fuzzy multi-objective linear programming. Fuzzy Sets Syst. 157, 1303–1316.

Liang, T.F., 2008. Fuzzy multi-objective production/distribution planning decisions with multi-product and multi-time period in supply chain. Comput. Ind. Eng. 55, 676–694.

Lim, C.H., Lam, H.L., 2014. Biomass demand-resources value targeting. Energy Convers. Manage. 87, 1202–1209.

Lim, M., Zhang, D., Goh, W., 2009. An iterative agent bidding mechanism for responsive manufacturing. Eng. Appl. Artif. Intell. 22 (7), 1068–1079.

Linthicum, D., 1999. Enterprise Application Integration. Addison-Wesley, Massachusetts, USA.

Lopez, C., Salmeron, J.L., 2012. Dynamic risks modelling in ERP maintenance projects with FCM. Inform. Sci. 256, 25–45.

Luis, A.M., Recio, G., 2014. Bi-criterion optimisation for configuring an assembly supply chain using Pareto ant colony meta-heuristic. Manuf. Syst. 33, 188–195.

Luis, A.M., Zhang, D.Z., 2011. Multi-objectiveant colony optimisation:A meta-heuristic approach to supply chain design. Int. J. Prod. Econ. 131, 407–420.

McKendry, P., 2002. Energy production from biomass next term (Part 1): overview of biomass. Bioresour. Tehnol. 83 (1), 37–46.

Makey, P., 1998. Enterprise Resource Planning. Butler Group Limited, Hull, UK.

Marinos, T., Zahir, I., 2002. Integrating Cross-Enterprise Systems: An Innovative Framework for the Introduction of Enterprise Application Integration. Information Systems Evaluation and Integration Group (ISEIG) Brunel University, Uxbridge,UK.

Mastrocinque, E., Yuce, B., Lambiase, A., Packianather, M., 2013. A multi-objective optimization for supply chain network using the bees algorithm. Swarm Evol. Comput. 5 (38), 1–11.

Moncayo-Martinez, L., Recio, G., 2014. Bi-criterion optimisation for configuring an assembly supply chain using Pareto ant colony meta-heuristic. Manuf. Syst. 33 (1), 188–195.

Moncayo-Martınez, L., Zhang, D., 2011. Multi-objective ant colony optimization: a meta-heuristic approach to supply chain design. Prod. Econ. 131 (1), 407–420.

Moyaux, T., Chaib-draa, B., D'Amours, S., 2004. The impact of information sharing on the efficiency of an ordering approach in reducing the bullwhip effect. Systems, Man, and Cybernetics.

Murphy, F., Devlin, G., Mc Donnell, K., 2014. Forest biomass supply chains in Ireland: a life cycle assessment of GHG emissions and primary energy balances. Appl. Energy. 116, 1–8.

Nagarajan, M., Sosic, G., 2008. Game-theoretic analysis of cooperation among supply chain agents: review and extensions. Eur. J. Oper. Res. 187 (3), 719–745.

Nagurney, A., Yu, M., 2012. Sustainable fashion supply chain management under oligopolistic competition and brand differentiation. Int. J. Prod. Econ. 135 (2), 532–540.

Nagy, A.B., Adonyi, R., Halasz, L., Friedler, F., Fan, L.T., 2001. Integrated synthesis of process and heat exchanger networks: algorithmic approach. Appl. Therm. Eng. 21, 1402–1427.

Ness, B., Urbel-Piirsalu, E., Anderberg, S., Olsson, L., 2007. Categorising tools for sustainability assessment. Ecol. Econ. 60 (3), 498–508.

Ng, W.P.Q., Lam, H.L., Yusup, S., 2013. Supply network synthesis on rubber seed oil utilisation as potential biofuel feedstock. Energy. 55, 82–88.

Ng, W.P.Q., Lam, H.L., 2014. A supply network optimization with functional clustering of industrial. Cleaner Prod. 71, 87–97.

Ng, W.P.Q., Promentilla, M.A., Lam, H.L., 2014. An algebraic approach for supply network synthesis. Cleaner Prod.Corrected Proof, (In Press).

Nwe, E., Adhitya, A., Halim, I., Srinivasan, R., 2010. Green supply chain design and operation by integrating LCA and dynamic simulation. Comput. Aided Chem. Eng. 28, 109–114.

Paksoy, T., Pehlivan, N.Y., Özceylan, E., 2012. Application of fuzzy optimization to a supply chain network design: a case study of an edible vegetable oils manufacturer. Appl. Math. Modell. 36, 2762–2776.

Peidro, D., Mula, J., Poler, R., Verdegay, J.L., 2009. Fuzzy optimization for supply chain planning under supply, demand and process uncertainities. Fuzzy Sets Syst. 160 (18), 2640–2657.

Penfield, P., 2008. Sustainability can be a competitive advantage, Material Handling Industry of America, <www.mhi.org/media/news/7056> (accessed 01.09.2014).

Pham, D., Ghanbarzadeh, A., 2005. The Bees Algorithm. Manufacturing Engineering Centre, Cardiff University, Cardiff, UK.

Porter, M., 1998. Cluster and the new economics of competition. Harvard Bus. Rev. November–December, 76 (6), 76–90.

POST—UK Parliamentary Office of Science and Technology, 2006. Carbon footprint of electricity generation. <www.parliament.uk/documents/upload/postpn268.pdf> (accessed 10.08.2014).

Pucker, J., Zwart, R., Jungmeier, G., 2012. Greenhouse gas and energy analysis of substitute natural gas from biomass for space heat. Biogas Bioenergy 38 (0), 95–101.

Ramezani, M., Bashiri, M., Tavakkoli-Moghaddam, R., 2013. A new multi-objective stochastic model for a forward/reverse logistic network design with responsiveness and quality level. Appl. Math. Modell. 37, 328–344.

Robinson, E.P., Satterfield, R.K., 1998. Designing distribution systems to support vendor strategies in supply chain management. Decis. Sci. 29, 685–706.

Sarkis, J., Zhu, Q., Lai, K., 2011. An organizational theoretic review of green supply chain management literature. Int. J. Prod. Econ. 130 (1), 1–15.

Sayigh, A., 1999. Renewable energy: the way forward. Appl. Energy 64, 15–30.

Silva, A., Sousa, J., Costa, S., Runkler, T., 2004. A multi-agent approach for supply chain management using ant colony optimization. In: Proceedings of International Conference on Systems, Man and Cybernetics. Hague, Netherlands, 2004.

Sheu, J.B., 2011. Bargaining framework for competitive green supply chains under governmental financial intervention. Transp. Res. Part E Logistics Transp. Rev. 47 (5), 573–592.

Sobel, M.J., Turcic, D., 2008. Risk Aversion and Supply Chain Contract Negotiation. Department of Operations Weatherhead School of Management Case Western Reserve University, Cleveland,Ohio.

Srivastava, S., 2007. Green supply-chain management: a state-of-the-art literature review. Int. J. Manage. Rev. 9 (1), 53–80.

Swaminathan, J., Smith, S., Sadeh, N., 1998. Modeling supply chain dynamics: a multiagent approach. Decis. Sci. 29 (3), 607–632.

Syam, S.S., 2002. A model and methodologies for the location problem with logistical components. Comput. Oper. Res. 29 (9), 1173–1193.

Syarif, A., Yun, Y., Gen, M., 2002. Study on multi-stage logistics chain network: a spanning tree-based genetic algorithm approach. Comput. Ind. Eng. 43 (1–2), 299–314.

Themistocleous, M., Irani, Z., OKeefe, R., 2001. ERP and application integration: exploratory survey. Bus. Process Manage. 7 (3), 195–204.

Truong, T.H., Azadivar, F., 2005. Optimal design methodologies for configuration of supply chains. Int. J. Prod. Res. 43 (11), 2217–2236.

Turan, P., Nimet, Y.P., Eren, Ö., 2012. Application of fuzzy optimization to a supply chain network design: a case study of an edible vegetable oils manufacturer. Appl. Math. Model. 36 (6), 2762–2776.

Vance, L., Cabezas, H., Heckl, I., Bertok, B., Friedler, F., 2013. Synthesis of sustainable energy supply chain by the P-graph framework. Ind. Eng. Chem. Res. 52 (1), 266–274.

V-Neumann, J., Morgenstern, O., 1944. Theory of Games and Economic Behaviour. Princeton University Press, Princeton, NJ.

Wang, S., Xia, H., Liu, F., 2002. Agent-based modelling and mapping of a manufacturing system. Mater. Process. Technol. 129 (1–3), 518–523.

Xu, R., Zhai, X., 2010. Analysis of supply chain coordination under fuzzy demand in a two-stage supply chain. Appl. Math. Modell. 34, 129–139.

Yan, H., Yu, Z., Cheng, T.C.E., 2003. A strategic model for supply chain design with logical constraints: formulation and solution. Comput. Oper. Res. 30 (14), 2135–2155.

Yu, Y., Liang, L., Xu, K.N., Wang, Z.Q., 2006. VMI integration considering pricing, production capacity and raw material procurement. Int. J. Logist. Res. Appl. 9 (4), 335–350.

Yuce, B., Packianather, M.S., Mastrocinque, E., Pham, D.T., Lambiase, A., 2013. Honey bees inspired optimization method: the Bees Algorithm. Insects. 4 (4), 646–662.

Yuce, B., Mastrocinque, E., Lambiase, A., Packianather, M.S., Pham, D.T., 2014. A multi-objective supply chain optimisation using enhanced Bees Algorithm with adaptive neighbourhood search and site abandonment strategy. Swarm Evol. Comput. 18, 71–82.

Yue, J.F., Austin, J., Wang, M.C., Huang, Z.M., 2006. Coordination of cooperative advertising in a two-level supply chain when manufacturer offers discount. Eur. J. Oper. Res. 168 (1), 65–85.

Zamarripa, M.A., Aguirre, A.M., Mendez, C.A., Espuna, A., 2013. Mathematical programming and game theory optimization-based tool for supply chain planning in cooperative/competitive environments. Chem. Eng. Res. Des. 91 (8), 1588–1600.

Zhang, D., Anosike, A., Lim, M., 2006. An agent-based approach for e-manufacturing and supply chain integration. Comput. Ind. Eng. 51 (2), 343–360.

Zhang, T., Zhang, D., 2007. An agent-based simulation of consumer purchase decision-making and decoy effect. Bus. Res. 60 (8), 911–922.

Zhang, X.Y., Huang, G.Q., 2010. Game-theoretic approach to simultaneous configuration of platform products and supply chains with one manufacturing firm and multiple cooperative suppliers. Int. J. Prod. Econ., 121–136.

Zhao, Y., Wang, S., Cheng, T.C.E., Yang, X., Hunag, Z., 2010. Coordination of supply chains by option contracts: a cooperative game theory approach. Eur. J. Oper. Res. 207 (2), 668–675.

Zhao, R., Neigbour, G., Han, J., McGuire, M., Deutz, P., 2012. Using game theory to describe strategy selection for environmental risk and carbon emissions reduction in the green supply chain. J. Loss Prev. Process Ind. 25 (6), 927–936.

Zhu, Q.H., Dou, Y.J., 2007. Evolutionary game model between governments and core enterprises in greening supply chains. Syst. Eng. Theory Pract. 27 (12), 85–89.

CHAPTER

Supply and demand planning and management tools toward low carbon emissions 13

Zainuddin Abdul Manan and Sharifah Rafidah Wan Alwi

Process Systems Engineering Centre (PROSPECT), Faculty of Chemical Engineering, Universiti Teknologi Malaysia, Johor, Malaysia

INTRODUCTION

Carbon dioxide (CO_2), methane (CH_4), and nitrous oxide (N_2O) gases are continuously emitted to the atmosphere as a result of various human activities. Carbon emission refers to the net emission of CO_2 as well as other greenhouse gases that are expressed in terms of CO_2 equivalents from a life cycle assessment (LCA) perspective (Weidema et al., 2008). These gases have been attributed to global warming and climate change. The key sectors contributing to carbon emissions include the energy supply, industry, transport, agriculture, forestry, water, and wastewater. Power generation plants under the energy sector and energy-intensive industries like steel mills, cement plants, and petroleum refineries contribute approximately 60% of the global CO_2 emissions (Table 13.1) (van Straelen et al., 2010). Stricter regulations regarding carbon and greenhouse gas emissions have driven companies, research organizations, and nongovernmental bodies to invest in developing technologies to reduce carbon emissions.

Numerous models have been developed and published regarding energy planning with consideration for CO_2 emissions. Mirzaesmaeeli et al. (2010) have developed a multiperiod mixed-integer linear programming (MILP) model for power generation planning of electric systems. The main objective of the model is to determine the optimal mix of energy supply sources and pollutant mitigation options that meet specified electricity demands and CO_2 emissions targets at minimum cost. Hashim et al. (2005) developed an MILP model and applied it to the existing Ontario power generation fleet and analyzed it from three different perspectives: economic; environmental; and integrated mode combining the objectives of the economic and environmental modes through the use of an external pollution index as a conversion factor from pollution to cost.

Table 13.1 Global Stationary CO_2 Emitters (van Straelen et al., 2010)

Process	Number of Sources	Emissions (10^6 t CO_2/y)
Fossil fuels		
Power	4,942	10,539
Cement production	1,175	932
Refineries	638	798
Iron and steel industry	269	646
Petrochemical industry	470	379
Oil and gas processing	N/A	50
Other sources	90	33
Biomass		
Bioethanol and bionenergy	303	91
Total	7,887	13,466

Pekala et al. (2010) proposed a model to handle two subcases of energy planning systems: optimum biofuels production with consideration of multiple footprints and deployment of carbon capture and storage (CCS) retrofit with concern for cost-effectiveness. Sadegheih (2010) developed a methodology for designing an optimal configuration for system transmission planning with carbon emission costs. The methodology involves the use of a mixed-integer programming model, a genetic algorithm, and simulated annealing to solve the power transmission network planning problem. Muis et al. (2010) developed a model for the optimal planning of electricity generation schemes for a nation to meet a specified CO_2 emission target. The model used an optimizer to select a scheme for renewable energy integration in the existing electricity generation setup, and this was applied to the Malaysian electricity sector. The aforementioned approaches directly involve the use of models in the electricity generation and transmission planning to achieve the minimum carbon emissions. The results of the work of Hashim et al. (2005) suggested that fuel balancing and fuel switching are effective ways to reduce carbon emissions. Atkins et al. (2010) proposed some potential carbon reduction schemes to include CCS, increase in renewable energy generation, and an extensive reduction in fossil fuel–based thermal energy generation. Fuel replacement, "end of pipe" treatment, and fuel reforming were other alternatives proposed by Clarke (2001).

Therefore, a pertinent question arises: what happens to the actual CO_2 emissions from other stationary emission sources outside the electricity sector? Power generation sites, refineries, cement plants, gas processing plants, and iron and steel plants are the predominant stationary CO_2 emission sources (Gale, 2002). For any country to meet its carbon emission reduction targets, all sources of carbon emissions must be considered, and plans regarding how to mitigate and reduce must be initiated and implemented.

The aforementioned mathematical techniques offer the advantage of computational effectiveness and the ability to handle problem dimensionality. However, they are less popular among policy-makers, government officials, and engineering practitioners because of the difficulty to set-up the problem models and to master the techniques. In addition, the mathematical approach typically provides designers with relatively less design insights.

In contrast, graphical techniques are typically easier to master and apply, and they are vital as visualization tools for systems planning and engineering design. However, they have limitations in dealing with problem dimensionality and often cannot guarantee a global optimal solution. The two approaches are complementary and are widely used to provide better understanding through visualization (graphical approaches) and are able to handle highly complex problems (mathematical modeling) (Manan and Alwi, 2007).

This chapter presents a set of graphical techniques based on the Pinch Analysis to assist decision makers during the planning, design, and retrofit of systems to achieve low carbon emissions. Pinch Analysis, which was originally developed to identify optimal energy utilization strategies for process plants (Hohmann, 1971), has been used to plan emission reduction in the electricity sector. The Pinch-based electricity planning approach was initially focused on the targeting and planning of electricity generation mix on the macro-scale (Atkins et al., 2010). Other approaches included the decomposition of total carbon footprint into material and energy-based components or, alternatively, into internal and external components (Tjan et al., 2010), locating the minimum CO_2-neutral and low-carbon sources for energy sector planning (Lee et al., 2009), and determining how aggregate carbon emissions targets for the power generation sector can be met while minimizing the need for retrofitting power plants (Tan et al., 2009). There are also recent works regarding carbon capture and sequestration planning that involve the integration of CO_2 sources emitted from power plants, oil refineries, and cement plants, with CO_2 sinks from geological storage sites based on graphical (Diamante et al., 2013) as well as algebraic techniques (Ooi et al., 2013). The recent work of Diamante et al. (2014) simultaneously considered injectivity constraints of every sink as well as time of availability of various sources and sinks.

In this chapter, opportunities for integrating carbon demand processes with CO_2 sources within an industrial complex (e.g., steel plants, cement factories, petroleum refineries, power plants, and other processes that emit CO_2 from their production process) are explored and analyzed. The use of source and demand curves (SDCs) based on Pinch Analysis provides vital visualization insights for the planning and retrofitting of systems toward achieving a low-carbon industrial sites and regions. The chapter also describes how the carbon SDCs can be used as a tool for planners and designers to systematically search for process change opportunities to holistically reduce carbon emissions in industrial sites.

CARBON PINCH ANALYSIS DESIGN METHOD

The first approach for the use of the Pinch Analysis technique for carbon-constrained energy planning was first developed by Tan and Foo (2007). Their method assumed that within a system there exists a set of energy sources, each with a specific carbon intensity characteristic of the fuel or the technology. At the same time, the system was also assumed to contain a set of demands, each with a specified carbon footprint limit. The aim of developing the approach was to determine the minimum amount of zero-carbon energy resources needed to meet the specified emission limits and to determine how the energy sources should be allocated to meet the different demands while complying with emission limits. This section, using the approach developed by Tan and Foo (2007), attempts to explain the method of generating the Energy Demand Composite Curve, the Energy Supply Composite Curve, targeting demand for zero-carbon energy, and effect of reduced emission limit on zero-carbon energy demand by using the hypothetical case of Tan and Foo (2007).

THE ENERGY DEMAND COMPOSITE CURVE

Table 13.2 adopted from the case study of Tan and Foo (2007) is used to illustrate how to plot the Composite Curves and how to use the concept of Pinch Analysis in targeting demand for zero-carbon energy and the effect of reduced emissions.

Table 13.2 provides the demands or sinks in primary energy equivalents (right side of the table). These values are the hypothetical energy demand distributed over three geographic regions, with each having its own expected energy usage and emission limits. To generate the Composite Curve, the energy demands of the sinks are first arranged in order of increasing emission factors. The emission factors are calculated by dividing the emission limits for each region by the corresponding energy consumption. These factors were found to be 20, 50, and 100 t CO_2/TJ for regions I, II, and III. The overall emission factor for the three regions can be calculated by dividing the total emissions limit by the total energy demand across all the regions; it was found to be 50 t CO_2/TJ for this case. A plot of cumulative energy on the horizontal axis against cumulative emissions on the

Table 13.2 Data for Case Studies (Tan and Foo, 2007)

Energy Resource	Emission Factor $C_{in, out}$ (tCO_2/TJ)	Available Resource S_i (TJ)	Energy Demand	Expected Consumption D_j (TJ)	Emission Limit, $D_jC_{in,j}$ (10^6 t CO_2)
Coal	105	600,000	Region I	1,000,000	20
Oil	75	800,000	Region II	400,000	20
Natural gas	55	200,000	Region III	600,000	60
Others[a]	0	>400,000			
Total		>2,000,000	Total	2,000,000	100

[a]*Refers to energy sources with zero or near-zero CO_2 emissions.*

vertical axis is then made. Figure 13.1 shows how the plot will look after plotting it for this case study.

The emission factor for each region can be read on the plot as the slope of each segment representing regions I, II, and III. The straight dotted line drawn from the origin represents all the demands for the three regions taken as a single demand without regard to their individual disaggregated emission limits.

THE ENERGY SUPPLY COMPOSITE CURVE

The procedure for constructing the Energy Supply Composite Curve is similar to that of the Energy Demand Composite Curve as demonstrated by Figure 13.2. The only exception is that the zero-carbon sources are not initially included when constructing the plot.

The Composite Curves are constructed in such a manner that their geometry reflected the cumulative nature of both emissions and energy. Both quantities can be read directly from the graphs.

AGGREGATED PLANNING WITH OVERALL EMISSION LIMITS (CASE 1)

The work illustrates a scenario in which it is assumed that the major concern for planning is only the total CO_2 emissions. In this case, the dotted line in Figure 13.1 is taken as the Energy Demand Composite Curve. Combining the two Composite Curves now provides the Pinch diagram in Figure 13.3.

The targeting is performed by the shifting Source Composite Curve horizontally to the right, until the two curves meet at the Pinch Point. The horizontal

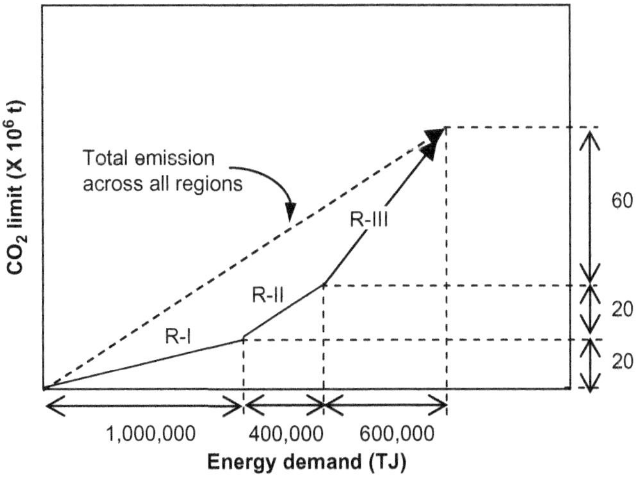

FIGURE 13.1

The Energy Demand Composite Curve (Tan and Foo, 2007).

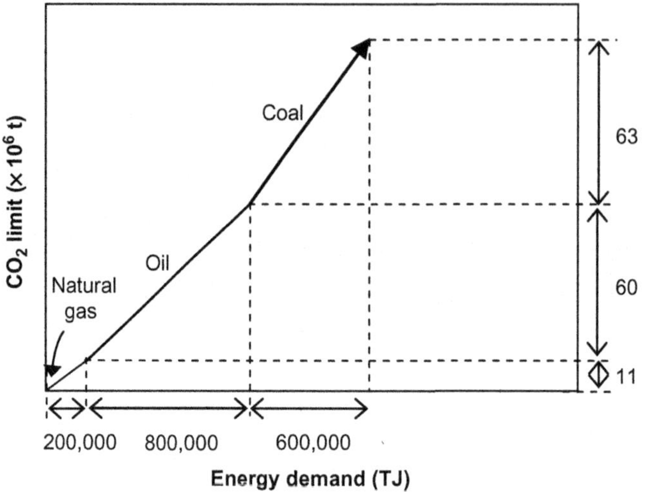

FIGURE 13.2

The Energy Supply Curve (Tan and Foo, 2007).

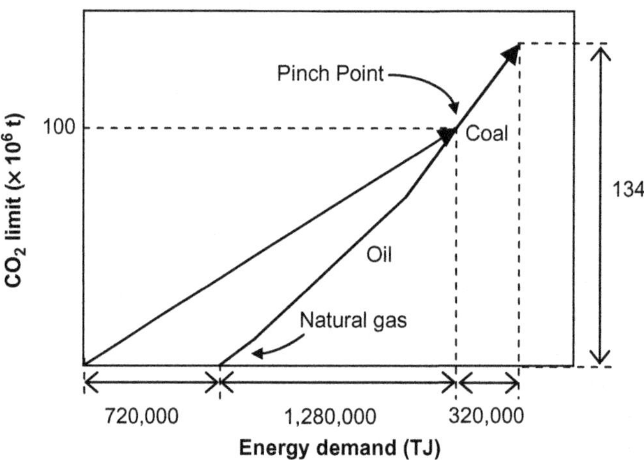

FIGURE 13.3

Targeting demand for zero-carbon energy (Case 1) (Tan and Foo, 2007).

distance that the Source Composite Curve was moved is now the amount of zero-carbon energy needed in the system. For this case it was found to be approximately 720,000 TJ.

In this particular planning scenario, total energy demand and combined emission limits of 2,000,000 TJ and 100 Mt of CO_2 were specified. Looking at the

Table 13.3 Energy Allocation for Cases 1 and 2

	Energy Resource				
	Coal	**Oil**	**Natural Gas**	**Others**	**Total**
Case 1					
Quantity (TJ)	280,000	800,000	200,000	720,000	2,000,000
Emissions (10^6 t CO_2)	29.4	60.0	11.0	0	100.4
Case 2					
Quantity (TJ)	90,000	800,000	200,00	910,000	2,000,000
Emissions (10^6 t CO_2)	9.5	60.0	11.0	0	80.5

plot, it can be seen that the coordinates of the Pinch Point correspond to these values. It can also be seen from the plot that there is an extension of the Source Composite Curve beyond that of the Demand Composite Curve. This represents the amount of excess energy that cannot be utilized if the set emission limits are to be met. The unusable energy in this case is represented by 320,000 TJ of coal. The resulting energy allocation for this planning scenario is given in Table 13.3.

From the plot and the resultant energy allocation table, it can be seen that the emission targets can only be achieved by omitting the portion of unusable coal; if it has to be utilized, then more zero-carbon energy will have to be used to compensate for the increased emission generated from the combustion of coal.

TARGETING EMISSIONS WITH FIXED ZERO-CARBON ENERGY SUPPLY

This scenario assumes that a fixed amount of zero-carbon energy (910,000 TJ) is available. The objective is to determine the lowest amount of CO_2 emissions that can be achieved. The introduction of the zero-carbon energy source results in a separation of the Source Composite Curve from the original Demand Composite Curve of case 1. The gap created between the two curves can be removed by moving the Demand Composite Curve downward (Figure 13.4).

The uppermost point of the shifted Demand Composite Curve now becomes the new emission limit. The overall emission limit for the three regions is then reduced by approximately 20% to 80 Mt of CO_2.

In this case, because there is a shift in the Pinch Point, the amount of unusable coal increases to approximately 510,000 TJ, corresponding to approximately 85% of the total available supply of coal. The utilization of coal could be increased by leaving either the natural gas or the oil unused instead, but the implication is that more zero-carbon energy must be utilized to compensate for the higher emission factor of coal. The energy allocation for this case is also presented in Table 13.3.

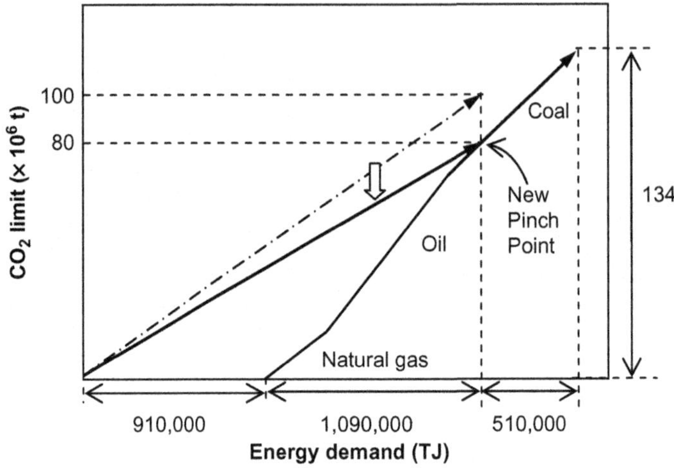

FIGURE 13.4

Effect of reduced emission limit on zero-carbon energy demand (Case 2) (Tan and Foo, 2007).

CARBON FOOTPRINT IMPROVEMENT BASED ON PINCH ANALYSIS

An approach to determine the strategies to reduce carbon footprints using the Pinch Analysis concept was developed by Tjan et al. (2010). The approach is based on segregating the total footprint of a company into material-based and energy-based components. The approach was developed to be a simpler form of LCA. The authors designed a graphical tool that combines the use of Pinch Analysis and the LCA technique to produce a Carbon Footprint Composite Curve. Depending on the application being considered, the segregation could be based on the internal and external components rather than the material-based or energy-based components. The components were then plotted as the Source Composite Curve using the approach of Tan and Foo (2007), with economic value on the horizontal axis and CO_2 emissions on the vertical axis. A benchmark carbon intensity goal was then set at a value lower than the plant's current total carbon intensity, and this served as the Demand Composite Curve.

The overall carbon intensity of the process can be determined by drawing a diagonal line from the origin to the end point of the Source Composite Curve and the slope of the line is equivalent to the overall carbon intensity of the process, combining both internal and external components (Tahara et al., 2005).

To determine strategies for reducing the carbon intensity from the current level to a desired value, the author considered two scenarios. Figure 13.5 shows a Carbon Footprint Composite Curve plot based on the internal and external components. In the first scenario, the external carbon intensity is lower than the

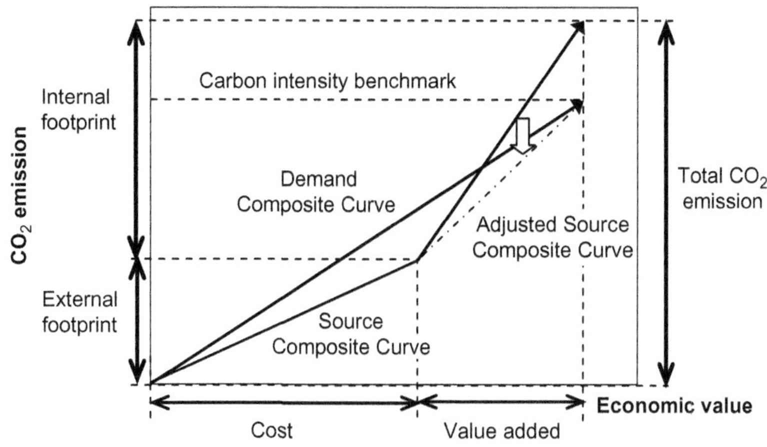

FIGURE 13.5

Carbon Footprint Composite Curves (Scenario 1) (Tjan et al., 2010).

internal carbon intensity, and the strategy that would be most appropriate is to reduce the internal carbon footprint of the plant. To visualize this graphically, a reduction in the slope of the internal carbon footprint segment of the Source Composite Curve until it touches the Demand Composite curve, and this gives an idea of how much of the internal carbon footprint needs to be reduced to achieve the set benchmark level. In the second scenario presented, the external carbon footprint has a much higher value than the internal carbon intensity. This gives a situation in which the Source Composite Curve lies entirely above the Demand Composite Curve, which is the benchmark value for the overall carbon intensity of the plant. Shifting the internal carbon footprint component diagonally along the external carbon footprint segment of the Source Composite Curve until the tip of the Composite Curve touches the carbon intensity benchmark as shown in Figure 13.6 results in the reduction of the external footprint of the product and corresponds to an increase in the internal footprint.

Practically, this approach offers companies with low internal carbon footprints a route through which they can best reduce their overall carbon intensity by reducing the use of external inputs per products. Tjan et al. (2010) further demonstrated their approach by implementing two industrial case studies to show the applicability of the approach.

CARBON EMISSION PINCH ANALYSIS

Given a fixed number of stationery CO_2 emission point sources (e.g., flue gas stacks releasing CO_2) and available carbon demands for an industrial park (IP),

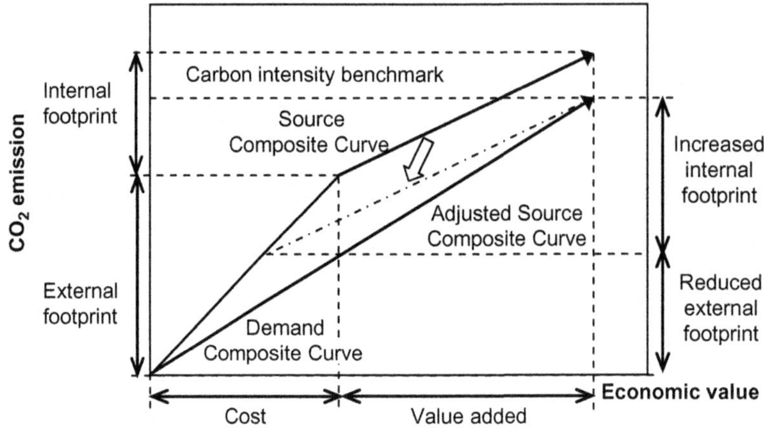

FIGURE 13.6

Carbon Footprint Composite Curves (Scenario 2) (Tjan et al., 2010).

the holistic planning and retrofitting of an IP to achieve the minimum carbon emission involves determining answers to the following questions.

1. What would be the:
   a. Maximum possible amount of carbon exchange?
   b. Minimum carbon emission target?
   c. Minimum fresh carbon supply?
2. What possible and practical carbon demands should be introduced?
3. How can emissions be minimized via process changes?
4. How can an optimal carbon allocation network be designed?

There are two possible cases for CO_2 planning and retrofitting of an IP:

- The retrofit case. This case involves *existing* IPs that include processes with fixed *pure* CO_2 demands.
- The planning case. This case involves IPs where processes with CO_2 demands are introduced to achieve the emission targets of the IPs. The economics of adding processes demanding pure CO_2 also need to be addressed.

The carbon emission Pinch Analysis (CEPA) introduced by Sadiq et al. (2012) is a holistic graphical methodology for IP carbon planning that addresses the aforementioned issues. There is also an algorithm methodology called the Generic Carbon Cascade Analysis introduced by Manan et al. (2014). For CEPA, the methodology first identifies potential carbon emission point sources to be integrated with carbon demands. Next, the maximum carbon exchange potential and the minimum carbon targets are established by prioritizing process changes options using the carbon management hierarchy (CMH) as a guide. Generic SDCs are used as a visualization tool for the carbon emission planning of new IPs and for retrofitting of existing IPs. A CEPA retrofit case is described next using an illustrative case study.

STEP 1: IDENTIFY THE CARBON SOURCES AND DEMANDS

The CO_2 point sources are identified first. Typical flue gas sources include furnace and boiler stacks, flare and purge systems, and drying towers. Depending on the type of fuel used, and the combustion efficiency, the flue gas sources may contain N_2, O_2, CO_2, CO, NO_X, and SO_X. Appropriate CO_2 demands are then introduced. The limiting data to be extracted from the source stream are the flue gas flow rate and the composition of CO_2 in the flue gas. The flow rate of gases that can satisfy the demand processes and the limiting CO_2 composition are the required data for CO_2 demands.

It may not be possible to reuse all of the available carbon sources, especially those containing small CO_2 concentrations, hazardous substances, and/or those that are remotely located within an IP. The capital and operating costs involved in recycling these carbon sources can be very high, such that they may not be cost-effective to be used for carbon exchange (van Straelen et al., 2010). Only sources that do not pose any geographical, safety, and cost constraints and that can reasonably contribute to the emission reduction of the IP in terms of quantity (flow rate) and quality (mass load) can be considered for possible carbon exchange. The remaining sources will be treated and released to the atmosphere as needed.

Once the CO_2 sources have been selected, the appropriate CO_2 demands that can accept the emission sources are introduced. This step involves a detailed survey as well as careful planning and selection of the appropriate and profitable CO_2-consuming processes to be introduced within the IP. Examples of such processes are syngas production, microalgae cultivation, enhanced oil recovery (EOR), methanol, and beverage production. Table 13.4 shows the data for CO_2 sources and demands for case study 1.

Table 13.4 Data for an Illustrative Case Study 1

Sources	F_T (t/h)	CO_2 Composition (%)	F_{CO2} (t/h)	F_{OG} (t/h)	Cum F_{CO2} (t/h)	Cum F_{OG} (t/h)
S_1	50	80	40	10	40	10
S_2	40	70	28	12	68	22
Demands						
D_1	45	85	38.25	6.75	38.25	6.75
D_2	35	65	22.75	12.25	61	19

STEP 2: TARGETING THE MAXIMUM CARBON EXCHANGE

Step 2 involves CO_2 targeting using the generic SDCs. Previous works dealing with mass, gas, and Property Pinch Analysis have plotted the mass flow rate of the main carrier fluid (e.g., gas streams such as nitrogen and hydrogen) versus mass load/concentration of contaminants or minor components present in a process stream (Wan Alwi et al., 2009). This approach does not apply to gaseous streams containing CO_2 because CO_2 may not be the main carrier fluid. A flue gas stream may contain CO_2 and a larger proportion of other gases, including N_2, O_2, CO, NO_x, and SO_x. Note that using the conventional plot of primary gas flow rate versus the mass load of contaminant for CO_2 targeting will yield the minimum *flue gas* instead of the minimum carbon emission target. Therefore, the generic SDC has been used in this work to suit the unique carbon exchange problem.

In case study 1, the SDC plots the cumulative flow rate of other gases (F_{OG} in the flue gas, apart from CO_2) versus the cumulative flow rate of CO_2 (F_{CO2}) for the sources and demands. The values of F_{OG}, cumulative F_{CO2}, and cumulative F_{OG} are shown in columns 5–7 of Table 13.4. To get the minimum fresh CO_2 and emission flow rate targets, the entire Source Curve is moved horizontally until it pinches the Demand Curve (Figure 13.7). The horizontal gap between the Demand and Source Curves at zero F_{OG} gives the minimum fresh CO_2 flow rate of 11.25 t/h. The overshoot of the Source Curve gives the minimum CO_2 emission flow rate of 18.25 t/h. Because the quantity of other gases is also quite high, the concentration is not represented in ppm. The gradient is the ratio between tons of other gases and tons of CO_2 gas. The Pinch Point for case study 1 is at the F_{CO2} value of 38.25 t/h.

Figure 13.1 shows that maximizing carbon exchange via integration of emission sources with carbon demands yields a potential emission reduction of 73%

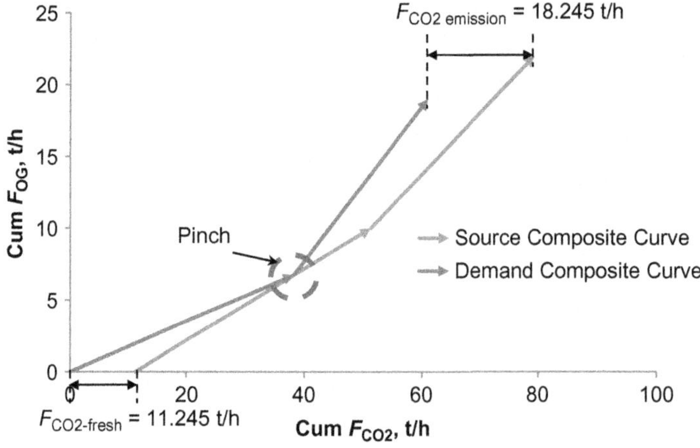

FIGURE 13.7

Maximum carbon exchange targets using SDC.

(i.e., from 68 to 18.25 t/h). The preceding approach for maximum carbon exchange targeting applies to the retrofit case when there are existing demands for pure CO_2 in the IP.

STEP 3: SETTING THE HOLISTIC MINIMUM CARBON TARGETS

Further reductions in the fresh CO_2 as well as the carbon emission targets obtained from step 2 can be achieved by exploring options for process changes using the CMH as a guide. The CMH and the heuristics to guide process planning to achieve the holistic minimum carbon targets are described next.

The Carbon Management Hierarchy (CMH)

Figure 13.8 shows the CMH that is used to guide a designer/planner to screen and prioritize process changes options to achieve the holistic minimum carbon targets for an IP. The hierarchy consists of four levels. Direct reuse at the top of the hierarchy is the most preferred level, and carbon sequestration at the bottom is the least preferred level.

Level 1: Direct reuse

CO_2 emission point sources may be directly reused (without any treatment) to satisfy various CO_2 demands. Note that the maximum carbon exchange target is given by the total flow rate of all emission point sources that can be directly reused and integrated with various demand processes.

For holistic carbon planning, the *direct reuse* should be the top-priority carbon management option. Note that a piece-meal implementation of process changes would typically favor the source elimination and reduction options (level 2) as the top-priority option. Figure 13.9 demonstrates the advantage of implementing the

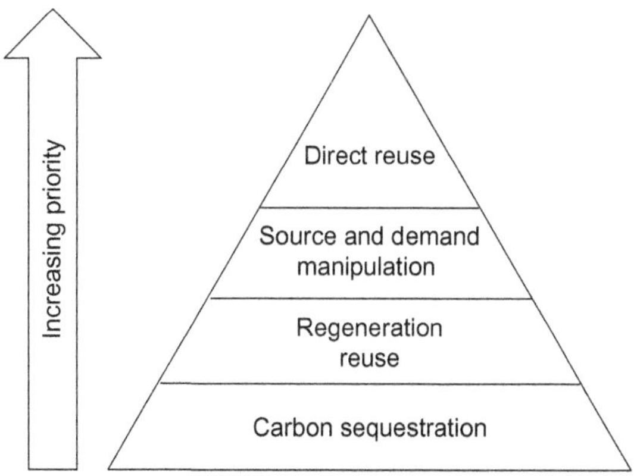

FIGURE 13.8

The Carbon Management Hierarchy (CMH).

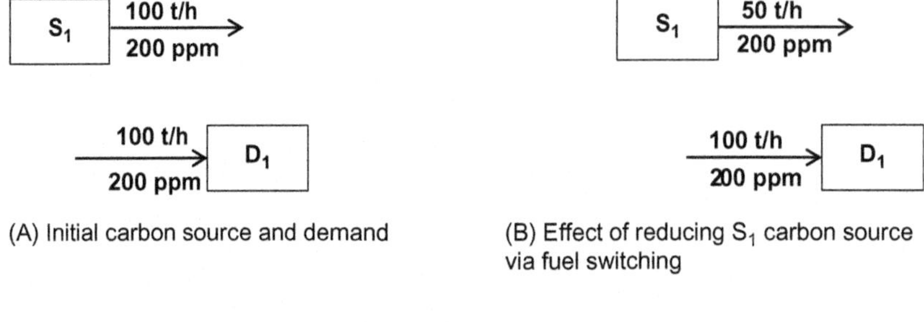

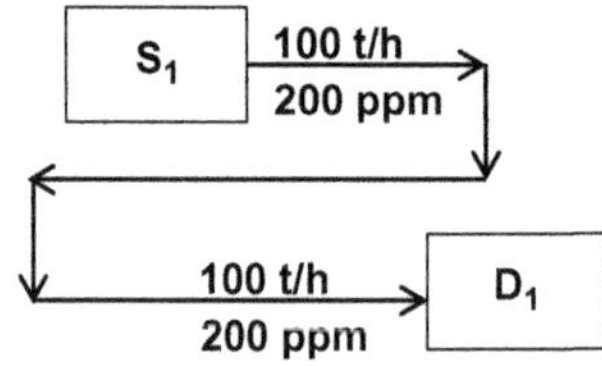

(C) Direct reuse of a carbon source to satisfy a carbon demand

FIGURE 13.9

Effect of source reduction (B) versus direct (C) on carbon targets as compared to the baseline (A).

direct reuse before the elimination option within a holistic and integrated context. Consider a 100 t/h carbon source and assume that microalgae that requires 100 t/h of carbon demand is also available. Figure 13.9B shows that fuel switching can only reduce the carbon source emission from 100 to 50 t/h while the demand is still maintained at 100 t/h. Instead, Figure 13.9C shows that direct reuse of the carbon emission source to cater to the microalgae CO_2 demand has managed to eliminate both the pure external carbon demand and the carbon emission source to 0 t/h. Therefore, to achieve the minimum carbon emission as well as fresh supply, the *direct reuse* should be the highest-priority process change option in the context of the holistic carbon planning.

Level 2: Source and demand manipulations

Level 2 of the CMH includes options for source elimination and reduction as well as demand manipulation (demand elimination, reduction, and generation) as described here.

Source elimination and reduction. Examples of source elimination and reduction options are fuel switching, reaction chemistry substitution, process parameter changes, and equipment modifications. Emission sources may be eliminated or reduced when these options are applied using heuristic H_1 and the SDC as guides:

> H_1: above the Pinch, apply process changes to eliminate or reduce carbon source streams.

Demand manipulation. Demand manipulation involves applying strategies to change the size of carbon demands. The flow rates of process demands may be changed to reduce emission or to eliminate the fresh CO_2 requirement, depending on the demand location relative to the Pinch Point, new demands may also be introduced. In overall demand planning, demand manipulation provides the degree of freedom for designers to further reduce emissions within an IP. Two heuristics are introduced to systematically guide demand manipulations:

H_2: increase the limiting flow rate or add new demand processes above the Pinch; above the Pinch, increase the limiting flow rates of demand processes to reduce CO_2 emissions.

H_3: reduce or eliminate demand processes below the Pinch; below the Pinch, fresh CO_2 can be reduced by reducing or eliminating demand processes, particularly for the CO_2 planning case. However, this option may not be applicable to the *retrofit case* when the fresh CO_2 demand exists and is an essential feature of an IP.

Level 3: Regeneration—reuse

The quality of CO_2 emission sources can be upgraded before being reused, for example, via chemical absorption using amine-based solvent (regeneration) according to the heuristic:

H_4: regenerate source stream to achieve purity higher than the Pinch purity.

The quality of CO_2 sources above the Pinch may be upgraded via regeneration across the Pinch. The regenerated source can then be integrated with a demand below the Pinch to reduce CO_2 emissions above the Pinch, as well as the fresh CO_2 requirement below the Pinch. Note that regeneration of CO_2 emission sources should only be performed when higher-priority carbon reduction strategies have been implemented.

Level 4: Carbon sequestration

CO_2 sequestration or CO_2 capture and storage should only be considered when all other higher-level CMH options have been explored. This is the least preferred carbon management strategy. Examples of CO_2 sequestration include tree planting and storage of CO_2 in exhausted oil wells and in deep saline aquifers.

STEP 4: MINIMUM CARBON NETWORK DESIGN

Designing the carbon allocation network to achieve the holistic minimum carbon target is the final step of the holistic carbon planning methodology. This is achieved by applying the resource allocation network design technique of Wan Alwi et al. (2009).

IP CASE STUDY

The step-wise application of the holistic carbon planning methodology in an IP is demonstrated in this section. The key processes listed in the preliminary IP master plan include a coal-fired power station, an ammonia production plant, a fermentation plant, and a bioenergy plant (biomass-based). To achieve a targeted carbon emission limit, the IP planner has the flexibility to further include profitable industries into the IP.

Step 1: Data extraction

Processes within the IP are assumed to operate 24 h/day, 330 d/y.

Source data extraction

The major CO_2 emission point sources extracted from the IP are shown in Table 13.5. The source data are arranged in descending order of CO_2 composition. The data taken are based on the work of Metz et al. (2005).

Carbon demands planning

Various CO_2-consuming processes were carefully studied, and four promising demand processes with the potential to reduce the CO_2 emissions of the IP were proposed. These include a beverage manufacturing plant consuming food-grade CO_2, methanol production plant, microalgae cultivation, and EOR. Detailed descriptions of the demand processes and their roles in reducing carbon emission within the IP are described next.

i. *Carbonated Beverage Manufacturing Plant*
   A carbonated beverage plant typically consumes CO_2 with purity as high as 99.9%. Carbonated beverage plants are usually built within IPs that feature refineries, petrochemicals, and gas processing plants because of the various sources of CO_2 that are readily available with a wide range of qualities. Treating these CO_2 sources on-site is relatively cheaper than buying food-grade CO_2 because transportation costs are eliminated.

Table 13.5 Data for CO_2 Sources for the IP Case Study

Sources	F_T (t/h)	CO_2 Composition (%)	F_{CO2} (t/h)	F_{OG} (t/h)	Cum F_{CO2} (t/h)	Cum F_{OG} (t/h)
Fermentation (S_1)	25	100	25	0	25	0
Power station: coal-fired boilers (S_2)	3,000	14	420	2,580	445	2,580
Ammonia production (S_3)	375	8	30	345	475	2,925
Bioenergy from biomass	563	8	45	518	520	3,443

ii. *Methanol Production Plant*
Methanol production using CO_2 from flue gas has been investigated by Mignard et al. (2003). Musabbeh and Shammari (2006) observed a 20% increase in methanol production in their feasibility study to enhance methanol production via CO_2 injection using CO_2 from flue gases. The approach managed to enhance the productivity and reduced the CO_2 emissions of the methanol production plant. The use of flue gas in CO_2 reforming resulted in an energy-efficient methanol plant and helped to optimize the synthesis gas composition (Aasberg-Petersen et al., 2008).

iii. *Microalgae Cultivation*
CO_2 fixation by photoautotrophic algal cultures can potentially curb the atmospheric release of CO_2 to help mitigate global warming (Ono and Cuello, 2003). High-rate algae ponds can potentially produce 17,000–28,000 L of transportation fuel per hectare per year (Putt et al., 2011). This estimate is based on an average specific productivity of 20 g of dry biomass per m^2 per day and a 300-day growing season with a production capacity of more 59 t of dry biomass per hectare per year (Brennan and Owende, 2010). Producing 100 t of algal biomass fixes 183 t of CO_2 (Chisti, 2008). Because the different algal species have been known to tolerate CO_2 and other components in flue gas (Wang et al., 2008), microalgal biomass production can potentially make use of the CO_2 released from the various stationary emission sources in an IP. Combining wastewater treatment with CO_2 fixation to cultivate microalgae for biodiesel production is therefore advantageous (Muñoz and Guieysse, 2006). In addition to providing a rich source of renewable biodiesel microalgae production can function as a secondary wastewater treatment process.

iv. *Enhanced Oil Recovery*
The CO_2 from emission point sources can also be reused in EOR. The use of CO_2 emissions from industrial processes in EOR is gaining wide acceptance because of its ability to reduce emissions while improving the recovery of oil. With CO_2-EOR, significant volumes of CO_2 emissions can be stored while increasing domestic oil production (Ferguson et al., 2009). Estimates of the world storage capacity through EOR range between 73.3 billion ($\times 10^9$) t CO_2 and 238.8 billion ($\times 10^9$) t CO_2 (Gaspar Ravagnani et al., 2009). The authors attributed the success of any EOR project to the global oil price, oil production, and capital expenditure. The relatively low cost and technological know-how associated with this approach make it more attractive (Lackner et al., 2010). Previous research also suggested that CO_2 storage in depleted oil and gas reservoirs has the least potential environmental risk (Anderson and Newell, 2003). This further highlights the advantages and feasibility of CO_2 emission reduction via EOR. The stability of stored CO_2 and the probability of its leakage after storage were investigated by van der Zwaan and Smekens (2009). They performed a detailed sensitivity analysis for the

Table 13.6 Data for CO_2 Demand Processes

Demands	F_T (t/h)	CO_2 Composition (%)	F_{CO2} (t/h)	F_{OG} (t/h)	Cum F_{CO2} (t/h)	Cum F_{OG} (t/h)
Beverage plant (D_1)	100	99	99	1	99	1
EOR (D_2)	500	80	400	100	499	101
Methanol production (D_3)	300	50	150	150	649	251
Microalgae cultivation (D_4)	200	10	20	180	669	431

CO_2 leakage rate in a geological formation. The analysis concluded that, to constitute a meaningful climate change mitigation option, a 0.1%/y leakage rate is acceptable for CCS.

The location of potential sites CCS and reuse for EOR are also important to consider. Promising potential sites include petroleum refineries, fertilizer and hydrogen production facilities, gas processing facilities, and power stations. Iijima and Kamijo (2003) conducted a detailed economic analysis of CO_2-EOR plants in the Middle East, focusing on the capacity of the CO_2 recovery unit, utility cost, pipeline cost, and other operational requirements. They found that the arrival cost of CO_2 at the oil field was slightly more than 1 USD/MSCF for CO_2 sources within 100–300 km of oil fields. It can be concluded that within this range of proximity of CO_2 sources to the oil fields, CO_2 recovered from flue gas sources can replace the injection of natural gas into oil reservoirs.

Table 13.6 presents the data for the demand processes under the carbon-planning scenario (as opposed to the retrofit scenario). The demand data were carefully planned and selected by considering the feasibility of using CO_2 from the IP point source emissions.

Step 2: Setting the holistic minimum carbon targets

The holistic minimum carbon targets were determined by implementing all the carbon reduction options using the CMH levels as a guide.

CMH Level 1: Direct reuse

Level 1 is the highest CMH level. This level can be referred to as the maximum carbon exchange targeting stage because it involves the direct reuse of carbon sources to satisfy carbon demands. The SDCs in Figure 13.10 were plotted using the IP source and demand data in Tables 13.5 and 13.6. The SDC Pinch Point occurs at cum F_{CO2} of 649 t/h and cum F_{OG} of 251 t/h. The minimum fresh CO_2 flow rate is 583.14 t/h, and the minimum CO_2 emissions is 434.14 t/h. Maximizing carbon exchange allows a reduction of up to 16.5% emissions from a total CO_2 flow rate of more than 520 t/h.

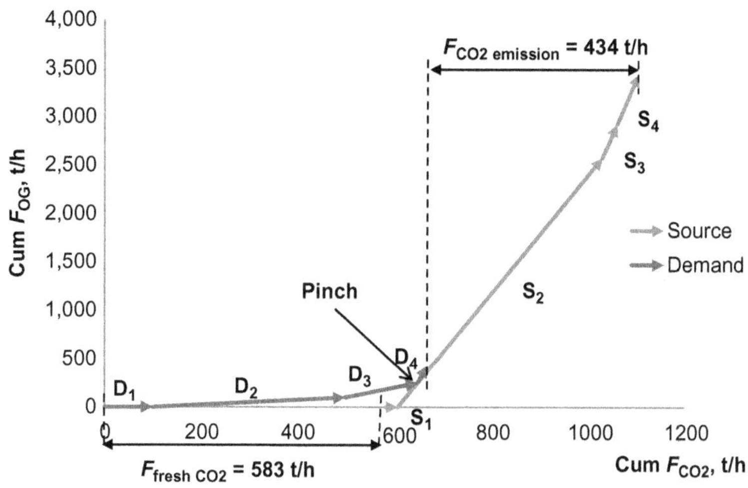

FIGURE 13.10

Targeting the maximum carbon exchange—CMH—direct reuse level.

Table 13.7 Data for CO_2 Sources for the IP Case Study after Reducing S_2

Sources	F_T (t/h)	CO_2 Composition (%)	F_{CO2} (t/h)	F_{OG} (t/h)	Cum F_{CO2} (t/h)	Cum F_{OG} (t/h)
Fermentation (S_1)	25	100	25	0	25	0
Power station: coal-fired boilers (S_2)	2,000	14	280	1,720	305	1,720
Ammonia production (S_3)	375	8	30	345	335	2,065
Bioenergy from biomass	563	8	45	518	380	2,583

CMH Level 2: Source and demand manipulations

Further reductions in fresh CO_2 as well as CO_2 emissions were achieved by elimination and reduction of the CO_2 sources, and by manipulation of the demands.

a. *Source Elimination and Reduction*
   Referring to the SDC in Figure 13.10, it can be seen that the point source emissions S_3–S_4 and part of S_2 are above the Pinch Point. According to heuristic H_1, apply process changes to eliminate or reduce carbon source streams above the Pinch. In line with heuristic H_1, the S_2 power station flue gas generation capacity from the coal-fired boiler can be reduced to 3,000–2,000 t/h (assuming the power plant has performed load-shifting operations). Table 13.7 presents the new source data after implementing this process change.

Figure 13.11 shows the SDC with the total CO_2 emissions reduced from 434.14 to 294.14 t/h after implementing heuristic H_1. The Pinch Point and the fresh CO_2 targets remained the same.

The preceding example demonstrates the importance of systematically implementing process changes based on the knowledge of the Pinch Point locations while using the holistic CMH framework as a guide to effectively reduce emissions. Inappropriate application of process changes may result in excess emissions instead of emissions reductions. For example, reducing the flow rate of an emission source below the Pinch Point will result in an increase in the fresh CO_2 target without any reduction in CO_2 emissions. In addition, opportunities to reduce both fresh CO_2 and CO_2 emissions may be missed if the direct reuse (level 1 of the CMH) were applied after source elimination and reduction strategies (level 2 of the CMH).

b. *Demand Manipulations*

CO_2 flow rate targets can be reduced further by manipulating CO_2 demands using heuristics H_2 and H_3.

- H_2: increase demand flow rate above the Pinch.

Above the Pinch, CO_2 emissions can be reduced by increasing the demand's limiting flow rate subject to constraints such as the market demand and the maximum flow rate of available sources. This rule is applicable for the IP retrofit as well as for planning/design cases.

Figure 13.11 shows that, above the Pinch, worthwhile process changes can be achieved by increasing the CO_2 demands for methanol production (D_3) and microalgae cultivation (D_4). However, the maximum flow rate of D_3 is limited to 300 t/h as per the market demand. However, the maximum

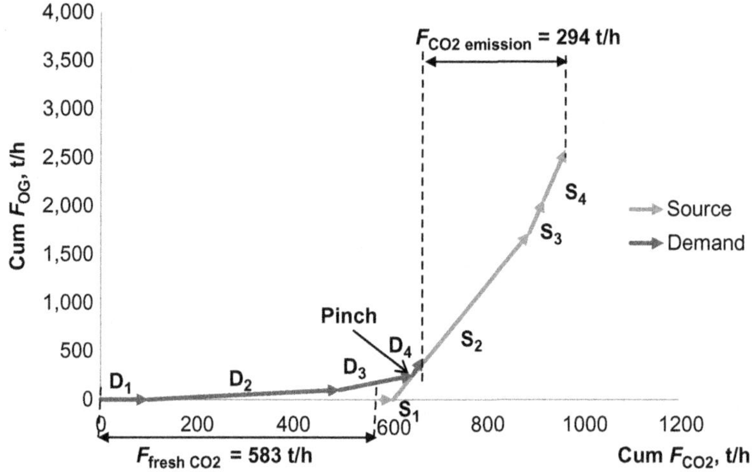

FIGURE 13.11

SDC after implementing source reduction process changes according to heuristic 1.

flow rate of D_4 is 300 t/h because of the limited space available for microalgae cultivation in the water treatment plant. Table 13.8 shows the new limiting data after demand manipulations.

The modified sources and demands data were plotted to yield the new SDC shown in Figure 13.12. Above the Pinch, increasing demand D_4 led to 3.4% reduction in the site CO_2 emissions (i.e., from 294.14 to 284.14 t/h). Because the changes only affect the demand above the Pinch, the Pinch Point and the fresh CO_2 target remained the same.

Below the Pinch, heuristic 3 can be used to guide demand manipulation under the IP planning case.

- H_3: reduce/eliminate demands below the Pinch.

Demand D_1 (beverage plant), which typically requires expensive food-grade CO_2, should be the prime candidate to be eliminated or reduced to decrease the investment cost. The limiting data after elimination of demand D_1 below the

Table 13.8 Data for CO_2 after Demand Manipulations of D_4

Demands	F_T (t/h)	CO_2 Composition (%)	F_{CO2} (t/h)	F_{OG} (t/h)	Cum F_{CO2} (t/h)	Cum F_{OG} (t/h)
Beverage plant (D_1)	100	99	99	1	99	1
EOR (D_2)	500	80	400	100	499	101
Methanol production (D_3)	300	50	150	150	649	251
Microalgae cultivation (D_4)	300	10	30	270	679	521

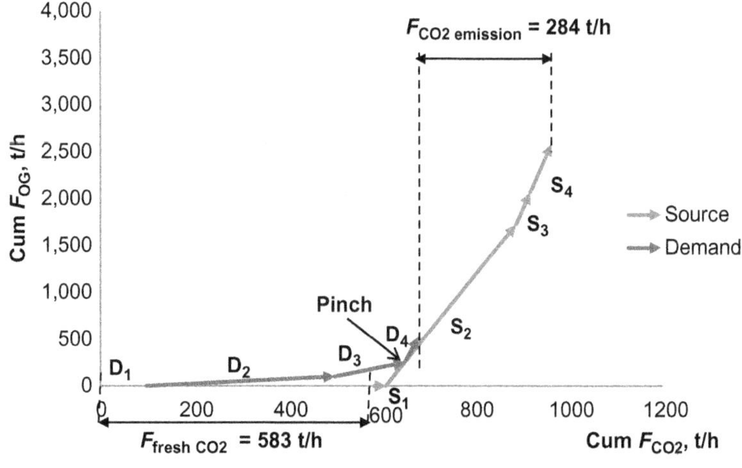

FIGURE 13.12

SDC after demand manipulations according to heuristic 2.

Table 13.9 Data for CO_2 after Demand Manipulations of D_1

Demands	F_T (t/h)	CO_2 Composition (%)	F_{CO2} (t/h)	F_{OG} (t/h)	Cum F_{CO2} (t/h)	Cum F_{OG} (t/h)
EOR (D_2)	500	80	400	100	400	100
Methanol production (D_3)	300	50	150	150	550	250
Microalgae cultivation (D_4)	300	10	30	270	580	520

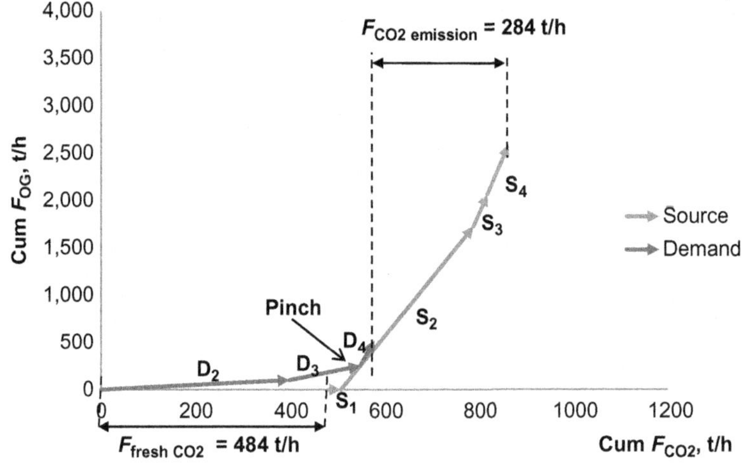

FIGURE 13.13

SDC after eliminating demand D_1 below the Pinch.

Pinch is shown in Table 13.9. D_1 elimination managed to reduce the fresh CO_2 target by 17%. Figure 13.13 shows the resulting SDC with the new Pinch Point at a CO_2 flow rate of 550 t/h, a new fresh CO_2 target of 484.30 t/h, and a new CO_2 emission target of 284.30 t/h.

CMH Level 3: Regeneration—reuse

The fresh CO_2 supplied by *regenerating* or upgrading a source stream above the Pinch so that it can be *reused* as fresh CO_2 below the Pinch. For the IP undergoing study, regenerating S_2 to 99% purity can reduce the targets for fresh CO_2 as well as for CO_2 emissions. The stream data after regeneration of the S_2 emission source from the cogeneration plant are shown in Table 13.10.

Figure 13.14 shows the SDC after S_2 has been regenerated and reused to meet the fresh CO_2 requirement below the Pinch. Regeneration (across the Pinch) has managed to reduce the fresh CO_2 requirement from 484.30 to 200.05 t/h. At the same time, CO_2 emissions were reduced from 284.30 to 30.05 t/h. To achieve the targeted reductions, an appropriate treatment unit has to be installed to concentrate S_2 to 99% purity of CO_2.

Table 13.10 Data for CO_2 Sources for the IP Case Study after Regenerating S_2

Sources	F_T (t/h)	CO_2 Composition (%)	F_{CO2} (t/h)	F_{OG} (t/h)	Cum F_{CO2} (t/h)	Cum F_{OG} (t/h)
Fermentation (S_1)	25	100	25	0	25	0
Power station: coal-fired boilers (S_2)—regenerated	283	99	280	3	305	3
Ammonia production (S_3)	375	8	30	345	335	348
Bioenergy from biomass	563	8	45	518	380	865

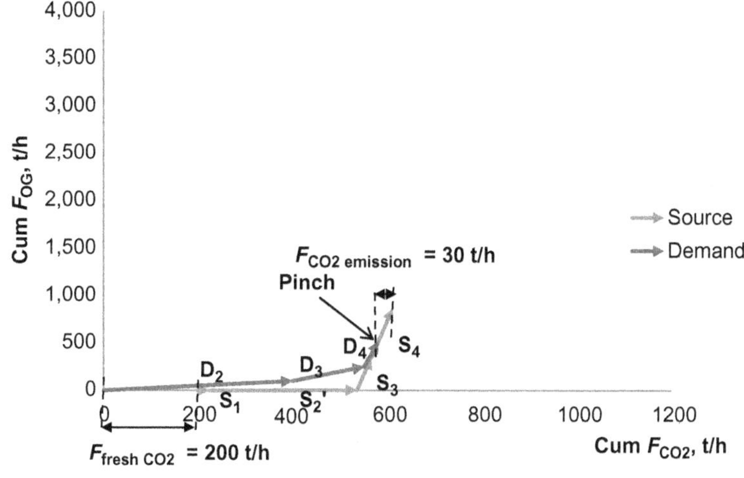

FIGURE 13.14

SDC after regeneration source S_2.

CMH Level 4: Carbon sequestration

Carbon sequestration involves the capture and storage of carbon emissions (or CCS) from the atmosphere via physical and biological processes. CCS should be considered only when all higher-level CMH options have been implemented. CCS strategies such as planting trees and CO_2 storage in deep saline aquifers or exhausted oil wells have been used. For the IP case study, tree planting was proposed as the strategy to sequester carbon emissions.

The national average for urban forest carbon sequestration density in the United States was found to be 25.1 versus 53.5 t/ha for forests (Nowak and Crane, 2002). A study of urban forests by Zhao et al. (2009) in Hangzhou, China, revealed that 1,328,166.55 t/y of carbon are sequestered at a sequestration rate of 1.66 t/ha/y of carbon. The typical maximum rate of carbon sequestration by well-stocked trees in forests and plantations occurs between the ages of 10 and 30 y. As an indication, at the age of 30 y, approximately 54.5–141.69 t of CO_2 are sequestered per hectare in forests with productivity levels ranging from low to high (Zhao et al., 2009).

For the IP undergoing study, 30.05 t/h of CO_2 are still emitted to the atmosphere after the CMH regeneration stage. This translates to 238,034 t CO_2/y of emissions. Assuming a forest with a low sequestration rate of 54.5 t/ha, a total forest area of 4,368 ha is required to enable the IP to sequester the 238,034 t CO_2 of annual emissions.

The SDC in Figure 13.15 shows the holistic minimum carbon targets achieved through implementation of all the CMH-guided process changes. The effects of

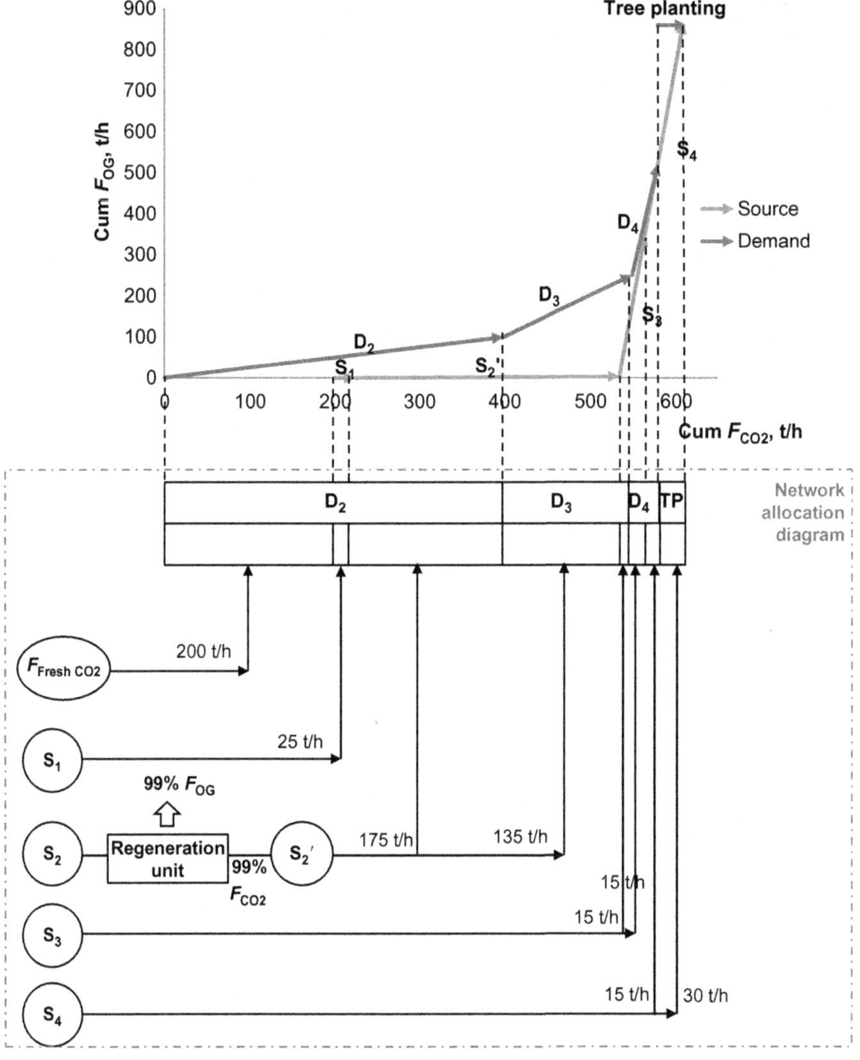

FIGURE 13.15

Final NAD for the IP.

the CMH-guided changes on the minimum CO_2 targets and the Pinch Point location are also shown.

Step 3: Carbon distribution network design

The final step involves designing a carbon distribution network to achieve the holistic minimum carbon targets. The network allocation diagram (NAD) proposed by Wan Alwi et al. (2009) was used for this purpose. The NAD technique involved the graphical mapping of CO_2 sources with CO_2 demands based on the CO_2 concentration and flow rate. The source demand mapping was performed within each SDC segment that appeared between vertical lines that marked the beginning of a new source or a demand line (Figure 13.15). The boxes below the SDC represent the quantity of CO_2 sources to satisfy the CO_2 demands. Figure 13.15 shows the carbon distribution network that yields the holistic minimum carbon targets for the IP. Tree planting was chosen as a measure to absorb the remaining 30.03 t/h CO_2 emissions.

CONCLUSION

The supply and demand planning and management tools as well as the end-of-pipe solution for carbon emission reduction from stationary point sources have been described in this chapter. The techniques presented for carbon supply and demand planning include carbon Pinch Analysis design method, carbon footprint improvement with Pinch Analysis, and CEPA.

REFERENCES

Aasberg-Petersen, K., Nielsen, C.S., Dybkjaer, I., Perregaard, J., 2008. Large Scale Methanol Production from Natural Gas. Haldor Topsoe, Lyngby, Denmark, p. 1–14.

Anderson, S., Newell, R., 2003. Prospects for Carbon Capture and Storage Technologies. Resources for the Future Press, Washington, DC, pp. 20–35.

Atkins, M.J., Morrison, A.S., Walmsley, M.R.W., 2010. Carbon Emissions Pinch Analysis (CEPA) for emissions reduction in the New Zealand electricity sector. Appl. Energy 87 (3), 982–987.

Brennan, L., Owende, P., 2010. Biofuels from microalgae—a review of technologies for production, processing, and extractions of biofuels and co-products. Renew. Sustainable Energy Rev. 14 (2), 557–577.

Chisti, Y., 2008. Biodiesel from microalgae beats bioethanol. Trends Biotechnol. 26 (3), 126–131.

Clarke, S.C., 2001. CO_2 management—a refiners perspective. Digital Refining. Retrieved from <www.digitalrefining.com> (accessed 18.04.2014).

Diamante, J.A.R., Tan, R.R., Aviso, K.B., Bandyopadhyay, S., Ng, D.K.S., Foo, D.Y., 2013. A graphical approach for pinch-based source-sink matching and sensitivity analysis in carbon capture and storage (CCS) systems. Ind. Eng. Chem. 52, 7211–7222.

Diamante, J.A.R., Tan, R.R., Foo, D.Y., Ng, D.K.S., Aviso, K.B., Bandyopadhyay, S., 2014. Unified pinch approach for targeting of carbon capture and storage (CCS) systems with multiple time periods and regions. J. Cleaner Prod. 71, 67–74.

Ferguson, R.C., Nichols, C., Leeuwen, T.V., Kuuskraa, V.A., 2009. Storing CO_2 with enhanced oil recovery. Energy Procedia. 1 (1), 1989–1996.

Gaspar Ravagnani, A.T.F.S., Ligero, E.L., Suslick, S.B., 2009. CO_2 sequestration through enhanced oil recovery in a mature oil field. J. Pet. Sci. Eng. 65 (3–4), 129–138.

Gale, J., 2002. Overview of CO_2 emissions sources, potential, transport and geographical distribution of storage possibilities. Proceedings of the workshop on CO_2 dioxide capture and storage, Regina, Canada, 18–21 November 2002, pp. 15–29.

Hashim, H., Douglas, P., Elkamel, A., Croiset, E., 2005. Optimization model for energy planning with CO_2 emission considerations. Ind. Eng. Chem. Res. 44 (4), 879–890.

Hohmann, E., 1971. Optimum Networks for Heat Exchange (PhD Thesis). University of Southern California, Los Angeles, CA, USA.

Iijima, M., Kamijo, T., 2003. Flue gas CO_2 recovery and compression cost study for CO_2 enhanced oil recovery. In: Gale, J., Kaya, Y. (Eds.), Greenhouse Gas Control Technologies—Sixth International Conference. Pergamon Press, Oxford, UK, pp. 109–114.

Lackner, K.S., Park, A.H.A., Miller, B.G., 2010. Eliminating CO_2 emissions from coal-fired power plants. Generating Electricity in a Carbon-Constrained World. Academic Press, Boston, MA, USA, pp. 127–173.

Lee, S.C., Sum Ng, D.K., Yee Foo, D.C., Tan, R.R., 2009. Extended pinch targeting techniques for carbon-constrained energy sector planning. Appl. Energy 86 (1), 60–67.

Manan, Z.A., Alwi, S.R.W., 2007. Water pinch analysis evolution towards a holistic approach for water minimization. Asia-Pac J. Chem. Eng. 2 (6), 544–553.

Manan, Z.A., Wan Alwi, S.R., Sadiq, M.M., Varbanov, P., 2014. Generic carbon cascade analysis technique for carbon emission management. Appl. Therm. Eng. 70 (2), 1141–1147.

Metz, B., Davidson, O., Coninck, H. d., Loos, M., Meyer, L., 2005. IPCC, 2005—Carbon Dioxide Capture and Storage. Cambridge University Press, Cambridge, UK, Chapter 2: Sources of CO_2. pp 77–101. Retrieved from <www.ipcc.ch> (accessed 18.04.2014).

Mignard, D., Sahibzada, M., Duthie, J.M., Whittington, H.W., 2003. Methanol synthesis from flue-gas CO_2 and renewable electricity: a feasibility study. Int. J. Hydrogen Energy 28 (4), 455–464.

Mirzaesmaeeli, H., Elkamel, A., Douglas, P.L., Croiset, E., Gupta, M., 2010. A multi-period optimization model for energy planning with CO_2 emission consideration. J. Environ. Manage. 91 (5), 1063–1070.

Muis, Z.A., Hashim, H., Manan, Z.A., Taha, F.M., Douglas, P.L., 2010. Optimal planning of renewable energy-integrated electricity generation schemes with CO_2 reduction target. Renewable Energy 35 (11), 2562–2570.

Muñoz, R., Guieysse, B., 2006. Algal-bacterial processes for the treatment of hazardous contaminants: a review. Water Res. 40 (15), 2799–2815.

Musabbeh, A.M.A., Shammari, S.M.A., 2006. Enhancing Methanol Production by CO_2 Injection. Saudi Methanol Company, Jubail, Saudi Arabia.

Nowak, D.J., Crane, D.E., 2002. Carbon storage and sequestration by urban trees in the USA. Environ. Pollut. 116 (3), 381–389.

Ono, E., Cuello, J.L. (2003). Selection of Optimal Microalgae Species for CO2 Sequestration. Alexandria, VA: Proceedings of Second Annual Conference on Carbon Sequestration, p. 1–7.

Ooi, R.E.H., Foo, D.C.Y., Ng, D.K.S., Tan, R.R., 2013. Planning of carbon capture and storage with pinch analysis techniques. Chem. Eng. Res. Des. 91 (12), 2721–2731.

Pekala, L.M., Tan, R.R., Foo, D.C.Y., Jezowski, J.M., 2010. Optimal energy planning models with carbon footprint constraints. Appl. Energy 87 (6), 1903–1910.

Putt, R., Singh, M., Chinnasamy, S., Das, K.C., 2011. An efficient system for carbonation of high-rate algae pond water to enhance CO_2 mass transfer. Bioresour. Technol. 102 (3), 3240–3245.

Sadegheih, A., 2010a. A novel formulation of carbon emissions costs for optimal design configuration of system transmission planning. Renewable Energy 35 (5), 1091–1097.

Sadiq, M.M., Manan, Z.A., Wan Alwi, S.R., 2012. Holistic carbon planning for industrial parks—a waste-to-resources process integration approach. J. Cleaner Prod. 33, 74–85.

Tahara, K., Sagisaka, M., Ozawa, T., Yamaguchi, K., Inaba, A., 2005. Comparison of "CO_2 efficiency" between company and industry. J. Cleaner Prod. 13 (13–14), 1301–1308.

Tan, R.R., Foo, D.C.Y., 2007. Pinch analysis approach to carbon-constrained energy sector planning. Energy 32 (8), 1422–1429.

Tan, R.R., Sum Ng, D.K., Yee Foo, D.C., 2009. Pinch analysis approach to carbon-constrained planning for sustainable power generation. J. Cleaner Prod. 17 (10), 940–944.

Tjan, W., Tan, R.R., Foo, D.C.Y., 2010. A graphical representation of carbon footprint reduction for chemical processes. J. Cleaner Prod. 18 (9), 848–856.

van der Zwaan, B., Smekens, K., 2009. CO_2 capture and storage with leakage in an energy-climate model. Environ. Model. Assess. 14 (2), 135–148.

van Straelen, J., Geuzebroek, F., Goodchild, N., Protopapas, G., Mahony, L., 2010. CO_2 capture for refineries, a practical approach. Int. J. Greenhouse Gas Control 4 (2), 316–320.

Wan Alwi, S.R., Aripin, A., Manan, Z.A., 2009. A generic graphical approach for simultaneous targeting and design of a gas network. Resour. Conserv. Recycl. 53 (10), 588–591.

Wang, B., Li, Y.Q., Wu, N., Lan, C.Q., 2008. CO_2 bio-mitigation using microalgae. Appl. Microbiol. Biotechnol. 79, 707–718.

Weidema, B.P., Thrane, M., Christensen, P., Schmidt, J., Løkke, S., 2008. Carbon footprint. J. Ind. Ecol. 12 (1), 3–6.

Zhao, M., Kong, Z.H., Escobedo, F.J., Gao, J., 2009. Impacts of urban forests on offsetting carbon emissions from industrial energy use in Hangzhou, China. J. Environ. Manage. 91 (4), 807–813.

CHAPTER

Setting a policy for sustainability: the importance of measurement

14

Richard C. Darton

Department of Engineering Science, University of Oxford, Oxford, UK

INTRODUCTION

THE NEED FOR MEASUREMENT

The influential report of the World Commission on Environment and Development—the Brundtland (1987) report—drew attention to the pressing need for people to enjoy an improved quality of life while not harming the prospects of future generations. Historically, the activities that have increased wealth and welfare for humanity have mostly been at the expense of the natural world. We have consumed resources and converted natural landscapes to agriculture and urban settlements, and in doing so we have produced "wastes" and land-use change at a rate that will not be feasible as population size and living standards increase. We need different, *sustainable* approaches to our provision and consumption of freshwater, food, energy, and other resources. Meeting this challenge requires our decisions to reflect a sustainability agenda. Setting and implementing policy for a sustainable future means that we must be able to choose development options that take us in the best direction for increased sustainability. Choosing the "best" option implies that we can compare, and thus measure, sustainability outcomes.

It may be argued in a particular case that, of the options available, none is completely sustainable. It may also be the case that one option is better in some regard (e.g., consumes less resources), whereas another is better in some other way (e.g., produces more human benefit). Although both of these difficulties will often be encountered, we cannot allow them to prevent choices being made using sustainability criteria. The alternative is that if we cannot decide that one option is more sustainable than another, then we will be unable to steer our activities in a sustainable direction, and our desire for continued improvement of the human condition for the whole population of the planet will not be attained.

Thus, the ability to assess the sustainability of a development option is central to applying the concept in practice. The assessment allows us to choose the "best" option (however we choose to define it) or to reject all options as insufficiently sustainable against an absolute standard (however we choose to define that). Because sustainability is a holistic quality with many different aspects, much attention has been given to the criteria that should be used, formulated in terms of sets of indicators that address the important factors (Pinter et al., 2005; Bell and Morse, 2008).

For example, in the commercial world, the Dow Jones Sustainability Index (DJSI) has been developed to monitor how companies treat economic, social, and environmental issues with regard to mitigating risk and exploiting opportunity (Searcy, 2009). The set of criteria (indicators) in the DJSI are designed to assess corporate sustainability, "a business approach that creates long-term shareholder value by embracing opportunities and managing risks deriving from economic, environmental, and social developments." This understanding of sustainability differs from that of the Brundtland report, illustrating the cardinal importance of being clear about the working definition of sustainability used in making any assessment. Dyllick and Hockerts (2002) pointed out that many companies have adopted eco-efficiency as a guiding principle, thereby neglecting impacts on human and social affairs. Any assessment based on a restricted understanding of sustainability is bound to result in only a partial assessment of sustainability.

The Global Reporting Initiative (GRI, 2002) has developed a framework to facilitate the reporting of sustainability performance in a consistent and comprehensive way to give such reports similar credibility to the financial reports that have long been standard practice in the world of audited accounts. The GRI guidelines include a set of indicators with broad coverage that demonstrates *performance* against *goals* for the organization. Thus, policy is built into the assessment. The indicators are generic in nature and, for an assessment that is more focused either on business sector-related concerns or on local conditions, the GRI is developing, respectively, sector-specific supplements and national annexes. An example of sector-specific sustainability metrics is the set developed by the Institution of Chemical Engineers (2002). These are consistent with the GRI approach and are intended for use by chemical manufacturing industry.

Dalal-Clayton and Sadler (2014) describe many such generic methods, and these are very useful in designing assessment approaches. But generic methods, when applied to specific cases, need to be interpreted and adapted in a systematic and transparent manner; otherwise, the resulting set of indicators may not be comprehensive in covering all relevant issues or may not relate to an acceptable definition of sustainability, or may be inadequate in some other way.

A GENERALIZED SYSTEM CONSTRUCT

In promoting sustainability, it is necessary to make assessments in a wide variety of circumstances. We might be interested in a large-scale application, such as

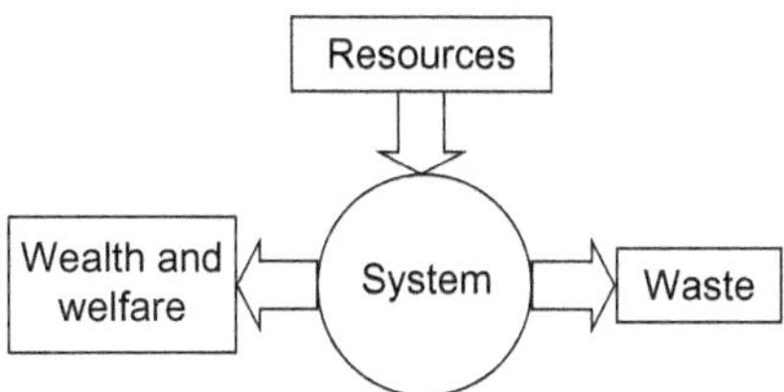

FIGURE 14.1

In this general view, the *system* consists of processes that convert *resources* to *wealth and welfare* (output that we desire) and *waste* (output that we do not want).

sustainability policy for a province or country, or a more local question involving domestic or commercial choices. The case may not be geographically constrained, as in the case of a global supply chain or business. When making an assessment, it is convenient to view the object being assessed as a *system* comprising *processes* that have sustainability *impacts*. It is through quantifying the impacts that we can see whether one system is more sustainable than another. Policy for changing the system can then be evaluated. Our most general view of a system is shown in Figure 14.1.

In many respects the system of Figure 14.1 resembles a heat engine, which is familiar to students of thermodynamics. A heat engine converts an amount of heat resource into a quantity of work with, according to the second law, an inevitable coproduction of waste heat. However, in the sustainability analogy of Figure 14.1, system operation involves a wide range of resources. In addition to energy we have to consider materials such as water and minerals, human and social input, and economic flows. The impacts of the system will be equally varied in nature, so our techniques of analysis must be sufficiently broadly based to capture them. Setting the system boundary is an important feature of the analysis, because it determines what activities, and thus impacts, must be included; this is a very well-known issue in this type of assessment in which subjective choice of system boundary can invalidate the results (Suh et al., 2004).

CAPITAL ASSETS OR STOCKS

Accounting for the sustainability impacts of the system conveniently uses the approach of the *triple bottom line* introduced by Elkington (1998). This is a well-known framing that views sustainable development as balancing the three "domains," "pillars," or "dimensions" of economy, environment, and society. The resources mentioned in Figure 14.1 are part of the capital stocks, and we can envisage any system activity as increasing, decreasing, or leaving unchanged the capital stock of each domain. This type of thinking about capital is similar to that used by the World Bank (2005) in its approach to sustainable development.

Economic capital includes funds available to purchase physical objects such as buildings, equipment, infrastructure, and the capital tied up in such objects, and also intangible assets like intellectual property. Environmental capital, sometimes called natural capital, is the value of the natural world comprising renewable and nonrenewable resources and also including the value in complex natural systems such as ecosystems with their interdependencies and genetic diversity and landscapes with their aesthetic appeal or cultural importance. The third type of capital is the combined human and social capital. Human capital is the intrinsic value of people according to their individual attributes such as knowledge and skills, resourcefulness, imagination, creativity, and physical strength. Social capital is the extra value that groups acquire through personal interactions and organizations. Group attributes typically arise from mutual support and trust, organizational efficiency, collaboration in eliminating social ills, altruistic behavior, positive feedback encouraging beneficial behavior and achievement. Institutional capital, which is sometimes considered separately, is taken in this work as part of social capital; this comprises the value in robust human institutions that promote activities and structures beneficial for sustainability.

Note that although it may be possible to measure economic capital in monetary units (using the US dollar, British pound, Euro, etc.), assigning monetary values to environmental capital and human and social capital raises fundamental problems. It is not easy to decide on a monetary value in the absence of a free market in which stocks can be bought and sold. For many people, valuing human or environmental assets in financial terms seems unethical in promoting purely utilitarian considerations. This has been discussed by Costanza et al. (1997) with reference to valuing nature as a provider of ecosystem services. These authors point out that refusing to consider the possibility of a monetary valuation does not solve the problem; decision-making will continue to occur in which monetary values are implicitly assumed, and these might be dangerously wrong, perhaps leading to the destruction of natural capital whose worth is simply not recognized. A similar argument is encountered in the human/social domain; assigning a monetary value to a human life may be morally repugnant, but in practice many policy decisions about allocation of resources (e.g., for health care, road safety) do imply such a valuation, even if it is not made explicit. In this chapter it is recommended to measure impacts on capital in terms that are appropriate to the particular impacts—these may be number of lives lost, mass of pollutant emitted, value added ($), kilometers traveled, number of jobs created, and so on. A great deal can then be said about the sustainability of the system, when a comprehensive set of data is available, using transparent methods of weighting and ranking different indicators. It may be possible, and sometimes useful, to attempt a final reconciliation in purely financial terms, but one has to be sure that the monetary valuations are well chosen, and that the richness of the original assessment is not lost to an extent that would distort subsequent decision-making.

REQUIREMENTS FOR A MEASUREMENT FRAMEWORK

The measurement framework will need to provide a way of analyzing the system to produce a set of indicators that represents the assessment of sustainability that we need. Dalal-Clayton and Bass (2002) emphasize that such a framework should be:

- Systemic—Different functions within the system are recognized and the relationships between them are appropriately accounted for.
- Hierarchical—The decomposition of high-level aspects into contributory, more detailed descriptions is performed consistently, avoiding overlap between components at the same level.
- Logical—Lower levels in the hierarchy contribute information about higher levels. Specifically, a higher level will specify why a particular feature is needed, and the lower level will specify which elements contribute to that feature.
- Communicable—We should be able to express the aims and methods simply, and they should be readily understandable and not too abstract or technical. Ease of communication will help support an important objective and will help ensure that the framework and its implementation will be transparent. Thus, stakeholders and other assessment practitioners should understand how the results were obtained and should easily adapt them to changed circumstances or alternative assumptions.

These are very general requirements, and they have been met by several reported sustainability assessment methodologies (Dalal-Clayton and Sadler, 2014). A framework constructed according to these requirements should generate a set of indicators that satisfy the heuristics outlined in Box 14.1.

THE PROCESS ANALYSIS METHOD

The process analysis method (PAM) was developed by Chee Tahir and Darton (2010a) for an agricultural industry application, the production of palm oil fruits. At its heart is the schematic representation of the system under review as a process, or set of processes, causing impacts (changes in either amount of quality) to the capital assets or stocks (Figure 14.1). In the analysis, an inventory is made of all the activities that cause a significant impact, and each impact is assessed to see what issues or effects arise for sustainability. As a check on the issues, we identify which stakeholders (impact receivers) are involved. For example, in a manufacturing operation, sale of product will give rise to a profit representing the added value, which increases the economic capital. An issue associated with profit generation is whether it is sufficient to reward the providers of economic capital who have invested in the operation. Unprofitable operations are a poor use of capital, so a useful indicator is one that addresses this, such as return on capital

BOX 14.1 HEURISTICS FOR A SET OF SUSTAINABILITY INDICATORS

1. Clear definition of what is to be assessed (e.g., factory operation, supply chain, industry sector) and purpose of assessment	*The system boundary must be specified* *A set of indicators that is not planned for some defined purpose is merely a collection of statistics*
2. The nature of a sustainable outcome is explicitly defined, and it addresses all three domains—economic, environmental, and human/social	*If the indicator set does not address all these aspects, then it may still be useful, but it will not provide a measure of sustainability*
3. Coverage—key aspects are included	*The indicator set should aim at completeness and at the specified level of detail. A systematic methodology makes it easier to spot omissions*
4. Data are available, preferably quantified empirical data	*Proxy indicators may be necessary for features that are identified as important, but for which data are lacking. Important indicators should not be suppressed because data are not (yet) available—the need for additional data should be reported*
5. Duplication and needless complication are avoided	*Double-counting, excessive detail, highly technical indicators, and a complex methodology obscure the message*
6. Excluded: parameters that are system characteristics but that do not measure any aspect of sustainability	*Statistics that only present interesting background should be kept separate from the sustainability indicator set*
7. Composite indicators are used, if appropriate	*Combining indicators to provide a single value can be extremely powerful if there is a good rationale for the combination, even though information is lost; the weighting of the various components should be explained*

used (perhaps measured in %). As shown in Figure 14.2, there is thus a chain of cause and effect that links the system process with the indicators and their values (metrics): *process* causes *impact* and creates *issues* described by *indicators* measured by *metrics*.

Applying the PAM in a particular case requires a number of distinct activities that build up the analysis: first the preliminary scoping and essential definitions, then the sustainability framework yielding indicators and metrics, and finally verification and modification using external input.

PRELIMINARY SCOPING AND ESSENTIAL DEFINITIONS

Overview and background

A thorough review of the system is made to identify the processes that will have a significant impact. A convenient approach is first to draw up schemes

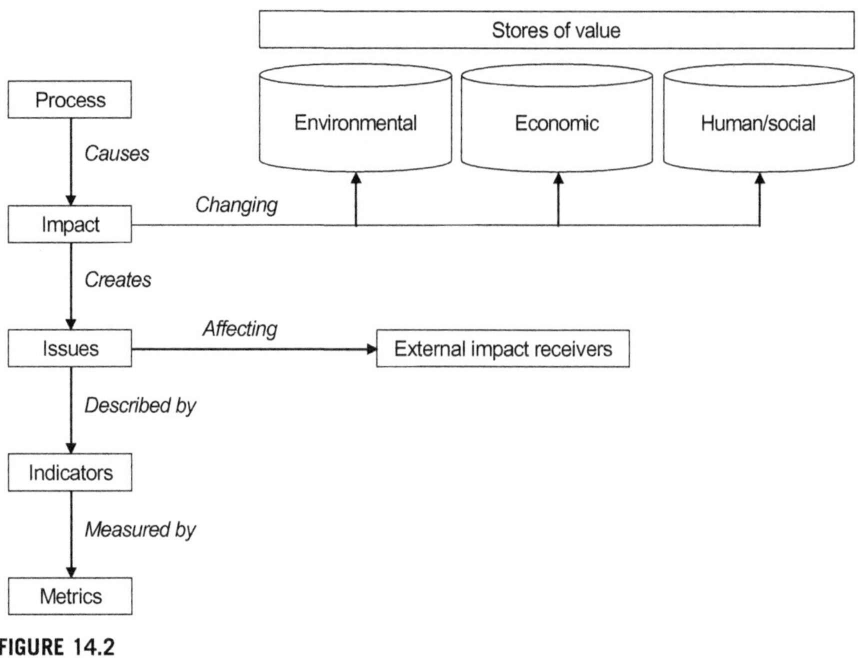

FIGURE 14.2

The PAM sustainability framework.

From Smith et al. (2013).

showing the flows of material and energy, and also the inputs and outputs of capital (economic, environmental, and human/social). From this scheme, the processes that cause impacts can be identified. The background is also reviewed to put the system in its context and to clarify the objectives of the analysis. This helps to identify stakeholders, issues associated with resource flows, and stakeholder priorities.

Defining the system boundary

The system boundary reflects the nature of the system and the purpose for which the analysis is being performed. It is essential that the system boundary is set to include all relevant effects, but not drawn so widely that extraneous activities are included that could confuse subsequent analysis. For an organization, this issue is discussed in detail by the Global Reporting Initiative (2005). The temporal scale is the period over which the impacts of the operations are considered and which must be sufficiently large to cover intergenerational effects (typically decades).

Defining sustainability

It is important that the analysis is based on an explicit definition of sustainable development. This chapter uses the well-known 1987 definition of Brundtland, which is "development, which meets the needs of the present without compromising the ability of the future generations to meet their own needs." Although a widely accepted and carefully crafted definition, it is very general and requires interpretation for specific situations. It helps transparency if this interpretation is made explicit in specifying what will be considered a sustainable outcome.

SUSTAINABILITY FRAMEWORK

Within the sustainability framework (Figure 14.2), a systematic analysis leads to selection of the indicator set and appropriate metrics.

Internal impact generators

The activities that have an impact on the three types of capital are termed internal impact generators (IIG). These relate to activities within the system (internal) and are typically activities or the policies that cause the activities to occur. The impacts are experienced by the owners or guardians of the capital.

External impact receivers

The owners or guardians of the capital impacted by the impact generators are termed external impact receivers (EIR). The EIR will be stakeholders, by definition. This is a useful check to establish that there are stakeholders who are experiencing an impact and, thus, that the impact is real and significant. Stakeholder interests in, and concern about, the issues arising from impacts play an important role in the assessment.

Issues

Impacts are characterized in terms of *issues*—stakeholder concerns about a particular impact. It is essential that the issues, which will be described by the indicators, do cover the extent and nature of the impact. Identifying the set of issues is crucial in the analysis because it leads directly to the indicator set. Consultation of stakeholders will often be required.

Indicators

Indicators are used to describe the issues. Help with identifying possible indicators can be obtained from previously published studies. The link between "issue" and "indicator" should be as simple and direct as possible.

Metrics

The metrics quantify the impact caused by the IIG. They should be backed by scientific or other quantifiable data and must be relevant and specific to the defined purposes. Schwarz et al. (2002) have noted that it will often be useful if the metrics are comprehensible to both technical and nontechnical audiences and point out the desirability of low cost and ease of data collection.

VERIFICATION AND MODIFICATION

To ensure that the indicators and metrics are applicable, it is necessary to verify and revise the indicators and metrics through fieldwork reviews and consultation with experts and stakeholders. This consultation will also inform the priorities and weightings that might be appropriate in presenting the results. Finally, a refined set of indicators and metrics is obtained that is both necessary and sufficient to monitor the sustainability performance of the system.

SOME EXAMPLES OF SUSTAINABILITY ASSESSMENT

THE OIL PALM CASE STUDY (CHEE TAHIR AND DARTON, 2010a,b)

The first case study was performed on an agricultural production process in Malaysia: the cultivation of oil palm in matured estates to supply palm fruit feedstock for edible oil production and potentially for biofuel. Fieldwork with stakeholders was undertaken to clarify the analysis and the issues. The study was intended to provide the industry and company management with an objective tool to judge the sustainability performance of oil palm estates, and to compare different locations in a benchmarking exercise. For this reason, it was convenient to view the impact generators as the set of policies used to manage the estates. These are given in Table 14.1. The impact generators are very general, and we can expect to see somewhat similar sets for any manufacturing process if the analysis has a similar objective. The specific impacts for the oil palm study were identified from the detailed overview of the system, which is the first step in the methodology. This underlines the importance of drawing up, at an early stage in the analysis, a thorough description or scheme for the system to make an inventory of all the significant activities. The system spatial boundary was set at the perimeter of the estate with all activities within this perimeter being included; in addition, supplier and contractor activities outside this boundary were also included, where these were considered to be integral to the business of the estate.

The framework analysis was conducted for each impact generator in Table 14.1. For example, within the *human/social domain* and for the IIG *management of social capital*, one chain leading to an indicator and metric is:

> *Management of social capital (process)* causes *employment opportunities (impact)* and creates *job prospects for local communities (issue)* described by *job creation (indicator)* and the metric *number of positions filled by locals.*

The whole set of approximately 22 impact generators resulted in 72 metrics specific to the oil palm case. In the case of product management, it was found that different aspects of the policies adopted had significant impacts in all three domains, and this is shown in Table 14.1. However, in other areas the impacts

Table 14.1 Environmental, Economic, and Social Domains

Environmental Domain	Economic Domain	Social Domain
Energy, land, water, transport, and biodiversity management		
Material management	Economic material management	
Waste management	Management of waste with economic worth	
Environmental product management	Economic product management	Social product management
Management of supplier and contractor environmental practices		
	Economic human (and social) capital management	Human (and social) capital management
	Economic utility and service management	
	Economic distribution to capital providers, suppliers and contractors, employees, government agencies, local communities	

Source: From Chee Tahir and Darton (2010b).

were experienced in only one or two domains. For example, policies on waste management clearly have a significant effect on the environment; however, they also impact the economic capital through the possibility of waste upgrading and recycling, which reduces cost per unit of output.

The study revealed the problem of contradiction between indicators. For example, effort must be taken to prevent damage to the environment by careless treatment of harvested palm fruit, but this effort presents a cost that reduces profitability. Thus, the requirement of good environmental product management seems to be opposed to that of good economic product management. The PAM only exposes such tensions that are encountered in the management of most undertakings and does not itself reconcile them. Managers have to consider where a good balance lies and perhaps what might be done to minimize or avoid the problem. To present both aspects is not double-counting, providing that the indicators really are representing different impacts.

The study found that two business perspectives could be used to point the way toward greater sustainability. These are:

- *Resource efficiency*, which measures how effectively the capital is used or created (change can occur in both the amount of capital and its quality); this perspective incorporates the judgment that enhancing the quality and preserving or increasing the extent of the capital is beneficial for sustainable development. Enhancing and preserving the capital, for example, will mean that more is available for future generations to enjoy (intergenerational equity).

- *Fairness in benefit*, which means both how fairly the benefit of using the capital is distributed and how fairly disbenefits are distributed. This perspective incorporates another judgment, which is that sustainable development is advanced by a fairer distribution of benefits that, in turn, promotes wider economic and social development. "Fairness" also applies to the distribution of undesirable outcomes or disbenefits such as pollution, resource depletion, or reduction in biodiversity.

These business perspectives provide a link between the priorities arising from the definition of sustainability and the system processes that cause impacts. We expect these perspectives to align with the corporate social responsibility (CSR) policy of the business. The need for CSR policy to specify the values that drive sustainability assessment of business operations has been stressed by Labuschagne et al. (2005).

THE CAR TRANSPORT SYSTEM (SMITH et al., 2013)

The car transport study was designed to examine how the sustainability of the UK car fleet had changed between two particular years (1995 and 2005). The two underlying questions were whether we could, in this way, tell whether motor car use in the United Kingdom is becoming more or less sustainable, and also whether the process analysis methodology could be applied to a distributed service-based system. Figure 14.3 is a top-level view of the analysis showing how the use of resources and the generation of outputs have impact on the three capital domains.

In this case, the desired output is mobility. To focus on the car fleet itself and omit questions regarding whether mobility could be more sustainably provided in some different way, the system boundary was drawn to include only the life cycle of the motor car in the United Kingdom. This included the import of vehicles and fuel into the United Kingdom, but the road system and questions of urban design and other transport modes were excluded. A sustainable transport system was defined as one that delivers for all socioeconomic groups and, at any given time, levels of mobility and safety of movement that society considers to be ideal. At the same time it maximizes positive impacts on society, while minimizing undesirable human/social impacts. Other aspects of the definition deal with the use of stock-dependent or nonrenewable resources, as suggested by the rules of Rennings and Wiggering (1997).

In this study, four IIG were identified as responsible for the significant impacts on the three domains of capital, as follows.

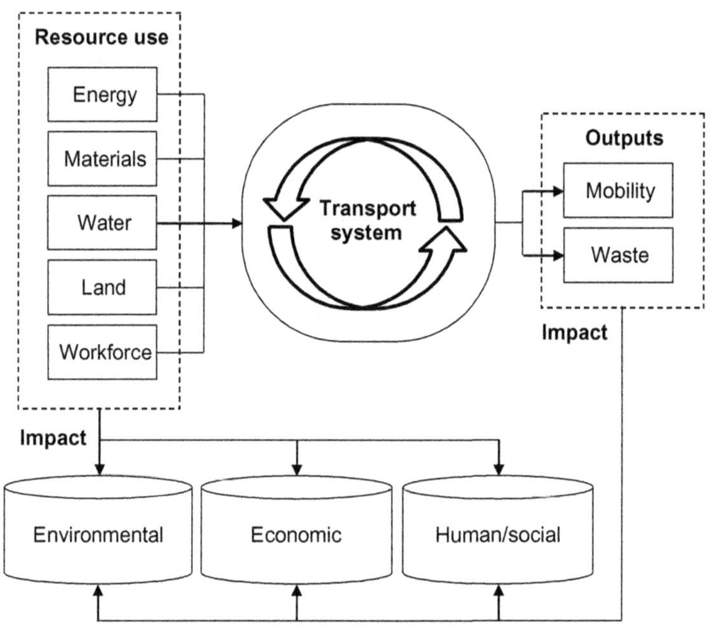

FIGURE 14.3

Top-level view of analysis.

From Smith et al. (2013).

User decisions

Users are not a homogeneous group, and the impact they have is an emergent property of a large number of individual decisions, on the amount of travel, on the type of vehicle driven, on driving style, and so on.

Government policy and regulation

These include regulations for manufacturers (such as emissions standards and safety), regulations for drivers (standards of maintenance and other safety-related issues but also environmental protection regulations such as rules for disposal), and the important issues of taxation and licensing.

Car manufacturer design, research, and development choices

Technology choice plays a key role because it greatly influences resource use and waste production, and it influences the quality of the mobility provided.

Car manufacturer sustainability policy and practices

The extent of the manufacturer's commitment to environmentally and socially sustainable practices influences the impacts of car manufacture and use.

It might be supposed that manufacturers respond to government pressures and consumer demand, and they do, of course. But in providing technology solutions and promoting greater sustainability, the manufacturers themselves present the other stakeholders with new options to choose. Sometimes the government stakeholder can only make policy and enforce a particular regulation if the appropriate technology is available to meet the required outcome (e.g., with respect to tail pipe emissions).

In this study no separate fieldwork was undertaken to determine stakeholder views, because there is a range of published material and excellent quality statistics available (though somewhat less for the earlier study year 1995). For example, the UK government publishes an annual survey of British Social Attitudes, which includes a substantial section on transport. Other data were available from a variety of published sources regarding road traffic accidents, motor car emissions to the environment, health impacts as a result of car-related pollution, and so on. It was somewhat surprising to find that these data are not regularly collated to provide an overview of the sustainability of the motor car, which is an important feature of our society. The initial system overview quickly revealed this importance: servicing a population of approximately 60 million, the car fleet in 2010 comprised nearly 30 million units, with vehicle-kilometers driven approaching 460 billion annually. This is an enormous benefit in terms of mobility provision, but naturally there are associated disbenefits in many areas, especially environment and human health.

Our study was designed to measure the changing sustainability of the car fleet by looking at sustainability impacts, good and bad. That is, it was not predicated on any particular view of transport policy and did not measure performance against goals. For example, no view was taken on how many journeys *should* be undertaken by a motor car nor on how quick they should be. In this, the study differed from many assessments of transport infrastructure, where policy goals have been established first and the assessment seeks to measure performance against these targets. As pointed out by Holden et al. (2013), the introduction of policy goals related to local- and project-level issues can cause the assessment to lose sight of really important issues on a global scale, such as long-term ecological sustainability, and issues of intragenerational and intergenerational equity. An assessment that starts from an accepted definition of sustainable development and applies that definition should capture such essential global issues.

ASSESSING THE SUSTAINABILITY OF ARSENIC MITIGATION TECHNOLOGY FOR DRINKING WATER (ETMANNSKI AND DARTON, 2014)

The PAM requires, in its final step, stakeholders and experts to be consulted to ensure that the analysis has correctly identified the significant issues that are of importance to stakeholders, and also to provide clarity on stakeholder preferences. In this study of arsenic mitigation technology for drinking water, the stakeholder (user) consultation was of central importance. Naturally occurring arsenic is found in groundwater in many countries, but it is particularly a problem in populous areas in the Bengal basin (India and Bangladesh), where approximately 100 million people are affected. The question addressed was why, when offered a technology to remove arsenic (a life-threatening contaminant) from their drinking water, do many people in the area either not adopt it at all or quickly give up using it? The technical solution, which may seem, to an outsider, to offer a sustainable solution to a terrible problem, is frequently found to be unsustainable. In this case, the spatial system boundary was taken to be that of each arsenic-affected community (approximately 100 households). A sustainable arsenic mitigation option was defined as one that satisfies the following requirements:

- Economic—For all stakeholders, their ability to use and/or provide the technology does not diminish with time because of financial constraints.
- Environmental—Maximum use is made of renewable or replaceable resources and materials; there is no production of waste to systematically degrade the environment.
- Sociocultural—The technology provides treated water acceptable in taste, color, and odor; the location and volume of treated water matches user needs; technology operation is easy for the user; application of the technology gains the trust and confidence of users; and in all aspects, it suits the level of awareness of the arsenic problem in the user community.

Fieldwork in West Bengal was undertaken in 2012. The most important element was the survey of 933 households to determine user perspectives. Ten different technologies were included in the survey. A full sustainability analysis was undertaken using the PAM, where the application of arsenic mitigation technology was taken to be the system of processes shown in Figure 14.4. This basically follows the life cycle of the technology, because we were interested in identifying which elements could be improved, to result in a more sustainable solution and particularly in public health improvement.

Analysis of the survey data showed that a number of issues were considered by users to be "very important," and a composite indicator could be created combining the key considerations of *trust* and *confidence*. We found that technologies enjoying the highest rating of trust and confidence were the most utilized. Trust is based on social relations and shared values, and confidence is an evidence-based belief in the successful outcome of a proposition (Earle and Siegrist, 2006). It was clear from these results that the way in which the technology is presented to the user is very important, because much depends on the elements of trust and confidence. As an illustration, there is no way for users themselves, in the rural setting, to determine if the water is safe; they must trust the operator (if it is a

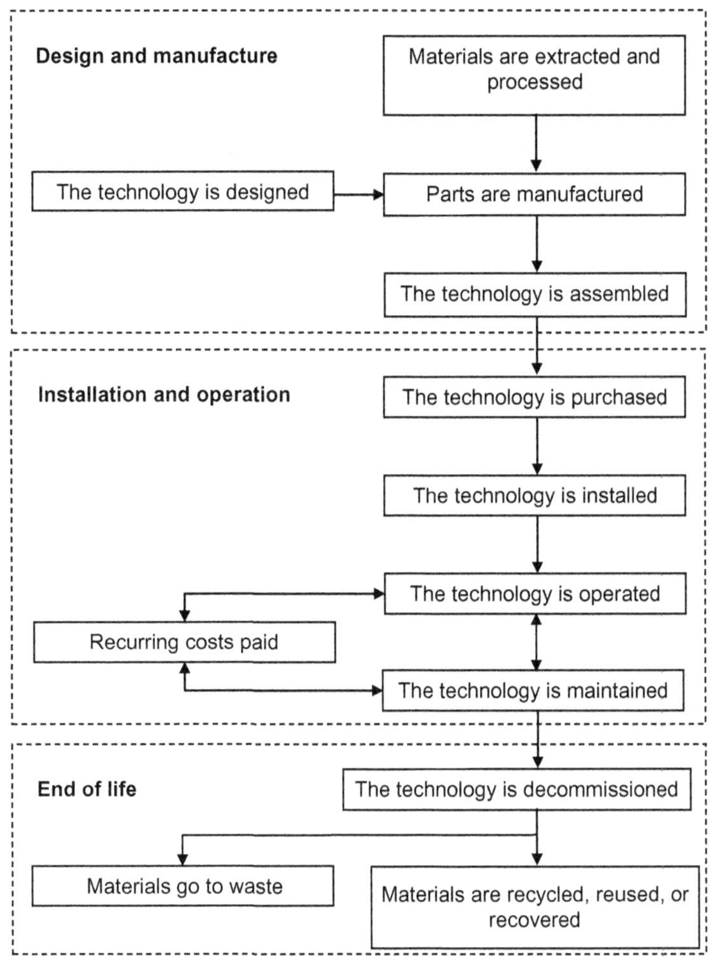

FIGURE 14.4

Arsenic mitigation as a system of processes.

From Etmannski and Darton (2014).

community-scale system), and believe what they have been told, and have confidence in the equipment. Trust and confidence are thus of key importance.

In addition to features related to trust and confidence, other factors that the user found to be very important were cost and the convenience to the user of the water provision. An issue that rated less highly with stakeholders but that emerged from the PAM analysis was the disposal of arsenic-rich waste (AsRW). This is often currently discarded in the local environment in an uncontrolled manner. Such disposal is regarded with widespread nonchalance, and indeed there are very few safe options available within the rural community. Nevertheless, this AsRW is potentially harmful and can lead to point-source recontamination. Users identified "understanding the effects of arsenic on health" as a very important issue, and general levels of understanding were often poor. For example, it is often believed that arsenic can be removed from the water by boiling it, or that arsenicosis can be "caught" like an infectious disease. It seems probable that better public awareness could lead to changed attitudes toward AsRW disposal, better disposal options, and adherence to stricter and safer disposal protocols. The PAM analysis also showed that a major challenge in achieving long-term sustainability is to plan for sustainable financing that can take over when charitable or government funding runs out.

SUMMARY AND CONCLUSIONS

The PAM interprets and applies the concept of sustainable development to the practical problem of assessing the sustainability of a system in operation. The method is systemic, hierarchical, logical, and communicable, and results in a set of indicators that aims to be comprehensive in its coverage of all three domains of sustainability. Because the selection of indicators for any particular application is based on an inventory of the system processes that give rise to sustainability impacts, the indicator set is tailored to that particular application; it is also based on a clearly stated definition of what constitutes a sustainable outcome. This contrasts with the situation sometimes encountered in which indicators have been selected in an unsystematic fashion from published lists. The PAM bears some similarity to the pressure-state-response (PSR) approach, which also considers cause and impact (OECD, 1993). However, PSR and its derivatives involve the modeling of this causal relationship and, thus, the full understanding of its complexity. For a large system with many internal and external interactions, the PSR approach becomes difficult and time-consuming because of the need to fully understand these relationships. The PAM observes the dependence of cause and effect but does not require a detailed explanation of this; it only seeks to characterize the magnitude of all significant sustainability impacts as system output. The cause–effect link is much less rigorously described by the PAM, but it is therefore easier to apply when the goal is to obtain an adequate comprehensive indicator set.

In this chapter we have shown how the method has been applied to three very different systems to enable the definition of indicator sets that can guide management and policy formulation toward improving sustainability. When the application is similar, we would expect to obtain similar indicator sets that will aid comparison and benchmarking. Nevertheless, each set should be tailored to the specific characteristics of the case study by following the steps in the method.

REFERENCES

Bell, S., Morse, S., 2008. Sustainability Indicators: Measuring the Immeasurable? Earthscan, London, UK.

Brundtland, G. (Ed.), 1987. Our Common Future: The World Commission on Environment and Development. Oxford University Press, Oxford, UK.

Chee Tahir, A., Darton, R.C., 2010a. The process analysis method indicators to quantify the sustainability performance of a business. J. Clean. Prod. 18 (16–17), 1598–1607.

Chee Tahir, A., Darton, R.C., 2010b. Sustainability indicators: using the process analysis method to select indicators for assessing production operations. Chem. Eng. Trans. 21, 7–12.

Costanza, R., dArge, R., de Groot, R., Farber, S., Grasso, M., Hannon, B., Limburg, K., Naeem, S., O'Neill, R.V., Paruelo, J., Raskin, R.G., Sutton, P., van den Belt, M., 1997. The value of the world's ecosystem services and natural capital. Nature 387 (6630), 253–260.

Dalal-Clayton, B., Bass, S., 2002. Sustainable Development Strategies: A Resource Book. Earthscan, London, UK.

Dalal-Clayton, B., Sadler, B., 2014. Sustainability Appraisal. Routledge, Abingdon, UK.

Dyllick, T., Hockerts, K., 2002. Beyond the business case for corporate sustainability. Bus. Strategy Environ. 11, 130–141.

Earle, T.C., Siegrist, M., 2006. Morality information, performance information, and the distinction between trust and confidence. J. Appl. Psychol. 36 (2), 383–416.

Elkington, J., 1998. Cannibals with Forks: The Triple Bottom Line of 21st Century Business. New Society Publishers, Stony Creek, CT, USA.

Etmannski, T.R., Darton, R.C., 2014. A methodology for the sustainability assessment of arsenic mitigation technology for drinking water. Sci. Total Environ. 488–489, 505–511. Available from: http://dx.doi.org/10.1016/j.scitotenv.2013.10.112.

Global Reporting Initiative, 2002. Sustainability Reporting Guidelines. Global Reporting Initiative, Amsterdam, The Netherlands.

Global Reporting Initiative, 2005. Boundary Protocol. Global Reporting Initiative, Amsterdam, The Netherlands.

Holden, E., Linnerud, K., Banister, D., 2013. Sustainable passenger transport: back to Brundtland. Transp. Res. Part A—Policy Pract. 54, 67–77. Available from: http://dx.doi.org/10.1016/j.tra.2013.07.012.

Institution of Chemical Engineers, 2002. The Sustainability Metrics: Sustainable Development Progress Metrics Recommended for use in the Process Industries. Institution of Chemical Engineers, Rugby, UK.

Labuschagne, C., Brent, A.C., van Erck, R.P.G., 2005. Assessing the sustainability performance of industries. J. Clean. Prod. 13 (4), 373–385.

OECD, 1993. OECD Core Set of Indicators for Environmental Performance Reviews, OECD Environment Monographs No. 83. OECD, Paris, France.

Pinter, L., Hardi, P., Bartelmus, P., 2005. Sustainable Development Indicators: Proposals for a way forward; UNDSD report. IISD, Winnipeg, Manitoba, Canada.

Rennings, K., Wiggering, H., 1997. Steps towards indicators of sustainable developments: linking economic and ecological concepts. Ecol. Econ. 20, 25–36.

Schwarz, J., Beloff, B., Beaver, E., 2002. Use sustainability metrics to guide decision-making. Chem. Eng. Prog. 58–63, 1 July.

Searcy, C., 2009. The Role of Sustainable Development Indicators in Corporate Decision-Making. IISD, Winnipeg, Manitoba, Canada.

Smith, T.W., Axon, C.J., Darton, R.C., 2013. A methodology for measuring the sustainability of car transport systems. Transp. Policy 30, 308–317. Available from: http://dx.doi.org/10.1016/j.tranpol.2013.09.019.

Suh, S., Lenzen, M., Treloar, G.J., Hondo, H., Horvath, A., Huppes, G., Jolliet, O., Klann, U., Krewitt, W., Moriguchi, Y., Munksgaard, J., Norris, G., 2004. System boundary selection in life-cycle inventories using hybrid approaches. Environ. Sci. Technol. 38 (3), 657–664.

World Bank, 2005. Focus on Sustainability 2004. Chapter 4, Our Commitment to Sustainable Development. World Bank, Washington, DC, USA.

CHAPTER

Sustainability assessments of buildings, communities, and cities 15

Umberto Berardi

Faculty of Engineering and Architectural Science, Ryerson University, Toronto, ON, Canada

INTRODUCTION

The increasing attention toward sustainability is pushing the construction sector and the governance of the built environment toward rapid changes. Policies, laws, and regulations around the world are asking the sector to adopt sustainable innovations in terms of products and processes to encourage a more sustainable built environment (Hellstrom, 2007). This attention for the building sector arises from its energy consumption and GHG emissions, which, in developed countries, account for 30% and 40% of the total quantities, respectively (IPCC, 2007). Forecasts of the EIA (2010) show that energy consumption in buildings is increasing at a slightly higher rate than those of the industrial and transportation sectors. However, according to the IPCC (2007), the building sector has the highest energy saving and pollution reduction potential, given the flexibility of its demands. IPCC showed that in countries that are not members of the Organization for Economic Cooperation and Development (non-OECD) and in economies in transition, potential CO_2 savings in buildings could be 3 and 1 $GtCO_2$-eq every year in 2030. Therefore, a possible reduction of almost 6 $GtCO_2$-eq every year is possible worldwide in the next 20 y if the building sector embraces sustainability. This highlights why sustainable buildings and communities are often considered a priority for a sustainable world (Butera, 2010; Larsson, 2010).

Climate change has raised concerns over the rapid depletion of the environment and its resources. International research has confirmed that the built environment is the most promising sector for a rapid transition to sustainability (GhaffarianHoseini et al., 2013). In this scenario, many examples of sustainable urban environments are showing the advantages of sustainability. Meanwhile, an increasing request for tools to assess their sustainability is recorded.

The assessment of sustainability of the built environment is an essential step toward its promotion (Crawley and Aho, 1999; Kibert, 2007; EPA, 2008) and recently (du Plessis and Cole, 2011). However, large difficulties exist creating useful and measurable assessment indicators (Mitchell, 1996). Sustainability assessments have been defined as the processes of identifying, predicting, and evaluating the potential impacts of different initiatives and alternatives (Devuyst, 2000). The possibility to assess both products and processes has often been considered particularly important for a sector as inertial and conflicting as that of the built environment (Winston, 2010; CIB, 2010).

The sustainability assessments were addressed with rating tools for buildings more than two decades ago in Europe and North America before diffusing worldwide (Häkkinen, 2007) and more recently (Sev, 2011; Berardi, 2012). In some way, these systems have promoted the commercial image of recent green buildings.

Although there is a high demand for and much attention given to green buildings, there is an increasing awareness that these are insufficient to guarantee sustainability of the built environment (Häkkinen, 2007; Cole, 2010). Recent literature has discussed the importance to go beyond the sustainability assessment of single buildings and to enlarge the assessment scale to communities and cities to meet all the different aspects of sustainability (Turcu, 2013). Cole (2011) clarified that a significant achievement in sustainability assessments has been the introduction of rating systems for the urban design. These increase the assessment scale and allow consideration of aspects not accounted for at the building scale. Examples of some aspects are the flows and the synergies between initiatives within the built environment and consequent social and economic effects of sustainability in the built environment (Berardi, 2011).

Sustainability assessments on community and city scales are proving to be much more than the summation of individual green elements, because the scaling-up results in complex interactions that significantly alter the results obtained on the building scale (Haapio, 2012). Requests to go beyond the building-centric approach in sustainability assessments have favored the discussion about new possible areas of sustainability assessment within the built environment (Berardi, 2013a,b,c). In fact, systems originally developed for buildings have been criticized for their inability to capture what makes a built environment sustainable for its citizens (Rees and Wackernagel, 1996). The rare consideration of criteria related to social and economic aspects of sustainability has often been underlined at the building scale, but many other limits of current sustainability assessment tools exist (Conte and Monno, 2012).

Previous considerations show that different scales are considered necessary to assess sustainability of the built environment. In fact, only cross-scale evaluations allow recognition that the whole urban environment has a prime role in social and economic sustainability and a huge impact on environmental sustainability (Mori and Christodoulou, 2012).

More than 50% of the world's population currently lives in urban areas, and this figure is expected to increase to 70% by 2050 (UN, 2008). In Europe, 75% of the population lives in urban areas, and by 2020 the number is expected to reach 80% (EEA, 2006). The importance of urban areas is also confirmed by the diffusion of megacities of more than 20 million people, which are gaining ground in Asia, Latin America, and Africa (Figure 15.1). As a result, most resources are currently consumed in the urban environment worldwide. This contributes to the

FIGURE 15.1

Views of the Bund of Shanghai in 1990 and 2010.

From Berardi (2013b).

economic and social importance of the urban areas, and also to their poor environmental sustainability. Their metabolism generally consists of the input of goods and the output of wastes with unavoidable externalities (Turcu, 2013).

Urban sustainability has attracted much criticism because urban areas rely on too many external resources. Promoting sustainability in the urban environment has been interpreted as reducing the impact of cities on the environment. However, other interpretations of urban sustainability have often promoted a more anthropocentric approach, according to which urban areas should respond to demand based on people's needs and focus on the quality of life and other social aspects of sustainability (Turcu, 2013).

Large and mega urban areas increase the difficulties in promoting sustainability and their consequent request for tools for sustainability assessments. New assessment systems have been created in the past few years to answer such requests (Sharifi and Murayama, 2013). Tanguay et al. (2010) and Berardi (2013c) have reviewed available indicators for measuring sustainability. They showed that most of the currently used indicators are characterized by a strong environmental approach. This is evident considering indices such as the ecological footprint, the water footprint, the environmental sustainability, and the environmental vulnerability. The preservation of natural resources is a key component in ensuring sustainability and, as a matter of fact, it is also a key dimension of people's well-being. Furthermore, people directly benefit from environmental assets and services because these allow them to satisfy basic needs and to enjoy leisure time (OECD, 2011a). Thus, it is commonly accepted that environmental sustainability has high effects on the social dimension of well-being (Vallance et al., 2011).

This chapter is based on the belief that measures of the different aspects of sustainability, including social and economic aspects, should be explicitly considered in sustainability assessment of the built environment at the different scales. Consequently, systems exclusively related to environmental and ecological assessments are not considered. The attention is focused on sustainability rating systems that adopt a multicriterion approach to consider the different dimensions of sustainability using the triple bottom line approach (Pope et al., 2004).

Multicriterion systems are gaining increasing attention because they are easily understood and allow a step implementation for each criterion (Berardi, 2012). In the systems considered, sustainability is generally evaluated by the summation of the results of different performances related to environmental, social, and economic aspects (Scerri and James, 2010). One of the limitations of multicriterion systems is their additional structure based on the sum of different evaluations. This and other limitations are considered here.

The chapter is organized in six different sections. The second section describes the framework of sustainability assessment in the built environment. The third, fourth, and the fifth sections review sustainability assessment systems for buildings, communities, and cities, respectively. In each section, the most known systems are presented and compared, and then the outcome of their applications is discussed. The final section presents concluding remarks and research trends.

FRAMEWORK OF SUSTAINABILITY ASSESSMENT

Before comparing sustainability assessment systems for the built and urban environment, it is helpful to clarify what is meant by sustainability, assessment, community, and city. In fact, lack of consensus on the definitions of these terms prevents the possibility of comparing existing rating systems. This section is largely based on a recently published article (Berardi, 2013a).

Sustainability is not a single and well-defined concept. At least 100 definitions have been given to this term, and new definitions are continually added, often clouding its concept (Hopwood et al., 2005). Sustainability has also been accused of being indefinable because every time a definition has been formulated, it has always left out some of the possible meanings (Robinson, 2004). The concept of sustainability dates back to the 1970s. Its theoretical framework evolved after the publication of "The Limits to Growth" and led to the famous definition proposed by the Brundtland Commission (WCED, 1987). In the 1990s, an intensive debate about different definitions and models of sustainability occurred.

The multitude of interpretations that the term sustainability has received indicates a resistance in the acceptance of a unique official definition and a preference to adapt this term to the context in which it is considered at any time (Martens, 2006). Paradoxically, the Sustainable Buildings and Climate Initiative (SBCI) of UNEP has declared that sustainability requires all the different interpretations that are often given to the term, because the concept of sustainability represents the synthesis of all of them (UNEP-SBCI, 2009).

The wide meaning of sustainability opens several options for the considerable criteria in sustainability assessments. Sustainability is time-dependent and socially dependent, and it has different interpretations for different people, with partial dependence on the point of view of the assessment (Martens, 2006) and more recently (Dempsey et al., 2011). These sources of uncertainty have contributed to the belief that several levels of sustainability exist, and that it is more useful to consider sustainability as a relative concept. The introduction of the concepts of strong and weak sustainability increased this belief; strong sustainability states that it is not possible to accept an exchange between environment and economy, whereas weak sustainability accepts their substitutability (Mori and Christodoulou, 2012). The resilience over temporal and spatial cross-scales has recently been used as a measure of relative sustainability (Mayer, 2008) and more recently (Barr and Devine-Wright, 2012).

If the definition of sustainability suffers from ambiguity, so does its assessment. A sustainability assessment can be defined as the process of identifying, measuring, and evaluating the potential impacts of alternatives for sustainability (Devuyst, 2000). Several sets of sustainability indicators have been developed so far, but none has emerged as a universal measure (Pope et al., 2004). Multicriterion rating systems focusing on environmental indicators have been proposed in different fields in the past 20 y. However, as increasing attention on sustainability is recorded among sociologists, economists, and politicians, new assessment indicators have been promoted.

Sustainability indicators have raised the debate about the way in which they were developed and used: from the top, initiated primarily by governments and based on expert input (expert-led), or from the bottom (citizen-led), drawing on local networks and involving citizens. The tensions between expert-led versus citizen-led systems of sustainability assessment recently seemed to be solved through the integration of the two approaches (Reed et al., 2006). Meanwhile, doubts about the objectivity of the assessment are still often raised. In fact, previous research has shown that the assessor, the point of view of the assessor, and time of assessment often play a prime role in the assessment results because they influence the considered criteria (Martens, 2006). Consequently, a transparent, objective, and plural (or promoted in a multiagent contest) assessment has recently been considered necessary.

The difficulties in performing sustainability assessments in the urban environment are greater because the object of the assessment is often an unbounded entity. Most of the systems presented here were originally proposed for the assessment of buildings. They answered the request of sustainable and green buildings and were intended as tools to give objectivity to their performance. At the same time, the increasing awareness of the limits of the assessment of buildings has led to developing systems for urban communities and cities. These systems are considered useful to assess the built environment in a more integrated way, and they have been proposed as tools for marketing as well as for planning issues. Nevertheless, these systems suffer from a lack of data and of exact definition of the boundaries of their assessed entities. An urban area can be identified in different ways in terms of land use, infrastructure, or people density (UN-Habitat, 2006); these criteria raise ambiguity about urban boundaries.

For example, the evaluation of transportation generally covers several communities, whereas population density may consider residents or workers. Different criteria have been used to define the boundaries of communities and cities, among which administrative criterion, population density, and economic characteristics are the most common (UN-Habitat, 2006). Urban sprawl is also increasing the confusion in establishing exact boundaries. Consequently, during sustainability assessment, attention must be given to external impacts (leakage effects) on areas beyond the assessed boundaries (Bithas and Christofakis, 2006).

Considering the scale of urban communities, high uncertainty exists regarding their dimensions. In the Haussmannian fabric, approximately $200 \times 200\ m^2$ represented the dimensions of a neighborhood; in South America, the grid is generally larger and it often reaches the dimensions of $400 \times 400\ m^2$, whereas in cities such as New York it is generally rectangular ($100 \times 200\ m^2$). Apart from geometrically planned cities, the boundaries of a community are generally difficult to establish. As a consequence, during sustainability assessments, they are often established only considering the area object of assessment. This criterion is often meaningless for sustainability.

The importance of the interactions between different parts of the built environment has been recognized as an unavoidable aspect of sustainability and has increased the request for assessments at scales larger than buildings (Berardi, 2011). However, it is widely recognized that evaluations of countries and regions are often far from capturing, influencing, and assessing sustainability of the daily practices of people (Mori and Christodoulou, 2012). Communities and cities are therefore considered the institutional and geographical levels closer to citizens where sustainability can efficiently be promoted and assessed. As a matter of fact, they represent the nearest natural environment, social network, and economic market around a citizen.

Urban areas are the lowest level where problems can be meaningfully resolved in an integrated, holistic, and sustainable way (Aalborg Charter, 1994). International policies have started focusing on sustainability assessments in urban areas and, therefore, the number of communities and cities experimenting sustainability assessments is increasing. This trend shows that sustainability assessments are recognized as tools for monitoring urban dynamics and land promotion. Many communities have developed their own sustainability assessment systems (Atkisson, 1996) and later (Corbiére-Nicollier et al., 2003).

Several frameworks of sustainability assessment indicators have been proposed (Bentivegna et al., 2002; Xing et al., 2009) and more recently (Mori and Christodoulou, 2012). Figure 15.2 is one of these and it covers the most important indicators, although it should be considered as a reference that allows the opportunity to contextualize different visions of sustainability by assigning different importance to each criterion. In particular, Figure 15.2 adopts the recent interpretation of the concept of sustainability as composed of four dimensions (instead of the classic three): environmental; economic; social; and institutional sustainability.

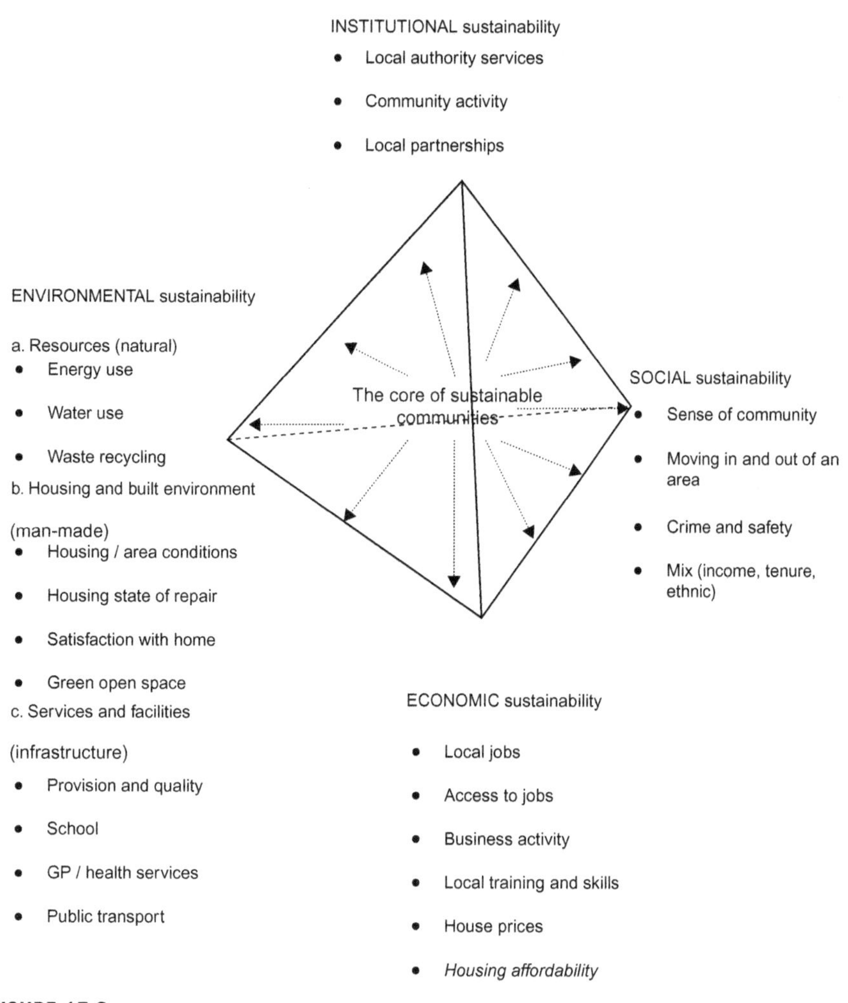

FIGURE 15.2

Framework for the sustainability assessment in the built environment (Turcu, 2013).

Indicators in Figure 15.2 are common to many sustainability assessment systems, although specific urban settings may require alterations in the frameworks. In this sense, using an adaptive approach to sustainability assessment indicators for specific local situations has generated a multitude of different systems (Reed et al., 2006). This trend is also briefly described here, especially because the request for comparability among the systems is a challenging topic in sustainability assessments. In some way, the goal of this chapter is to compare the most diffused multicriteria systems for sustainability assessment of the built environment.

SYSTEMS FOR SUSTAINABILITY ASSESSMENT OF BUILDINGS

According to many studies, the sustainability assessment is considered necessary to increase the diffusion of sustainable buildings (Cheng et al., 2008). Unfortunately, the construction sector is unfamiliar with performance measurements and, although many assessment systems already exist worldwide (Figure 15.3), their diffusion is still low in absolute terms. Sustainability building certification programs and rating systems are used worldwide, with the only exceptions being Africa (except South Africa) and Latin America (except Brazil). Sustainability measurements, in the building sector, are capturing much attention worldwide, rapidly moving from fashionable certifications to current practices. In 2010, 650 m^2 obtained a sustainability certification throughout the world, with projections for 1,100 m^2 in 2012 and for more than 4,600 m^2 in 2020 (Bloomc and Wheelock, 2010).

Sustainability assessment has reached early adopters and is increasing, and the subject is becoming common in specialized press and journals (Bloomc and Wheelock, 2010), but, even in active countries it is not yet a common practice (McGraw-Hill Construction, 2008). According to innovation diffusion theories, communication is generally the most important element for the introduction of a new paradigm. In this sense, sustainability assessments represent the framework and communication labels for sustainable constructions.

FIGURE 15.3

Sustainability assessment systems around the world (in gray, countries with adopted systems).

The increasing number of certified buildings shows that awareness for sustainability is increasing. Moreover, the assessment scale allowed by many rating systems, which permit defining several sustainability grades, has shown a trend toward higher sustainability levels in the past few years (Berardi, 2012). For example, few buildings among first assessed projects were rated as LEED platinum (best rating) buildings from 1999 to 2002. Then, in 2003, some buildings were rated platinum; currently, this rating is often common among sustainable buildings (Bloomc and Wheelock, 2010).

It is often unclear how to categorize and recognize sustainable buildings and how to measure their sustainability (Steurer and Hametner, 2013). After the energy crisis in the 1970s, regulations promoted energy consumption limits for buildings around the world. As a result, energy consumption evaluation became the first *ante litteram* measure of sustainability for buildings. Meanwhile, sustainability consciousness evolved to the point at which the energy consumption is considered just one among many other parameters.

The complexity of a building suggests a multidisciplinary approach in sustainability assessment (Langston and Ding, 2001). This is also because buildings cannot be considered assemblies of raw materials, but they are generally high-order products that incorporate different technologies assembled according to unique processes on-site (Ding, 2008). The sustainability of a building therefore should be evaluated for every subcomponent, for integration among them in functional units and assembled systems (e.g., the air conditioning system, the envelope), and for the building in its entirety.

A first, but partial, approach to sustainability assessments is through the sustainability evaluation of building products. A similar approach for environmental evaluations is internationally established for many kinds of products. For example, ISO 14020 (2000) defines three environmental labels: the eco-certification of environmental labels (type I); the self-declared environmental claims (type II); and the environmental declarations (type III). Among these, type III is the most common label for building products. However, environmental declarations of products are rarely performed by manufacturers, and environmental product declarations (EPD) in the building sector are slowly diffusing (McGraw-Hill Construction, 2008).

Product eco-certification assessment systems have been developed in different countries with labels such as the American Green Seal, the European Eco-Label, the French NF Environment Mark, the German Blue Angel, and the Japanese Eco Mark. Moreover, specific evaluations for building products exist, especially for timber and concrete-based ones. The aforementioned labels have a binary evaluation and indicate a sustainable product without the ability of measuring its greenness. Since 2011, the new European Construction Products Directive states that a sustainable resource use evaluation is part of the assessment for the CE mark (305/2011, CPR). This implies a larger diffusion of environmental assessments for construction products, at least in Europe. Energy labels of equipment (e.g., heat pumps) represent another way of assessing building sustainability. However, the adoption of certified sustainable materials is not sufficient to obtain a sustainable building because the complexity of this requires a holistic and integrated evaluation (Ding, 2008).

In this sense, product labels and certifications are considered a database for starting a sustainability analysis. However, the building and construction sector is a complex input–output sector, where the material flux is difficult to standardize and, rarely, a priori programmed (Cole, 1998). Some researchers have started promoting assessments that look at buildings as processes of people's satisfaction. This means to look at user requests that evolve through occupancy and that change the way in which buildings behave, too. The dynamic perspective is also supported by considering that local parameters such as weather and international phenomena (e.g., climate change or cultural shifts) continually influence the operational needs of the building. Moreover, buildings are constructed according to client's requests. These aspects prevent buildings from being manufacture-standardized products. Finally, construction stakeholders constitute a variegated network of subjects and differences among them imply several possible points of view in sustainability assessments (Cole, 1998) and later (de Blois et al., 2011; Albino and Berardi, 2012). The next section presents some of the most common systems for sustainability assessment of buildings.

DESCRIPTION OF THE SYSTEMS

According to ISO 15392 (2008), sustainability in construction includes considering sustainable development in terms of its three primary dimensions (economic, environmental, and social) while meeting the requirements for technical and functional performance. In 2008, the Building Research Establishment found that more than 600 sustainability assessment rating systems for buildings worldwide had been created (BRE, 2008). Meanwhile, new systems are continually proposed, whereas the most diffused ones receive an update yearly.

This evolving situation has led to new standards for increasing the comparability among systems, such as "Sustainability in building construction—Framework for methods of assessment of the environmental performance of construction works—Part 1: Buildings" (ISO 21931-1, 2010) and "Sustainability of construction works—Sustainability assessment of buildings—General framework" (ISO 15643-1, 2010).

Systems for sustainability assessment span from energy performance evaluation to multidimensional quality assessments. Hastings and Wall (2007) grouped existing systems into:

- Cumulative energy demand (CED) systems, which focus on energy consumption
- Life cycle analysis (LCA) systems, which focus on environmental aspects
- Total quality assessment (TQA) systems, which evaluate the different dimensions of sustainability (ecological, economical, and social aspects)

This division should not be strictly considered because many assessment systems do not fit perfectly into one category. CED systems are often monodimensional and aim at measuring sustainability of the building through energy-related indicators. LCA systems measure the impact of the building on the environment by assessing the emission of one or more chemical substances related to the building construction and operation. LCA can have one or more evaluation parameters, whereas TQA systems are multidimensional because they assess several parameters. The first two categories of systems have a quantitative approach to the assessment, whereas a TQA system generally has a qualitative or quantitative approach for different criteria.

CED systems

CED systems measure and evaluate the energy consumption of the building. Energy is furnished to buildings to cover needs such as heating, ventilation, air conditioning, water heating, lighting, entertainment, and telecommunications. The specification of the energy request is of primary importance because CED systems can refer to just some of these consumptions (often, just heating and hot water consumption) or they can consider all needs without distinction regarding the final use.

CED systems evaluate the energy consumption over a time unit that generally corresponds to 1 y. However, monthly or semiannual evaluations have been proposed (Marszal et al., 2011). Energy consumption for residential buildings in developed countries at middle latitudes assumes values of hundreds of kWh/m^2 net floor surface per year (kWh/m^2y); e.g., heat consumption of traditional US buildings is 300 kWh/m^2y on average (Butera, 2010). Energy consumption in European buildings is generally lower than in the United States, but still much higher than current technological achievable standards, with values of 108 kWh/m^2y in Greece, 113 kWh/m^2y in Italy, 172 kWh/m^2y in Germany, 178 kWh/m^2y in Poland, and 261 kWh/m^2y in the United Kingdom (Butera, 2010).

Referring to traditional buildings, operating energy consumption dominates the building energy demand with 80% of the total during the life cycle of the building (Suzuki and Oka, 1998). A small energy percentage is consumed for material manufacture and transportation, construction, and demolition. Consequently, energy-saving policies have typically given attention to operation energy performance only (EC, 2010). However, energy consumption standards in new buildings are largely decreasing under the pressure of more stringent requirements (Figure 15.4). This trend reconsiders the way in which CED systems assess the sustainability of buildings.

In the United States, zero-energy buildings (or ZEB) are discussed in the Energy Independence and Security Act (EISA, 2007), whereas the recast of the European Energy Performance of Buildings Directive (EC, 2010) has established that all new private buildings should be ZEB after 2020, whereas public buildings are required to achieve this standard by 2018. A ZEB can be defined as a building with a very high level of energy efficiency, so that the overall annual primary

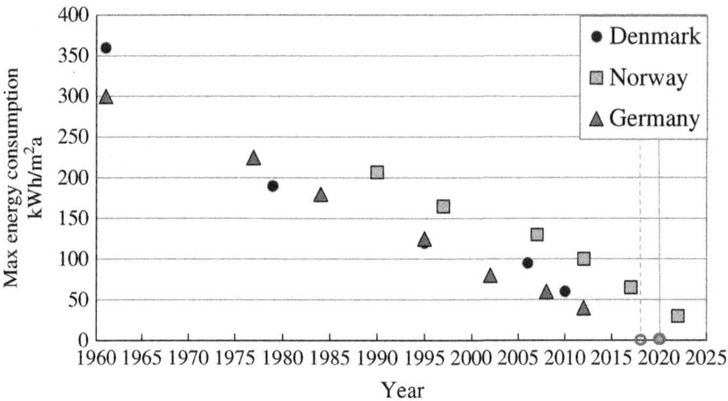

FIGURE 15.4

Energy requirements in some European building codes over the years (data taken from national regulations) in kWh/m^{2}y and zero-energy standard forced by the 31/2010 directive in 2020 for private buildings and in 2018 for public buildings.

energy consumption is equal to the on-site energy production from renewable energy sources. A universally accepted definition of ZEB is still lacking, and several proposed methodologies for ZEB calculations differ for the metric of the analysis (energy, CO_2 emission, costs), the balancing time, and the type of energy use considered in the assessment (Marszal et al., 2011). Table 15.1 reports some recently proposed definitions of ZEB.

As highly efficient buildings are built, the energy needs during construction and demolition processes, together with the embodied energy in construction materials, become relatively more significant. Hernandez and Kenny (2010) have defined the life cycle zero-energy building (LC-ZEB) concept for energy consumption equity in a whole-life perspective. A life cycle evaluation of energy use implies enlarging time and space boundaries in the assessment (Suzuki and Oka, 1998) and represents the trend for energy-based sustainability assessment in the construction sector.

Overall, CED systems adopt a monodimensional analysis that considers the energy flux only. Apart from an energy analysis, some researchers have accounted for other measurement units, such as exergy or emergy. Exergy is the maximum useful work that brings the system into a heat reservoir equilibrium, whereas emergy is the available solar energy directly and indirectly used in a transformation. These units of measurement are related to thermodynamic principles of resource use and may be more appropriate than energy to evaluate building consumption (Marszal et al., 2011), although energy data are more common in literature. The limits of monodimensional analyses have led to promote systems that consider more assessment criteria.

Table 15.1 Definitions of Net ZEB by Order of Appearance

Author	Definition
Gilijamse (1995)	A zero-energy house is defined as a house where no fossil fuels are consumed, and annual electricity consumption equals annual electricity production. Unlike the autarkic situation, the electricity grid acts as a virtual buffer with annually balanced delivers and returns.
Iqbal (2004)	A zero-energy home is one that optimally combines commercially available renewable energy technology with the state-of-the-art energy efficiency construction techniques. A zero-energy home may or may not be grid-connected. In a zero-energy home, annual energy consumption is equal to the annual energy production.
Charron (2005)	Homes that utilize solar thermal and solar photovoltaic (PV) technologies to generate as much energy as their yearly load are referred to as net zero-energy solar homes (ZESH).
Torcellini et al. (2006)	A ZEB is a residential or commercial building with greatly reduced energy needs through efficiency gains such that the balance of energy needs can be supplied with renewable energy technology.
EISA (2007)	A net zero-energy commercial building is a high-performance commercial building designed, constructed, and operated: (i) to require a greatly reduced quantity of energy to operate; (ii) to meet the balance of energy needs from sources of energy that do not produce greenhouse gases; (iii) to act in a manner that will result in no net emissions of greenhouse gases; and (iv) to be economically viable.
Laustsen (2008)	Net ZEB are buildings that over the course of a year are neutral, meaning that they deliver as much energy to the supply grids as they use from the grid. Seen in these terms, they do not need any fossil fuels for heating, cooling, lighting, or other energy uses, although they sometimes draw energy from the grid.
EC, European Commission (2010)	The nearly zero or very low amount of energy required should be covered to a very significant extent by energy from renewable sources, including energy from renewable sources produced on-site or nearby.
Hernandez and Kenny (2010)	An LC-ZEB is one in which the primary energy used in the building in operation plus the energy embodied within its constituent materials and systems, including energy-generating ones, over the life of the building is equal to or less than the energy produced by its renewable energy systems over their lifetime.
Lund et al. (2011)	A ZEB combines highly energy-efficient building designs, technical systems, and equipment to minimize the heating and electricity demand with on-site renewable energy generation typically including a solar hot water production system and a rooftop PV system. A ZEB can be off-grid or on-grid.

Source: *From Kibert (2012) and Berardi (2013a,b,c).*

LCA systems

Several systems have been developed for the environmental assessment of manufactured products according to a life cycle perspective, such as environmental risk assessment (ERA), material flow accounting (MFA), input–output analysis (IOA), and LCA. These systems generally breakdown products and processes into elementary parts.

LCA is the most commonly used of these systems. It divides a building into elementary activities and raw materials to assess the environmental impact over a life cycle from manufacture and transportation to deconstruction and recycling (Seo et al., 2006). LCA is a robust methodology refined on the basis of manufacturing sector experiences. It consists of four phases (ISO 14040, 2006): the goal and definition phase; the life cycle inventory; the life cycle impact assessment; and the improvement assessment phase. LCA systems allow the comparison of products on the basis of the same functional quality. This describes the quality of a product service as well as its duration (e.g., square meter of a building element with a substitution rate of 50 y).

The scientific rigor of LCA is inherent to assessments from cradle-to-grave phases, although it is limited by uncertainties in collecting data relating to building processes. LCA diffusion in the building sector is limited by a lack of information (Seo et al., 2006). In fact, the specificities of the construction processes require data for every building material in any region. Databases have been created for LCA evaluations and implemented in specifically designed software in several geographic areas: BEES in the United States, BEQUEST and ENVEST in England; SIMAPRO and Eco-Quantum in the Netherlands; Ecoinvent in Switzerland; and GaBi in Germany. However, these databases are only valid for assessments in a specific region. The United Nations Environment Program's Sustainable Buildings and Climate Initiative (UNEP-SBCI, 2010) has recently adopted the common carbon metric. This system allows emissions from buildings around the world to be consistently assessed and compared. The assessment reports the carbon intensity, which is the evaluation in weight equivalent of carbon dioxide emitted per square meter per year ($kgCO_2e/m^2y$). The assessment is mainly based on the operational consumption, but it can be extended to the whole life cycle of the building.

An obstacle for LCA diffusion is its specialist structure: outputs of LCA systems are represented by environmental impacts expressed through chemical substances, which are not easily understood by construction sector stakeholders (Langston and Ding, 2001). Another limit that has been discussed is related to the fact that LCA systems do not consider social and economic impacts. To fit this limit, some studies combine the disaggregation analysis necessary for an LCA with an evaluation of economic costs. Such an approach is interesting for the building sector because life cycle cost (LCC) analysis represents a familiar paradigm to construction stakeholders. Combined LCA–LCC can be useful to evaluate environmental and economic aspects in life terms by assigning a price to the different chemical elements. For example, BEES and GaBi systems already permit the selection of cost-effective environmentally preferable products.

TQA systems

TQA systems aim at considering the three aspects of sustainability of buildings: environmental issues such as GHG emission and energy consumption; economic aspects such as investment and equity; and social requirements such as accessibility and quality of spaces. The most common TQA systems are the multicriteria systems. They are largely increasing the attention for sustainable assessment of buildings because they are well related to market interests and stakeholders' culture (Newsham et al., 2009). Multicriteria systems base the evaluation on criteria measured by several parameters and compare real performances with reference ones. Each criteria has a certain amount of available points over the total assessment, whereas the evaluation of sustainability comes from summing the results of assessed criteria. Multicriteria systems are generally easy to understand and can be implemented in steps for each criteria. Moreover, step implementation is allowed during the analysis; in fact, these systems enable the assessment of the building at several stages, from the concept design to the final construction, and can be used during construction. A critical aspect of multicriteria systems is their additional structure (Hahn, 2008).

COMPARISON BETWEEN SYSTEMS

Several multicriteria systems exist worldwide. Because many are just the adaptation of more famous ones to a regional level or for specific scopes, only the most adopted systems are considered here. These are BREEAM, LEED, CASBEE, SBTool, and Green Globes (Berardi, 2012). Other famous rating systems are the Australian Building Greenhouse Rating (ABGR), the Green Home Evaluation Manual (GHEM), the Chinese Three Star, the US Assessment and Rating System (STARS), and the South African sustainable building assessment tool (SBAT).

The United Kingdom was the first country to release a multicriteria system for sustainability assessment before this concept entered into the agenda of international policies with the Rio Conference. The British Building Research Establishment Environmental Assessment Method (BREEAM) was planned at the beginning of the 1990s by the British Research Establishment and was released in 1993. The system has a large diffusion in the United Kingdom, where almost 20,000 buildings have been certified (Figure 15.5), and several hundreds of thousands have registered for assessment since it was first launched in 1990. Since 2009, as a consequence of the worldwide attention garnered for this system, an international version has been released and BREEAM has released versions for Canada, Australia, and Hong Kong. The system is differentiated for 11 building typologies and its evaluations are expressed as percentage of successful over total available points: 25% for pass classification; 40% for good; 55% for very good; 70% for excellent; and 85% for outstanding. The evaluation categories are management, health and well-being, energy, transport, water, materials, land use, ecology, pollution, and innovation.

The most well-known TQA system is LEED (Leadership in Energy and Environmental Design), which was released in 1998 by the US Green Building Council (USGBC). LEED allocates points to incentivize building project teams to comply with requirements that address the social, environmental, and economic outcomes identified by USGBC. Points are allocated through a weighting process whereby a credit receives one or more LEED points based on each credit's relative effectiveness. The more effective a building is at addressing the goals of the system, the more points it receives. The fourth version of this system was released at the end of 2013. This system is currently available for several building typologies (Table 15.2). There are six evaluation categories to obtain the 100 possible

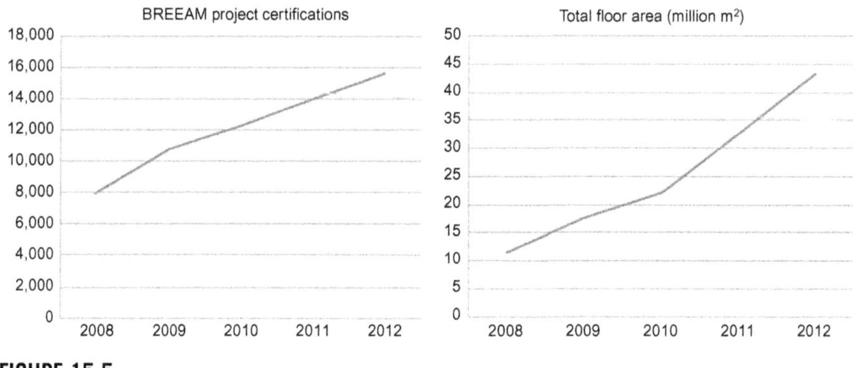

FIGURE 15.5

Diffusion of BREEAM certifications toward the years.

From BRE (2014).

Table 15.2 Current Available LEED Systems

Building Design + Construction	Building Operations + Maintenance
New construction	Existing buildings
Core and shell	Schools
Schools	Retail
Retail	Hospitality
Hospitality	Data centers
Data centers	Warehouses and distribution centers
Warehouses and distribution centers	
Health care	**Interior Design + Construction**
	Commercial interiors
	Retail
	Hospitality
Homes	**Neighborhood Development**
Homes and multifamily low rise	Plan
Multifamily mid-rise	Built Project

points of the standard: location and transportation; sustainable site; water efficiency; energy and atmosphere; material and resources; and indoor environment quality. A weighting approach in LEED was introduced in version 3 (2009); the basic approach is that each of the LEED credits are independently evaluated along each impact category in a matrix-style format with credits as rows, impact categories as columns, and associations between credits and impact categories as individual cells. For each cell, an association between credit and impact category is determined and given a weight that depends on the relative strength of that association (i.e., credit outcome weighting). The weighting is established according to the goals to which a LEED project should answer: reverse contribution to global climate change; enhance individual human health and well-being; protect and restore water resources; protect, enhance, and restore biodiversity and ecosystem services; promote sustainable and regenerative material resources cycles; build a greener economy; and enhance social equity, environmental justice, and community quality of life. LEED points accumulated are divided in the following categories: at least 40 points for certified buildings; 50 for silver; 60 for gold; and 80 for platinum. Although released in the United States, GBC has diffused worldwide over the years, and recently the World GBC has opened regional chapters in Europe, Africa, America, and Asia. At the end of May 2013, almost 60,000 buildings were registered for certifications, of which almost 50,000 were in the United States, whereas the registered LEED projects were 1,156 in China, 808 in the United Arab Emirates, 638 in Brazil, and 405 in India. Current requests for new certifications for buildings are pending in 110 countries (Berardi, 2012).

CASBEE (Comprehensive Assessment System for Building Environmental Efficiency) is a Japanese rating system developed in 2001 that is also available in English. CASBEE covers a family of assessment tools based on a life cycle evaluation: predesign; new construction; existing buildings; and renovation (CASBEE, 2010). This system is based on the concept of closed ecosystems and considers two assessment categories, building performance and environmental load. Building performance covers criteria such as indoor environment, quality of services, and outdoor environment, whereas environmental loads cover criteria such as energy, resources and materials, reuse and reusability, and off-site environment. By relating the previous two main criteria, CASBEE results are presented as a measure of eco-efficiency on a graph with environmental loads on one axis and quality on the other, so that sustainable buildings for CASBEE have the lowest environmental loads and highest quality. Two hundred buildings have been certified with this system, although the number is rapidly increasing.

At the end of the 1990s, the Sustainable Building Council promoted an internationalization of rating systems under the leadership of Natural Resources Canada (NRC). Toward this initiative a common protocol, SBMethod, was developed. Using the general scheme, several countries have proposed national versions of this system, such as Verde in Spain, SBTool PT in Portugal, and SBTool CZ in the Czech Republic. In Italy, this protocol was implemented in 2000 as SBTool IT before evolving in ITACA. Ten Italian regions have already adopted modified versions of this system to more aptly cover regional specificities. In 2005, adapting the Canadian version of BREEAM, the Green Globe Initiative (GBI) launched a new rating system known as Green Globes. Criteria for this include project management, site, energy, water, indoor environment, resource, building materials, and solid waste.

A critical aspect of multicriteria systems regards the selection of criteria and the weights given to them. These elements show which aspects of building performance are given more consideration in sustainability assessments. Figure 15.6 shows weights assigned by these five systems grouping the criteria of each into seven main categories. The selection of these categories was based on main sustainability building aspects (Langston and Ding, 2001) and later (Berardi, 2012): site selection, energy efficiency, water efficiency, materials and resources, indoor environmental quality, and waste and pollution. The category "others" contains criteria that do not fit into the other six categories. For LEED, version 2 was used for this comparison. Credits assigned in the LEED system for low-emitting materials were assigned to the IEQ category; however, they could also be assigned to the waste and pollution category. Management and innovation criteria have been included in the category "others." For example, LEED assigns 7% of its credits to innovations, BREEAM has 15% for construction management, and Green Globe has 12.5% for project management. Moreover, in the category "others," there are points given by CASBEE for mitigation and off-site solar energy, and by GBTool for the cultural perception of sustainability.

It is interesting to note that energy efficiency among assessment systems in Figure 15.6 is always considered the most important category (weight average among the six systems, 25.5%), followed by IEQ (17.7%), waste and pollution (15.9%), sustainable site (13.2%), and materials and resources (11.5%).

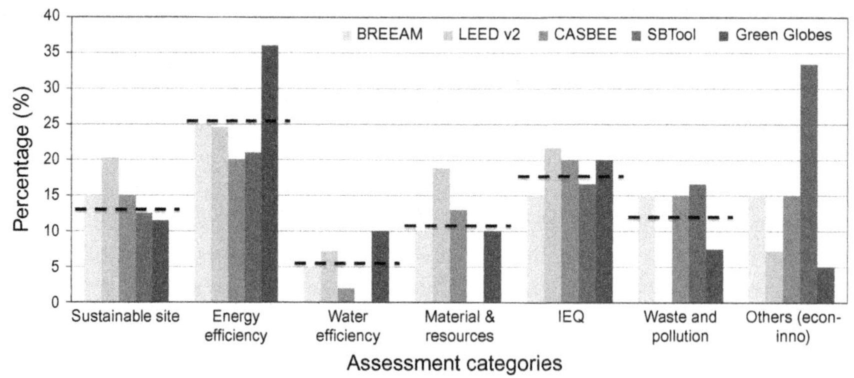

FIGURE 15.6

Comparison of the weight assigned by five sustainable assessment systems for building, grouping the respective criteria into seven categories.

Readapted from Berardi (2012).

The Green Globes assigns a higher percentage of its assessment weight to the energy-efficient (36%). This is established by the inclusion of criteria that are not presented in other systems, such as the correct size energy-efficient system or energy-efficient transportation.

Above-averages have no rigorous meaning, standard deviations among systems are high, and percentages may change if other versions of the systems are considered. However, studies have shown similar structures among sustainability rating systems (Smith et al., 2006). Finally, it should be remembered that evaluation criteria and weights comprise just one of the ways to compare systems. Fowler and Rauch (2006) compared the aforementioned systems for other properties (applicability, usability, communicability), again finding some similarities. Differences among the systems have led to the creation of the Sustainable Building Alliance to establish common evaluation categories and to improve comparability among systems (Berardi, 2013b).

Many studies have discussed the limits of rating systems. Unscientific criteria selection has been criticized by Rumsey and McLellan (2005) and by Schendler and Udall (2005). Bower et al. (2006) stated the lack of overall life cycle perspective in evaluations is a critical aspect of TQA systems. The US National Institute of Standards and Technology analyzed the LEED system from an LCA perspective, and this is not a completely reliable sustainability assessment system (Scheuer and Keoleian, 2002). From Figure 15.6, it is clear that in the selection of assessment criteria, environmental aspects in existing systems receive much more attention than economic and social aspects (Sev, 2009). Recently, some multicriteria rating systems more closely related to a TQA have been released. For example, the Deutsche Gesellschaft für Nachhaltiges Bauen (DGNB), available since 2009, aims at evaluating sustainability through the quality of the building; economic aspects emerge explicitly and, in the category of technical quality, paradigms such as performance, durability, ease of cleaning, as well as dismantling and recycling are considered, too. More attention is given to social aspects than in other rating systems. Functional aspects such as space efficiency, safety, risk of hazardous incidents, handicap accessibility, suitability for conversion, public access, and art and social integration are also considered. Nine-hundred thirteen projects have been assessed through this system, and this number is increasing rapidly.

CHARACTERISTICS OF CERTIFIED BUILDINGS

In the present section, assessment results with a sustainability rating system are used as a proxy variable to analyze the characteristics of certified buildings. In fact, the results of constructed buildings can be useful to understand the state of the art of sustainability assessment of buildings. This section reports results more extensively described by Berardi (2012).

Sustainable rating systems reviewed previously are voluntary standards, and the adoption of which is often motivated by signaling reasons. This means that the owner of the building decides to perform a sustainability assessment to communicate something to the market and the public (Mlecnik et al., 2010). According to King and Toffel (2007) and Berardi (2012), signaling and intrinsic benefits are mixed together when sustainable rating systems are used. In their analysis, this clearly emerged from the decreasing number of buildings that obtained a larger number of credits than the minimum number for a given certification level. Buildings generally aim at an established certification level and rarely show higher performance than the minimum ones for the given certification level.

Although there is space for improvement in LEED (Bower et al., 2006) and later (Hahn, 2008; Newsham et al., 2009), this system is the most diffused system worldwide; hence, it was chosen for the following analysis. A sample of 490 buildings was selected in the GBC database from already-built projects assessed with version 2 of LEED. The sample was composed of buildings that had allowed diffusion of their evaluation data. Selected buildings belonged to several typologies, with a majority being commercial (52%) and residential (30%) buildings. The time of construction was very similar among buildings, from 2002 to 2009. Figure 15.7 shows points earned on average over the total possible points. The data suggest several considerations:

- Sustainable sites is an important category in the overall evaluation; however, assessed buildings reach less than 50% of the available points on average. The selection of a sustainable site is often influenced by property possibilities, municipal policies, and previous land uses, making a free selection difficult.
- Energy and atmosphere is the category with the largest number of points, but with the lowest rate of successful points over possible points (38%).

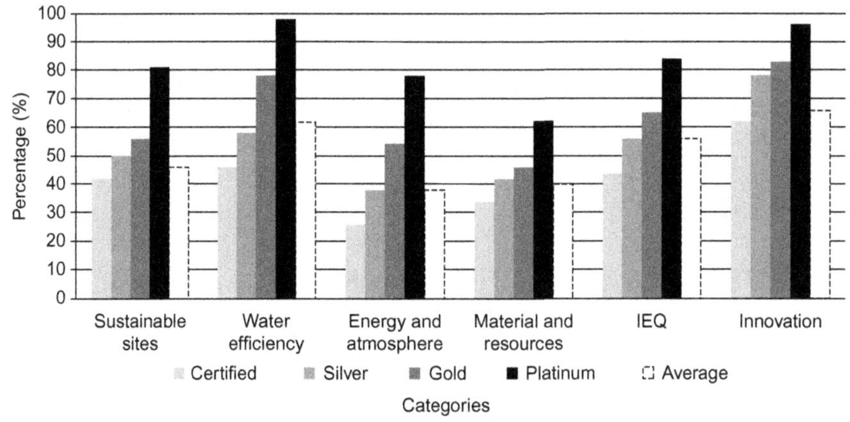

FIGURE 15.7

Earned points over the total possible points in each assessment category for different classes of 490 LEED-rated buildings (Berardi, 2012).

- Indoor environmental quality is the second category for available points but the first for contributing to the total score, and earned points on average are 56% of available points.
- Water efficiency receives only a few points in the standard, despite its importance for a sustainable building. The most probable reason for this is that few actions can lead to a significant efficiency in the use of this resource and, in fact, buildings obtained 62% of the available points.
- Material and resources category has a considerable number of available points, but effectively earned ones are few, with an average of 40%.
- Innovation and design process category has a low number of available points and, on average, buildings are successful in this category (66%).

Figure 15.7 also represents the percentages for buildings of different classes. In platinum buildings, the percentage of earned points in the energy and atmosphere category increases with respect to other classes of buildings, becoming the most contributing category to the overall score in absolute value. However, if compared with the total available points in this category, obtained points have a lower percentage than in other categories. The material and resources category is characterized by the high percentage of points. The high percentage of success in the innovation category can be justified by the freedom the LEED system allows for points in this category. Moreover, it is interesting to look at the results for the water efficiency category; the importance of this resource, together with the ease of building systems for water harvesting, suggest that water efficiency can be reached independently from the rate of certification. The comparison between achieved points in silver and gold buildings shows that the improvement in the assessment is slightly influenced by the material and resources category. Conversely, a larger improvement occurs in the energy and atmosphere category.

Figure 15.8 disaggregates the statistics in Figure 15.7 by representing the earned points for any of the 69 criteria in LEED version 2. Figure 15.7 shows which points are more often reached. In the indoor environmental quality category, criteria from IEQ 1.0 to 5.0 are earned by a high percentage of buildings in any class; these criteria correspond to the air-monitoring system, an increase in ventilation, management of air quality during construction, use of low-emitting materials, and control of pollutant source. This suggests that sustainable buildings have learned how to achieve good indoor quality or, on the contrary, that the required target levels are in line with or below the common practice of sustainable buildings.

Energy-related criteria are among the less achieved ones. In particular, the percentage of buildings with renewable energy production is low for any class of buildings, with only 1% of certified buildings in the selected sample able to produce 20% of energy from renewable sources (E&A 2.3). High energy performance (E&A 1) is partially achieved, and many buildings make only limited choices toward optimization: high success rates for E&A 1.1 and 1.2 (optimize energy performance through lighting power and lighting controls) but low success

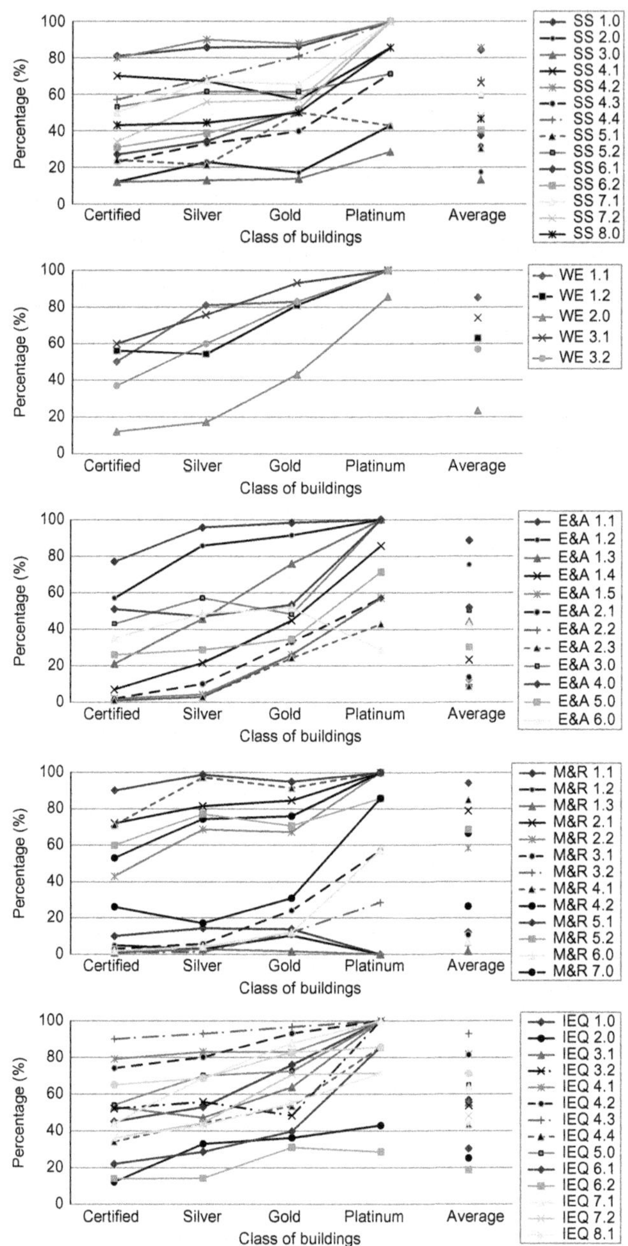

FIGURE 15.8

Percentages of earned points over total points in several categories of the LEED system in buildings of different classes.

From Berardi (2012).

rates for E&A 1.3, 1.4, and 1.5 criteria, which are related to energy saving of HVAC, equipment, and appliances.

Urban and brownfield redevelopment criteria (SS 2.0, 3.0) have low success rates, confirming that the possibility of selecting a "more sustainable" land occurs after the priority of construction. On the contrary, criteria regarding alternative transportation (Public Transportation Access SS 4.1 and Bicycle Storage and Changing Rooms SS 4.2) have a high success rate. A similar discourse is valid for other criteria in the sustainable site category, such as the mitigation of the heat island effect (SS 7.2).

In the water efficiency category, water use reduction has a high percentage of success among all certification levels with values that, in certified buildings, go from 60% for 20% reduction in water use (WE 3.1) to 37% for 30% reduction (WE 3.2). In contrast, the implementation of innovative wastewater technologies (WE 2.0) represents a complicated target for best-rated buildings.

Finally, criteria in the material and resources category have different behavior. In fact, high successful percentages are reached for construction waste management (M&R 2.1, 2.2) and use of local and regional materials (M&R 5.1, 5.2) in any class of buildings. In contrast, other criteria in this category show a low success rate even in platinum buildings, and among these are criteria for adoption of building reuse materials (M&R 1.1, 1.2, 1.3) and rapidly renewable materials (M&R 6.0).

LIMITS AND TRENDS IN SUSTAINABILITY ASSESSMENT OF BUILDINGS

As described in the previous section, single and multidimension systems for sustainability assessment of buildings exist. However, assessments through a single dimension have received much criticism (Nijkamp et al., 1990). An increasing awareness of externalities, risk, and long-term effects have led to a larger diffusion for multicriteria systems (Janikowski et al., 2000). Available multicriteria systems have been accused of a lack of completeness because they neglect some criteria; e.g., they rarely take into account the economic and social dimensions of sustainability (Ding, 2008). Moreover, by neglecting the evaluation of economic aspects, current systems allow and incentivize an additive logic, which has been largely criticized. The importance of economic and social evaluations has recently emerged in defining systems for developing countries (Gibberd, 2005). Limits of sustainability assessments suggest that more complete rating systems are necessary to assess the multidimensional aspects of sustainability.

A comprehensive approach to the evaluation has led to design systems that require more detailed information. For example, the last version of GBTool comprises more than 120 criteria. However, the complexity has been pointed out as a limit for the diffusion of current rating systems (Mlecnik et al., 2010). In fact, when sustainability rating systems are perceived as too complex by building stakeholders, then sustainable practices are slowly adopted (Albino and Berardi, 2012). A balance between completeness in coverage and simplicity of use is necessary to spread sustainability assessment systems. The larger diffusion of multicriteria TQA systems than LCA ones is probably because of their simplicity and checklist structure. In fact, although LCA analysis is often more rigorous than that of multicriteria systems, they are still complex and their diffusion is limited to a few specialists. The importance of simple systems is also emerging as a factor in making them useful as design tools by introducing them when only preliminary information is available. The review among some TQA systems has shown a trend for whole-life perspective analysis as the assessment is moving to cover the operation phases and sometimes the dismantling phase.

An open aspect of sustainability rating systems regards possible regional adaptations in assessment criteria. The experience of BREEAM and SBC-ITACA, similar to many other experiences, shows that regions are adapting the original system to local characteristics and priorities with regional criteria. It is evident that sustainability evaluation needs site adaptations to fit sustainable requirements with contextual aspects. This approach is shared more and more worldwide.

An important trend in sustainability assessment is seen in the increasing attention to the neighborhood and construction site. First, assessment systems considered the building a manufactured product, and evaluated it almost in isolation. However, the importance given to the surrounding site is largely increasing; e.g., available points for sustainable sites have increased from 15% to 23% from version 2.2 to version 3 of LEED. The next section shows how the rating systems have recently evolved to consider aggregation of buildings, neighbors, and communities.

SUSTAINABILITY ASSESSMENT OF URBAN COMMUNITIES

In this section, the most internationally well-known systems for sustainability assessment of urban communities through multicriterion ratings are described and compared. The considered systems were selected for their established worldwide diffusion and resonance with the help of institutions and organizations actively involved in promoting their use. This criterion resulted in neglecting the systems that have been developed for specific cases and communities (Cartwright, 2000) or the indicators developed within research projects (Xing et al., 2009). However, these last references were considered for the framework of comparison and for the discussion about the critical aspects of the selected systems.

DESCRIPTION OF THE SYSTEMS

Current systems for communities were developed within the past years as an evolution of sustainability assessments systems for buildings. They maintain the logic and structure of the analogous building assessment systems, but they have gone beyond the building scale by redefining the assessment criteria. The most well-known systems are BREEAM Communities (Com), CASBEE for Urban Development (UD), and LEED for Neighborhood Development (ND). In addition to their worldwide adoption, they were selected because BREEAM, CASBEE, and LEED protocols have already reached a significant diffusion for sustainability assessments of buildings, and they are quickly diffusing to communities.

In 2009, the British Building Research Establishment launched BREEAM Com. This system received an update in 2012. The new version is simpler, less prescriptive, and better-aligned with the planning process with respect to the original one. BREEAM Com applies the BREEAM methodology to the community level and can be used for both new and regeneration development projects. BREEAM Com assesses the environmental, social, and economic impacts of a community. The assessment criteria are divided into eight categories: climate and energy; resources; place shaping; transport and movement; ecology and biodiversity; buildings; business and economy; and community (BREEAM, 2009). At this time, eight projects have been certified under BREEAM Communities, and another 18 are currently registered and undergoing assessment, with the size of development ranging from 2 to 179 ha.

CASBEE UD results, similarly to those for the analogous system for buildings, are presented as a measure of eco-efficiency on a graph with loads on one axis and quality on the other. Sustainability for CASBEE corresponds to the lowest environmental loads and the highest quality and performance. The performance criteria consider aspects as the natural environment, quality of services, and the contribution to the local environment, whereas the environmental load covers aspects related to the impact on local environment, social infrastructure, and energy and material consumptions. CASBEE UD was developed in 2007 to assess urban areas (CASBEE, 2007), and since then it has officially been used as a self-assessment tool in many projects. CASBEE UD partially promotes local stakeholders' engagement in the choice of the weighting coefficients assigned to different criteria. This often uses a qualitative assessment.

LEED for ND is a system developed by the USGBC in partnership with Congress for the New Urbanism (CNU) and the Natural Resources Defense Council (NRDC). It was developed in 2009 for the United States, and it was later applied in Canada and China; versions for other countries are currently being developed (LEED, 2009). LEED ND places emphasis on the site selection, design, and construction elements that bring buildings and infrastructures together into a neighborhood and relates this to its landscape and regional context. Smart location, neighborhood pattern, eco-design, green infrastructure, and buildings are the main categories considered by LEED ND (LEED, 2009). Moreover, LEED ND rates innovative practices and regional priorities as the sustainable features of a community.

COMPARISON BETWEEN THE SYSTEMS

The systems described in the previous section are compared according to the assessment criteria. First, it is important to clarify the dimensions of an urban community in the different systems. BREEAM allows consideration of sizes from 10 units (small projects) to 6,000 units (large projects) as an urban community; however, it also considers bespoke projects of more than 6,000 units after confirmation by the British Research Establishment. LEED suggests considering communities with an extension less than 1.3 km^2 and suggests dividing the project if the surface exceeds this value (LEED, 2009). For example, the pilot projects that have been assessed with LEED ND have an average project size of 1.2 km^2 and a median size of 0.12 km^2. In particular, the smallest size was 687 m^2, whereas the largest was approximately 51.8 km^2 (LEED, 2009). No indication regarding the dimensions of an urban community is presented in CASBEE UD. These data confirm that the dimension of a community has been particularly heterogeneous, ranging from a single building to a medium city.

Each sustainability assessment system bases its evaluation on several parameters whose rates are generally obtained comparing real performances with referenced ones (benchmarks). However, a few criteria are evaluated by looking at the presence of an element. For example, LEED ND enables earning 1 point for the presence of a bicycle network without going into the assessment of its characteristics. Points earned in this simple way are rarer than those in the other systems.

In multicriterion systems, each criterion has a certain weight over the total assessment, and the overall sustainability evaluation comes from the weighted sum of the results for all the criteria. A fundamental aspect in multicriterion systems is the selection of the criteria. Unfortunately, reasons for the choice of the criteria are not explicitly discussed by any of the responsible agencies (Haapio, 2012). Table 15.3 reports the main categories of the assessment criteria that are used by the three systems considered here. Similarities between the main categories of the different systems exist. For example, all consider the sustainability of the land in terms of ecology and natural environments. However, other sustainability aspects are considered in only some systems. BREEAM Com, for example, attributes more importance to business opportunities, whereas social aspects such as the history, tradition, and culture preservation are only considered in CASBEE UD.

Using the respective manuals, the assessment criteria of each system were divided into seven main categories. These categories were chosen according to both the structure of the systems and the requirements of a sustainable urban community (Kellett, 2009). Selected categories were sustainable land (sustainable planning, design and buildings, microclimates), location (previous land use, reduction of sprawl), transportation (pedestrian, bicycle, or public transportation), resource and energy (use and selection of materials, waste management, energy production, and efficiency), ecology (biodiversity), economy and business (employment and new opportunities), and well-being (quality of life).

Table 15.3 Main Categories in the Three Sustainability Assessment Systems Considered in This Chapter

	CASBEE for Urban Development		
BREEAM Communities	**Performance**	**Load**	**LEED for ND**
Climate and energy: reducing the contribution to climate change through energy efficiency and passive design	Natural environment, microclimates, and ecosystems	Environmental impact on microclimates, façades, and landscape	Smart location and linkage: emphasizes development of preferred urban areas, brownfield redevelopment, bicycle network or housing and jobs proximity, and wetlands and water bodies conservation
Resources: emphasizes sustainable and efficient use of resources through construction management	Service functions for the designated area	Social infrastructure	Neighborhood pattern and design: focuses on walkable streets, public transportation access, reduction of car-dependency compact development, and also mixed-use neighborhood centers and mixed-income communities
Place shaping: provides a framework for the design and layout of local area	Contribution to the local community: history, culture, scenery, and revitalization	Management of the local environment	Green infrastructure and buildings: decreasing environmental impact caused by construction and maintenance of buildings and infrastructure; energy and water efficiencies are mainly emphasized; some attention to waste management
Transportation: focuses on sustainable public transportations, walking, and cycling			Innovation and design process
Community: encourages integrating the community with surrounding areas and emphasizes mixed use			Regional priority
Ecology and biodiversity: aims at conserving the site ecological value by reducing pollution			
Business: aims at providing opportunities for local businesses			
Buildings: focuses on the sustainability performance of buildings			

Content analysis was used to attribute the criteria of each system to the previous categories. However, the organization of assessment criteria into the previous seven categories resulted in some difficulties because the systems were not easily and fully accessible and criteria among systems did not perfectly overlap.

Figure 15.9 depicts the categories with a high frequency of occurrence in the varied systems. The results show that great importance is assigned to the sustainable use of the land, ecological measures, and sustainable transportations. On the contrary, economic themes are scarcely considered. The average weights among systems for each category were 33% for sustainable land, 9% for location, 13% for transportation, 16% for resources and energy, 21% for ecology, 3% for economy and opportunity, and 5% for well-being.

Surprisingly, existing systems assign a low weight to energy and resource-related topics. This is significantly different from the sustainability assessment of buildings where energy-related criteria represent the most influencing category (Berardi, 2012). For example, the weight assigned to the criteria related to building energy efficiency, solar orientation, on-site renewable energy sources, and district heating and cooling was only 8 points more than the possible 110 in LEED ND. Together with the attribution of criteria in the seven categories, the analysis showed the level of detail and the difference among criteria in each system. The number of criteria used by each system is particularly different. For example, transportation is assessed through 11 criteria within BREEAM Com (over the 51 parameters used by this system), whereas fewer criteria are considered in the other two systems. This also corresponds to different weights that transportation criteria have: 22% of total weight in BREEAM Com; 8% in CASBEE UD; and 15% in

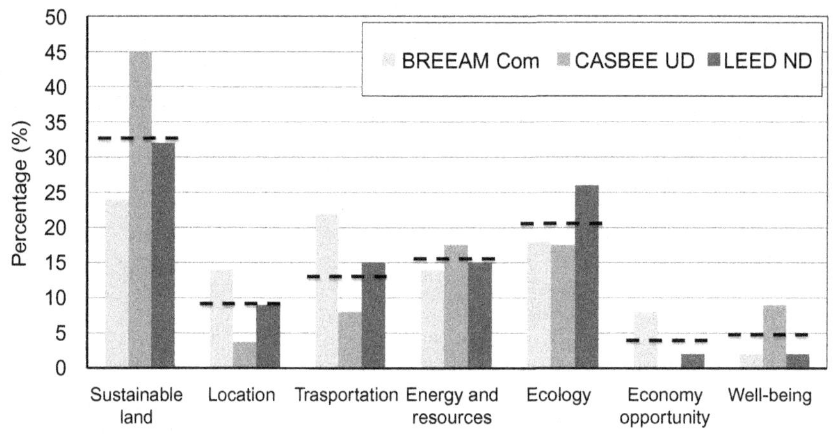

FIGURE 15.9

Comparison of the weight given by three sustainability assessment systems for urban communities grouping the assessment criteria into seven categories (Berardi, 2013c).

LEED ND. In particular, LEED ND is more focused on the promotion of a compact design of the community, favoring walkable streets. This reminds us of the different status and types of public transportation in the United Kingdom and United States where previous systems have been developed. As a consequence, US communities are first interested in increasing compactness to accommodate walkable streets and, second, they consider the public transportation (Berardi, 2013c).

The different weights given to the location category are probably influenced by the different scopes and applications of the systems. LEED ND aims to be applied in new urban communities, whereas BREEAM Com is mainly focused on interventions of rehabilitation. This different field of application has probably discouraged assigning a high weight to location in BREEAM Com, because this category is generally satisfied in rehabilitation projects.

The well-being category has received a low average weight in Figure 15.9. This is also related to the fact that only assessment parameters that exclusively referred to social aspects of the quality of life have been considered in this category. However, other parameters, such as bike transportation or urban microclimate, play a fundamental role for the quality of life.

CHARACTERISTICS OF CERTIFIED COMMUNITIES

In this section, a sample of certified communities with LEED ND system is used for a detailed examination of the characteristics of sustainable communities. LEED ND was chosen because there are a large number of assessed communities versus other rating systems. Forty-two communities assessed with LEED ND were considered. Table 15.4 gives the percentage of total and mean earned points of each assessment criteria. This allows consideration of the frequency of successfully achieved points and the criteria most influential on the final rate. The data suggest several considerations. In the smart location and linkage category, the selection of a sustainable site was often difficult because it was influenced by property possibilities and municipal policies. Moreover, the selection of a brownfield redevelopment was uncommon, whereas many communities preferred reduced automobile-dependence categories (Berardi, 2013c).

The neighborhood pattern and design category had the largest number of points within LEED ND. Criteria in this category showed significant differences in their success rates; criteria such as the diversity of uses or the presence of walkable streets were commonly satisfied in assessed projects, whereas other criteria, such as the affordability of rental housing, the restoration of habitat, or wetlands and water bodies, were uncommon.

The green construction and technology category showed that sampled communities were able to minimize site disturbance, reduce water use, and plan a storm water management. However, they became particularly unsuccessful with criteria such as building reuse and adaptive reuse, solar orientation, on-site energy generation, wastewater management, contaminant reduction, and district heating and cooling. This suggests that energy and resource efficiency is still difficult to

Table 15.4 Criteria of LEED ND Version 3 and Percentages of Total and Mean Earned Points Over the Total Possible Points Among the Assessment Categories in a Sample of Assessed Communities (Berardi, 2013c)

	Max. Available Points	% of Communities That Earned Points	Mean Earned Points
Smart location and linkage	**27**		**16.8**
Smart location	Prerequisite		
Imperiled species and ecological communities	Prerequisite		
Wetland and water body conservation	Prerequisite		
Agricultural land conservation	Prerequisite		
Floodplain avoidance	Prerequisite		
Preferred locations	10	96	6.9
Brownfield redevelopment	2	45	0.9
Locations with reduced automobile dependence	7	92	4.0
Bicycle network and storage	1	77	0.8
Housing and jobs proximity	3	85	2.3
Steep slope protection	1	66	0.7
Site design for habitat or wetlands and water body conservation	1	56	0.6
Restoration of habitat or wetlands and water bodies	1	36	0.3
Long-term conservation management of habitat	1	34	0.3
Neighborhood pattern and design criteria	**44**		**25.6**
Walkable streets	Prerequisite		
Compact development	Prerequisite		
Connected and open community	Prerequisite		
Walkable streets	12	96	6.1
Compact development	6	87	2.9
Mixed-use neighborhood centers	4	99	3.3
Mixed-income diverse communities	7	80	5.3
Reduced parking footprint	1	73	0.7
Street network	2	80	1.3
Transit facilities	1	63	0.6
Transportation demand management	2	55	0.8
Access to civic and public spaces	1	89	0.9
Access to recreation facilities	1	88	0.9
Visitability and universal design	1	63	0.6
Community outreach and involvement	2	90	0.9

Table 15.4 Criteria of LEED ND Version 3 and Percentages of Total and Mean Earned Points Over the Total Possible Points Among the Assessment Categories in a Sample of Assessed Communities (Berardi, 2013c) *Continued*

	Max. Available Points	% of Communities That Earned Points	Mean Earned Points
Local food production	1	27	0.4
Tree-lined and shaded streets	2	40	0.4
Neighborhood schools	1	40	0.3
Green construction and technology	**29**		**13.3**
Certified green building	Prerequisite		
Minimum building energy efficiency	Prerequisite		
Minimum building water efficiency	Prerequisite		
Construction activity pollution prevention	Prerequisite		
Certified green buildings	5	86	2.9
Building energy efficiency	2	73	1.5
Building water efficiency	1	40	0.6
Water-efficient landscaping	1	40	0.6
Existing building reuse	1	18	0.2
Historic resource preservation and adaptive use	1	18	0.2
Minimize site disturbance during construction	1	82	0.8
Stormwater management	4	86	2.9
Heat island reduction	1	75	0.8
Solar orientation	1	32	0.3
On-site renewable energy sources	3	30	0.3
District heating and cooling	2	15	0.2
Infrastructure energy efficiency	1	22	0.2
Wastewater management	1	22	0.2
Recycled content in infrastructure	1	53	0.5
Solid waste management infrastructure	1	53	0.5
Light pollution reduction	1	64	0.6

achieve on the community scale, and that communities aiming at sustainability certification do not rigorously pursue solutions within this category. Moreover, results in Table 15.4 suggest that sustainable communities are generally able to reduce the impact of their material uses, although this ability is shown by selecting new virgin materials rather than by using recycled ones.

LIMITS OF SUSTAINABILITY ASSESSMENT OF COMMUNITIES

This section discusses the following limits of sustainability; assessment systems of communities; assessment of a weak sustainability and lack of an appropriate assessment of social, environmental, and economic sustainability; static sustainability assessment; and minimal adaptability and lack of stakeholders' engagement.

Assessment of a weak sustainability

Although the number of criteria that are considered in the sustainability assessment systems for communities is generally high (51 criteria in BREEAM Com, 80 in CASBEE UD, and 59 in LEED ND), every system is dominated by an environmentally biased approach. The analysis has also revealed that every system lacks integration of the different aspects and criteria but follows a strict additional approach. This weakness was recognized as a possible incentive to the promotion of weak sustainability (Berardi, 2013c). Bourdic and Salat (2012) criticized existing systems because they only assess the different criteria by comparison to benchmark values, and they stated that there is no quantitative evidence that a high-rated community emits less carbon or is more sustainable than a lower-rated one. Considering the often untransparent process for the selection of the criteria, their weights, and their benchmarks, previous criticisms are difficult to overcome. Moreover, the aggregated level of assessment, which synthesizes the evaluation in one single rate, reduces the ability to obtain a robust and transparent output (Mori and Christodoulou, 2012).

An important leakage in existing systems involves the social features of sustainability. Although it is unavoidable that a sustainable community should promote social relationships and well-being of citizens, the analyzed systems poorly assess the importance of social life and the sense of citizenship. They misrepresent one of the main reasons for urban life. The low importance of social aspects is caused by the adoption of an approach that considers almost exclusively the physical and material properties of the built environment. On the contrary, the new interpretations of sustainability and the increasing awareness that the built environment is more than the physical space should lead to consideration of social criteria (Bond and Morrison-Saunders, 2011).

Vallance et al. (2011) discussed the importance for the built environment to assume the status of the locus of a community, and they emphasized the importance of assessing the sense of community. This suggests that other criteria have to be considered. A proposal for a set of social indicators has been given by Albino and Dangelico (2012). Based on a review of country-relevant well-being indicators, they proposed three main domains of well-being criteria: material well-being; quality of life; and social inclusion. Moreover, they indicated 10 relevant dimensions with 45 indicators:

- Material well-being: income and wealth, employment, and housing
- Quality of life: health, education, work–life balance, political well-being, and safety
- Social inclusion: social cohesion and equity

Several of these indicators have been introduced in a recently proposed tool for city (CASBEE), which considers crime prevention and quality of housing for the living environment, educational, cultural, medical, and child care services for social services, and information pressure for social vitality (Murakami et al., 2011). Obviously, the application of previous criteria raises difficulties such as the lack of benchmarking data. A possible solution may be represented by parameters that compose the Better Index Life indicator, for which national open source values are available (OECD, 2011b). However, further studies are necessary to evaluate social sustainability parameters on the community scale.

Another limitation in existing systems regards the misrepresentation of economic sustainability. Figure 15.9 showed that low importance is given in current systems to the ability to promote business and economic opportunities within a sustainable development. Reasons for this can be related to the strong attention to economic aspects historically given in the evaluations of development. However, local businesses and new economic activities are critical for sustainable communities. Consequently, sustainability assessment systems should be able to take these into account, decouple, and promote both economic growth and environment protection. Among the dimensions of the social sustainability assessment proposed by Albino and Dangelico (2012), few also referred to economic sustainability. For example, in the category of material well-being, they consider the household net income, the employment rate, and the percentage of homeless people. Moreover, criteria referring to socioeconomic aspects within an urban community have to be taken into account to prevent decoupling between these two aspects.

Another limit in the current systems for urban communities refers to environmental protection measurements. Given the high consumption of resources in urban environments, sustainable communities should increase their focus on reusing materials and their circulation in closed-loop cycles of production, recovery, and remanufacture (Mang and Reed, 2012). This should result in more eco-friendly practices. Existing sustainable assessment systems rarely focus on a lifestyle in harmony with nature. Criteria aiming to integrate physical, functional, and emotional properties of a community in a holistic perspective would help promote a more integrated idea of sustainability and would overcome the eco-technocratic trend that can be seen in Table 15.3.

Limits of static sustainability assessment

In current systems, the assessment is a process realized once at the beginning of the urban community development. However, recent definitions of sustainability have encouraged looking at the assessment as a moving target and have shown that doing so only once is not sufficient (Brandon and Lombardi, 2005).

Moreover, existing systems are seldom considered later, especially because the sustainability assessments are often promoted by developers alone. The static assessment prevents looking at the trends in the evolution and performance of a community (Berardi, 2013c). Instead, continuous evaluations should be incentivized in a way that sustainability assessments become an interactive process that could be used to map the evolution of the urban development (Lowe, 2008). This means that it is particularly important to monitor progress through a continuous check (Innes and Booher, 2000).

The importance of dynamic evaluations is also related to the time dependence of sustainability and to the changes of the requisites and benchmarks that sustainability requires (Martens, 2006) and later (Berardi, 2011). Criteria that consider the evolution of a community should be introduced in the assessments.

Limits of adaptability and stakeholders' engagement

The analysis of existing systems has shown a link between the assessment systems and the context in which they have been developed. This relationship limits the use of these systems in other countries, unless the criteria are modified to consider specific culture and laws of those countries (Haapio, 2012). BREEAM has recently made available different regional weightings to account for different priorities. Similarly, the Canadian Green Building Council has developed Canadian Alternative Compliance Paths to apply the LEED ND rating system. This process of adaptation of the assessment systems aims to contextualize new assessment criteria. A larger adaptability of the systems should be encouraged to use the systems in countries besides those for which they have been formulated, especially considering the rapid urbanization processes in developing countries.

An important opportunity for adapting the systems is offered by stakeholders' engagement (Albino and Berardi, 2012). Mathur et al. (2008) and Mascarenhas et al. (2009) have conceptualized the importance of stakeholders' engagement in the development of sustainability assessment systems.

All the systems compared previously do not promote communities' engagement adequately. A new definition of the implementation steps of these systems is desirable to build a participative context that stimulates citizens and increases their awareness toward sustainability measures in their community (Berardi, 2013a). Citizen-led experiences of sustainability assessment of communities are able to measure indicators that are better suited to individual happiness within the community (Morse and Fraser, 2005). In fact, indicators developed from the bottom have often been proven to be successful for measuring the level of community activity, satisfaction with local area, and perception of community spirit (Hardi and Zdan, 1997). Research in this sense has emphasized the importance of sharing knowledge through a transparent process of citizens' involvement to define and prioritize indicators (Thomson et al., 2010).

As Reed et al. (2006) found, the local context only becomes visible when the indicators are checked through the lens of local citizens, because these are needed to unpack area-specific and hidden local conditions. Integration between expert-led and citizen-led indicators and assessment criteria is therefore necessary to synthesize different aspects. Finally, the reader should not forget that the systems compared in this section are often promoted by an urban developer, whereas sustainability assessment systems should be program tools that, after having been contextualized and used for a community in the development stages, can be used for continuous evaluations.

SYSTEMS FOR SUSTAINABILITY ASSESSMENT OF CITIES

DESCRIPTION OF THE SYSTEMS

Sustainability implies scale dependence of the attributes. Previous sections have shown the limits of evaluating sustainability at the level of one or more buildings. In fact, the ways in which a building connect and depend on surroundings would be better evaluated on larger scales. The importance of the interaction between buildings and infrastructures (grids, roads, public transportation, parks, etc.) has increasingly been recognized as an unavoidable aspect of sustainability. In fact, although the sustainability has often been performed on a small-scale level, it is clear that connections with city services have to be taken into account. Moreover, although the spacial dependence of sustainable development makes uncertain which is the most appropriate scale in sustainability assessments, the boundaries of a community are often too weak and meaningless in terms of sustainability.

The increasing awareness of the role played by public infrastructures and services for a sustainable built environment comes together with the increasing awareness that cities, more than international agreements and national policies, are the leading actors in addressing sustainability (Spiekermann and Wegener, 2003; Reed 2007). Cities are close to citizens and their actions are often more purposeful than generic international agreements and plans because cities have direct control over the natural environment, the social condition of the population, and the economic activities of the community. Cities have the power to influence the design daily by authorizing the development of new area or modifying public services.

In 1994, in Aalborg, few cities signed the Charter of European Cities and Towns Towards Sustainability (Aalborg Charter, 1994). This affirms that cities are the largest unit capable of initially addressing the architectural, social, economic, political, natural resource, and environmental imbalances damaging the modern world and the smallest scale at which problems can be meaningfully resolved in an integrated, holistic, and sustainable fashion. Citics have recognized that sustainability is neither a vision nor an unchanging state, but rather a creative, local, balance-seeking process extending into all areas of local decision-making. This means that the sustainability assessment should continually evolve following the paths of the city modifications. The number of cities that have adopted this charter is continually increasing and has more than 2,000 cities (Figure 15.10). Several other agreements have been signed after that, such as the Leipzig Charter in 2007, the ICLEI network plan in 2008, and the C40 or the World Sustainable Capitals in 2010 (Berardi, 2011). Meanwhile, the Covenant of Mayors is representing the mainstream European movement involving local authorities, voluntarily committing to increasing energy efficiency and use of renewable energy sources in their territories to meet and exceed the European Union 20% CO_2 reduction objective by 2020. This last agreement is, so far, leading almost 6,000 cities in Europe to measure the level of their environmental sustainability. Similar initiatives are increasing at an exponential rate worldwide.

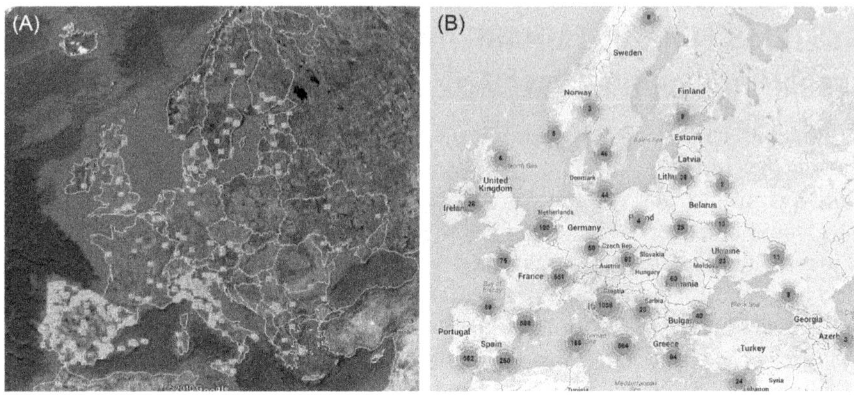

FIGURE 15.10

European cities that have signed the Aalborg Charter as of December 1, 2011.

From www.aalborgplus10.dk (A). Covenant of Mayors signatories at the end of 2013. From www.covenantofmayors.eu/index_en.html (B).

In this section, a review of major sustainability systems for sustainability assessment of cities is provided, discussing their structures, scales, and evaluation methods. In the past few years, a challenging number of sustainability assessment systems for cities have been proposed. These systems often differ significantly because they reflect the many possible approaches to the creation of sustainable development (Tanguay et al., 2010). Table 15.5 gives general information about systems.

The European Common Indicators (ECI) was launched by the Environment Commissioner Margot Wallström at the Third European Conference on Sustainable Cities in January 2001. After the launch of the ECI initiative, 80 local authorities signed the agreement on the adoption of the ECI for monitoring progress toward sustainability and reporting back to the European level and actively taking part in the testing phase and process that commenced after adoption, aiming at developing and helping build this new monitoring tool on the basis of practical experiences. Since 2008 the participating cities have been given the possibility to share and compare their data. The systems are used each year for nominating the Green Capital of Europe, an initiative focused on monitoring environmental sustainability at the local level. A set of 10 environmental sustainability indicators have been developed together with the methodologies for collecting the data for each indicator (European Common Indicators, 2003). Ten local sustainability indicators were identified through a bottom-up process: availability of public open areas and services; children's journeys to and from school; citizen's satisfaction with the local community; local contribution to global climate change; local mobility and passenger transportation; noise pollution; products promoting sustainability; quality of the air; sustainable land use; sustainable management of the local authority; and local enterprises.

Table 15.5 General Information About Some Sustainability Assessment Systems for Cities

Indicator System	Time of Initiation	Promoter	# of Indicators
European Common Indicators	2001	Funded by the European Commission, the Italian Ministry of Environment and Territory, and the Italian National Environmental Protection Agency (APAT) Project partners included Ambiente Italia, Eurocities, and Legambiente	10
Estidama	2007	The Abu Dhabi Urban Planning Council (UPC)	65
CASBEE-City	2008	Japan Green Build Council (JaGBC)/ Japan Sustainable Building Consortium (JSBC)	36
Global City Indicators	2008	Global City Indicators Facility with support from the World Bank, the Inter-American Development Bank, the University of Toronto, and the Government of Canada	63
Urban Sustainability Index	2010	McKinsey & Company, Columbia University, and Tsinghua University's School of Public Policy and Management	18
Indicators for Sustainable Development Goals	2013	Leadership Council of the Sustainable Development Solutions Network	100

The Global City Indicators Program (GCIP) is a decentralized, city-led initiative that enables cities to measure, report, and improve their performance and quality of life, facilitate capacity building, and share practices through an easy-to-use web portal (GCIP, 2009). The program seeks the improvement of big urban challenges such as poverty reduction, economic development, climate change, and the creation and maintenance of inclusive and peaceful cities and defines a set of standardized indicators that are essential to measure performance, capture trends and developments, and support cities in becoming global partners. The GCIP is suitable and applicable for all cities regardless of their size. However, at the present time, cities with more than one million people are targeted to reach a critical mass. Given the lack of a standardized definition of a "city," the unit of measurement used is the municipality. The GCIP also accommodates and aggregates data from metropolitan areas. The GCIP is organized into two broad categories: city services (which include services typically provided by city governments and other entities) and quality of life (which includes critical contributors to overall quality of life). The two categories are structured around 18 themes (Table 15.6).

Table 15.6 Global City Indicators Themes Program

City Services Themes		Quality of Life Themes	
Education	Energy	Civic engagement	Shelter
Finance	Fire and emergency response		
Governance	Health	Economy	Social equity
Recreation	Safety		
Solid Waste	Transportation	Environment	Technology and innovation
Water	Wastewater		

Source: *From GCIP (2009).*

The Leadership Council of the Sustainable Development Solutions Network (SDSN) has recently launched the Action Agenda for Sustainable Development with 10 goals and 30 targets. A series of 100 Indicators for Sustainable Development Goals (ISDG) has followed to map out operational priorities for the post-2015 development agenda (ISDG, 2013). The indicators are arranged across the 10 goals for cross-cutting thematic issues such as: ending extreme poverty including hunger; achieving development within planetary boundaries; ensuring effective learning for all children and youth for life and livelihood; achieving gender equality, social inclusion, and human rights; achieving health and well-being at all ages; improving agriculture systems and raising rural prosperity; empowering inclusive, productive, and resilient cities; curbing human-induced climate change and ensuring sustainable energy; securing biodiversity and ensuring good management of water, oceans, forests, and other natural resources; and transforming governance and technologies for sustainable development.

Estidama, which means "sustainability" in Arabic, is the initiative proposed to transform Arabic cities into a model of sustainable urbanization. Its aim is to create more sustainable communities, cities, and global enterprises, and to balance the four pillars of Estidama. This rating system aims to address the sustainability of a given development throughout its life cycle from design to operation. The system is organized into seven categories that are fundamental to more sustainable development (Estidama, 2010): integrated development process; natural systems; liveable communities; precious water; resourceful energy; stewarding materials; and innovating practice.

The Urban Sustainability Index (USI) was built to evaluate how cities in developing countries, particularly Chinese cities, are confronting the challenge of balancing environmental sustainability and growth. The indicators were drawn from data readily available from these cities. These indicators are spread across five categories that encompass environmental sustainability as well as a city's overall standard of living: basic needs; resource efficiency; environmental health; built environment; and commitment to future sustainability (The Urban Sustainability Index, 2011).

CASBEE-City represents one of the most advanced sustainability assessment systems on a city level. It combines the quality of a city with its environmental load; from the ratio of these two parameters, the system calculates the built environment efficiency. CASBEE-City uses the scalability of CASBEE beyond individual buildings to assess the whole environmental performance with a triple bottom line approach. The system implements the concept of environmental efficiency and allows evaluation of a city from two aspects: decreasing negative environmental load (L) emitted outside the city and improving environmental quality (Q) and activities inside the city.

The environmental load emitted from a city focuses on the emissions of GHG as a result of city activities. In CASBEE-City, GHG are calculated by combining the CO_2 from energy sources (in industrial, residential, commercial, transportation, and energy conversion sectors), the industrial processes, the waste disposal sector, the emission from the agriculture sector, and other sources of GHGs. Moreover, in CASEBEE-City, the environmental load also takes into account environmental load reduction and the support to other regions for reducing CO_2 missions. In this way, the system considers several cross-boundaries effects (Figure 15.11).

In CASEBEE-City, the quality of a city measures the environmental, social, and economic situations and improvements in citizens' and other stakeholders' activities and quality of life. Almost 30 items are considered for the assessment. For example, the system considers the industrial vitality (amount of gross regional product and change in number of employees), the amount of economic exchanges (number of people visiting the city and efficiency of public transportation) and the financial viability (tax revenues, outstanding local bonds), the living environment (with indicators such as crime prevention, disaster preparedness, and quality of housing), social services (educational, cultural, medical, and child care services), and social vitality (population change attributable to migration or to change in births/deaths, or pressure toward an information society).

COMPARISON BETWEEN THE SYSTEMS

The systems described in the previous section are compared here. First, Table 15.5 gives that there is a great difference in the number of indicators defined by each system as a consequence of different levels of details among the systems. Although sustainable development goals proposes 100 indicators for evaluating the sustainability, the ECI are summarized with 10.

It is also noticeable that the systems have implemented different templates for categorizing the indicators. Although ECI has not defined any set of categories, ISDG has used 16 main issues that, in a cross-cutting combination with the main goals of the system, create the assessment indicators. CASBEE seems to have a different concept in classification of the indicators that is derived from the

Major category	Middle category	Subcategory
L1. GHG emissions	L1.1 CO2 from energy sources	L1.1.1 Industrial sector*
		L1.1.2 Building (residential) sector
		L1.1.3 Building (commercial) sector
		L1.1.4 Transportation sector
		L1.1.5 Energy conversion sector*
	L1.2 Industrial processes*	
	L1.3 Waste disposal sector	
	L1.4 Agriculture sector*	
	L1.5 Three gases (HFCs, PFCs and SF6)	
L2. Environmental load reduction and CO2 absorption	L2.1 Low-carbon energy sources	
	L2.2 CO2 sinks	
L3. Support to other regions for reducing CO2 emissions	L3.1 Domestic trade, etc.	

Note: There are two calculation methods for assessing environmental load: "emitter-pays principle" and "beneficiary-pays principle." The latter method is applied to items marked with an asterisk; GHG emissions from the producing area are deducted and reallocated evenly to consuming areas across the country.

Major category	Middle category	Subcategory
Q1. Environment	Q1.1 Nature conservation	Q1.1.1 Ratio of natural and agricultural land use
	Q1.2 Local environmental Quality	Q1.2.1 Air
		Q1.2.2 Water
		Q1.2.3 Noise
		Q1.2.4 Chemicals
	Q1.3 Resource recycling	Q1.3.1 Recycling rate of general waste
	Q1.4 Environmental measures	Q1.4.1 Efforts and policies to improve the environment and biodiversity
Q2. Society	Q2.1 Living environment	Q2.1.1 Adequate quality of housing
		Q2.1.2 Adequate provision of parks and open spaces
		Q2.1.3 Adequate sewage systems
		Q2.1.4 Traffic safety
		Q2.1.5 Crime prevention
		Q2.1.6 Preparedness for natural disaster
	Q2.2 Social services	Q2.2.1 Adequacy of education services
		Q2.2.2 Adequacy of cultural services
		Q2.2.3 Adequacy of medical services
		Q2.2.4 Adequacy of child-care services
		Q2.2.5 Adequacy of services for the disabled
		Q2.2.6 Adequacy of services for the elderly
	Q2.3 Social vitality	Q2.3.1 Rate of population change due to births & deaths
		Q2.3.2 Rate of population change due to migration
		Q2.3.3 Progress toward an information society
		Q2.3.4 Efforts and policies for vitalizing society
Q3. Economy	Q3.1 Industrial vitality	Q3.1.1 Amount equivalent to gross regional-products per capita
		Q3.1.2 Ratio of change in the number of employees
	Q3.2 Economic exchanges	Q3.2.1 Index equivalent to the number of people visiting the city
		Q3.2.2 Efficiency of public transportation
	Q3.3 Financial viability	Q3.3.1 Tax revenues
		Q3.3.2 Outstanding local bonds

FIGURE 15.11

Criteria in CASBEE-City.

From CASBEE (2011).

different interpretation of sustainability. CASBEE performs the evaluation after having divided every assessment in two categories, environmental quality and environmental load.

Figure 15.12 shows the percentage of indicators allocated to nine categories by each assessment system for cities: natural environment; built environment; mobility; water and waste management; energy; economy; well-being and culture; innovation; and governance. These categories were chosen according to both structure of the systems and the requirements of a sustainable city. The organization of indicators into fully separated categories resulted in some difficulties because some indicators can be classified in more than one category, especially when the categories by themselves had some overlaps. For example, the issues related to watershed management could be considered in water management and natural environment fields. Despite all the conflicts that existed in the classification of indicators, the figure gives a general view of the way that varied issues have received attention by each system.

According to Figure 15.12, there is variety in the issues that each system covers. In general, different systems adopted different approaches according to the context they were designed for. For example, Estidama or USI have been designed for cities in developing countries, and the chart reveals that most of their attention is assigned to the water and waste management or energy issues, rather than well-being and culture, which are emphasized more by the other systems. The GCIP and ISDG systems that were developed in a more international *consensus* apply a general view to the different sustainability fields.

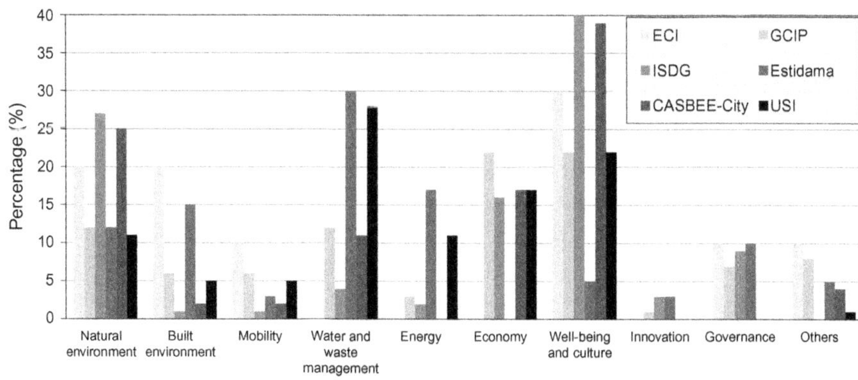

FIGURE 15.12

Comparison of the weight given by different sustainability assessment systems for cities. Urban communities grouping the assessment criteria into nine categories.

DISCUSSION AND CONCLUSIONS

An important trend in sustainability assessment is the increasing attention given to the network of the assessed elements of sustainability (the single buildings within their neighborhoods and scales up to the city or higher). First, assessment systems considered the building as a manufactured product evaluated in isolation. Then, the importance given to the surrounding site largely increased. This is because the scale dependence of sustainability in the built environment is unavoidable. The interactions between elements of the built environment make the sustainable attribute context-related and promote the evaluation of criteria that mirror the strong sustainability of the built environment, more than the presence of single elements or objects.

The dependence on the scale has shown the importance of enlarging the spatial cross-boundaries in the evaluation of sustainability. When sustainability assessment at the community level was introduced, it seemed to represent the minimum unit of analysis for a complete evaluation, especially for the social and economic dimensions of sustainability. Later, assessors started considering criteria that could be solved at the level of a city. Among these criteria were:

- To adhere to ethical standards during development by ethical trading throughout the supply chain and by providing safe and healthy work environments and conditions
- To provide a mix of type zones
- To integrate development in the local context, conserve local heritage, and culture
- To guarantee access to local infrastructure and services to all citizens
- To involve all interested parties through a collaborative approach
- To provide social and cultural value over time and for all the people

Most of the systems reviewed in previous sections were created for developed countries. Then, they started to being applied to developing countries, where the construction sector is showing high rates pushed by the urbanization process and the increased population. It noteworthy that the potential application of sustainability assessment systems in developing or in less developed countries represents a key opportunity for any sustainability assessment systems.

This chapter has presented and discussed different systems for the sustainability assessment of urban communities. A clarification of the concepts behind the sustainability assessment has been achieved by looking at the possible meanings of sustainability, assessment, community, and city. Most well-known systems have been considered, including BREEAM, CASBEE, and LEED. Assessment systems have mainly been compared considering their assessment criteria.

This chapter has shown several limits of the available systems, because they lack appropriate assessment of social, economic, and environmental sustainability. The comparison among systems has revealed that they generally promote a weak sustainability and accept that economic development can reduce natural capitals. This reduces their capability to measure sustainability in the long term.

The discussion has also shown that the assessments should support sustainable business as well as the well-being of people and environmental protection. This means that sustainable development needs higher cross-scale considerations. This also corresponds to the use of current sustainability assessment systems as tools for monitoring urban transition. Because the chapter mainly focused on top-down systems, the limits of adaptability of existing systems to different countries have been considered and the need to redefine and adapt the assessment criteria through citizens' engagement has been discussed. Finally, statistical data of assessed buildings and communities have been reported to understand a few limits and difficulties actually encountered in the promotion of sustainability.

No buildings, community, or city can achieve sustainability on its own, but a sustainable built environment should help continue sustainable use of the global hinterland. Every urban area disrupts ecosystems but, at the same time, the concentration of populations and consumption has its benefits regarding global sustainability.

REFERENCES

Aalborg Charter, 1994. Aalborg Charter, European Conference on Sustainable Cities & Towns in Aalborg, Denmark.

Albino, V., Berardi, U., 2012. Green buildings and organisational changes in Italian cases studies. Bus. Strategy Environ. 21 (6), 387–400.

Albino, V., Dangelico, R.M., 2012. Green economy principles applied to cities: an analysis of best performers and the proposal of a set of indicators. Proceedings of the IFKAD, Matera, Italy. pp. 1722–1739.

Atkisson, A., 1996. Developing indicators of sustainable community: lessons from sustainable Seattle. Environ. Impact Assess. Rev. 16 (4–6), 337–350.

Barr, S., Devine-Wright, P., 2012. Resilient communities: sustainabilities in transition. Local Environ. Int. J. Justice Sustainability 17 (5), 525–532.

Bentivegna, V., Curwell, S., Deakin, M., Lombardi, P., Mitchell, G., Nijkamp, P., 2002. A vision and methodology for integrated sustainable urban development: BEQUEST. Build. Res. Inf. 30 (2), 83–94.

Berardi, U., 2011. Beyond sustainability assessment systems: upgrading topics by enlarging the scale of assessment. Int. J. Sustainable Build. Technol. Urban Dev. 2 (4), 276–282.

Berardi, U., 2012. Sustainability assessment in the construction sector: rating systems and rated buildings. Sustainable Dev. 20 (6), 411–424.

Berardi, U., 2013a. Clarifying the new interpretations of the concept of sustainable building. Sustainable Cities Soc. 8 (1), 72–78.

Berardi, U., 2013b. Moving to Sustainable Buildings Paths to Adopt Green Innovations in Developed Countries. Versita, London, UK.

Berardi, U., 2013c. Sustainability assessments of communities through rating systems. Environ. Dev. Sustainability 15 (6), 1573–1591.

Bithas, K.P., Christofakis, M., 2006. Environmentally sustainable cities: critical review and operational conditions. Sustainable Dev. 14 (3), 177–189.

Bloomc, E., Wheelock, C., 2010. Green Building Certification Programs, Pike Research Report 2Q.

Bond, A.J., Morrison-Saunders, A., 2011. Re-evaluating sustainability assessment: aligning the vision and the practice. Environ. Impact Assess. Rev. 31 (1), 1–7.

Bourdic, L., Salat, S., 2012. Building energy models and assessment systems at the district and city scales: a review. Build. Res. Inf. 40 (4), 518–526.

Bower, J., Howe, J., Fernholz, K., Lindburg, A., 2006. Designation of environmentally preferable building materials—fundamental change needed within LEED. Dovetail Partner Report.

Brandon, P.S., Lombardi, P., 2005. Evaluating Sustainable Development in the Built Environment. Blackwell, Oxford, UK.

BRE, Building Research Establishment, 2008. A Discussion Document Comparing International Environmental Assessment Methods for Buildings. BRE, Glasgow, UK.

BRE, Building Research Establishment, 2014. BBREEAM in numbers. Available at: <www.breeam.org/page.jsp?id = 559> (accessed 22.07.2014).

BREEAM Communities, 2009. Technical Manual. BREEAM Communities Assessor Manual Development Planning Application Stage SD5065B, BRE Global Ltd. Glasgow, UK.

Butera, F.M., 2010. Climatic change and the built environment. Adv. Build. Energy Res. 41, 45–75.

Cartwright, L.E., 2000. Selecting local sustainable development indicators: does consensus exist in their choice and purpose? Planning Pract. Res. 15, 65–78.

CASBEE for Cities, 2011. Technical Manual. Institute for Building Environment and Energy Conservation, Japan.

CASBEE for New Construction, 2010. Technical Manual. Institute for Building Environment and Energy Conservation, Japan.

CASBEE for Urban Development, 2007. Technical Manual. Institute for Building Environment and Energy Conservation, Japan.

Charron, R., 2005. A review of low and net-zero energy solar home initiatives. Nat. Resour. Canada, 1–8.

Cheng, C., Pouffary, S., Svenningsen, N., Callaway, M., 2008. The Kyoto Protocol, the CDM and the Building & Construction Industry. A report for the UNEP sustainable buildings and construction initiative, Paris, France.

CIB, Conseil International du Bâtiment, 2010. Towards Sustainable and Smart-Eco Buildings. Summary Report on the EU-Funded Project Smart-ECO Buildings in the EU, Rotterdam, The Netherlands.

Cole, R.J., 1998. Emerging trends in building environmental assessment methods. Build. Res. Inf. 261, 3–16.

Cole, R.J., 2010. Building environmental assessment in a global market. Int. J. Sustainable Build. Technol. Urban Dev. 1 (1), 11–14.

Cole, R.J., 2011. Environmental issues past, present & future: changing priorities & responsibilities for building design. Proceedings of the SB11 Conference, Helsinki, Finland.

Conte, E., Monno, V., 2012. Beyond the buildingcentric approach: a vision for an integrated evaluation of sustainable buildings. Environ. Impact Assess. Rev. 34, 31–40.

Corbiére-Nicollier, T., Ferrari, Y., Jemelin, C., Jolliet, O., 2003. Assessing sustainability: an assessment framework to evaluate Agenda 21 actions at the local level. Int. J. Sustainable Dev. World Ecol. 10, 225–237.

CPR 305, 2011. Construction Products Regulation CPD. European Commission, Brussels, Belgium.

Crawley, D., Aho, I., 1999. Building environmental assessment methods: application and development trends. Build. Res. Inf. 27 (4/5), 300–308.

de Blois, M., Herazo-Cueto, B., Latunova, I., Lizarralde, G., 2011. Relationships between construction clients and participants of the building industry: structures and mechanisms of coordination and communication. Archit. Eng. Des. Manage. 7 (1), 3–22.

Dempsey, N., Bramley, G., Power, S., Brown, C., 2011. The social dimension of sustainable development: defining urban social sustainability. Sustainable Dev. 19 (5), 289–300.

Devuyst, D., 2000. Linking impact assessment and sustainable development at the local level: the introduction of sustainability assessment systems. Sustainable Dev. 8 (2), 67–78.

Ding, G.K.C., 2008. Sustainable construction—the role of environmental assessment tools. J. Environ. Manage. 86 (3), 451–464.

du Plessis, C., Cole, R.J., 2011. Motivating change: shifting the paradigm. Build. Res. Inf. 39 (5), 436–449.

EC, European Commission, 2010. Directive 2010/31/CE of the European Parliament. Energy Performance of Buildings Directive. Official journal of the European Communities, Brussels, Belgium.

EEA, European Environmental Agency, 2006. Urban sprawl in Europe: the ignored challenge, Report 10. <www.eea.europa.eu/publications/eea_report_2006_10> (accessed 22.08.2014).

EIA, Energy Information Administration, 2010. International Energy Outlook 2010. U.S. Department of Energy. Washington DC <www.eia.doe.gov/oiaf/ieo/highlights.html> (accessed 22.08.2014).

EISA, 2007. Energy Independence and Security Act of 2007: December 21, US Government, Washington, DC. <www.gpo.gov/fdsys/pkg/PLAW-110publ140/content-detail.html> (accessed 22.08.2014).

EPA, Environmental Protection Agency, 2008. Green Building Strategy—defines green building and explains EPA's strategic role in facilitating the mainstream adoption of effective green building practices.

Estidama, 2010. Abu Dhabi Urban Planning Council. <estidama.org/template/estidama/docs/PCRS%20Version%201.0.pdf> (accessed 30.11.2014).

European Common Indicators, 2003. Ambiente Italia Research Institute. <www.cityindicators.org/Deliverables/eci_final_report_12-4-2007-1024955.pdf> (accessed 30.11.2014).

Fowler, K.M., Rauch, E.M., 2006. Sustainable Building Rating Systems Summary. Pacific Northwest National Laboratory. U.S. Department of Energy, PNNL-15858.

GCIP, The Global City Indicators Program, 2009. World Bank Urban Development Unit. <openknowledge.worldbank.org/handle/10986/10244> (accessed 30.11.2014).

GhaffarianHoseini, A., Dahlan, N.D., Berardi, U., GhaffarianHoseini, A., Makaremi, N., GhaffarianHoseini, M., 2013. Sustainable energy performances of green buildings: a review of current theories, implementations and challenges. Renewable Sustainable Energy Rev. 25, 1–17.

Gibberd, J., 2005. Assessing sustainable buildings in developing countries—the sustainable building assessment tool SBAT and the sustainable building lifecycle SBL. Proceedings of the World Sustainable Building Conference, Tokyo, Japan, pp. 1605–1612.

Gilijamse, W., 1995. Zero-energy houses in the Netherlands. In: Proceedings of Building Simulation '95, Madison, WI, pp. 276–283.

Haapio, A., 2012. Towards sustainable urban communities. Environ. Impact Assess. Rev. 32 (1), 165–169.

Hahn, T.J., 2008. Research and solutions: LEED-Ing away from sustainability: toward a green building system using nature's design. Sustainability 1 (3), 196–201.

Häkkinen, T., 2007. Assessment of indicators for sustainable urban construction. Civ. Eng. Environ. Sys. 24 (4), 247–259.

Hardi, P., Zdan, T., 1997. Assessing Sustainable Development: Principles in Practice. International Institute for Sustainable Development, Winnipeg, Manitoba, Canada.

Hastings, R., Wall, M., 2007. Sustainable Solar Housing, Vol. 1—Strategies and Solutions. Earthscan, London, UK.

Hellstrom, T., 2007. Dimensions of environmentally sustainable innovation: the structure of eco-innovation concepts. Sustainable Dev. 15 (3), 148–159.

Hernandez, P., Kenny, P., 2010. From net energy to zero energy buildings: defining life cycle zero energy buildings LC-ZEB. Energy Build. 42 (6), 815–821.

Hopwood, B., Mellor, M., O'Brien, G., 2005. Sustainable development: mapping different approaches. Sustainable Dev. 13 (1), 38–52, <sustainabledevelopment.un.org/content/documents/3233indicatorreport.pdf> (accessed 30.11.2014).

Indicators for Sustainable Development Goals, 2013. The Leadership Council of the Sustainable Development Solutions Network.

Innes, J., Booher, D., 2000. Indicators for sustainable communities: a strategy building on complexity theory and distributed intelligence. Planning Theory Pract. 1 (2), 173–186.

IPCC, Intergovernmental Panel for Climate Change, 2007. Summary for Policymakers, Climate Change.. Cambridge University Press, New York, NY, USA, IPCC WG1 Fourth Assessment Report.

Iqbal, M., 2004. A feasibility study of a zero energy home in Newfoundland. Renewable Energy 292, 277–289.

ISO standard 14020, 2000. Environmental labels and declarations—general principles.

ISO standard 14040, 2006. Environmental management—life cycle assessment—principles and framework.

ISO standard 15392, 2008. Sustainability in building construction—general principles.

ISO standard 15643-1, 2010. Sustainability of construction works—sustainability assessment of buildings—Part 1: General framework.

ISO standard 21931-1, 2010. Sustainability in building construction—framework for methods of assessment for environmental performance of construction works—Part 1: Buildings.

Janikowski, R., Kucharski, R., Sas-Nowosielska, A., 2000. Multi-criteria and multi-perspective analysis of contaminated land management methods. Environ. Monit. Assess. 601, 89–102.

Kellett, R., 2009. Sustainability Indicators for Computer-based Tools in Community Design, 1. Canada Mortgage and Housing Corporation, Ottawa, Canada.

Kibert, C.J., 2007. The next generation of sustainable construction. Build. Res. Inf. 356, 595–601.

King, A.A., Toffel, M.W., 2007. Self-regulatory Institutions for Solving Environmental Problems: Perspectives and Contributions from the Management Literature. Governing the Environment: Interdisciplinary Perspectives. Delmas, M., Young O. (Eds.).

Langston, C.A., Ding, G.K.C., 2001. Sustainable Practices in the Built Environment, second ed. Butterworth-Heinemann, Oxford, UK.

Larsson, N., 2010. Rapid GHG reductions in the built environment under extreme conditions. Int. J. Sustainable Build. Technol. Urban Dev. 1 (1), 15–21.

Laustsen, J., 2008. Energy Efficiency Requirements in Building Codes, Energy Efficiency Policies for New Buildings. OECD/IEA, Paris, France.

LEED for Neighborhood Development, 2009. Manual, The US Green Building Council.

Lowe, I., 2008. Shaping a sustainable future—an outline of the transition. Civ. Eng. Environ. Sys. 25 (4), 247–254.

Lund, H., Marszal, A., Heiselberg, P., 2011. Zero energy buildings and mismatch compensation factors. Energy Build. 43 (7), 1646–1654.

Mang, P., Reed, B., 2012. Designing from place: a regenerative framework and methodology. Build. Res. Inf. 40 (1), 23–38.

Marszal, A.J., Heiselberg, P., Bourrelle, J.S., Musall, E., Voss, K., Sartori, I., Napolitano, A., 2011. Zero energy building—a review of definitions and calculation methodologies. Energy Build. 43 (4), 971–979.

Martens, P., 2006. Sustainability: science or fiction? Sustainability Sci. Pract. Policy 2 (1), 36–41.

Mascarenhas, A., Coelho, P., Subtil, E., Ramos, T.B., 2009. The role of common local indicators in regional sustainability assessment. Ecol. Indic. 10 (3), 646–656.

Mathur, V.N., Price, A.D.F., Austin, S., 2008. Conceptualizing stakeholder engagement in the context of sustainability and its assessment. Constr. Manage. Econ. 26 (6), 601–609.

Mayer, A.L., 2008. Strengths and weaknesses of common sustainability indices for multidimensional systems. Environ. Int. 34 (2), 277–291.

McGraw-Hill Construction, 2008. Key Trends in the European and U.S. Construction Marketplace, SmartMarket Report. McGraw-Hill Construction, New York, NY, USA.

Mitchell, G., 1996. Problems and fundamentals of sustainable development indicators. Sustainable Dev. 4, 1–11.

Mlecnik, E., Visscher, H., van Hal, A., 2010. Barriers and opportunities for labels for highly energy-efficient houses. Energy Policy 38 (8), 4592–4603.

Mori, K., Christodoulou, A., 2012. Review of sustainability indices and indicators: towards a new City Sustainability Index CSI. Environ. Impact Assess. Rev. 32 (1), 94–106.

Morse, S., Fraser, E.D.G., 2005. Making 'dirty' nations look clean? The nation state and the problem of selecting and weighting indices as tools for measuring progress towards sustainability. Geoforum 36, 625–640.

Murakami, S., Kawakubo, S., Asami, Y., Ikaga, T., Yamaguchi, N., Kaburagi, S., 2011. Development of a comprehensive city assessment tool: CASBEE-City. Build. Res. Inf. 39 (3), 195–210.

Newsham, G.R., Mancini, S., Birt, B.J., 2009. Do LEED-certified buildings save energy? Yes, but.... Energy Build. 41 (8), 897–905.

Nijkamp, P., Rietveld, P., Voogd, H., 1990. Multicriteria Evaluation in Physical Planning. North-Holland, New York, NY, USA.

OECD, Organisation for Economic Co-operation and Development, 2011a. How's Life? Measuring Well-Being. OECD Publishing, Paris.

OECD, Organisation for Economic Co-operation and Development, 2011b. Your better life index. <oecdbetterlifeindex.org/> (accessed 30.11.2014).

Pope, J., Annandale, D., Morrison-Saunders, A., 2004. Conceptualising sustainability assessment. Environ. Impact Assess. Rev. 24 (6), 595–616.

Reed, B., 2007. Shifting from 'sustainability' to regeneration. Build. Res. Inf. 356, 674–680.

Reed, M.S., Fraser, E.D.G., Dougill, A.J., 2006. An adaptive learning process for developing and applying sustainability indicators with local communities. Ecol. Econ. 59, 406–418.

Rees, W., Wackernagel, M., 1996. Urban ecological footprints: why cities cannot be sustainable—and why they are a key to sustainability. Environ. Impact Assess. Rev. 16, 223–248.

Robinson, J., 2004. Squaring the circle: on the very idea of sustainable development. Ecol. Econ. 48 (4), 369–384.

Rumsey, P., McLellan, J.F., 2005. The green edge—the green imperative. Environ. Des. Constr., 55–56.

Scerri, A., James, P., 2010. Accounting for sustainability: combining qualitative and quantitative research in developing 'indicators' of sustainability. Int. J. Social Res. Method 13 (1), 41–53.

Schendler, A., Udall, R., 2005. LEED is Broken; Let's Fix It. Grist Environmental News & Commentary.

Scheuer, C.W., Keoleian, A., 2002. Evaluation of LEED Using Life Cycle Assessment Methods. National Institute of Standard and Technologies, NIST GCR 02-836.

Seo, S., Tucker, S., Ambrose, M., Mitchell, P., Wang, C.H., 2006. Technical Evaluation of Environmental Assessment Rating Tools, Research and Development Corporation, Project PN05, 1019.

Sev, A., 2009. How can the construction industry contribute to sustainable development? A conceptual framework. Sustainable Dev. 173 (3), 161–173.

Sev, A., 2011. A comparative analysis of building environmental assessment tools and suggestions for regional adaptations. Civ. Eng. Environ. Sys. 28 (3), 231–245.

Sharifi, A., Murayama, A., 2013. A critical review of seven selected neighborhood sustainability assessment tools. Environ. Impact Assess. Rev. 38, 73–87.

Smith, T., Fischelein, M., Suh, S., Huelman, P., 2006. Green Building Rating Systems—A comparison of the LEED and Green Globes in the US Proc. Carpenters Industrial Council, USA.

Spiekermann, K., Wegener, M., 2003. Modelling urban sustainability. Int. J. Urban Sci. 7, 47–64.

Steurer, R., Hametner, M., 2013. Objectives and indicators in sustainable development strategies: similarities and variances across Europe. Sustainable Dev. 21 (4), 224–241.

Suzuki, M., Oka, T., 1998. Estimation of life cycle energy consumption and CO_2 emission of office buildings in Japan. Energy Build. 28 (1), 33–41.

Tanguay, G., Rajaonson, J., Lefebyre, J., Lanoie, P., 2010. Measuring the sustainability of cities: an analysis of the use of local indicators. Ecol. Indic. 10, 407–418.

The Urban Sustainability Index, 2011. Urban China initiative. <www.urbanchinainitiative.org/wp-content/uploads/2012/05/McKinsey-2011-Urban-Sustainability-Index.pdf> (accessed 30.11.2014).

Thomson, C.S., El-Haram, M.A., Emmanuel, R., 2010. Managing the knowledge flow during sustainability assessment. Proc. Inst. Civ. Eng. Urban Des. Planning 163 (2), 67–78.

Torcellini, P., Pless, S., Deru, M., Crawley, D., 2006. Zero Energy Buildings: A Critical Look at the Definition. National Renewable Energy Laboratory NREL, Golden, CO, NREL/CP-550-39833.

Turcu, C., 2013. Re-thinking sustainability indicators: local perspectives of urban sustainability. J. Environ. Planning Manage. 56 (5), 695–719.

UN, United Nations, 2008. World urbanization prospects: the 2007 revision population database. <esa.un.org/unup/> (accessed 30.11.2014).

UNEP-SBCI, United Nations Environment Programme—Sustainable Buildings & Climate Initiative, 2009. Buildings and Climate Change: a Summary for Decision-Makers. UNEP DTIE Sustainable Consumption & Production Branch, Paris, France.

UNEP-SBCI, United Nations Environment Programme—Sustainable Buildings & Climate Initiative, 2010. Common Carbon Metric CCM. UNEP DTIE Sustainable Consumption & Production Branch, Paris, France.

UN-Habitat, United Nations Human Settlements Programme, 2006. The State of the World's Cities Report 2006/2007, London, UK, Earthscan.

Vallance, S., Perkins, H.C., Dixon, J.E., 2011. What is social sustainability? A clarification of concepts. Geoforum 42 (3), 342–348.

WCED, World Commission on Environment and Development, 1987. Our Common Future, The Report of the Brundtland Commission. Oxford University Press, Oxford, UK.

Winston, N., 2010. Regeneration for sustainable communities? Barriers to implementing sustainable housing in urban areas. Sustainable Dev. 18 (6), 319–330.

Xing, Y., Horner, R.M.W., El-Haram, M.A., Bebbington, J., 2009. A framework model for assessing sustainability impacts of urban development. Account. Forum 33, 209–224.

Index

Note: Page numbers followed by "*b*" "*f*" and "*t*" refer to boxes, figures and tables, respectively.

F

Q

R

S

W

Z

Printed in the United States
By Bookmasters